MICROBIAL BIOFILMS

Advances in Biotechnology and Bioengineering

MICROBIAL BIOFILMS

CHALLENGES AND ADVANCES IN METABOLOMIC STUDY

Series editor

HEMEN SARMA
Department of Botany, Bodoland University, Rangalikhata, Deborgaon, Kokrajhar (BTR), Assam, India

Edited by

SANKET JOSHI
Oil and Gas Research Center, and Central Analytical and Applied Research Unit, Sultan Qaboos University, Muscat, Oman

DIBYAJIT LAHIRI
Department of Biotechnology, University of Engineering and Management, Kolkata, West Bengal, India

RINA RANI RAY
Department of Biotechnology, Maulana Abul Kalam Azad University of Technology, Kolkata, West Bengal, India

MUBARAKALI DAVOODBASHA
School of Life Sciences, B.S. Abdur Rahman Crescent Institute of Science and Technology, Chennai, Tamil Nadu, India

ELSEVIER

ACADEMIC PRESS
An imprint of Elsevier

Academic Press is an imprint of Elsevier
125 London Wall, London EC2Y 5AS, United Kingdom
525 B Street, Suite 1650, San Diego, CA 92101, United States
50 Hampshire Street, 5th Floor, Cambridge, MA 02139, United States
The Boulevard, Langford Lane, Kidlington, Oxford OX5 1GB, United Kingdom

ISBN: 978-0-323-95715-1

For information on all Academic Press publications visit our website at https://www.elsevier.com/books-and-journals

Publisher: Stacy Masucci
Acquisitions Editor: Kattie Washington
Editorial Project Manager: Sam Young
Production Project Manager: Fahmida Sultana
Cover Designer: Christian J. Bilbow

Typeset by TNQ Technologies

Dedication

All the editorial members would like to dedicate the book to the editors Fahmida Sultana and Samuel Young for their constant help in processing the book.

Dr. Dibyajit Lahiri dedicates the book to his paternal grandfather Late Asim Lahiri, grandmother Mrs. Ashoka Lahiri, maternal grandfather Late Prodyout Kumar Guha, grandmother Late Ila Guha, father Mr. Debasish Lahiri, mother Madhumita Lahiri, father-in-law Dilipkumar Dey, mother-in-law Mimi Dey, sister Dr. Debasmita Bhattacharya, brother-in-law Mr. Baisampayan Bhattacharya, niece Ms. Purbita Bhattacharya, nephew Kiaan Bhattacharya, and my wife Mrs. Ankita Dey (Lahiri).

Contents

5. From symbiosis to dysbiosis in gut-consequence includes metabolic syndrome

Nithyakalyani Mohan and Anusha Sunder

6. Metabolomic study of biofilm-forming natural microbiota of skin biofilm

Gokul Shankar Sabesan and Annie Jeyachristy Sam

7. Metabolomic study of biofilm-forming natural microbiota of vaginal biofilm

Ashok Kumar Krishna Kumar, Jayanthi Malaiyandi, Andrea Mary Francis, and Vidya Ramachandran

8. Metabolome analysis for host–microbiota interactions

Sreejita Ghosh, Dibyajit Lahiri, Moupriya Nag, and Rina Rani Ray

9. Multispecies metabolomics interactions resulting in the development of resistance

Bhavdip Anavadiya, Dhaval Acharya, Meenu Saraf, and Dweipayan Goswami

10. Natural phytochemicals: a potential alternative antibiofilm agent

Kanchanlata Tungare, Ameyota De, Shreya Das, Vaishnavi Kesarkar, Mustansir Bhori, and Renitta Jobby

11. Medical devices—associated biofilm infections and challenges in treatment

Sreejita Ghosh, Dibyajit Lahiri, Moupriya Nag, and Rina Rani Ray

12. NMR-based metabolomics study of microbial biofilm

Bandita Dutta, Dibyajit Lahiri, Moupriya Nag, Ashmita Samanta, and Rina Rani Ray

13. Isotope labeling LC-MS for metabolomics of biofilm study and tracer-based biofilm metabolomics analysis

Yuvaraj Sivamani, Karan Narasimha Murthy, Harshit Sajal, and Sumitha Elayaperumal

14. Lipidomics profiling of microbial biofilm

Gowtham Kumaraguru, Saruniyadevi Moorthy, Ashok Kumar Krishna Kumar, and Jayanthi Malaiyandi

15. Functional metabolomics approaches in determining the inhibition of biofilm

Bandita Dutta, Dibyajit Lahiri, Moupriya Nag, and Rina Rani Ray

16. Reprogramming the metabolomics of biofilms

Harshit Sajal, Yuvaraj Sivamani, Nimisha Joly Chennakadan, Yadhu Krishna Menon Kakkanattu, and Sumitha Elayaperumal

17. Dysbiosis of microbiome: a risk factor for cancer, metabolic and inflammatory diseases

Yuvaraj Sivamani, Harshit Sajal, and Sumitha Elayaperumal

18. Strategies to reduce microbial biofilm in medical prosthesis and other devices

Juhita Dhar, Anupriya Das, Soumyadip Biswas, Talaha Nishat Ahmed, Sanjana Ghosh, and Arup Kumar Mitra

19. Targeting microbial biofilms using genomics-guided drug discovery

Damayanti Chakravarty, Dipabarna Bhattacharya, Sayak Ganguli, and Upal Das Ghosh

20. Biofilms: a sociomicrobiological nexus

Lynn D'Lima and Manju Phadke

21. In silico and in vivo methods for designing antibiofilm agents against *Pseudomonas aeruginosa* and *Staphylococcus aureus*

Saichandana Tamba, Aravind Panicker, Renitta Jobby, Satish Kumar Rajasekharan, Karthi Shanmugam, and Vinothkannan Ravichandran

22. Biofilm: a threat to medical devices

Jignasa H. Bera, Leyon Selvin Raj A., Sishir Gang, and Dhara N. Patel

23. Biofilm formation in acute and chronic respiratory infections caused by nosocomial gram-negative bacteria

Shreya Karmakar, Soumyadeep Mukherjee, Natasha Mary Joseph, Archana Jena Priyadarshini, Vinothkannan Ravichandran, Satish Kumar Rajasekharan, and Renitta Jobby

List of contributors

Dhaval Acharya Department of Microbiology, B N Patel Institute of Paramedical and Sciences, Anand, Gujarat, India

Bhavdip Anavadiya Department of Microbiology & Biotechnology, University School of Sciences, Gujarat University, Ahmedabad, Gujarat, India

Ankita Banerjee Department of Microbiology, St. Xavier's College (Autonomous), Kolkata, West Bengal, India

Jignasa H. Bera Department of Medical Laboratory Technology, Bapubhai Desaibhai Patel Institute of Paramedical Sciences (BDIPS), Charotar University of Science and Technology, Changa, Gujarat, India

Dipabarna Bhattacharya University of Helsinki, Helsinki, Finland

Mustansir Bhori School of Biotechnology and Bioinformatics, D Y Patil Deemed to be University, Belapur, Maharashtra, India

Soumyadip Biswas Department of Microbiology, St. Xavier's College (Autonomous), Kolkata, West Bengal, India

Damayanti Chakravarty Cell and Molecular Biology Laboratory, The University of Southern Mississippi, Hattiesburg, MS, United States

Nimisha Joly Chennakadan Department of Biotechnology and Bioinformatics, JSS Academy of Higher Education and Research, Mysore, Karnataka, India

Lynn D'Lima Department of Microbiology, S.I.E.S College of Arts, Science & Commerce (Autonomous), Mumbai, Maharashtra, India

Anupriya Das Department of Microbiology, St. Xavier's College (Autonomous), Kolkata, West Bengal, India

Shreya Das School of Biotechnology and Bioinformatics, D Y Patil Deemed to be University, Belapur, Maharashtra, India

Ameyota De School of Biotechnology and Bioinformatics, D Y Patil Deemed to be University, Belapur, Maharashtra, India

Sylvia Denis Department of Microbiology, St. Xavier's College (Autonomous), Kolkata, West Bengal, India

Juhita Dhar Department of Microbiology, St. Xavier's College (Autonomous), Kolkata, West Bengal, India

Bandita Dutta Department of Biotechnology, Maulana Abul Kalam Azad University of Technology, Haringhata, West Bengal, India

Sumitha Elayaperumal Department of Biotechnology and Bioinformatics, JSS Academy of Higher Education and Research, Mysore, Karnataka, India

Andrea Mary Francis Department of Human Genetics, Sri Ramachandra Institute of Higher Education and Research, Chennai, Tamil Nadu, India

Sishir Gang Department of Nephrology, Muljibhai Patel Urological Hospital, Nadiad, Gujarat, India

Sayak Ganguli Department of Biotechnology, St. Xavier's College, Kolkata, West Bengal, India

Sreejita Ghosh Department of Biotechnology, Maulana Abul Kalam Azad University of Technology, Haringhata, West Bengal, India

Upal Das Ghosh Department of Microbiology, Bidhannagar College, Kolkata, West Bengal, India

Sanjana Ghosh Department of Microbiology, St. Xavier's College (Autonomous), Kolkata, West Bengal, India

Dweipayan Goswami Department of Microbiology & Biotechnology, University School of Sciences, Gujarat University, Ahmedabad, Gujarat, India

C.R. Hemalatha Bioinformatics—SRET, SRIHER (DU), Chennai, Tamil Nadu, India

Renitta Jobby Amity Institute of Biotechnology, Amity University Maharashtra, Mumbai, Maharashtra, India; Amity Centre of Excellence in Astrobiology, Amity University Maharashtra, Mumbai, Maharashtra, India

Natasha Mary Joseph Department of Microbiology, School of Life Sciences, Pondicherry University, Puducherry, India

Anupma Jyoti Kindo Department of Microbiology, SRMC & RI, SRIHER (DU), Chennai, Tamil Nadu, India

Shreya Karmakar Department of Botany, Institute of Science, Banaras Hindu University, Varanasi, Uttar Pradesh, India

Vaishnavi Kesarkar School of Biotechnology and Bioinformatics, D Y Patil Deemed to be University, Belapur, Maharashtra, India

Ashok Kumar Krishna Kumar Department of Biotechnology, School of Life Sciences, Vels Institute of Science Technology and Advanced Studies, Chennai, Tamil Nadu, India; Department of Lifescience, University of Greenwich, London, United Kingdom

Gowtham Kumaraguru Department of Biotechnology, Vels Institute of Science Technology and Advanced Studies, Chennai, Tamil Nadu, India; Plant Biotechnology Lab, Department of Biotechnology, Bharathiar University, Coimbatore, Tamil Nadu, India

Dibyajit Lahiri Department of Biotechnology, University of Engineering and Management, Kolkata, West Bengal, India

Jayanthi Malaiyandi Department of Biotechnology, School of Life Sciences, Vels Institute of Science Technology and Advanced Studies, Chennai, Tamil Nadu, India; Department of Biotechnology, Vels Institute of Science Technology and Advanced Studies, Chennai, Tamil Nadu, India

Yadhu Krishna Menon Kakkanattu Department of Biotechnology and Bioinformatics, JSS Academy of Higher Education and Research, Mysore, Karnataka, India

Arup Kumar Mitra Department of Microbiology, St. Xavier's College (Autonomous), Kolkata, West Bengal, India

Nithyakalyani Mohan Apollo Hospital and Research Foundation, Apollo Hospitals, Chennai, Tamil Nadu, India

Saruniyadevi Moorthy Department of Biotechnology, Vels Institute of Science Technology and Advanced Studies, Chennai, Tamil Nadu, India; MTA Infotech, Varanasi, Uttar Pradesh, India

Soumyadeep Mukherjee Department of Botany, Institute of Science, Banaras Hindu University, Varanasi, Uttar Pradesh, India

Karan Narasimha Murthy Department of Biotechnology and Bioinformatics, JSS Academy of Higher Education and Research, Mysore, Karnataka, India

Moupriya Nag Department of Biotechnology, University of Engineering and Management, Kolkata, West Bengal, India

Talaha Nishat Ahmed Department of Microbiology, St. Xavier's College (Autonomous), Kolkata, West Bengal, India

Aravind Panicker Amity Institute of Biotechnology, Amity University Maharashtra, Mumbai, Maharashtra, India

Dhara N. Patel Department of Medical Laboratory Technology, Bapubhai Desaibhai Patel Institute of Paramedical Sciences (BDIPS), Charotar University of Science and Technology, Changa, Gujarat, India

Manju Phadke Department of Microbiology, S.I.E.S College of Arts, Science & Commerce (Autonomous), Mumbai, Maharashtra, India

Archana Jena Priyadarshini Post Graduate Department of Microbiology, St. Xavier's College (Autonomous), Kolkata, West Bengal, India

Leyon Selvin Raj A. Department of Medical Laboratory Technology, Bapubhai Desaibhai Patel Institute of Paramedical Sciences (BDIPS), Charotar University of Science and Technology, Changa, Gujarat, India

Satish Kumar Rajasekharan Department of Food Sciences, Institute of Postharvest Technology and Food Sciences, Agricultural Research Organization (ARO), The Volcani Center, Rishon LeZion, Israel

Vidya Ramachandran Department of Microbiology, School of Life Sciences, Vels Institute of Science Technology and Advanced Studies, Chennai, Tamil Nadu, India

Vinothkannan Ravichandran Centre for Drug Discovery and Development (CD3), Amity Institute of Biotechnology (AIB), Amity University Maharashtra, Mumbai, Maharashtra, India

Rina Rani Ray Department of Biotechnology, Maulana Abul Kalam Azad University of Technology, Haringhata, West Bengal, India

Drisha Roy Department of Biotechnology, University of Engineering and Management, Kolkata, West Bengal, India

Gokul Shankar Sabesan Faculty of Medicine, Manipal University College Malaysia, Melaka, Malaysia

Harshit Sajal Department of Biotechnology and Bioinformatics, JSS Academy of Higher Education and Research, Mysore, Karnataka, India

Ashmita Samanta Department of Biotechnology, University of Engineering and Management, Kolkata, West Bengal, India

Annie Jeyachristy Sam Faculty of Medicine, Universiti Kuala Lumpur Royal College of Medicine Perak, Ipoh, Perak, Malaysia

Meenu Saraf Department of Microbiology & Biotechnology, University School of Sciences, Gujarat University, Ahmedabad, Gujarat, India

Karthi Shanmugam School of Chemical and Biotechnology, SASTRA University, Thanjavur, Tamil Nadu, India

Shahitha Sikkander Department of Microbiology, Muthayammal College of Arts and Science (Autonomous), Namakkal, Tamil Nadu, India

Yuvaraj Sivamani Department of Pharmaceutical Chemistry, Cauvery College of Pharmacy, Mysore, Karnataka, India

Suhana Sultana Department of Microbiology, St. Xavier's College (Autonomous), Kolkata, West Bengal, India

Anusha Sunder Apollo Hospital and Research Foundation, Apollo Hospitals, Chennai, Tamil Nadu, India

Saichandana Tamba Amity Institute of Biotechnology, Amity University Maharashtra, Mumbai, Maharashtra, India

Kanchanlata Tungare School of Biotechnology and Bioinformatics, D Y Patil Deemed to be University, Belapur, Maharashtra, India

Nagasinduja Venkatesan Department of Microbiology, Muthayammal College of Arts and Science (Autonomous), Namakkal, Tamil Nadu, India

Preface

Biofilms are adherent group of sessile microbial communities, characterized by irreversible cells that adhere to the biotic or abiotic surfaces using pili and extracellular polymeric substances. Biofilm-related infections are reported to be responsible for ~65%–80% of all infections, posing a serious health issues. Acute infections caused by pathogenic microbial cells that comprise bacteria and fungi have been studied extensively for the last few decades. Most of the research works in the field of microbial pathogenesis focus on various types of acute infections. The mechanism of biofilm formation varies from species to species associated with it. The common feature recognized about the formation of biofilm is the extracellular polymeric substances or extracellular matrix that holds the cells together. Quorum Sensing (QS) is reported to provide the competence to the pathogen to perceive the microbial cell numbers and synchronize their behavior by cell-to-cell signaling using small molecules known as autoinducers, which include small peptides, acylhomoserine lactones (AHLs), and quinolones. These autoinducers cause aggregations of the biofilm-forming cells with one another by secretion of extracellular polymeric substances and are majorly responsible for the expression of virulence in pathogens. Thus, it is quite imperative to understand how QS works in the host environment. Plenty of research studies have been performed since last decades, and have been focused on the characterization of microbial biofilms comprising genomics and postgenomics functional approaches. But at recent times, differential profiles including proteomic and metabolomics approaches are used for the characterization of the biofilm. The proteomic profile helps in dissecting the complexity of microbial communities by analyzing the expression of the protein, function, modification, and interactions over temporal scales. Metabolomics refers to the study of different cell products that help in understanding the physiological state of microorganisms. Each of this analysis reveals extensive amount of information which combines the complementary techniques that could simply contribute in understanding the development of biofilm. With the stated objective, this book was designed to address various technical and application details of microorganisms and their biofilms in the light of novel approaches for identification, understanding the routes of infections, and possible treatments.

This book aims to bring together contributions from different experts to significantly advance the knowledge of this research field. It will be potentially useful for both clinicians and researchers who are researching on biofilms. On this perspective, this book deals with the metabolomic characteristics of bacterial biofilm. This book will bring forth various analytical techniques which can be achieved in analyzing the metabolomics of the biofilm. It will also focus on the various types of metabolomics studies of biofilm formation such as oral biofilm and biofilm by various nosocomial organisms. This book will shed light into the world of microbial biofilm, its related infection, and pathogenesis. This book also aims to provide insight into how alternate strategies can be achieved by the use of bacteriophages or other novel compounds bringing about disintegrating the biofilm.

We have taken utmost care to make sure that the information provided by contributors is without any error, and in case if you find any, then please pardon us, as it is unintentional. We are thankful to all contributors, for timely submission of their chapters. We would like to especially thank the Elsevier team for their continuous support for this book: Stacy Masucci, Kattie Washington, Sam Young, Fahmida Sultana, and Christian J. Bilbow. We would also like to kindly acknowledge our respective universities, for their support and encouragement to carry out such scholarly work.

Editors: Sanket Joshi
Dibyajit Lahiri
Rina Rani Ray
MubarakAli Davoodbasha

CHAPTER

1

Microbial biofilms: issues and challenges

Nagasinduja Venkatesan and Shahitha Sikkander

Department of Microbiology, Muthayammal College of Arts and Science (Autonomous), Namakkal, Tamil Nadu, India

1. History and introduction

Antonie Van Leeuwenhook in the 17th century observed "animalcules" on some living and dead matter. He even invented these "animalcules" on his own teeth as deposits. These are the dental plaques of bacteria. This is the reliable affirmation of the survival of microbial biofilm. Apart from cohesive nature, the higher resistance of biofilms to disinfectants such as chlorine was registered by Characklis in 1973. The term biofilm was coined by Costerton in 1978, and also he insisted about the biofilm importance to the world. Now we roughly exemplify these biofilms as the microbial colonies clung to a foundation and are wrapped inside an extracellular polymeric substance (EPS) composed by the microbial cells. Biofilms may articulate on ample surfaces, which include aquatic systems, living tissues, medical devices, and water piping system. A huge number of microbes form biofilms in aqueous environments. This can occur in both legitimate and manmade environments under distinct conditions that may occur in a variety of agents. The term biofilm is becoming familiar nowadays. In a simpler way, tooth plaque that is causing tooth decay, clogs in kitchen sink or drains in a house, and the slippery rocks in water bodies are some of the examples to understand about biofilm easily. The biofilms are nothing but it is a composition of living microbes, which may endure as a society or confederation. These biofilms are alive with an entangled social formation, which shields them and also permits them to flourish (Percival et al., 2011).

2. Denotation of biofilm

A biofilm is denoted as an association of bacteria confined in an autonomous construction of exopolysaccharide matrix, which attaches to a living or nonliving substance. These are the

Microbial Biofilms
https://doi.org/10.1016/B978-0-323-95715-1.00003-0

agglomeration of microbial cells that are analogous with the surface in an inevitable manner. It affords supportive circumstances for the motility of genetic material from one cell to another (Donlan, 2002). An organized association of microbial cells solidly hooked to a surface and ingrained in a pattern that comprises of EPS is known as biofilm. Biofilm formation can be evident in many surfaces inclusive of tissues, medical devices, piping in the industries, or potable water system or event in aquatic systems. Eminently complicated biofilms are experienced in water system that composes corrosion products, diatoms from fresh water, clay particles, and filamentous bacteria. On the other hand, biofilm that is formed on medical devices evidenced to contain cocci or single microbe and the associated extracellular polymeric matrix (EPS). National Institute of Health has openly disclosed that around 80% of microbial infections are due to the involvement of biofilm. Biofilms that occur in medical devices such as catheters, prosthetic joints, and pacemakers result in biofilm-associated infections. Some clinically important biofilm infections are shown in Fig. 1.1 (Muhammad et al., 2020).

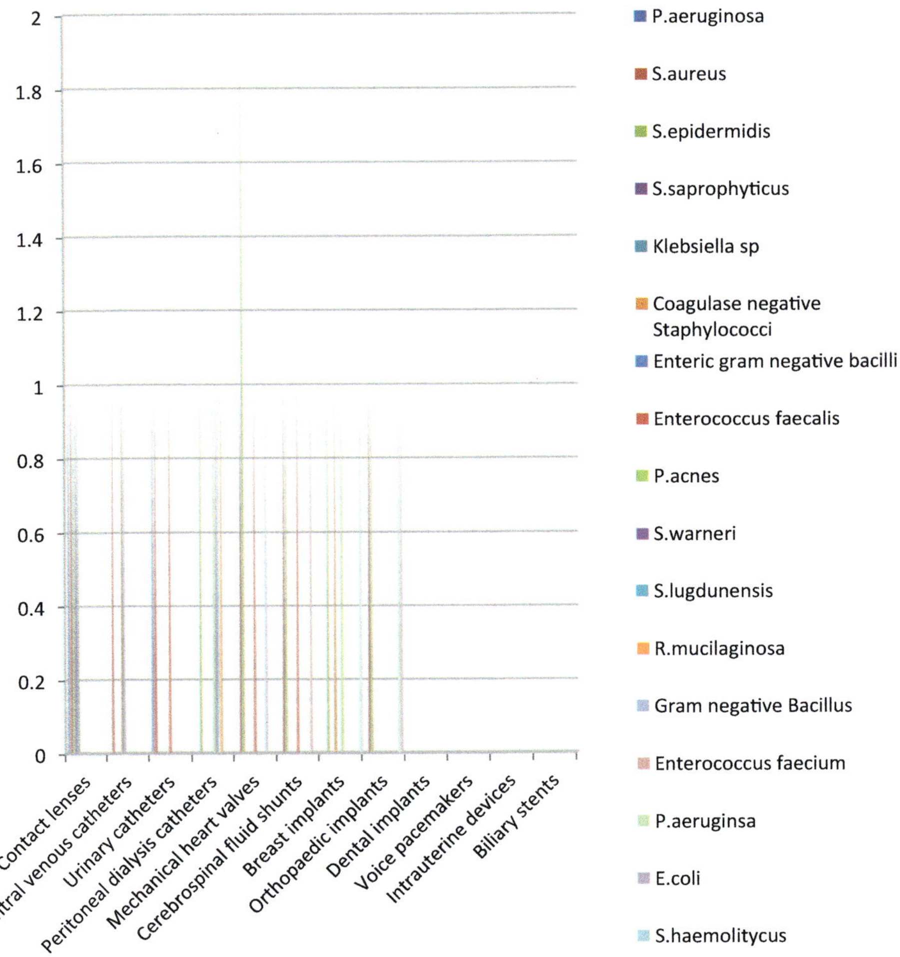

FIGURE 1.1 Clinically significant biofilm infections. *Source Muhammad, M. H., Idris, A. L., Fan, X., Guo, Y., Yu, Y., Jin, X., Qiu, Y., Guan, X., & Huang, T. (2020). Beyond risk: Bacterial biofilms and their regulating approaches. Frontiers in Microbiology, 11(928), 1–20.*

3. Biofilm matrix

Microorganisms produce an EPS such as proteins that include enzymes (<1–2%), 1% of DNA, 1%–2% of polysaccharides, and <1% of RNA along with water, which makes up to 97% and is hence considered as the important part of biofilm and is liable for the nutrient flow inside the matrix of biofilm. The two prime components of biofilm matrix are (1) water channel that helps in transport of nutrients and (2) intact packed cells without visible pores in it. These bacterial biofilms are usually ahead the entry of antibiotics and disinfectants and even escape the immune system of humans. Microbes that form biofilm adorned potential to cope up and offset the antibiotics that impact in treatment prolongation. Some genes are switched on biofilm-forming bacteria, which stimulate the articulation of stress genes that in turn change phenotypes resistant by some changes such as cell density, temperature or nutrition, osmolarity, and pH. Discrete components of biofilms are represented in Fig. 1.2, which imply the biofilm principle and what makes it resistant to a number of environmental factors (Jamal et al., 2015).

A single group of microbe can form a biofilm, but in attributes, biofilms relatively and consistently dwell a blend of mass species of bacteria along with fungi, yeasts, algae, protozoa, and other microbes together with abiotic detritus and erosion products. Microbial cells that are present in biofilms are nested inside an endogenously formed extracellular matrix (ECM) that inhere biomolecules such as proteins, polysaccharides, and nucleic acids (Okuda et al., 2018). The configuration and network of EPS may differ based on the microbial type, resident prune pressure, presence of substrates, and the environment of the host. The synthesis of EPS, its dimensional management, and grouping will conflict based on the single or multispecies of microorganisms present in the biofilm. So far, a disparate array of biomolecules has been evidenced, which can be compiled into two dominant groups: (1) cell surface

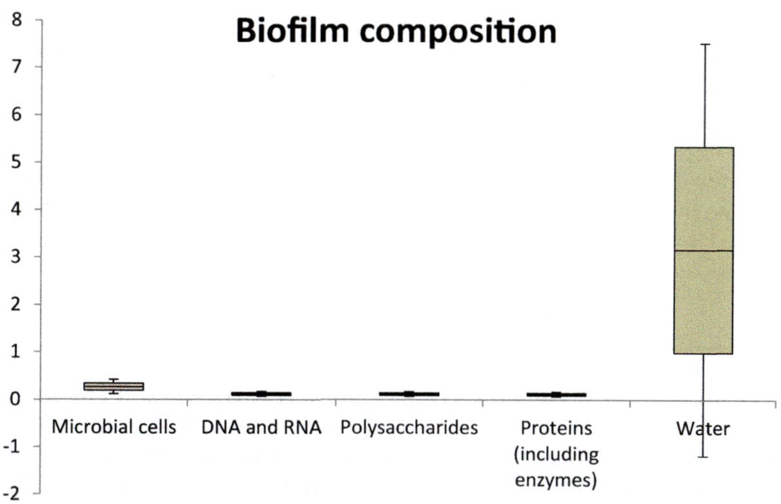

FIGURE 1.2 Biofilm composition. *Source Muhsin Jamal, Ufaq Tasneem, Tahir Hussain and Saadia Andleeb. 2015. Bacterial biofilm: Its composition, formation and role in human infections. Research Reviews: Journal of Microbiology and Biotechnology. 4(3): 1–14.*

associated and (2) extracellularly secreted (Karygianni et al., 2020). Biofilm formed by *Staphylococcus aureus* and *Staphylococcus epidermidis*, the components are extracellular DNA (eDNA), proteins and polysaccharide intracellular adhesion (PIA).

In different staphylococcal strains, the biofilm component and contribution varies. PIA synthesis plays a vital role in the staphylococcal biofilm, but there are some strains that are PIA independent. Apart from this, the proteins that are present in ECM also differ based on strains. For example, in *Pseudomonas aeruginosa*, minimum of three noticeable alginate exopolysaccharides seem to provide help for construction and development of biofilm. High alginate synthesis is seen in mucoid strains with uneven biofilm. The involvement of alginate is depicted in the basic biofilm formation, which is also liable for the strength of mature biofilm. Alginate biosynthesis genes are not expressed in nonmucoid strains, and Pel and Psl genes are intricate in the biofilm enactment. Biofilm formation is inhibited in *P. aeruginosa* by DNase I that shows eDNA is necessary for the biofilm to be established in the initial stage. Currently, it has been reported that an ECM that comprises nucleic acid, proteins, and β-glucan is synthesized by *Haemophilus influenzae* when it forms biofilm. Added to this, eDNA plays a prime component of ECM and is indispensable in maintaining the biofilm (Okuda et al., 2018).

4. Network of biofilm

When a precise microbe hooks up with facade of some matter in humid surroundings and commences to replicate, biofilms are formed. The appendage is initiated by the secretion of slimy, mucilaginous substance. Biofilm can scheme on any conceivable facade, viz., plastics, metals, medical implants, natural materials, contact lenses, kitchen counters, swimming pool/hot tub walls, animal andand human tissues, and many more. Microorganisms to sustain in the environmental stresses such as dehydration, UV radiation, inadequate nutrients, utmost pressure, intense temperature, high salinity, increased pressure, antimicrobial agents, and disinfectants form biofilm. The proceedings of bacterial biofilms are intricate. It is broadly considered that the initiation of biofilm formation is a reversible appendage of bacteria onto a facade dogged with the irreversible attachment, generally assisted by clinging bacterial structures and short-range interactions. EPS production supports the reversible attachment. Followed by this, they form a formulated structure entangled in an EPS matrix. At the end, bacterial cells leave the mature biofilm and diffuse into the external surface so that it can colonize novel niches. The biofilm formation phases are shown in Fig. 1.3 (Muhammad et al., 2020).

5. Phases in biofilm formation

Generally biofilm formation takes place by four steps: (1) attachment of the microbe to a surface, (2) formation of microcolony, (3) maturation, and (4) dissemination of microbes that may help to colonize new surface. The origin of biofilm formation is the early point of association of the microbe with a facade, which can be reversible repeatedly delineated by the cells that are inclined polarly. Following this, the microbe then begins to scheme a monolayer by producing an ECM to protect them. This matrix is composed of structural proteins, extracellular polysaccharides, nucleic acids, and cell debris, which are together known as

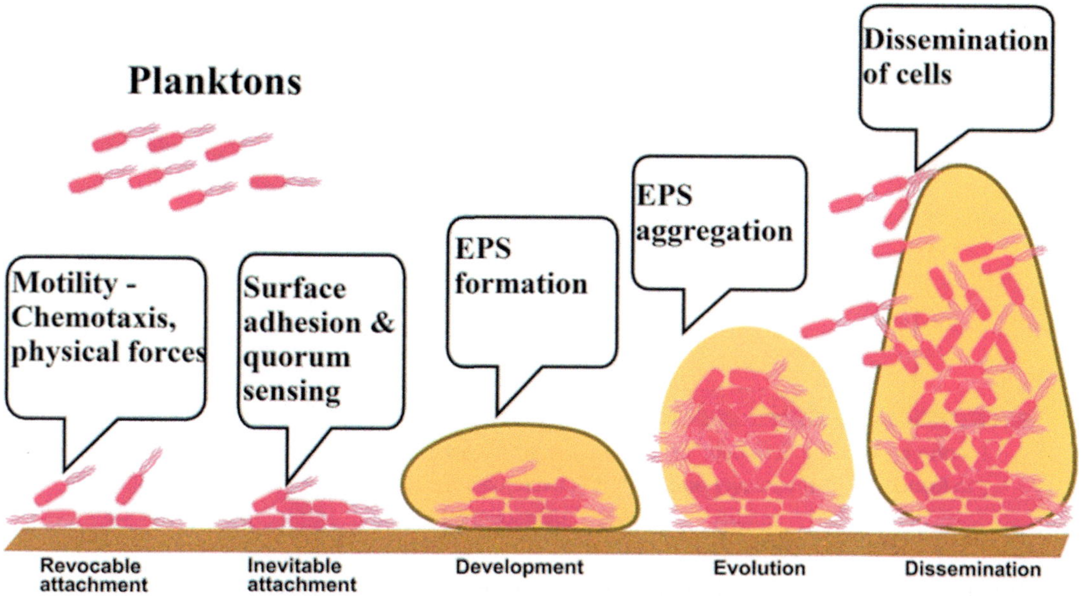

FIGURE 1.3 The five main phases leading to the development and dispersal of biofilm. *Source Muhammad, M. H., Idris, A. L., Fan, X., Guo, Y., Yu, Y., Jin, X., Qiu, Y., Guan, X., & Huang, T. (2020). Beyond risk: Bacterial biofilms and their regulating approaches.* Frontiers in Microbiology, 11*(928), 1–20.*

EPS. The extracellular DNA (eDNA) monopolizes these germinal processes of matrix formation followed by which structural proteins and polysaccharides proceed with the further steps. Formation of microcolonies is seen at this stage that display momentous growth, thereby helping quorum sensing/cell signaling process. In a number of gram-negative species, cyclic di-GMP (c-di-GMP) plays as a crucial intracellular signaling molecule. The biofilm structure is not a homogenous slime monolayer, but it is made up of heterogenous substances with water channels that transport oxygen and nutrients to the cells within them. They also entangle different kind of particles such as minerals, components from host including RBCs, platelets, and fibrin. High c-di-CMP is displayed in the biofilm state that stimulates the biofilm formation and also suppresses swimming motility by the flagella. A three-dimensional growth mechanism is evident in biofilm that makes the attachment irreversible from this stage. At the final stage, in a mature biofilm, few cells initiate to separate and discharge into the external environment as planktons to initiate a new cycle for the formation of biofilm (Armbruster & Parsek, 2018). The schematic representation of these steps involved in biofilm formation is shown in Fig. 1.4.

6. Role of biofilm

Microbes living in biofilm benefit many preferences such as escaping from a predator, dryness, risk to antimicrobial substances, enhanced procurement of nutrition, etc. Both beneficial and opportunistic pathogenic bacteria are supported by biofilms as their site of endurance,

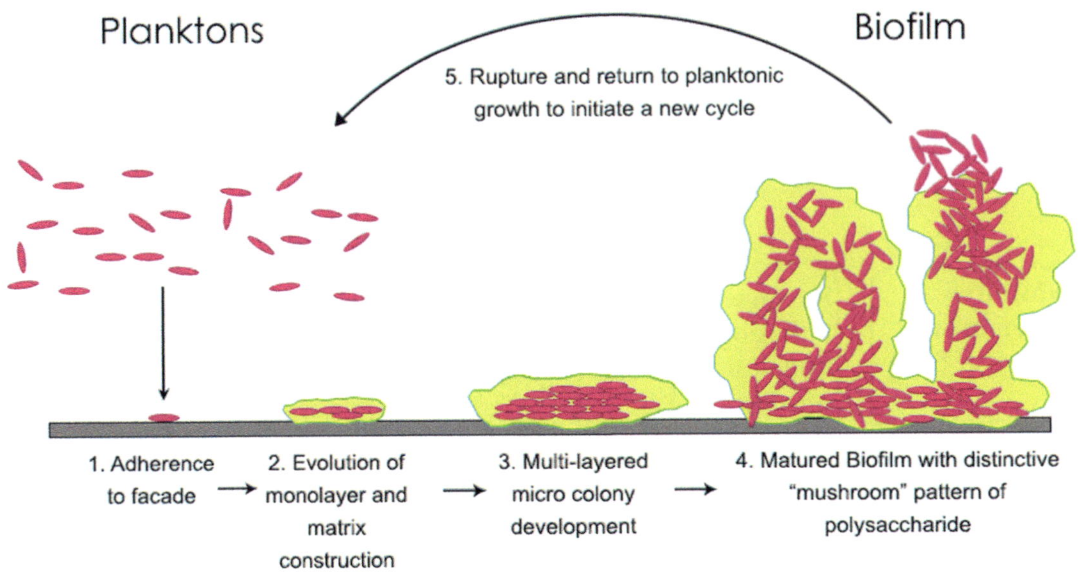

FIGURE 1.4 Graphical representation of biofilm formation. *Source Birte Hollmann, Mark Perkins, Dean Walsh. Biofilms and their role in pathogenesis. British society for Immunology.*

thereby preserving them and expanding the probable microbe for their survival in the environment. Streambed biofilms are also treated as valuable indicators for the quality of water and ecological status. In an aquatic environment such as lakes and rivers, it is really a peculiar concern to determine the reaction of biofilm against the anthropogenic pollution where the chemical compounds such as metals, drugs used as medicine, etc. are discharged into the water bodies, which are rising indefinitely from the wastewater treatment plants and industrial effluents, agricultural wastes, etc. Bacteria cells in biofilms are 1000 times low susceptible to antimicrobial agents than those present as planktons. In a biofilm, the entry of either nutrients or the antimicrobials is mediated by their diffusion through the matrix formed in the biofilm. Due to this, restriction in entry of substances results in the poor transport of materials; thereby making those resistant to antimicrobials, disinfectants, or even the inactivation of antimicrobial substance may also occur. Many researches had proved the resistance to antimicrobials by the organisms in biofilm because of this poor penetration of substances. Hence, on these aspects, biofilms are considered as thoughtful universal health concern as they are able to abide antibiotics, defense systems of the host, and even stress in the environment, which makes them causing continual chronic infections (McCarty et al., 2012).

Biofilms sometimes also play a positive role like *S. epidermidis*, which is the commensal bacterium blocking the establishment of pathogens in the skin by stimulating the host cell immune defense mechanism, thereby preventing the adhesion of the pathogens. Yet biofilms are mostly associated with the pathogenic infection causing microbes in humans, plants, and animals. Biofilm formation highly is evident in food and food processing environments. The formation of biofilm in such places results in risk of food safety. Biofilm formation in food industries may be due to the inadequate and indecisive cleaning practices followed there, which results in the bacterial colonization. Proper sanitization will be a challenge in biofilms as it makes the microbial cells endure tolerance to antimicrobials and disinfectants. Apart from this, biofilm formation and its endurance in several surfaces, which may be associated to medical, food, or even other sectors, would make them act as stockpile of numerous infectious pathogens. Discrete microbes can grow in food and form biofilm in food matrixes and utensils, which results in the contamination of the final product in the food industry as biofilms are hard to eradicate. It also serves for the transmission of disease, and also the shelf life of the products is also affected. Along with this, most predominantly biofilm makes the microbes resistant to the antimicrobials and disinfectants, which can affect the health and economy in global perspective (Abebe, 2020).

7. Supremacy of microbes in biofilm

In a biofilm, microbes possess intermicrobial interaction/cell signaling, which helps in the design of community with fine structure that permits cells to precede the metabolic activities, thereby increasing thorough function of the community. Bacteria that resist inside the biofilm unveil phenomenal resistance against stress in their surroundings, particularly to antibiotics. It is also said that microbial cells dispersed from a biofilm possess a high potential for the cause of cytotoxicity, mortality, and virulence than the planktons. The main cause of this resistance is based on many factors such as follows: (1) polysaccharide matrix prevents the penetration of antimicrobials, (2) inside the biofilm matrix, nongrowing cells may present or cells that generate response to stress in an unfavorable conditions may also be present; (3) high solidity of the bacterial cells that for mature biofilm that exhibit definite growth; (4) up/downregulation of drug targets; (5) the discharge of intracellular drugs by means of the transporter proteins and pumps; (6) stress response in the cells is changed by the upregulation of the resistance genes; (7) presence of persisters which are the cells that exhibit resistant phenotype. Persisters are formed spontaneously in the biofilm. They are dormant in nature, inactive metabolically, and do not divide like regular cells; apart from this, they are extensively recalcitrant to antibiotics and disinfectants. Biofilm mechanisms are complex and multifactorial (Santos et al., 2018). Biofilm formation in teeth is known as dental biofilms that exhibit symbiotic interaction between the organisms, which results in coadhesion and coaggregation of the bacteria and promotes colonization of bacteria over the teeth coated with saliva and accomplishing transitory and contiguous formation of completely organized biofilm communities. Sometimes, these dental biofilms also support the human systems by nutritional synergy by permitting the organism in the breakdown of complex salivary components. Quorum sensing is chiefly seen in microbial biofilms that help in regulation and development of the biofilm. It also provides other endeavor, which includes symbiosis, fruiting bodies or spore formation, production of bacteriocin, genetic competence, pathogenesis, and virulence (Li & Tian, 2016).

8. Detrimental effect of biofilm

Every year, infections caused by biofilm associated infection and their burden involved in health and services are poorly understood. In the United States, the nosocomial infections alone that are caused by bacteria and fungi associated with biofilm are estimated to be around 80%. There are three main groups that cause biofilm-related illness: they are (1) biofilm disease related to device, (2) chronic biofilm disease that is not related to device, and (3) malfunctioning of device due to biofilm disease. The snag in biofilm-related diseases is due to their recalcitrant nature, and treatment difficulties make it endure as the dominant obstacle throughout the world. Heavy doses of antibiotics and sometimes even a combination of many drugs for lengthened period of time are used to treat the disease caused by biofilm-associated disease. Such a treatment procedure can lead to undesirable consequences to the patient and also may lead to traumatic removal or replacement of the medical devices. This biofilm infection is also a major reason for high mortality rates. Hospitalized patients under medical devices are prone to chronic conditions by biofilm-associated microbial infections and extend the hospital stay with high medical costs to both private institutions and public globally. This biofilm-associated infections will also induce (1) malfunction of device, (2) biomaterial degradation, and (3) infections caused by diverse microbes. It also acts as a cause of systemic and local infections for patients admitted in intensive care unit. One such example is in patients affected by respiratory infection and immunocompromised in nature caused by *Candida* sp., when biofilm can support the growth of ventilator-associated bacterial pneumonia caused by *S. aureus* and *P. aeruginosa*. The polymicrobial biofilm that causes dental plaque on teeth surfaces, when not controlled properly, will cause periodontitis and dental caries. Main organisms involved in dental caries are streptococci, lactobacilli, and actinomycetes. Formation of biofilms on medical device–related infections is seen in inserted biomaterials that are implanted in human body that includes joint implants, catheters, etc. Biofilm in catheters may aggravate urinary tract infections. Likewise, biofilm formed in joint implants may result in further surgery for the removal or replacement of the implants. Many such examples that are caused by biofilm are urinary tract infections by *Escherichia coli*, *S. aureus* causing catheter infections, and *Haemophilus influenzae* that leads to child middle ear infections, gingivitis, and dental plaque are difficult to treat and may also be relapsing. Apart from this, *S. aureus* causing endocarditis, joint prostheses, and heart valve and *P. aeruginosa* that cause infections in patients with cystic fibrosis are also included in this list. The threatening fact in such biofilm infections is increasing mortality and morbidity. One most common acute ear infection is otitis media; in children, it may also be caused by biofilms; followed by this, biofilm-associated disease includes bacterial endocarditis that is affected in inner heart surfaces and heart valve, lung disease such as cystic fibrosis, *Legionella* biofilms that escapes from air and water from cooling or heating systems causes Legionnaire's disease, etc. Some nosocomial infections are also caused by biofilm microbes. The colonization of microbes in the form of biofilm is seen even in many household surfaces such as toilets, sinks, cutting boards, countertops in bath and kitchen. Poor sanitization and disinfection is the main reason for this biofilm formation in household surfaces that may lead to illness with the pathogens.

Biofilms are also seen to colonize in dead tissues and even inside the living tissues. They also develop in inanimate things associated with hospitalization such as prosthetic heart valves, pacemakers, contact lenses, CSF shunts, and breast implants. Though biofilm formation may be done by both gram-negative and gram-positive bacteria, the frequent organisms involved are *P. aeruginosa*, *S. epidermidis*, and *S. aureus*. A report says that the two-third of infections in indwelling devices is caused by staphylococcal species. *P. aeruginosa* colonizes and forms biofilm inside the surfaces of metal pipes at hospitals water systems. It also causes pulmonary infections in many patients. Periodontitis is not only seen in gums, but also it damages soft tissues and bones that support teeth, which are due to *Pseudomonas aerobicus* and *Fusobacterium nucleatum*. *P. aeruginosa* biofilm is also seen in chronic wound infections. When pathogenic microbes attach to dead bone, it leads to chronic osteomyelitis, which is also a biofilm infection.

The challenges caused by biofilm-associated pathogens are not only seen in animals and human beings but also present in plants. The pierce disease (citrus canker) that occurs in grapevines is caused by *Xanthomonas citri* biofilm. *Xylella fastidiosa* is one of the substantial biofilm producers and will block the plant vasculature and cause pierce's disease of grapevine. Other examples are the bean leaves where brown spot disease is caused by biofilm-forming *Pseudomonas syringae*; short-term mortality in *Ocimum basilicum* and *Arabidopsis* roots is occurred by *P. aeruginosa* biofilm; an important plant pathogen name *Ralstonia solanacearum* colonizes in the xylem vessels by causing wilt disease and also forming biofilm.

Biofilm challenges are also experienced in food industries. Inside food industries, biofilms are formed with or without contacting the foods. About 60% of foodborne outbreaks are due to biofilms. Hence, it implies a serious risk in food safety. Here, in food industry, major contaminants are harbored from equipment, surrounding air and surfaces of food. Later, biofilms that grow on these environments will cause food spoilage, resulting in serious health risk to public and environment along with high economic consequences. *Listeria monocytogenes* tops the spoilage organism and foodborne infection by forming biofilm. Biofilms are also a reason for severe technical challenges of food industry, which forbid the heat flow across equipment surfaces, boost up the frictional resistance at the surfaces, and develop the surface corrosion rate, resulting in efficient production loss. The contamination risk of equipment and instruments and contamination of food by bacteria are directly mediated by biofilm-associated pathogenic microorganisms (Muhammad et al., 2020).

Inside the drinking water systems, one of the most prevailing microbial growths is by the formation of biofilms. The significant trouble in the distribution systems of drinking water is this biofilm formation. This contaminated water when consumed will cause disease outbreak that is liked to human diseases. Some bacteria that form biofilm in drinking water systems are *P. aeruginosa, Legionella pneumophilia, Campylobacter jejuni, Aeromonas hydrophila, Klebsiella pneumoniae*, and *Mycobacteria*. Biocorrosion of the water channel pipes is also caused by the biofilm microbes, as they escape from the biofilm matrix, which they form on the inner surface of the pipelines. This may also affect the water quality by changing the color, odor, taste, and turbidity and even reducing heat exchange efficacy. Metal corrosion by biofilm-producing bacteria chiefly belongs to sulfur-oxidizing/reducing bacteria, iron reducers/oxidizers, and manganese oxidizers. Together with all these reasons, biofilm-forming bacteria change the quality of drinking water affecting the pipelines and make them unsafe (Chan et al., 2019).

9. Mechanism of action in a biofilm network

Biofilm supports the microbial network in many ways, and some are discussed in the following.

9.1 Glycocalyx (capsules)

This makes the integral part of the biofilm with the thickness ranging from 0.2 to 1.0 μm, which is present in both gram-positive and gram-negative bacteria. The attachment of the biofilm to a solid surface is made possible by glycocalyx with the help of hydrogen bonds, electrostatic, and Van Der Waal forces, thereby creating adhesion and cohesion and also helping in biofilm maturation. Due to the regulated and flexible composition of glycocalyx, the growth of the pathogenic bacteria is supported by making host sustain in the intense environmental conditions. Different environmental conditions influence the components such as polysaccharides and glycoproteins present in the capsules of biofilm. Glycocalyx matrix also supports the bacteria to resist against antimicrobial agents. Antibacterial molecule, up to 25% of weight of the glycocalyx layer, is accumulated inside it. Biocide transport is limited by the matrix adsorption sites and also serves as an adherent to exoenzymes. Some antibacterial agents motility is inhibited by exoenzymes as it acts as the substrate source metabolite degradation of biocide; finally, it delays the susceptible drug activity.

9.2 Resistance mediated by enzymes

Enzymes act as a catalyst in the conversion of bactericide into a nontoxic material, thereby providing resistance to biofilm. Heavy metals such as cadmium, nickel, mercury, silver, antimony, zinc, copper, cobalt, lead, etc., aromatic compounds, and phenolic compounds are toxic in nature, which is degraded by few bacterial species. Ions and genes resistance to metals are detoxified by enzymatic reduction. A wide spectrum of resistant phenotype is enhanced by the presence of heavy metals.

9.3 Rate of proliferation and metabolism

Inside the biofilm, metabolic activity and growth rate of the microbial population are altered by the availability of oxygen and nutrients. The bacterial activity and growth inside the biofilm was influenced by the metabolic product and substrate concentration. At the fringe of a biofilm, nutrients and oxygen enhance the cells metabolic activity, thereby encouraging the bacterial growth. Added to this, cell growth is very slow inside the biofilm due to the meagre diffusion of the nutrients, thereby exorbitant to the metabolic activity. Antimicrobial agents attack the bacteria that are metabolically active, but those that are in sluggish state bacteria are becoming resistant, thereby gain protection against these biocides action. The availability of oxygen inside the biofilm rules the bacterial metabolism.

9.4 Paradox of cell endurance

Persisters are chiefly seen in biofilms that show tolerance against antimicrobials and even make it as multidrug resistant. In a bacterial community, these persisters are restrained by their stages in the growth. These types of cells are swiftly proliferated and sustain in the existence of antimicrobials only in the lethal dose. Persisters are chiefly produced in the stationary phase, which also supports the biofilm resistance. Some studies reported that when stationary phase is diluted, it declines the persisters. The protein responsible for the multidrug resistant is synthesized by the persisters by targeting the antibiotic targets (Fig. 1.5).

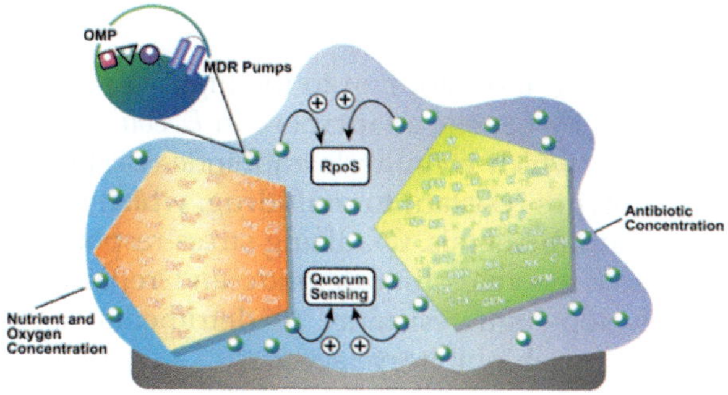

FIGURE 1.5 Drug resistance in biofilm. *Source Thien-Fah C. Mah and George A. O'Toole. Mechanisms of biofilm resistance to antimicrobial agents. TRENDS in Microbiology. 2001;9(1):34–39.*

9.5 Quorum sensing

The biofilm maturation is controlled by quorum sensing in many bacterial species. The synthesis of degradative enzyme regulation in a biofilm matrix is persuaded by the quorum sensing systems. Biofilms become susceptible to kanamycin; when there is a deficiency in the quorum sensing, this leads to the formation of thin biofilm and reduced EPS production and is known as deficient biofilm. Hence, from this, it is evident that quorum sensing system related to biofilm indirectly or directly influences the environmental stress.

9.6 Responding to stress

Resistance to cellular stress is developed by the stress response that happens in a biofilm. Cell envelope formation and fimbriae is controlled by the stress response in *E. coli* and *Salmonella enteritis* over *Typhimurium.* The precautionary factor for the cellular damage is the stress response. Microbial cells anchored in a biofilm matrix upsurge its resistance to antimicrobials due to the alteration of gene expresses that occurred by the general stress response.

9.7 Metabolism of microbes in a biofilm

The intrusion of biofilm-specific growth inside a biofilm will also interpret the biofilm resistance. The bacterial response to antimicrobials changes substantially based on the cells physiological state and habitat. The envelope of bacterial cell is altered in the scarcity of nutrients and also affects the barrier composition. When a biofilm is divulged to the inhibitory concentrations of antimicrobials, the resistant cell community adapts to the phenotypic changes. One such example is *E. coli* explicit resistant to H_2O_2 or UV light when it undergoes a stress of starvation or heat. But when the antimicrobials are removed, the resistant phenotypes are lost. Hence, it shows that when nutrition in a biofilm is narrowed, the growth of the cells is slowed and it gets into starvation state. Upon exposure with bactericides in a biofilm will lead to deficit in their respiratory action because of their activity adjacent to the biofilm–bulk fluid interface. When the cells grow in the media that contains nutrients for high growth rate, nongrowing cells are less susceptible to a diverse antimicrobials.

9.8 Genetic adaptation

The reduction in susceptibility and adoption to the highly protective recognizable phenotypes are seen inside a biofilm by genetic adaptation. In *E. coli*, various gene expressions are controlled by global regulator, which is multiple antibiotic resistance (*mar*) operons; it helps in the development of phenotypes resistant to multidrug, which includes resistant to organic solvents, disinfectants, and antimicrobials. Oxidizing stress response is seen in many fermentative bacteria that synthesize enzymes that degrade and repair oxidants. These cells that exhibit stress response develop high resistance to detrimental factors in short period of time when they are exposed to subinhibitory factors.

9.9 Envelope structure

When the cell envelope which is the outer membrane structure is modified, the penetration of antimicrobials inside the cell is inhibited, thereby making the cell resistant. Lipopolysaccharide membrane that is present as the outer coving of the cell prohibits the entry of hydrophilic antimicrobials, but the hydrophobic agents are exempted by the outer membrane protein. The antimicrobial resistance by mycobacterial cell wall is very imperative. Any changes in the protein profile of outer membrane cover of the cell cause change in the response to antimicrobials.

9.10 Efflux pumps

Survival of the bacteria in extreme conditions and against which include the presence of antimicrobials is mediated by the efflux systems. These systems provide acquired and intrinsic resistance to diverse or same group of antimicrobial agents. Hence, when efflux pump is overproduced, it helps the bacteria by exerting multidrug resistance character. Along with this, efflux pumps also serve by providing some other mechanisms such as antibiotic inactivation and modification of the target. Biofilm on wound surface is formed by the *Acinetobacter baumannii* that is present in the hospital environment and possesses multidrug resistance. Many researchers have said that plasmids that are responsible for resistance to antibiotics, biocides, dyes, and detergents seem to possess efflux pump encoding genes. When a bacterial biofilm is exposed in minimal concentration of antibiotics such as chloramphenicol, tetracycline, and few xenobiotics such as chlorinated phenols and salicylate, it triggers the efflux pumps and multidrug resistance expressions. One such example is *P. aeruginosa* mutant strain, which has high expression of efflux pump and plays an important role in the descending of susceptible nature to antibiotics by the biofilms formed by them.

10. Antibiotic resistance by biofilm

As discussed before, various reasons are responsible for the resistance to antibiotics that are exerted by a biofilm community. Some are incomplete or slow penetration of antimicrobials, change in environmental chemical compositions, and the population of microbes in a biofilm. These properties also make the biofilm a multicellular nature and exhibit resistance against antibiotics, finally resulting in the treatment failure. The key underlying factor for the resistance to antibiotics by biofilms is their multicellular nature. There are many ongoing researches in the production of biofilm as multicellular developmental process. The functioning of biofilm as multicellular system is mediated by the EPS by holding together the bacterial cells and making the environment heterogenous inside the biofilm. In a well-organized development of biofilm, inter- and intracellular signaling takes place. Upregulation and downregulation of genes and proteins mediate the attachment of bacteria in a surface. Followed by this, quorum sensing regulates the maturation of biofilm. Antibiotic resistance is mainly due to the multicellular nature of the biofilm, so when it is disrupted, the efficacy of the antimicrobials along with host defense mechanism increases and helps in the swift treatment and recovery from the persistent infection. A serious risk is imposed to medical treatment of pathogenic infections to humans, which are caused by this biofilm formation and their multicellular nature.

11. Antibiofilm approaches

Biofilm-associated diseases are difficult to treat due to their increase in antibiotic resistance nature in the bacterial communities. Regular antimicrobial chemotherapy is not effective in evacuating the microbial population situated in the ore of the biofilm, thereby making the position more worsen throughout the world. Hence, to defeat the bacterial biofilms, alternative methods and antibiofilm agents are being used. Different groups of compounds act as antibiofilm molecules, which inhibit the formation of biofilms that are from natural sources while some are synthetic or chelating agents. The formation of biofilm in *S. aureus* and *E. coli* is inhibited by the mature biofilm dispersal by some naturally synthesized small molecules such as D-amino acids and polyamine norspermidine. In a biofilm dispersal method, these molecules are implemented as the antibiofilm agents. It has been shown that NAC and tween 20, which act as antibiofilm molecule when combined with antibiotics, are efficient against nonpigmented biofilms. Tween 20 is highly efficient than NAC in mycobacterial biofilm. Biofilm degradative enzymes are used for the degradation of biofilms: they are Dispersin B, DNase I, and α-amylase. When the structural component of biofilm is degraded, it enhances the inflow of antibiotics inside the matrix, thereby making them more efficient. So these biofilm degrading enzymes will inhibit the formation of biofilm and also denature the mature biofilms of many bacteria such as *S. aureus*, *Vibrio cholerae*, and *P. aeruginosa*. In other cases, the quorum sensing signaling cascade is confused as it controls the biofilm formation. Quorum sensing signals of bacteria are disrupted by halogenated furanone, which is secluded from marine algae named *Delisea pulchra*. Some of the plant extracts such as garlic, ginseng, etc. attenuate the quorum sensing signals of bacteria. Azithromycin and usnic acid also contain inhibitory activity against fungal and bacterial biofilms. The activity of antibiotics is enhanced in the *P. aeruginosa* biofilms by the dispersion of signaling nitric oxide molecule. Many bioactive components are synthesized by bacteria and actinomycetes that exhibit antibiofilm properties. Some natural products such as 4-phenylbutanoic acid express promising antibiofilm activity on both gram-positive and gram-negative bacteria (Sharma et al., 2019).

Synthetic halogenated furanone is a secondary metabolite synthesized in *Dilsea pulchra*, which is a marine alga from Australia, which has the ability to interfere with bacterial signaling process and swarm cells motility. These compounds inhibit the aggregation in the surface with the compatible bacteria. It targets the *rhl* system that is responsible for the quorum sensing and also intrudes the *P. aeruginosa* biofilm matrix; hence, it collapses the genes associated to quorum sensing of biofilm maturity. The structure of biofilm is also changed by this molecule; thereby, removal of bacteria from the community takes place that leads to the biomass loss from the biofilm matrix. Some of the polyphenol components such as tannic acid, EGCG, and ellagic acid also have the same property and inhibit biofilm formation. But these compounds are required in large concentrations than furanones. A flavonoid named quercetin, which influences the quorum sensing and is used as the antibiofilm agent. The adherence mechanism during the biofilm formation is reduced by this compound by the inhibition of alginate production. Also production of EPS is reduced; thereby, poor attachment of bacteria occurs. Curcumin from *Curcuma longa* is a phytochemical that also exhibits similar antibiofilm properties.

The EPS of the biofilm acts as the protective layer for the microbial communities inside the biofilm matrix against the antimicrobials. Hence, EPS disruption results in the release of microbial cells from the biofilm to the antimicrobial agents. Few enzymes such as DNases and polysaccharide lyases disrupt the EPS. Enzymes DNase I and Dispersin B are the chief antibiofilm agents. eDNA seen inside the structure of biofilm is digested by DNase I and β 1-6 N-acetylglucosamine (PNAG); an EPS substance that helps in the bacterial attachment is cleaved by the Dispersin B, which is a glucoside hydrolase. EPS layer situated in the medical devices is also disrupted by this. The effectiveness of these substances is high when combined with the antimicrobial agents for bactericidal effect inside EPS.

The biofilm formation is constrained by creating a rift in the peptidoglycan layer and occurs in the cell wall of many bacteria. Tannic acid is one such substance that prevents the biofilm formation without affecting the growth of bacteria in *S. aureus*. It depends on the lytic transglycosylase, which is an immune dominant staphylococcal antigen A (Isa A); thereby, breakdown of peptidoglycan takes place. Transglycolases are lysozyme-like enzymes. The extracellular level of Isa A is enhanced by tannic acid, thereby inhibiting biofilm formation. Epigallocatechin gallate is a polyphenol that binds with peptidoglycan that interferes docking phase in the formation of biofilm (Roy et al., 2018).

It is a multistep process that includes disintegration of EPS by changing the cell physiology. This disassembly is carried out in many bacterial species by extracellular enzyme and surfactant production, which causes both solubilization and degradation of components that help in adhesion in a biofilm matrix. DNases, proteases, and surfactants also act as matrix-degrading gene product. Biofilm matrix disassembly is also carried out by the synthesis of an extracellular protease. Endogenous mediator of biofilm disassembly process is the nuclease enzyme. It is effective in separating cells from biofilm matrix. Likewise, ethyl pyruvate also possesses commendable antibiofilm activity.

Antimicrobial peptides are also an effective antibiofilm agent. These are positively charged molecules that penetrate the lipid bilayer and solubilize the aquatic environment. They bind electrostatically with lipopolysaccharides. The devitalization of lipid head groups are made possible by forming numerous pores and disrupting the cellular integrity.

The antimicrobial mode of action also changes on the alteration in the membrane permeability. This leads to the cytoplasmic membrane disruption by formation of pore; thereby, efflux of intracellular materials occurs. Lantibiotics are one such peptide molecules that are synthesized by ribosomes of gram-negative bacteria that act as the antibiofilm agent. One of the well-known lantibiotic is nisin; it forms complex with lipid molecules and inhibits biosynthesis of the cell wall. It also induces cytoplasmic membrane permeability by forming pores in short time. Another example is subtilin, which is similar to nisin and releases the cytoplasmic solutes from *Bacillus subtilis* and *Staphylococcus simulans*.

An important factor for the bacteria to survive in biofilm is cell division. Accumulation of silver in the intracellular vacuoles damages the plasma membrane and alters the electric potential and prevents cell division. Cytoplasmic proteins are inhibited by some antimicrobial peptides that have a role in cell division and survival. Negatively charged cell membranes are disrupted by a natural polymer known as chitosan as it is cationic in nature. Some antimicrobial peptides are able to penetrate outer membrane, there by inhibiting the DNA and protein synthesis, which are the basic components of the biofilm. Some exopolysaccharide also inhibits biofilm formation. There are some molecules that act as antibiofilm molecules, but their mechanisms are not known clearly. Few such examples are fisetin and esculetin, which are secondary metabolites that also inhibit biofilm.

12. Conclusion

Formation of biofilm takes place in well-regulated and persistent events. These are seen in many places and may be natural or manmade environments. The bacterial ability to form biofilm on surfaces is treated as hazardous destructive issue in many fields associated to medical devices, food industry, water treatment industries, pharmacy, healthcare services, and many more. Biofilm infections are considered dangerous as they develop resistance against prevailing antibiotics, thereby extending the hospitalization time and expenditure. There is a need of propitious treatment strategies for biofilm-associated infections. The biofilm formation is becoming a global issue. Formation of biofilm in an environment is mediated by a series of steps, making the microbial community stronger and resistant against antimicrobials, disinfectants, and environmental stress factors. Microbial population inside a biofilm matrix is benefited in many ways such as making them resistant to antimicrobials, which are carried out in many ways such as formation of EPS, cell endurance, quorum sensing, stress tolerance, genetic adaptation, efflux pumps, etc. Hence the control and dissemination of microbial biofilms is much needed for effective treatment and recovery of biofilm-related infections. This can be made effective by the use of antibiofilm strategies such as quorum quenching compounds and their combination with antimicrobials, dispersion of EPS in biofilm, disruption of biofilm, peptidoglycan, lipopolysaccharide, membrane permeability changes, cell division inhibition, and many more. Since the colonization of biofilms is evident everywhere especially in medical devices, their control measures have to be taken in a more effective way to avoid complications. This is because biofilm microbes are more resistant than the planktonic microbial population. The biofilm formation method, growth, and its resistance mechanisms have been researched by many scientists to get an effective treatment against biofilm-associated microbes. Much more research is needed in this perspective for the control of these types of microbial communities.

References

Abebe, Gedif Meseret (2020). The role of bacterial biofilm in antibiotic resistance and food contamination. *International Journal of Microbiology*, 1–10.

Armbruster, Catherine R., & Parsek, Matthew R. (2018). New insight into the early stages of biofilm formation. *Proceedings of the National Academy of Sciences of the United States of America, 115*(17), 4317–4319.

Chan, S., Pullerits, K., Keucken, A., Persson, K. M., Paul, C. J., & Radstrom, P. (2019). Bacterial release from pipe biofilm in a full-scale drinking water distribution system. *NPJ Biofilms Microbiomes, 5*, 9.

Donlan, R. M. (2002). Biofilms: Microbial life on surfaces. *Emerging Infectious Diseases, 8*(9), 881–890.

Jamal, Muhsin, Tasneem, Ufaq, Hussain, Tahir, & Andleeb, Saadia (2015). Bacterial biofilm: Its composition, formation and role in human infections. *Research Reviews: Journal of Microbiology and Biotechnology, 4*(3), 1–14.

Karygianni, L., Ren, Z., Koo, H., & Thurnheer, T. (2020). Biofilm matrixome: Extracellular components in structured microbial communities. *Trends in Microbiology, 28*(8), 668–681.

Li, Y., & Tian, X. (2016). Microbial interactions in biofilms: Impacts on homeostasis and pathogenesis. In D. Dhanasekaran, & N. Thajuddin (Eds.), *Microbial biofilms—importance and applications*. London: IntechOpen.

McCarty, S. M., Cochrane, C. A., Clegg, P. D., & Percival, S. L. (2012). The role of endogenous and exogenous enzymes in chronic wounds: A focus on the implications of aberrant levels of both host and bacterial proteases in wound healing. *Wound Repair Regen, 20*, 125–136.

Muhammad, Musa Hassan, Idris, Aisha Lawan, Fan, Xiao, Guo, Yachong, Yu, Yiyan, Jin, Xu, Qiu, Junzhi, Guan, Xiong, & Huang, Tianpei (2020). Beyond risk: Bacterial biofilms and their regulating approaches. *Frontiers in Microbiology, 11*(928), 1–20.

Okuda, Ken-ichi, Ryuichi, Nagahori, Yamada, Satomi, Sugimoto, Shinya, Sato, Chikara, Sato, Mari, Tadayuki, Iwase, Hashimoto, Kazuhiro, & Yoshimitsu, Mizunoe (2018). The composition and structure of biofilms developed by propionibacterium acnes isolated from cardiac pacemaker devices. *Frontiers in Microbiology, 9*(182), 1–12.

Percival, S. L., Malic, S., Cruz, h., & Williams, D. W. (2011). Introduction to biofilms. In S. Percival, D. Knottenbelt, & C. Cocchrane (Eds.), *Biofilms and veterinary medicine* (p. 6). Springer series on biofilms.

Percival, S. L., Suleman, L., Vuotto, C., & Donelli, G. (2015). Healthcare-associated infections, medical devices and biofilms: Risk, tolerance and control. *Journal of Medical Microbiology, 64*(Pt 4), 323–334.

Roy, R., Tiwari, M., Donelli, G., & Tiwari, V. (2018). Strategies for combating bacterial biofilms: A focus on antibiofilm agents and their mechanisms of action. *Virulence, 9*(1), 522–554.

Santos, A. L. S. D., Galdino, A. C. M., Mello, T. P., Ramos, L. S., Branquinha, M. H., Bolognese, A. M., Columbano Neto, J., & Roudbary, M. (2018). What are the advantages of living in a community? A microbial biofilm perspective. *Memórias do Instituto Oswaldo Cruz, 113*(9), e180212.

Sharma, D., Misba, L., & Khan, A. U. (2019). Antibiotics versus biofilm: An emerging battleground in microbial communities. *Antimicrobial Resistance and Infection Control, 8*, 76.

Further reading

Allewell, Norma M. (2016). Introduction to biofilms thematic minireview series. *Journal of Biological Chemistry, 291*(24), 12527–12528.

Birte Hollmann, Mark Perkins, Dean Walsh. Biofilm and their role in pathogenesis. Bitesized immunology. British Society of Immunology. 1–4.

Bogino, Pablo C., Mercedes Oliva, Maria de las, Sorroche, Fernando G., & Giordano, Walter (2013). The role of bacterial biofilms and surface components in plant-bacterial associations. *International Journal of Molecular Sciences, 14*, 15838–15859.

Gebreyohannes, Gebreselema, Nyerere, Andrew, Bii, Christine, & Sbhatu, Desta Berhe (2019). Challenges of intervention, treatment, and antibiotic resistance of biofilm-forming microorganisms. *Heliyon, 5*(8), e02192.

Hobley, Laura, Harkins, Catriona, MacPhee, Cait E., & Stanley-Wall, Nicola R. (2015). Giving structure to the biofilm matrix: An overview of individual strategies and emerging common themes. *FEMS Micriobiology Review, 39*, 649–669.

Li, Yuanzhe, Xiao, Peng, Wang, Yilin, & Yu, Hao (2020). Mechanisms and control measures of mature biofilm resistance to antimicrobial agents in the clinical context. *ACS Omega, 5*, 22684–22690.

Mann, Ethan E., & Wozniak, Daniel J. (2012). *Pseudomonas* biofilm matrix composition and niche biology. *FEMS Microbiology Reviews, 36*, 893–916.

Singh, S., Singh, S. K., Chowdhury, I., & Singh, R. (2017). Understanding the mechanism of bacterial biofilms resistance to antimicrobial agents. *The Open Journal of Microbiology, 11*, 53–62.

Vasdev, Kavita, Dewasthale, Sakshi, & Mani, Indra (2018). Microbial biofilm: Current challenges in health care industry. *Journal of Applied Biotechnology and Bioengineering, 5*(3), 160–164.

Wickramasinghe, Nirmani N., Hlaing, Mya M., Ravensdale, Joushua T., Coorey, Ranil, Scotr Chandry, P., & Dykes, Gary A. (2020). Characterization of the biofilm matrix composition of the psychrotrophic, meat spoilage pseudomonads. *Scientific Reports, 10*, 16457.

CHAPTER

2

Genetics of microbial biofilm development

Ashmita Samanta[1], *Drisha Roy*[1], *Dibyajit Lahiri*[1], *Rina Rani Ray*[2] and *Moupriya Nag*[1]

[1]**Department of Biotechnology, University of Engineering and Management, Kolkata, West Bengal, India;** [2]**Department of Biotechnology, Maulana Abul Kalam Azad University of Technology, Haringhata, West Bengal, India**

1. Introduction

At solid—liquid, solid—air, liquid—liquid, and liquid—air interfaces in a variety of settings, biofilms are organized and structured aggregates made up of microbial populations. According to recent studies, biofilm formations seen on the mucosal surfaces of numerous organs, medical devices, and wounds are responsible for approximately 80% of human infections. The biofilms can be made up of a single microbial layer of cells or cocultures embedded in an exopolymeric matrix, which is a highly hydrated matrix. The matrix includes key components such as microbial biopolymers such as polysaccharides and proteins along with glycoproteins, nucleic acids, and lipids (Carradori et al., 2020). Because microorganisms associated with biofilms have drastically decreased susceptibility to antimicrobial treatments, biofilms are very important for public health. This sensitivity could be inherited (since the biofilm formed naturally) or learned (due to the transfer of extrachromosomal elements to susceptible organisms in the biofilm). Because standard microdilution investigations concentrate on the behavior of planktonic (suspended) rather than biofilm (surface-associated) organisms, they are unable to determine whether biofilms are sensitive to antimicrobial medicines. Instead, direct susceptibility testing must be done against organisms found in biofilms, especially in situations that closely resemble in vivo conditions (Donlan, 2001).

In contrast to plasmids, transposons, and mutations, the mechanism of antimicrobial resistance and immune response in biofilm-related infections is strictly connected to (1) physical and chemical penetration barriers; (2) stress response activation; (3) nonstandard microbial development and form; (4) metabolic resistance that is active and linked to a tough bacterial

Microbial Biofilms
https://doi.org/10.1016/B978-0-323-95715-1.00001-7

response; and (5) emergence of new phenotypes (biofilm-related) as (persisters). Planktonic cells with no genetic resistance can be killed by antibiotics, but as they develop into biofilms, they can become 1000 times more resistant to the same treatment options (Weigelt et al., 2021). The removal of devices that have been contaminated with biofilm is the most efficient method for treating a number of illnesses, including bloodstream infections and urinary tract infections. Clinical decision-making could be affected if blood samples and catheter tip samples are collected and assessed in alternative ways or if the limitations of traditional medications for treating biofilm-associated disorders are better understood (Donlan, 2001).

Biofilms are made up of 75%–95% extracellular polymeric substance (EPS) and only 5%–25% bacteria. EPS components from the host ECM may be tweaked or created independently. In contrast to biofilms, which are free-floating microorganisms in an aquatic environment, planktonic bacteria are the traditional focus of microbiology research. The transition of planktonic bacteria to a biofilm lifestyle could be hazardous to a host because biofilms are notorious for resisting host defenses and drugs (Weigelt et al., 2021). Chronic wounds (60%) exhibit biofilms more frequently than acute wounds (6%). Biofilms are more frequently present in chronic wounds, which is one of the theories explaining why this results in slower healing. It has been established that biofilms prevent proper epidermal differentiation after healing, interfere with host immunological response, limit wound contracture and epithelialization, and induce prolonged inflammation. In contrast, the environment created by chronic wounds promotes the establishment of biofilm. It is unclear whether biofilms cause chronic wounds or the opposite, while both hypotheses are likely true. Even in some acute wound models where delayed healing is not present, biofilms may prevent or delay epidermal barrier regeneration and increase the risk of ulcer formation (Roy et al., 2014; Wu et al., 2019).

Biofilm growth and dispersal are closely controlled processes driven by genetic and environmental factors. Current studies suggest that quorum sensing (QS), bis-(3′-5′)cyclic diguanosine monophosphate (c-di-GMP), and short RNAs (sRNAs) are the main regulators of microbial biofilms, at least in some gram-negative species (Wolska et al., 2016).

2. Quorum sensing

Small, self-produced signal molecules, known as autoinducers, are the building blocks of QS, which is assumed to be a special language utilized for intercellular communication. When the number of bacteria and autoinducers reaches a critical mass, the bacteria start to notice this and react by activating or repressing the target genes (Wolska et al., 2016). Genes that are regulated by QS may make up 10%–15% of the bacterial genome. QS systems are essential for the growth and transmission of microbial biofilms. Although not engaged in attachment or the initial stages of biofilm formation, these systems are crucial for the later stages of the development of the biofilms and serve as the key moderators of biofilm dispersal (Davies et al., 1998).

The discovery that *Pseudomonas aeruginosa* produces at least two extracellular signals involved in cell-to-cell communication and cell density–dependent expression of many secreted virulence factors suggests that cell-to-cell signaling may play a role in the differentiation of *P. aeruginosa* biofilms, much like it does in the development of specialized structures in fruiting bacteria such as *Myxococcus*. In addition to two LasIR and RhlIR systems based on N-acylated homoserine lactone (AHL) and one based on the *Pseudomonas* quinolone signal, this bacterium contains at least three QS systems (PQS). AHL sensor/transcriptional regulator lasR, an autoinducer lasI required for the synthesis of N-(3-oxo-dodecanoyl)-L-homoserine lactone, and an autoinducer rhlI required for the synthesis of N-(butanoyl)-L-homoserine are present in both AHL systems (C4-AHL). OMV biogenesis is brought on by the 2-heptyl-3-hydroxy-4-quinolone, a sensing molecule in the PQS system, which is carried by OMVs. The hierarchy of control between the three systems is as follows: RhlR negatively regulates PQS, while LasR positively regulates RhlR and PQS (Pearson et al., 1994; Wolska et al., 2016). *P. aeruginosa* has two cell-to-cell signaling systems: the lasR-lasI and rhlR-rhlI systems. The lasI gene product controls the synthesis of the diffusible extracellular signal N-(3-oxododecanoyl)-L-homoserine lactone ($3OC_{12}$-HSL). When $3OC_{12}$-HSL concentrations are sufficient, the lasR product, a transcriptional regulator, activates a number of pathogenic strains, including lasI and the rhlR-rhlI system (Davies et al., 1998).

Rhamnolipids are essential for maintaining the channels in mushroom-shaped structures that allow for the proper distribution of nutrients and oxygen as well as the removal of waste products during the final stages of biofilm formation. QS signaling regulates the synthesis of rhamnolipids. The stimulation of rhamnolipid synthesis in the center of biofilm mushroom caps is consistent with QS's regulation role. Overproduction of these biosurfactants leads to the dispersion and separation of the biofilm from the surface. QS is also implicated in the release of a sizable amount of eDNA when a bacterial subpopulation goes through autolysis at a late stage of biofilm formation. Although the precise method by which PQS regulates autolysis is yet unknown, this is a widely known fact (Boles et al., 2005; Davey & O'toole, 2000; Lequette & Greenberg, 2005).

3. Genetic control of biofilm formation

Numerous substances, both organic and inorganic, have an impact on bacterial QS systems, which in turn affect the formation of biofilms. Two of them—vanillin and cinnamic acid—were even shown to support the formation of biofilm due to their capacity to boost AHL synthesis (Kalia, 2013). The ideal QS inhibitors should fulfill the following criteria: (1) they must be stable, low-molecular-weight compounds; (2) their activity must be highly specific and safe for eukaryotic hosts; and (3) they must not obstruct the basic metabolic processes that can act as targets for the development of drug resistance. A variety of QS inhibitors can also improve the potency of already prescribed antibiotics and restore the function of the immune system. For instance, it has been proven by Brackman et al. that vancomycin or clindamycin works better against *Staphylococcus aureus* when combined with the QS inhibitor hamamelitannin.

P. aeruginosa: 34% of the 73 biofilm-regulated genes that encode putative proteins with ambiguous functions were found to be associated with the development of *P. aeruginosa* biofilms. Additionally, when biofilms were subjected to tobramycin, the differential expression of 20 genes was observed, 12 of which were categorized as genes encoding putative proteins with unknown activities. The substantial number of uncharacterized genes/proteins that are differentially expressed in biofilms may not be surprising given that a sizable portion of sequenced bacterial genomes (25%–40%) contain hypothetical genes or genes of unknown function (Whiteley et al., 2001).

RNA-seq was used to analyze the transcriptomes of *P. aeruginosa* PAO1 cells cultured planktonically and as biofilms studied by (Kaleta et al., 2022). Nearly one-third of the genome, or 1812 genes, were found to express themselves differently in biofilms under test conditions. Furthermore, genes that encode previously undiscovered (hypothetical) proteins were responsible for a third of these alterations in *P. aeruginosa*'s transcriptome. It also shows how little is still known about how biofilms form because several of the genes encoding previously uncharacterized proteins have not been linked to biofilm developmental pathways before. An investigation of qRT-PCR was performed on a subset of the genes identified by RNA-seq as being differentially expressed to confirm the differential transcript abundance following biofilm formation. WspA, YfiB, and RocS1 are three signal sensors (receptors) that detect the presence of c-di-GMP. Additionally, *P. aeruginosa* has five phosphodiesterases (BifA, DipA, RocR, MucR, and Nbd) and five diguanylate cyclases (WspR, YfiN, SadC, RoeA, and SiaD). The best characterized chemosensor, WspA, recognizes growth on surfaces and activates phosphorylate WspR, a similar diguanylate cyclase. When WspR is phosphorylated, it clusters, moves about inside the cell, and exhibits increased cyclase activity (Güvener & Harwood, 2007).

The fundamental c-di-GMP signaling module of *P. aeruginosa* consists of five parts: (1) sensors of environmental signals, (2) enzymes that are involved in the production and breakdown of c-di-GMP, (3) particular effectors, such as riboswitches or proteins, are both allosterically controlled by c-di-GMP, (4) targets include exopolysaccharide synthase and secretion apparatus, flagellar basal body, promoter DNA, enzymes, and cellular structures, and (5) molecular output generated by effectors upon c-di-GMP activation (Hengge, 2009).

In *Escherichia coli*: DNA markers were used to compare the expression of genes in *E. coli* biofilms that had been cultivated for 40 h at varied flow rates in a silicone flow chamber to those of planktonic cells in the stationary phase. In the biofilm cells, 4.5% of the genome was repressed and 9% of the genome was activated, which resulted in a transfer of approximately 600 genes overall. When the transcriptional pattern of these cells was contrasted to that of exponentially developing cells, only 230 genes were found to be differentially expressed in biofilm cells, with 4.8% of them being upregulated and 0.5% being downregulated (Schembri et al., 2003). Only 79 genes, or 1.84% of the *E. coli* genome, showed significantly different expression patterns while the organism was growing a biofilm as opposed to plankton. The expression of three genes that are involved in adhesion and autoaggregation, many of which code for structural proteins including OmpC, OmpF, and OmpT, lpxC (which codes for a protein involved in the synthesis of lipid A), and slp, increased in biofilms (encoding an outer-membrane lipoprotein induced after carbon starvation). Two of these genes, *slp* and *ompC*, have recently been associated with the formation of *E. coli* biofilms on abiotic surfaces (Sauer, 2003).

Curli, fimbriae, and Antigen 43 (Ag43) have all been linked to the production of microcolonies and autoaggregation in *E. coli*. Surface-attached bacterial microcolonies grow into highly organized community structures as a result of a complicated series of molecular events involving numerous variables, many of which are likely unique to various bacterial species. The systems of cell-to-cell signaling that keep track of population density are involved in controlling community organization. Movement over the target surface to locations with higher nutrition availability is made possible by motility and chemotaxis. By enabling species to coexist in a cooperative, symbiotic way, metabolic interactions between various organisms contribute to the growth of microcolonies. Exopolysaccharide production is frequently linked to the development of intricate depth and three-dimensional structure, which likely increases resistance to antibacterial drugs (Hasman et al., 2000; Schembri et al., 2003).

***S. aureus*:** *S. aureus* is a major etiological agent in biofilms, which causes a significant share of hospital-acquired illnesses. A biofilm goes through numerous stages during its formation, including adhesion, development, maturation, and preservation of the mature biofilm. As a result, the bacterial cells inside a biofilm undergo changes in gene expression and transcription in response to stressors, as well as demands for metabolism and QS throughout different phases of biofilm growth (Schembri et al., 2003; Whiteley et al., 2001). Nassar et al. (2021) compared how the gene expression of *S. aureus* planktonic cells and their biofilm counterparts changed as the cells matured. At both biofilm formation stages, the expression of six genes—*rocD, gudB, sucC, grpE, hrcA*, and *clpC*—was increased. This implied that these six genes were critical for various stages of biofilm development. *rocD* and *gudB* are two genes that are involved in the metabolism and production of arginine, one of the amino acids. Ornithine–oxoacid transaminase and glutamate dehydrogenase are encoded by the genes *rocD* and *gudB*, respectively. The metabolism of glutamate, glutamine, and ammonia is regulated by the enzyme glutamate dehydrogenase (Nassar et al., 2021).

Compared with planktonic controls, the expression of *clpC* is increased in *S. aureus* biofilms (Becker et al., 2001). The *clpC* ATPase, which is encoded by this gene, is involved in a variety of processes, including the breakdown of misfolded proteins. Frees et al. illustrated that *clpC* was necessary for *S. aureus* to grow under thermal stress and was also necessary for the development of biofilms. It also functions as the *clpP* protease's ATPase partner, which was discovered to be necessary for *Pseudomonas fluorescens clpC*'s biofilm production. This *clpC*'s involvement in many stages of biofilm development makes it a viable target for antibiofilm therapy.

Enterobacterales, *E. coli* IMP, and *Klebsiella pneumoniae*: Members of the Enterobacterales, such as *E. coli* and *Klebsiella p.*, are the most frequent causes of urinary tract diseases in the population and in healthcare settings as well as the most serious lower pulmonary and bloodstream disorders. For gram-negative bacteria, the autoinducer-2 (AI-2) system is a component of the QS system, and AI-2 molecules are the collective name for the luxS gene products. In *E. coli* and *K. pneumoniae*, certain genes have been connected to the AI-2 system, and it has been shown that the repression of the AI-2 QS system is what causes the reduction of biofilm and the expression of the virulence genes. Additionally, it has been shown that in low-nutrient settings, luxS produces AI-2 molecules that can instruct other cells to form a biofilm (Beehan et al., 2015).

Members of the Enterobacterales, which are also involved in the formation of biofilms, include *E. coli* and *K. pneumoniae*. In hospitals and among internal patients, growth in biofilms promotes bacterial survival, raising the risk of sporadic nosocomial pathogens. Multidrug-resistant CREs, such as *E. coli* IMP and *K. pneumoniae* NDM-1, have been regarded as a significant health challenge due to their association with the common community- and hospital-acquired infections. In actuality, CRE outbreaks have already been noted globally (Al-Bayati & Samarasinghe, 2022). *E. coli* IMP and *K. pneumoniae* NDM-1, which have the blaIMP and blaNDM-1 resistance genes, respectively, both demonstrate high biofilm formation in comparison with the nonresistant control strains. In earlier studies of multidrug-resistant Enterobacteriaceae, similar conclusions about a faster rate of biofilm growth were reached. In addition to high biofilm development in *E. coli* IMP and *K. pneumoniae* NDM-1, the strongest overexpression of the resistance genes, bla IMP and bla NDM-1, was also observed at 24 h, when biofilm formation was at its best. This might be because these resistance genes have the power to influence how biofilms are formed. It has been demonstrated that resistance genes in particular plasmids can regulate the emergence of biofilms in *K. pneumoniae* (Ghanbarzadeh Corehtash et al., 2015).

***Vibrio cholerae*:** Cholera is an acute diarrheal disease with life-threatening complications that affects 2.9 million people annually and kills an estimated 95,000 people in the 69 countries where it is endemic. It is caused by a facultative bacterial infection called *V. cholerae*. Violent *V. cholerae* enters the body when food or liquid that has been contaminated is ingested. Due to the conditional expression of multiple well-known virulence factors, including the cholera toxin and the toxin coregulated pilus, once inside, it starts to colonize the small intestine and cause illness (TCP). Our knowledge of the biological functions, regulatory mechanisms, and biochemical composition of *V. cholerae* biofilms has greatly improved during the past 30 years. Adhesion, maturation, and dispersion are the three phases of biofilm development (Pombo et al., 2022).

V. cholerae multiplies in a sessile form after attachment and secretes the components of the extracellular matrix. The majority of this matrix is composed of the *Vibrio* polysaccharide (VPS), structural proteins *RbmA*, *RbmC*, and *Bap1*, and extracellular DNA (Yildiz et al., 2004). The extracellular biofilm matrix is primarily composed of VPS, which accounts for well over 50% of the total mass. The bulk of biofilm-relevant gene clusters, including the *vps* gene clusters, are negatively regulated by QS, which is triggered by high concentrations of the second messenger bis-(3–5′)-cyclic dimeric guanosine monophosphate (c-di-GMP) (Hammer & Bassler, 2009; Pombo et al., 2022). Cells spread out from the top layers of biofilms as they develop and go back to being planktonic, where they can colonize new surroundings. This process is less understood, but recent research has identified several components that are involved in *V. cholerae* biofilm dispersal, including proteins related to c-di-GMP signaling, enzymes that digest matrix, flagellar motility modulators, two-component regulators, and extracellular nucleases.

In *Bacillus subtilis*: Gram-positive, spore-forming *B. subtilis* is a common bacterium in both marine and terrestrial habitats. Ecophysiology suggests that the ability of *B. subtilis* to form biofilms on plant roots provides the cells with access to an abundant supply of nutrients contained in root exudates and osmostress-relieving chemicals. Biofilm development on roots is a crucial requirement for *B. subtilis* to promote plant growth and protect plants from bacteria that cause pathogens (Bremer et al., 2022). The three loci (*epsA-epsO*, *tapA-sipW-tasA*, and *bslA*) that encode the enzymes and proteins needed to create the extracellular matrix are plays important role in the biofilm formation of *B.subtilis*. The exopolysaccharide component of the matrix is synthesized by enzymes that are encoded by the 15 genes that make up the *epsA-epsO* operon. TasA, an extracellular protein that forms long fibers, is encoded by the *tapA-sipW-tasA* operon, and *tapA* functions as TasA's anchoring and assembly protein. As they are secreted across the cytoplasmic membrane, the N-terminal signal peptides of TapA and TasA are digested by *sipW*, a type I signal peptidase. The mono-cistronic *bslA* gene produces bslA, a hydrophobin-like biofilm surface layer protein (Arnaouteli et al., 2021; Vlamakis et al., 2013).

The transcription factor response regulator protein Spo0A, which is always present, is essential for *B. subtilis'* ability to differentiate into distinct cell types. Different kinases have the ability to phosphorylate Spo0A (Spo0AP) to encourage DNA binding. The phosphorylation of Spo0A is mediated by five histidine kinases with partially overlapping roles. The final kinase, KinE, is a cytoplasmic protein, while the other four, KinA, KinB, KinC, and KinD, are lodged in the cytoplasmic membrane. No single Kin kinase is entirely responsible for the formation of biofilms because each one's individual contribution to this differentiation process fluctuates in response to diverse growth conditions. Although the triggers for biofilm development are not fully known, KinC is crucial to the process. Surfactin, a lipopeptide involved in motility and generated by *B. subtilis* as well as other bacteria, controls its kinase activity. The main cellular cation, potassium, leaks out when surfactin enters the cytoplasmic membrane. As a result of this mechanism's stimulation of KinC activity, cellular Spo0AP levels rise (Bremer et al., 2022).

Mycobacterium tuberculosis: When it infects the human body, the bacterium that causes tuberculosis, *M. tuberculosis (M. tb)*, creates biofilms. The fact that biofilms make it so much more challenging to treat the disease is probably one of the reasons tuberculosis is the worst bacterial infection in the world. Biofilms are becoming more well recognized as an important bacterial growth form in their natural environments. Most people do not believe that TB is a biofilm infection. However, *M. tb* cells grown in vitro spontaneously group together and secrete extracellular matrix (ECM), indicating a propensity to form biofilms. In human tissues after TB infection, *M. tb* aggregates have long been discovered in autopsy research. Different *M. tb* biofilm indicators have been discovered in human postmortem samples and TB animal models in more recent investigations (Smith et al., 2022).

Due to the ability of *M. tb* cells to proliferate within biofilms and resist immune responses, the existence of *M. tb* biofilms during TB infection is of significant practical importance (Smith et al., 2022). The genetic reasons for *M. tb* biofilm formation have been investigated through the use of candidate gene approaches and the phenotypic characterization of deletion, knockdown, and overexpression mutants. *M. tb* biofilms and biofilm indicators have been discovered in clinical specimens. The mutations are rejected by selection because they interfere with regulatory systems that normally restrict biofilm formation to specific environmental conditions. Regulations would limit any potential adverse impacts of biofilm growth, such as the metabolic costs of matrix production and immunostimulation (Canetti, n.d.).

In *Enterococcus faecalis*: Gram-positive, opportunistic microorganisms called enterococci are the second-leading cause of infections acquired in hospitals (HAIs). *E. faecalis* and *Enterococcus faecium* are the members of the *Enterococcus* genus that are isolated from human infections the most frequently, with *E. faecalis* being more frequently found in HAI. *E. faecalis* can cause life-threatening endocarditis, bacteremia, wound infections, and infections connected to medical devices, such as catheter-associated urinary tract infections.

E. faecalis is frequently isolated with other pathogenic bacteria from polymicrobial biofilm-associated illnesses and causes multidrug-resistant, life-threatening infections. Transposon mutagenesis and conventional labor- and time-intensive allelic-exchange techniques are the only genetic tools that are now available to analyze complicated interactions in mixed microbial communities. Built on the streptococcal dCas9 platform, the *E. faecalis*-inducible CRISPRi system effectively silences both single and multiple genes. This approach can be used to investigate the essentiality of a gene and silence genes implicated in biofilm development and antibiotic resistance. Unusually, this method may be used to disturb already-formed biofilms and is optimized to research genes crucial for biofilm initiation, maturation, and maintenance. This approach will be useful for quickly and effectively looking into a variety of complex enterococcal biology-related topics (Afonina et al., 2020).

4. Genes required for various stages of biofilm formation in different bacteria

Since DNA is the building block of evolution, it has only one real function, which is to duplicate and preserve its own unique genetic code. An edge in reproductive fitness will protect a person's genetic makeup. The same rules hold for infectious agents such as viruses and plasmids that must be passed from one creature to another. These agents would disappear if they did not have a means of reproduction as they are unable to manufacture their own genetic material. Because of this, plasmids and bacteriophages have both created techniques for maintaining themselves inside a host bacterium as well as spreading to other bacteria (Jefferson, 2004). A biofilm is an ideal environment for horizontal genetic material exchange. The closeness encourages plasmid DNA conjugation and absorption by plasmid-capable bacteria as well as quick phage dissemination. Plasmids and phages have therefore developed strategies to cause their host to transition to the biofilm mode of growth, enabling them to spread to healthy bacteria and, on occasion, beyond the species barrier (Cvitkovitch et al., 2003).

Through horizontal gene transfer, the bacteria may also directly profit from the exchange of antibiotic resistance determinants within biofilms. It has been suggested that competence factor expression in *Streptococcus gordonii* has a role in the interchange of genetic material in biofilms as both a cause and an effect of biofilm formation. Regardless of whether the primary goal of competence factors is to assimilate external DNA to increase their genetic diversity or just to use it as a source of nutrition, the biofilm offers the ideal environment for the interchange of genetic material (Cvitkovitch et al., 2003; Loo et al., 2000) (Tables 2.1 and 2.2).

The first descriptions of specific genes that are up-or downregulated in biofilm bacteria were generated using transcriptional *lacZ* reporter-gene fusions, which gave rise to the hypothesis that bacterial adhesion causes the expression of a collection of genes that results in a biofilm phenotype. It was shown in a genome-wide screen in *E. coli* employing random chromosomal insertions of a *lacZ* reporter gene fusion construct that significant parts of the bacterial genome may be engaged in or influenced by biofilm production (Costerton & Lewandowski, n.d.; Dagostino et al., 1991). *algC*, the gene necessary for alginate synthesis, has a function in *P. aeruginosa* biofilms, and *csgA*, the gene which is encoded by a protein that is involved in the synthesis of colanic acid, is implicated in aggregation in *E. coli*. The formation of gram-positive biofilms also depends on EPS production. A glucosyltransferase known as glucan-binding protein GbpA has been linked to *Streptococcus mutans*' production of sucrose-dependent polysaccharides and the development of biofilms. Additionally, the b-1-6-linked poly-*N*-acetylglucosamine polymer known as PNAG or PIA is synthesized by the gene products encoded by the intercellular adhesin locus (icaADBC) in *S. aureus* and *Staphylococcus epidermidis* (polysaccharide intercellular adhesin) (Jefferson, 2004).

TABLE 2.1 Genes essential for the biofilm formation.

Genes	Protein/Function	Species	References
icaADBC	Intercellular adhesin synthesis	*Staphylococcus aureus, Staphylococcus epidermidis*	Otto (2012)
atlE	Autolysin/adhesin	*S. epidermidis*	Otto (2012)
bopABCD	Biofilm on plastic surfaces operon	*Enterococcus faecalis*	Hufnagel et al. (2004)
esp	Surface proteins of enterococcal	*E. faecalis*	Toledo-Arana et al. (2001)
agn43	Protein from the antigen in aggregating	*Escherichia coli*	Kjaergaard et al. (2000)
asI	Synthesis of the quorum sensing signal $3OC_{12}$-HSL	*Pseudomonas aeruginosa*	Davies et al. (1998)

TABLE 2.2 Gene expression in biofilms *up* or *down* compared with planktonic cells.

Genes		Proteins/Function	Regulation	Species	Reference
Adhesion	wcaB	Colanic acid synthesis	*Up*	*Escherichia coli*	Prigent-Combaret & Lejeune (1999)
	csgA	Curli	*Up*	*E. coli*	Jefferson (2004)
	clfA	Clumping factor/fibrinogen binding	*Down*	*Staphylococcus aureus*	Wolz et al. (2002)
	scaA	Coaggregation	*Down*	*Streptococcus gordonii*	Gilmore et al. (2003)
Quorum sensing	*pA4296*	Probable two-component response regulator	*Down*	*Pseudomonas aeruginosa*	Whiteley et al. (2001)
	comD, E	Competence factors	*Up*	*Streptococcus mutans*	Li et al. (2002)

Only five genes were found to be crucial for *P. aeruginosa* biofilms using in vivo expression technology, and these genes were later verified by insertional mutation. A mutation in PA2247 (2-oxoisovalerate dehydrogenase encoded), a homolog of *E. coli ubiB* (production of ubiquinone), has not been thoroughly characterized, and mutation in PA5065, a homolog of *E. coli ubiB*, was lethal. Planktonic growth was unaffected by mutations in three additional genes, PA0240 (putative porin encoded), PA3710 (aliphatic alcohol dehydrogenase encoded), and PA3782 (an AraC-like transcriptional regulator encoded), but these defects resulted in abnormal biofilm development in static and flowing systems (Finelli et al., 2003; Sauer, 2003). Intriguingly, in competition studies, all three *P. aeruginosa* mutants performed less well than the parent strain: after 5 days of development in culture, they made up less than

0.0001% of all biofilm cells, indicating a role for the altered genes in the formation of sessile communities. It was determined that the discovered novel genes were crucial for the creation, growth, and fitness of biofilms even though they had no effect on planktonic growth (Finelli et al., 2003). None of the genes discovered in this study was found in *E. coli* (Schembri et al., 2003) or *P. aeruginosa* (Whiteley et al., 2001) DNA microarray analyses that were conducted in the past.

5. Biomathematical modeling: the genetic basis of biofilm development

Promoting good biofilms and avoiding unhealthy ones require an understanding of the mechanisms behind biofilm creation, growth, and removal. It requires both experimental and modeling efforts to comprehend the complex process of biofilm formation, which is regulated by a number of biological, physical, and chemical factors. Important biofilm features such as cell numbers, cell viability, biofilm shape and EPS structure, nutrition profile, and genetic information may all be easily measured qualitatively or quantitatively through experiments (Zhang, 2017). The 1-D models are unable to demonstrate how lateral gradients affect the development of a biofilm structure or how pores, for example, contribute to total biofilm transformation (van Loosdrecht et al., 2002). To successfully use and regulate biofilms in industrial and medical contexts, biofilm models are the best way to comprehend the fundamental principles governing biofilm development, composition, structure, and function. There are many different types of mathematical models, ranging from extremely basic empirical correlations to complex and computationally costly algorithms that characterize the shape and activity of three-dimensional (3D) biofilms. Typically, the area of interest is separated into three sections: the bulk liquid, the biofilm itself, and the border layer (Mattei et al., 2018). Once again, mathematical modeling is an effective tool for understanding biofilms as a varied community, as well as the evolution of various species and their genetic origins. For scientists working on biofilm modeling, detailed experimental data present both possibilities and obstacles (Zhang, 2017).

The examination of mathematical models for biofilms is a topic with an abundance of literature. An outstanding overview of the mathematical modeling of biofilms is provided by the IWAtask group. The book outlines the fundamental procedures for developing a mathematical model. The many types of models include analytical models (A), pseudo-analytical models (PA), one-dimensional numerical models (N1), and multidimensional numerical models (N2 and N3). Wang and Zhang present a chronological analysis of numerous biofilm models developed during the 1980s and the early 2000s (Wang & Zhang, 2010). Models are categorized into four major categories according to their dimensionality, handling of diffusion, and complexity including physics, chemistry, and biological effects: continuum models in one dimension, coupled biofilm–fluid models, aggregate models with diffusion limits, and continuum–discrete diffusion models. According to the development paradigm, there must also be connections between groups of biofilm routes in a genetic network that is hierarchically organized and links between shape and function. It is still not possible to state that pathways controlling the structural development of an *E. coli* biofilm are directly coupled with pathways for the formation of a subpopulation of cells with higher tolerance to colistin. Folkesson et al. provided important evidence in this direction by showing that *E. coli* biofilms formed by F-plasmid-containing cells were more structured and had higher tolerance to colistin than cells without the F-plasmids (Folkesson et al., 2008). Zhang et al. created a theory for the study and forecasting of the spatial and temporal patterns of gene and protein expression inside microbial biofilms (Zhang et al., 2013). The hypothesis takes into account mRNA or protein production, microbial expansion, biomass advection, and gene transcript or protein turnover. Case studies show how the theory may predict microbial behaviors in biofilms that are qualitatively different from those of planktonic cells and repeat a variety of spatial patterns. Both Maximum entropy method (MaxEnt) and trait space model have been used to forecast the relative abundance of species for plant communities (Laughlin et al., 2012; Shipley et al., 2006). Both approaches use statistical methods and consist of three key elements: an underlying trait distribution, a performance filter that assesses the fitness of traits in diverse contexts, and a projection of the performance filter down some environmental gradient.

The dynamics of two-component types—dissolved (substrates and metabolic products) and particulate—are represented by all biofilm models (active and inert biomass, EPS). A transit mechanism, a consumption and growth mechanism, and a loss mechanism are the three main parts of these models, on average (Klapper & Dockery, 2002). The movement of dissolved materials within the biofilm matrix is governed by diffusion. It is essential for the growth of biofilms because the concentrations of nutrients and products regulate the rates of microbial responses, and all processes that result in an increase in volume are powered by nutrient availability (Picioreanu, 2000a).

The first to create a thorough quantitative model framework for biofilm formation was Picioreanu (Picioreanu, 2001; Picioreanu et al., 2000). The following factors are taken into consideration in these biofilm models:

- Convection: By resolving the Navier–Stokes equations for fluid flow, this is taken into consideration. At the moment, only nonturbulent flow conditions make this practical.
- Diffusion: Fick's law may be resolved to describe diffusion.
- Reaction: Standard reaction kinetics may be used to determine the chemical or microbiological conversion processes given the local concentrations from convection and diffusion processes are understood.
- Biofilm growth: The preceding step's freshly created biomass has to be disseminated. However, in two- and three-dimensional biofilm models, this procedure is not as simple as it is in one-dimensional biofilm models.
- Detachment: The first to mention a potential separation process based on the force generated by the liquid flow (Picioreanu, 2001). The biofilm's mechanical stress is computed, and when it rises over its tensile strength, the biofilm ruptures, resulting in detachment.

However, more recent experimental research has shown a great diversity in the structure, capabilities, and characteristics of biofilms, inspiring the creation of mathematical models that aim to capture this biological, ecological, and physical complexity. As a result, over time, the idea of biofilm has undergone tremendous alteration. The mechanical/physical features of biofilms and their interactions with the underlying fluid flow, in terms of deformation and detachment, have been the subject of extensive research over the past 10 years (Mattei et al., 2018).

Biofilm models have become increasingly complicated over the past few decades and have gradually integrated a vast array of physicochemical and biological activities. However, based on biomass representation, they may be divided into two major groups (Böl et al., 2013):

- Continuum models: These models regard biomass as a continuum and are based on the population-averaged behavior of several functional groups; hence, they do not directly account for the behavior of a single microbe. The equations for 1D models have the form:

$$\partial D/\partial t + \partial J/\partial z = R$$

[z = space coordinate; t = time variable; D = 1D property; J = 1D property flux; R = net property production rate]

- Discrete models: These models are often referred to as bottom-up models because the biofilm structure is not provided as an input to the model. Instead, the biomass units' actions and interactions with one another and the environment give rise to the complex morphology of biofilms. Discrete models have been classified into three categories based on the biomass representation and the preferred spreading mechanism: cellular automaton (CA) models, hybrid differential—discrete CA models, and individual-based models (IbMs).

6. CRISPR-based functional analysis of biofilm-forming genes in bacteria

The first CRISPRs (clustered, regularly interspaced short palindromic repeats) were found in *E. coli* in 1987 (Ishino et al., 1987). The direct DNA repetitions in CRISPRs range in size from 21 to 48 base pairs, and they are separated by similar-length nonrepetitive spacers (Barrangou, 2013). The CRISPR-Cas systems are made up of four components: tracrRNA (for Cas9), leader sequences, Cas proteins, and CRISPR areas (Karvelis et al., 2013; Liu et al., 2022). In most archaea and many bacteria, the CRISPR-Cas (CRISPR-associated) system is extensively distributed and serves as a defense mechanism against the invasion of foreign nucleic acids produced from plasmids, viruses, and phages (Jinek et al., 2012; Makarova & Koonin, 2015). The CRISPR-Cas system is divided into two basic classes, which branch out into six primary kinds and at least 33 subtypes (Makarova et al., 2020) based on the principles and effector module design. The spacer region is included in the CRISPR RNA (crRNA), which aids Cas proteins in identifying and cleaving foreign genomic elements (Marraffini, 2015). To guide this cleavage, a trans-activating crRNA (tracrRNA) must base-pair with the repeat region of the crRNAs to generate a mature duplex RNA (Chylinski et al., 2013; Fonfara et al., 2014).

In *P. fluorescens:* In gram-negative bacteria, the well-known GacA/S TCS regulates the expression of genes involved in QS, stress responses, biofilm formation, and virulence

(Gooderham & Hancock, 2009; Tahrioui et al., 2013). The GacA/S system is composed of the membrane-bound sensor histidine kinase GacS and the related response regulator GacA16. In proteobacteria such as *Pseudomonas* and *Halomonas*, inactivation by mutation of either GacA or GacS has a major effect on the production of EPS, secondary metabolites, and iron homeostasis (Brencic et al., 2009; Hassan et al., 2010; Heeb & Haas, 2001). A study conducted by Marie-Francoise Noirot-Gros modified the CRISPR interference (CRISPRi) technology to employ it in *P. fluorescens* species. The research revealed that CRISPRi may be utilized to investigate the detailed phenotypic traits (motility, cell shape, and biofilm production) in three biologically and genetically diverse species, SBW25, WH6, and Pf0-1. The CRISPRi system was modified for *P. fluorescens* SBW25, and tests on the SBW25, WH6, and Pf0-1 strains showed that it was effective at silencing genes. *P. fluorescens'* numerous species also give researchers enough opportunity to properly and comprehensively study these phenotypic traits. The study also discovered unanticipated traits connected to the production of extracellular matrix and complex biofilm architecture. The study also revealed how the PFLU1114 protein functions to successfully prevent SBW25 species from developing biofilms. The effects of CRISPRi range from the silencing of the two-component sensor kinase GacS, which affects mobility and biofilm formation, to the silencing of genes involved in cytokinesis and morphogenesis, the silencing of genes involved in c-di-GMP signaling, which likely affects other genes in operons and the silencing of swarming motility. Additionally, a considerable swarming loss was produced by the expression of the gacST guide. Compared with the strain utilized as a control that had no gRNA, GacS silencing caused notable biofilm formation defects. It was demonstrated that the amplitude of the inhibition was dependent on the DNA location and strand that the gRNA targeted as well as the concentration of aTc inducer by demonstrating that these properties may be leveraged to vary the amount of inhibition. We decided to construct gRNAs that bind to the start of the ORFs and are complementary to the protein-coding strand because of the significant suppression of gene transcription that these properties induce (30-fold in SBW25) and the 8.7-fold decrease of protein expression that they produce. According to these findings, dCas9 induction stops transcriptional elongation, which leads to efficient gene silencing.

In *E. coli*: In a study done by Azna Zuberi (Zuberi et al., 2017), CRISPRi technology is introduced to inhibit biofilm production in the AK-117 strain of *E. coli* by targeting the luxS gene. The synthase luxS is responsible for producing autoinducer-2 (AI-2) and subsequently controlling the preliminary phase of biofilm synthesis. Here is the first investigation into CRISPRi-mediated suppression in bacterial biofilms. A mutant endonuclease, dCas9, was added to the complementary sgRNA that was created. The luxS gene is then silenced using this modified form of the CRISPR system. The effect of the luxS gene on biofilm inhibition was examined using the crystal violet assay, the XTT reduction assay, and scanning electron microscopy (SEM). The relative qRT-PCR outcomes of knockdown strains also showed a dramatic decrease in luxS gene expression. The crystal violet test was used to gauge the decrease in biofilm formation brought on by the downregulation of the luxS gene in the knockdown strains AK-LV1, AK-LV2, and AK-LV3. The growth of biofilm was reduced the most in AK-LV3, then in AK-LV2, and then in AK-LV1. To assess the metabolic efficiency and survival of produced knockdown strains, the converted level of XTT was measured. Viability rates for the luxS knockdown strains derived from AK-117 (AK-LV1, AK-LV2, and AK-LV3) were 82%, 81%, and 69%, respectively. The effect of luxS gene silencing brought on by CRISPRi on AK-117 biofilms was examined using SEM. While the treated samples displayed distinct cell colonies, the cells in the control sample were visibly immersed in their self-synthesized matrix. The AK-LV3 version exhibited the greatest biofilm formation inhibition, whereas the AK-LV1 variant had the lowest levels.

In *Salmonella enterica serovar Typhimurium*: In a study done by Nandita Sharma et al. (sharma et al., n.d.), by deleting the CRISPR-Cas genes crisprI, crisprII, crisprI, and cas op, *Salmonella enterica serovar Typhimurium* was used to study the regulation of biofilm by CRISPR-Cas. One cas operon, two CRISPR arrays (CRISPR-I and CRISPR-II), and two CRISPR arrays makeup *Salmonella*'s type I-E CRISPR-Cas system. It has been established that this mechanism regulates the QS method used by *Salmonella enterica* subspecies *enterica* serovar Enteritidis to monitor biofilm formation. Additionally, it regulates the production of outer membrane proteins by the serovar Typhi, which has an impact on the growth of biofilms and bile resistance. Inhibition of the curli (csgA) and flagellar (fliC, flgK) genes prevented the formation of surface biofilms in knockout strains. The study also showed that the CRISPR-Cas system differently controlled the creation of surface-attached and pellicle biofilms by changing the expression of genes relevant to biofilms. To understand the temporal variations in biofilm formation in the knockout strains, the study assessed the expression of key biofilm components including flagella, cellulose, LPS, and curli. They were assessed for motility using a swarming experiment. Swarming rates were at least 20% lower in all knockdown strains compared with wild-type (WT). The outcomes demonstrate that all four knockout strains generate lesser curli. It was observed from the results that levels of bacterial biomass were elevated and its corresponding components in the CRISPR-Cas knockout strains cause greater pellicle production.

In *Streptococcus agalactiae*: In a Study conducted by Meng Nie et al. (Nie et al., 2022), it was discovered that CRISPR deficiency lowered hypervirulent piscine *Streptococcus agalactiae* strain GD201008-00 capacity to adhere, invade, and form biofilms by increasing the synthesis of CPS. The strain GD2008-001, which caused a streptococcosis epidemic in Guangdong Province in 2008, was discovered in tilapia with meningoencephalitis (Liu et al., 2012). A transcriptional regulator referred to as PcpsA, whose repressive action resulted in the cps operon's transcription being suppressed, as shown by RNA degradation experiments. The promoter PcpsA's suppressed activity eventually causes *S. agalactiae* to exhibit increased levels of virulence-related traits (invasion, adhesion, and synthesis of biofilm). It was determined from earlier studies that this bacterium only has one type II-A CRISPR-Cas system (Ma et al., 2018) and that deletion of CRISPR resulted in greatly increased expression of the genes cpsB, cpsC, and cpsJ associated with CPS production (Dong et al., 2021). In this study, we found that the synthesis of CPS was negatively regulated by crRNA, which enhanced the bacteria's capacity to cling, to penetrate, and to form biofilms on brain endothelial cells. The downregulation of the cps operon by CRISPR contributes negatively to the generation of CPS, according to quantitative reverse transcription-PCR (qRT-PCR) data.

In *Cedecea neteri*: The CRISPR/Cas9 gene editing technology, a powerful tool for altering bacterial genomes, facilitates research into host–microbe interactions. The *C. neteri* symbiont from *Aedes* mosquitoes was employed in the research by Shivanand Hegde et al. (2019) to get an altered outer membrane protein A (ompA) gene. Using the Scarless Cas9-Assisted Recombineering method, the researchers altered the ompA gene of the nonmodel *C. neteri* (Reisch & Prather, 2015). The study also revealed the employment of integrated antibiotic resistance genes and fluorescent marker protein into the bacterial genome using the CRISPR/Cas9 system, to showcase the gene's functionality in both in vitro and in vivo conditions. The researchers used CRISPR/Cas9 gene editing technology to create knockout variants in a *C. neteri* (a gut symbiont of *Aedes* mosquitoes) to examine the role of bacterial genetics in biofilm development and gut colonization. A monoassociation *C. neteri* ompA gene deletion reduced bacterial colonization of mosquitoes following infection.

In various bacteria: CRISPR-Cas systems influence a number of bacterial properties, including bacterial biofilm, QS, pathogenicity, and antibiotic resistance. Numerous virulence factors, including comDE, gtfC, vicR, smu0630, ace, esp, hyl, gelE, and asa1, are linked to the expression of CRISPR-Cas systems (*E. faecalis*). The genes cas3 and cas1 either directly or indirectly contribute to and have an impact on the development of bacterial biofilm. This mechanism of reciprocal modulation between CRISPR-Cas systems and QS systems involves a large number of genes, including cdpR, LasR, SmaI, SmaR, and SsoPox. CRISPR-Cas systems limit the proliferation of drug-resistant plasmids (Wu et al., 2022).

According to research, strains without the CRISPR1 or CRISPR3 loci synthesized more biofilms than bacteria carrying those loci did (Tong et al., 2017), and without CRISPR-Cas systems, *E. faecalis* root canal isolates produced more biofilm and had more prominent periapical lesions. The interaction between the CRISPR-Cas systems and the genes that regulate bacterial biofilm formation is demonstrated by this research.

The active CRISPR-Cas system, which is mediated by mutant heat-stable nucleoid structuring protein present in a variety of gram-negative bacterial species and suppresses transcription of target genes, greatly reduces *E. coli*'s capacity to produce biofilms. In Stx2 phage lysogen, *E. coli* has much higher phage resistance (Fu, Li, et al., 2017).

Since the protospacer-adjacent motif (PAM) in the genome of *P. aeruginosa* is necessary for the CRISPR-dependent elimination of biofilm features, endogenous targeting could employ the same mechanism as antiphage immune defense. The lysogenic phage DSM3 rendered WT *P. aeruginosa* PA14 incapable of forming biofilms. The observations suggest that CRISpr and cas genes are closely related to the loss of biofilm and that biofilm may be restored by deleting four of the six cas genes in *P. aeruginosa* type I−F or by interfering with CRISPR2 (CRISPR array 2) (Zegans et al., 2009).

To characterize the mechanisms of CRISPR-Cas in managing biofilms, the impacts of the technology are evaluated at the single Cas level. A cas3 gene deletion mutant of *S. mutans* UA159 showed lower biofilm development and weaker competitiveness against (cocultured) *Streptococcus sanguinis* in the vicinity of fluoride via upregulation of virulence genes (comD, gtfC, EvicR, and smu0630) (Tang et al., 2019). This study demonstrates how the cas3 gene of *S. mutans* may positively influence the development of biofilms and fluoride resistance.

According to research on *Serratia* sp. ATCC39006 strain (Fu, Su, et al., 2017; Patterson et al., 2016), the main QS mechanisms that control biofilm growth may have an impact on how CRISPR-Cas influences biofilm formation. In marine bacterium, *Chromobacterium violaceum* and gram-negative bacteria such as *P. aeruginosa*, there is a downregulating effect caused by the QS interfering enzyme lactonase SsoPox. These findings suggest that when bacteria reach large densities, QS systems regulate the cas genes, increasing the integration of foreign DNA snippets and amplifying the CRISPR-Cas-mediated resistance to phage infiltration as well as other genomic components.

7. Conclusion

It is known that the gene expression patterns of biofilms are different from those of planktonic bacteria, but it is still unclear what the genetic underpinnings of biofilm formation are. It was shown that the production of a significant number of proteins varied depending on the stage of biofilm growth. Numerous proteins were also found to express differently after 1 day of biofilm development, and when compared with the expression pattern in planktonic cells, the protein expression profile in developed biofilm cells altered the most (Sauer, 2003). The development of biofilms is followed by an increase of biofilm-related genes in the *E. coli* IMP and *K. pneumoniae* NDM-1 strains (Al-Bayati & Samarasinghe, 2022). While QS systems primarily control biofilm dispersal in *S. aureus* and *P. aeruginosa*, QS is essential for the growth of biofilms in *V. cholerae*, and it has been found that reduced cell density promotes this process. For *V. cholerae*, particularly prominent are the interactions between the QS and c-di-GMP regulatory networks (Wolska et al., 2016). Combining CLSM and qPCR has demonstrated a relationship between the expression profile of the selected gene cluster for biofilm, QS, stress survival, and antibiotic resistance and the emergence of phenotypic biofilms. Therefore, this morphological examination as well as the expression levels of genes during the biofilm formation may be used to study the diversity in the modulation of these kinds of genes in these significant CRE pathogens.

References

Afonina, I., Ong, J., Chua, J., Lu, T., & Kline, K. A. (2020). Multiplex CRISPRi system enables the study of stage-specific biofilm genetic requirements in *Enterococcus faecalis*. *MBio, 11*(5), e01101–e01120. https://doi.org/10.1128/mBio.01101-20

Al-Bayati, M., & Samarasinghe, S. (2022). Biofilm and gene expression characteristics of the carbapenem-resistant Enterobacterales, *Escherichia coli* IMP, and *Klebsiella pneumoniae* NDM-1 associated with common bacterial infections. *International Journal of Environmental Research and Public Health, 19*(8), 4788. https://doi.org/10.3390/ijerph19084788

Arnaouteli, S., Bamford, N. C., Stanley-Wall, N. R., & Kovács, Á. T. (2021). Bacillus subtilis biofilm formation and social interactions. *Nature Reviews Microbiology, 19*(9), 600–614. https://doi.org/10.1038/s41579-021-00540-9

Barrangou, R. (2013). CRISPR-cas systems and RNA-guided interference: CRISPR-cas systems and RNA-guided interference. *Wiley Interdisciplinary Reviews: RNA, 4*(3), 267–278. https://doi.org/10.1002/wrna.1159

Becker, P., Hufnagle, W., Peters, G., & Herrmann, M. (2001). Detection of differential gene expression in biofilm-forming versus planktonic populations of *Staphylococcus aureus* using micro-representational-difference analysis. *Applied and Environmental Microbiology, 67*(7), 2958–2965. https://doi.org/10.1128/AEM.67.7.2958-2965.2001

Beehan, D. P., Wolfsdorf, K., Elam, J., Krekeler, N., Paccamonti, D., & Lyle, S. K. (2015). The evaluation of biofilm-forming potential of *Escherichia coli* collected from the equine female reproductive tract. *Journal of Equine Veterinary Science, 35*(11–12), 935–939. https://doi.org/10.1016/j.jevs.2015.08.018

Böl, M., Ehret, A. E., Bolea Albero, A., Hellriegel, J., & Krull, R. (2013). Recent advances in mechanical characterisation of biofilm and their significance for material modelling. *Critical Reviews in Biotechnology, 33*(2), 145–171. https://doi.org/10.3109/07388551.2012.679250

Boles, B. R., Thoendel, M., & Singh, P. K. (2005). Rhamnolipids mediate detachment of *Pseudomonas aeruginosa* from biofilms: Rhamnolipid-mediated biofilm detachment. *Molecular Microbiology, 57*(5), 1210–1223. https://doi.org/10.1111/j.1365-2958.2005.04743.x

Bremer, E., Hoffmann, T., Dempwolff, F., Bedrunka, P., & Bange, G. (2022). The many faces of the unusual biofilm activator RemA. *BioEssays, 44*(5), 2200009. https://doi.org/10.1002/bies.202200009

Brencic, A., McFarland, K. A., McManus, H. R., Castang, S., Mogno, I., Dove, S. L., & Lory, S. (2009). The GacS/GacA signal transduction system of *Pseudomonas aeruginosa* acts exclusively through its control over the transcription of the RsmY and RsmZ regulatory small RNAs. *Molecular Microbiology, 73*(3), 434–445. https://doi.org/10.1111/j.1365-2958.2009.06782.x

Canetti, G. (n.d.). *The J. Burns Amberson Lecture*. 17.

Carradori, S., Di Giacomo, N., Lobefalo, M., Luisi, G., Campestre, C., & Sisto, F. (2020). Biofilm and quorum sensing inhibitors: The road so far. *Expert Opinion on Therapeutic Patents, 30*(12), 917–930. https://doi.org/10.1080/13543776.2020.1830059

Chylinski, K., Le Rhun, A., & Charpentier, E. (2013). The tracrRNA and Cas9 families of type II CRISPR-Cas immunity systems. *RNA Biology, 10*(5), 726–737. https://doi.org/10.4161/rna.24321

Costerton, W., & Lewandowski, Z. (n.d.). *Microbial biofilms*. 35.

Cvitkovitch, D. G., Li, Y.-H., & Ellen, R. P. (2003). Quorum sensing and biofilm formation in Streptococcal infections. *Journal of Clinical Investigation, 112*(11), 1626–1632. https://doi.org/10.1172/JCI200320430

Dagostino, L., Goodman, A. E., & Marshall, K. C. (1991). Physiological responses induced in bacteria adhering to surfaces. *Biofouling, 4*(1–3), 113–119. https://doi.org/10.1080/08927019109378201

Davey, M. E., & O'toole, G. A. (2000). Microbial biofilms: From ecology to molecular genetics. *Microbiology and Molecular Biology Reviews, 64*(4), 847–867. https://doi.org/10.1128/MMBR.64.4.847-867.2000

Davies, D. G., Parsek, M. R., Pearson, J. P., Iglewski, B. H., Costerton, J. W., & Greenberg, E. P. (1998). The involvement of cell-to-cell signals in the development of a bacterial biofilm. *Science, 280*(5361), 295–298. https://doi.org/10.1126/science.280.5361.295

Dong, Y., Ma, K., Cao, Q., Huang, H., Nie, M., Liu, G., Jiang, M., Lu, C., & Liu, Y. (2021). CRISPR-dependent endogenous gene regulation is required for virulence in piscine *Streptococcus agalactiae*. *Emerging Microbes & Infections, 10*(1), 2113–2124. https://doi.org/10.1080/22221751.2021.2002127

Donlan, R. M. (2001). Biofilm Formation: A clinically relevant microbiological process. *Clinical Infectious Diseases, 33*(8), 1387–1392. https://doi.org/10.1086/322972

Finelli, A., Gallant, C. V., Jarvi, K., & Burrows, L. L. (2003). Use of in-biofilm expression technology to identify genes involved in *Pseudomonas aeruginosa* biofilm development. *Journal of Bacteriology, 185*(9), 2700–2710. https://doi.org/10.1128/JB.185.9.2700-2710.2003

Folkesson, A., Haagensen, J. A. J., Zampaloni, C., Sternberg, C., & Molin, S. (2008). Biofilm induced tolerance towards antimicrobial peptides. *PLoS One, 3*(4), e1891. https://doi.org/10.1371/journal.pone.0001891

Fonfara, I., Le Rhun, A., Chylinski, K., Makarova, K. S., Lécrivain, A.-L., Bzdrenga, J., Koonin, E. V., & Charpentier, E. (2014). Phylogeny of Cas9 determines functional exchangeability of dual-RNA and Cas9 among orthologous type II CRISPR-Cas systems. *Nucleic Acids Research, 42*(4), 2577–2590. https://doi.org/10.1093/nar/gkt1074

Fu, Q., Li, S., Wang, Z., Shan, W., Ma, J., Cheng, Y., Wang, H., Yan, Y., & Sun, J. (2017). H-NS mutation-mediated CRISPR-cas activation inhibits phage release and toxin production of *Escherichia coli* Stx2 phage lysogen. *Frontiers in Microbiology, 8*. https://doi.org/10.3389/fmicb.2017.00652

Fu, Q., Su, Z., Cheng, Y., Wang, Z., Li, S., Wang, H., Sun, J., & Yan, Y. (2017). Clustered, regularly interspaced short palindromic repeat (CRISPR) diversity and virulence factor distribution in avian *Escherichia coli*. *Research in Microbiology, 168*(2), 147–156. https://doi.org/10.1016/j.resmic.2016.10.002

Ghanbarzadeh Corehtash, Z., Khorshidi, A., Firoozeh, F., Akbari, H., & Mahmoudi Aznaveh, A. (2015). Biofilm Formation and virulence factors among *Pseudomonas aeruginosa* isolated from burn patients. *Jundishapur Journal of Microbiology, 8*(10). https://doi.org/10.5812/jjm.22345

Gilmore, K. S., Srinivas, P., Akins, D. R., Hatter, K. L., & Gilmore, M. S. (2003). Growth, development, and gene expression in a persistent *Streptococcus gordonii* biofilm. *Infection and Immunity, 71*(8), 4759–4766. https://doi.org/10.1128/IAI.71.8.4759-4766.2003

Gooderham, W. J., & Hancock, R. E. W. (2009). Regulation of virulence and antibiotic resistance by two-component regulatory systems in *Pseudomonas aeruginosa*. *FEMS Microbiology Reviews, 33*(2), 279–294. https://doi.org/10.1111/j.1574-6976.2008.00135.x

Güvener, Z. T., & Harwood, C. S. (2007). Subcellular location characteristics of the *Pseudomonas aeruginosa* GGDEF protein, WspR, indicate that it produces cyclic-di-GMP in response to growth on surfaces. *Molecular Microbiology*. https://doi.org/10.1111/j.1365-2958.2007.06008.x

Hammer, B. K., & Bassler, B. L. (2009). Distinct sensory pathways in *Vibrio cholerae* el tor and classical biotypes modulate cyclic dimeric GMP levels to control biofilm formation. *Journal of Bacteriology, 191*(1), 169–177. https://doi.org/10.1128/JB.01307-08

Hasman, H., Schembri, M. A., & Klemm, P. (2000). Antigen 43 and type 1 fimbriae determine colony morphology of *Escherichia coli* K-12. *Journal of Bacteriology, 182*(4), 1089–1095. https://doi.org/10.1128/JB.182.4.1089-1095.2000

Hassan, K. A., Johnson, A., Shaffer, B. T., Ren, Q., Kidarsa, T. A., Elbourne, L. D. H., Hartney, S., Duboy, R., Goebel, N. C., Zabriskie, T. M., Paulsen, I. T., & Loper, J. E. (2010). Inactivation of the GacA response regulator in Pseudomonas fluorescens Pf-5 has far-reaching transcriptomic consequences: P. Fluorescens GacA transcriptomic fingerprint. *Environmental Microbiology, 12*(4), 899–915. https://doi.org/10.1111/j.1462-2920.2009.02134.x

Heeb, S., & Haas, D. (2001). Regulatory roles of the GacS/GacA two-component system in plant-associated and other gram-negative bacteria. *Molecular Plant-Microbe Interactions, 14*(12), 1351–1363. https://doi.org/10.1094/MPMI.2001.14.12.1351

Hegde, S., Nilyanimit, P., Kozlova, E., Anderson, E. R., Narra, H. P., Sahni, S. K., Heinz, E., & Hughes, G. L. (2019). CRISPR/Cas9-mediated gene deletion of the ompA gene in symbiotic Cedecea neteri impairs biofilm formation and reduces gut colonization of *Aedes aegypti* mosquitoes. *PLOS Neglected Tropical Diseases, 13*(12), e0007883. https://doi.org/10.1371/journal.pntd.0007883

Hengge, R. (2009). Principles of c-di-GMP signalling in bacteria. *Nature Reviews Microbiology, 7*(4), 263–273. https://doi.org/10.1038/nrmicro2109

Hufnagel, M., Koch, S., Creti, R., Baldassarri, L., & Huebner, J. (2004). A putative sugar-binding transcriptional regulator in a novel gene locus in *Enterococcus faecalis* contributes to production of biofilm and prolonged bacteremia in mice. *The Journal of Infectious Diseases, 189*(3), 420–430. https://doi.org/10.1086/381150

Ishino, Y., Shinagawa, H., Makino, K., Amemura, M., & Nakata, A. (1987). Nucleotide sequence of the iap gene, responsible for alkaline phosphatase isozyme conversion in *Escherichia coli*, and identification of the gene product. *Journal of Bacteriology, 169*(12), 5429–5433. https://doi.org/10.1128/jb.169.12.5429-5433.1987

Jefferson, K. K. (2004). What drives bacteria to produce a biofilm? *FEMS Microbiology Letters, 236*(2), 163–173. https://doi.org/10.1111/j.1574-6968.2004.tb09643.x

Jinek, M., Chylinski, K., Fonfara, I., Hauer, M., Doudna, J. A., & Charpentier, E. (2012). A programmable dual-RNA–guided DNA endonuclease in adaptive bacterial immunity. *Science, 337*(6096), 816–821. https://doi.org/10.1126/science.1225829

Kaleta, M. F., Petrova, O. E., Zampaloni, C., Garcia-Alcalde, F., Parker, M., & Sauer, K. (2022). A previously uncharacterized gene, PA2146, contributes to biofilm formation and drug tolerance across the ɣ-Proteobacteria. *Npj Biofilms and Microbiomes, 8*(1), 54. https://doi.org/10.1038/s41522-022-00314-y

Kalia, V. C. (2013). Quorum sensing inhibitors: An overview. *Biotechnology Advances, 31*(2), 224–245. https://doi.org/10.1016/j.biotechadv.2012.10.004

Karvelis, T., Gasiunas, G., Miksys, A., Barrangou, R., Horvath, P., & Siksnys, V. (2013). CrRNA and tracrRNA guide Cas9-mediated DNA interference in *Streptococcus thermophilus*. *RNA Biology, 10*(5), 841–851. https://doi.org/10.4161/rna.24203

Kjaergaard, K., Schembri, M. A., Ramos, C., Molin, S., & Klemm, P. (2000). Antigen 43 facilitates formation of multispecies biofilms. *Environmental Microbiology, 2*(6), 695–702. https://doi.org/10.1046/j.1462-2920.2000.00152.x

Klapper, I., & Dockery, J. (2002). Finger Formation in biofilm layers. *SIAM Journal on Applied Mathematics, 62*(3), 853–869. https://doi.org/10.1137/S0036139900371709

Laughlin, D. C., Joshi, C., van Bodegom, P. M., Bastow, Z. A., & Fulé, P. Z. (2012). A predictive model of community assembly that incorporates intraspecific trait variation. *Ecology Letters, 15*(11), 1291–1299. https://doi.org/10.1111/j.1461-0248.2012.01852.x

Lequette, Y., & Greenberg, E. P. (2005). Timing and localization of rhamnolipid synthesis gene expression in *Pseudomonas aeruginosa* biofilms. *Journal of Bacteriology, 187*(1), 37–44. https://doi.org/10.1128/JB.187.1.37-44.2005

Li, Y.-H., Tang, N., Aspiras, M. B., Lau, P. C. Y., Lee, J. H., Ellen, R. P., & Cvitkovitch, D. G. (2002). A quorum-sensing signaling system essential for genetic competence in *Streptococcus mutans* is involved in biofilm formation. *Journal of Bacteriology, 184*(10), 2699–2708. https://doi.org/10.1128/JB.184.10.2699-2708.2002

Liu, Y., Pinto, F., Wan, X., Yang, Z., Peng, S., Li, M., Cooper, J. M., Xie, Z., French, C. E., & Wang, B. (2022). Reprogrammed tracrRNAs enable repurposing of RNAs as crRNAs and sequence-specific RNA biosensors. *Nature Communications, 13*(1), 1937. https://doi.org/10.1038/s41467-022-29604-x

Liu, G., Zhang, W., & Lu, C. (2012). Complete genome sequence of Streptococcus agalactiae GD201008-001, isolated in China from Tilapia with meningoencephalitis. *Journal of Bacteriology, 194*(23), 6653. https://doi.org/10.1128/JB.01788-12

Loo, C. Y., Corliss, D. A., & Ganeshkumar, N. (2000). *Streptococcus gordonii* biofilm formation: Identification of genes that code for biofilm phenotypes. *Journal of Bacteriology, 182*(5), 1374–1382. https://doi.org/10.1128/JB.182.5.1374-1382.2000

van Loosdrecht, M. C. M., Heijnen, J. J., Eber, H., Kreft, J., & Picioreanu, C. (2002). *Mathematical modelling of biofilm structures* (Vol 12).

Ma, K., Cao, Q., Luo, S., Wang, Z., Liu, G., Lu, C., & Liu, Y. (2018). *Cas9* enhances bacterial virulence by repressing the *regR* transcriptional regulator in Streptococcus agalactiae. *Infection and Immunity, 86*(3), 17–005522. https://doi.org/10.1128/IAI.00552-17

Makarova, K. S., & Koonin, E. V. (2015). Annotation and classification of CRISPR-cas systems. In M. Lundgren, E. Charpentier, & P. C. Fineran (Eds.), *CRISPR* (Vol 1311, pp. 47–75). Springer New York. https://doi.org/10.1007/978-1-4939-2687-9_4

Makarova, K. S., Wolf, Y. I., Iranzo, J., Shmakov, S. A., Alkhnbashi, O. S., Brouns, S. J. J., Charpentier, E., Cheng, D., Haft, D. H., Horvath, P., Moineau, S., Mojica, F. J. M., Scott, D., Shah, S. A., Siksnys, V., Terns, M. P., Venclovas, Č., White, M. F., Yakunin, A. F., … Koonin, E. V. (2020). Evolutionary classification of CRISPR–cas systems: A burst of class 2 and derived variants. *Nature Reviews Microbiology, 18*(2), 67–83. https://doi.org/10.1038/s41579-019-0299-x

Marraffini, L. A. (2015). CRISPR-Cas immunity in prokaryotes. *Nature, 526*(7571), 55–61. https://doi.org/10.1038/nature15386

Mattei, M. R., Frunzo, L., D'Acunto, B., Pechaud, Y., Pirozzi, F., & Esposito, G. (2018). Continuum and discrete approach in modeling biofilm development and structure: A review. *Journal of Mathematical Biology, 76*(4), 945–1003. https://doi.org/10.1007/s00285-017-1165-y

Nassar, R., Hachim, M., Nassar, M., Kaklamanos, E. G., Jamal, M., Williams, D., & Senok, A. (2021). Microbial metabolic genes crucial for *S. aureus* biofilms: An insight from re-analysis of publicly available microarray datasets. *Frontiers in Microbiology, 11*, 607002. https://doi.org/10.3389/fmicb.2020.607002

sharma, N., Das, A., Raja, P., & Marathe, S. A. (n.d.). *The CRISPR-Cas System differentially regulates surface-attached and pellicle-biofilm in Salmonella enterica serovar Typhimurium*. 39.

Nie, M., Dong, Y., Cao, Q., Zhao, D., Ji, S., Huang, H., Jiang, M., Liu, G., & Liu, Y. (2022). CRISPR contributes to adhesion, invasion, and biofilm formation in Streptococcus agalactiae by repressing capsular polysaccharide production. *Microbiology Spectrum, 10*(4), e02113–e02121. https://doi.org/10.1128/spectrum.02113-21

Otto, M. (2012). Molecular basis of Staphylococcus epidermidis infections. *Seminars in Immunopathology, 34*(2), 201–214. https://doi.org/10.1007/s00281-011-0296-2

Patterson, A. G., Jackson, S. A., Taylor, C., Evans, G. B., Salmond, G. P. C., Przybilski, R., Staals, R. H. J., & Fineran, P. C. (2016). Quorum sensing controls adaptive immunity through the regulation of multiple CRISPR-cas systems. *Molecular Cell, 64*(6), 1102–1108. https://doi.org/10.1016/j.molcel.2016.11.012

Pearson, J. P., Gray, K. M., Passador, L., Tucker, K. D., Eberhard, A., Iglewski, B. H., & Greenberg, E. P. (1994). Structure of the autoinducer required for expression of *Pseudomonas aeruginosa* virulence genes. *Proceedings of the National Academy of Sciences, 91*(1), 197–201. https://doi.org/10.1073/pnas.91.1.197

Picioreanu, C. (2000a). *1. Introduction to biofilm structure* (Vol 57).

Picioreanu, C. (2001). Two-dimensional model of biofilm detachment caused by internal stress from liquid flow. *Biotechnology and Bioengineering, 72*(2), 14.

Picioreanu, C., van Loosdrecht, M. C. M., & Heijnen, J. J. (2000). A theoretical study on the effect of surface roughness on mass transport and transformation in biofilms. *Biotechnology and Bioengineering, 68*(4), 355–369. https://doi.org/10.1002/(SICI)1097-0290(20000520)68:4<355::AID-BIT1>3.0.CO;2-A

Pombo, J. P., Ebenberger, S. P., Müller, A. M., Wolinski, H., & Schild, S. (2022). Impact of gene repression on biofilm formation of *Vibrio cholerae. Frontiers in Microbiology, 13*, 912297. https://doi.org/10.3389/fmicb.2022.912297

Prigent-Combaret, C., & Lejeune, P. (1999). [4] Monitoring gene expression in biofilms, *310. Methods in enzymology* (pp. 56–79). Elsevier. https://doi.org/10.1016/S0076-6879(99)10006-5

Reisch, C. R., & Prather, K. L. J. (2015). The no-SCAR (Scarless Cas9 assisted recombineering) system for genome editing in *Escherichia coli. Scientific Reports, 5*(1), 15096. https://doi.org/10.1038/srep15096

Roy, S., Elgharably, H., Sinha, M., Ganesh, K., Chaney, S., Mann, E., Miller, C., Khanna, S., Bergdall, V. K., Powell, H. M., Cook, C. H., Gordillo, G. M., Wozniak, D. J., & Sen, C. K. (2014). Mixed-species biofilm compromises wound healing by disrupting epidermal barrier function: Preclinical burn wound biofilm infection. *The Journal of Pathology, 233*(4), 331–343. https://doi.org/10.1002/path.4360

Sauer, K. (2003). The genomics and proteomics of biofilm formation. *Genome Biology, 5*.

Schembri, M. A., Kjaergaard, K., & Klemm, P. (2003). Global gene expression in *Escherichia coli* biofilms: Gene expression in *E. coli* biofilms. *Molecular Microbiology, 48*(1), 253–267. https://doi.org/10.1046/j.1365-2958.2003.03432.x

Shipley, B., Vile, D., & Garnier, É. (2006). From plant traits to plant communities: A statistical mechanistic approach to biodiversity. *Science, 314*(5800), 812–814. https://doi.org/10.1126/science.1131344

Smith, T. M., Youngblom, M. A., Kernien, J. F., Mohamed, M. A., Fry, S. S., Bohr, L. L., Mortimer, T. D., O'Neill, M. B., & Pepperell, C. S. (2022). Rapid adaptation of a complex trait during experimental evolution of *Mycobacterium tuberculosis*. *ELife, 11*, e78454. https://doi.org/10.7554/eLife.78454

Tahrioui, A., Quesada, E., & Llamas, I. (2013). Genetic and phenotypic analysis of the GacS/GacA system in the moderate halophile Halomonas anticariensis. *Microbiology, 159*(Pt_3), 462–474. https://doi.org/10.1099/mic.0.061721-0

Tang, B., Gong, T., Zhou, X., Lu, M., Zeng, J., Peng, X., Wang, S., & Li, Y. (2019). Deletion of cas3 gene in Streptococcus mutans affects biofilm formation and increases fluoride sensitivity. *Archives of Oral Biology, 99*, 190–197. https://doi.org/10.1016/j.archoralbio.2019.01.016

Toledo-Arana, A., Valle, J., Solano, C., Arrizubieta, M. J., Cucarella, C., Lamata, M., Amorena, B., Leiva, J., Penadés, J. R., & Lasa, I. (2001). The enterococcal surface protein, esp, is involved in *Enterococcus faecalis* biofilm formation. *Applied and Environmental Microbiology, 67*(10), 4538–4545. https://doi.org/10.1128/AEM.67.10.4538-4545.2001

Tong, Z., Du, Y., Ling, J., Huang, L., & Ma, J. (2017). Relevance of the clustered regularly interspaced short palindromic repeats of *Enterococcus faecalis* strains isolated from retreatment root canals on periapical lesions, resistance to irrigants and biofilms. *Experimental and Therapeutic Medicine*. https://doi.org/10.3892/etm.2017.5205

Vlamakis, H., Chai, Y., Beauregard, P., Losick, R., & Kolter, R. (2013). Sticking together: Building a biofilm the Bacillus subtilis way. *Nature Reviews Microbiology, 11*(3), 157–168. https://doi.org/10.1038/nrmicro2960

Wang, Q., & Zhang, T. (2010). Review of mathematical models for biofilms. *Solid State Communications, 150*(21–22), 1009–1022. https://doi.org/10.1016/j.ssc.2010.01.021

Weigelt, M. A., McNamara, S. A., Sanchez, D., Hirt, P. A., & Kirsner, R. S. (2021). Evidence-based review of antibiofilm agents for wound care. *Advances in Wound Care, 10*(1), 13–23. https://doi.org/10.1089/wound.2020.1193

Whiteley, M., Bangera, M. G., Bumgarner, R. E., Parsek, M. R., Teitzel, G. M., Lory, S., & Greenberg, E. P. (2001). Gene expression in *Pseudomonas aeruginosa* bio®lms. *Binding Protein, 413*, 5.

Wolska, K. I., Grudniak, A. M., Rudnicka, Z., & Markowska, K. (2016). Genetic control of bacterial biofilms. *Journal of Applied Genetics, 57*(2), 225–238. https://doi.org/10.1007/s13353-015-0309-2

Wolz, C., Goerke, C., Landmann, R., Zimmerli, W., & Fluckiger, U. (2002). Transcription of clumping factor A in attached and unattached *Staphylococcus aureus* in vitro and during device-related infection. *Infection and Immunity, 70*(6), 2758–2762. https://doi.org/10.1128/IAI.70.6.2758-2762.2002

Wu, Y.-K., Cheng, N.-C., & Cheng, C.-M. (2019). Biofilms in chronic wounds: Pathogenesis and diagnosis. *Trends in Biotechnology, 37*(5), 505–517. https://doi.org/10.1016/j.tibtech.2018.10.011

Wu, Q., Cui, L., Liu, Y., Li, R., Dai, M., Xia, Z., & Wu, M. (2022). CRISPR-Cas systems target endogenous genes to impact bacterial physiology and alter mammalian immune responses. *Molecular Biomedicine, 3*(1), 22. https://doi.org/10.1186/s43556-022-00084-1

Yildiz, F. H., Liu, X. S., Heydorn, A., & Schoolnik, G. K. (2004). Molecular analysis of rugosity in a *Vibrio cholerae* O1 El Tor phase variant: Regulation of rugosity. *Molecular Microbiology, 53*(2), 497–515. https://doi.org/10.1111/j.1365-2958.2004.04154.x

Zegans, M. E., Wagner, J. C., Cady, K. C., Murphy, D. M., Hammond, J. H., & O'Toole, G. A. (2009). Interaction between bacteriophage DMS3 and host CRISPR region inhibits group behaviors of *Pseudomonas aeruginosa*. *Journal of Bacteriology, 191*(1), 210–219. https://doi.org/10.1128/JB.00797-08

Zhang, T. (2017). Modeling biofilms: From genes to communities. *Processes, 5*(4), 5. https://doi.org/10.3390/pr5010005

Zhang, T., Pabst, B., Klapper, I., & Stewart, P. S. (2013). General theory for integrated analysis of growth, gene, and protein expression in biofilms. *PLoS One, 8*(12), e83626. https://doi.org/10.1371/journal.pone.0083626

Zuberi, A., Misba, L., & Khan, A. U. (2017). CRISPR interference (CRISPRi) inhibition of luxS gene expression in *E. coli*: An approach to inhibit biofilm. *Frontiers in Cellular and Infection Microbiology, 7*, 214. https://doi.org/10.3389/fcimb.2017.00214

CHAPTER

3

Introduction on genomic analysis of biofilms

C.R. Hemalatha[1] and Anupma Jyoti Kindo[2]

[1]Bioinformatics—SRET, SRIHER (DU), Chennai, Tamil Nadu, India; [2]Department of Microbiology, SRMC & RI, SRIHER (DU), Chennai, Tamil Nadu, India

1. Introduction

Biofilms have always been a part of us and all around us from time immemorial. Combating medical biofilms remains to be a daunting experience confronting researchers and doctors alike. Though traditionally a single antibacterial agent or in combinations have been used to little effect for almost half a century, a remedy to completely remove these biofilms seems to be still in research. With the advent of drug target discoveries, scientists have been able to narrow down on molecules, which are the actual causes of thriving biofilms.

The National Institutes of Health (NIH) has estimated that up to 80% of human infectious diseases are biofilm related (Minutes of the National Advisory, 1997). More than 99% of bacteria found in nature exist in these stable, persistent biofilms. There is no single way to tackle biofilm formation, as microbes find newer ways of combating the drug very easily and quickly. They either use biofilm-inducing genes or dormancy genes initiating biofilm formation.

2. Biofilm formation

Costerton et al. (1978) gave the definition of biofilms as "polymicrobial community of adherent organisms with an extracellular polysaccharide matrix of their own making." Biofilm can form in any favorable environment that has proper nutrient conditions. The surfaces for attachment of cells can be abiotic such as metal, glass, plastic, medical implants, and stainless steel or biotic such as epithelial cells, human skin, and animal tissues. Apart from the environmental conditions such as temperature, pH, and ionic force of the medium, there

Microbial Biofilms
https://doi.org/10.1016/B978-0-323-95715-1.00024-8

are repulsive electrostatic and hydrodynamic forces in a liquid environment that inhibit the biofilm formation.

One of the first steps in biofilm formation, which is critical, is adherence (Budzik & Schneewind, 2006), followed by microcolony formation, maturation, and dispersal. Primary adhesion (An et al., 2020) involves balance of forces such as electrostatic, hydrophobic interactions, steric hindrance, van der Waal's forces, and hydrodynamic forces. Secondary adhesion involves specific bacterial adhesion molecules, attachment, and maintenance.

Bacteria-forming biofilms acquire smooth transition abilities from a planktonic to a sedentary biofilm life, which requires well-coordinated interaction of genes involved in forming biofilms. Recent DNA microarray analyses of biofilms showed that many genes exhibited differential expression, which may point out to the role of genetic makeup in biofilm formation (Beloin et al., 2004; Ren et al., 2004; Schembri et al., 2003). However, the agreement for differential gene expression is limited among these studies, probably reflecting differences in experimental conditions as well as the nature of biofilm itself (Junker et al., 2007). The environment within the biofilm is heterogenous, and biofilm formation a dynamic process (Domka et al., 2007). Furthermore, the different expression of some genes may be due to differences in growth of planktonic cells used as control (Lazazzera, 2005). Genetic analysis revealed that surface structures such as flagella and specific outer membrane adhesins, type 1, and curli fimbriae of *Escherichia coli* are important for biofilm formation, though they are not indispensable (Prigent-Combaret et al., 2000; Reisner et al., 2003).

3. Extracellular polymeric substances

According to Valle et al., 2012, "If biofilms can be metaphorically called a 'city of microbes,' (Watnick & Kolter, 2000), the EPS represents the 'house of the biofilm cells'." The EPS determine the immediate conditions of life of biofilm cells living in this microenvironment by affecting porosity, density, water content, charge, sorption properties, hydrophobicity, and mechanical stability (Flemming & Wingender, 2002).

EPS are biopolymers of microbial origin in which biofilm-producing microorganisms (archaea, bacteria, and eukarya) are entrenched. EPS comprises polysaccharides, a wide variety of proteins, glycoproteins, and glycolipids, and in some cases, surprising amounts of extracellular DNA (e-DNA). Unfortunately, it remains a substantial challenge to provide a complete biochemical profile of most EPS samples. It is often difficult to purify EPS matrix constituents apart from other components such as cells or other macromolecules transiently associated with the EPS (Nielsen & Jahn, 1999). The EPS biopolymers are highly hydrated and form a matrix, which keeps the biofilm cells together and retains water. This matrix interacts with the environment, e.g., by attaching biofilms to surfaces and providing nutrients for biofilm organisms.

4. Quorum sensing

Quorum sensing is a term used to describe intercellular signaling in bacteria. Although several quorum sensing systems are known, perhaps the two most thoroughly described systems are the acyl-homoserine lactone (acyl-HSL) systems of many gram-negative species and the peptide-based signaling systems of many gram-positive species (Bassler, 2002; Fuqua et al., 2001; Parsek & Greenberg, 2005). Microbiologists studying aspects of the bacterial community encounter information regarding two social phenomena exhibited by bacteria, quorum sensing, and biofilm development. These two topics have been inextricably linked, possibly because biofilms and quorum sensing represent two areas in which microbiologists focus on social aspects of bacteria. In addition, we believe that these two aspects of bacterial behavior represent a small part of the social repertoire of bacteria. Bacteria exhibit many social activities and they represent a model for dissecting social behavior at the genetic level. Owing to this fact, Sturme et al. introduce the term "sociomicrobiology," meaning the connection between quorum sensing and biofilms (Sturme et al., 2002).

5. Antibiotic resistance

One of the major reasons for antibiotic resistance by biofilms is the extracellular polysaccharide matrix. An example of poor penetration is *Klebsiella pneumoniae* (Anderl et al., 2000) biofilm resistance to ampicillin and ciprofloxacin. One of the examples of good glycopeptide penetrance is rifampin and vancomycin through a *Staphylococcus epidermidis* (Dunne, 1993) biofilm. Though 99% of bacteria are killed at bacteriocidal doses, 1% cells survive even in the face of greatly increased concentrations as in the case of multidrug tolerance in *Escherichia coli* (Keren et al., 2004) biofilms.

6. Central line-associated bloodstream infections

Central line-associated bloodstream infections (CRBSIs) have always been the point of attention in recent years. These infections are responsible for many cases of deaths and have ultimately raised the healthcare costs. Central venous catheters (CVCs) are progressively used in hospitals to manage seriously ill patients. CRBSIs occurring in the intensive care unit (ICU) are common occurrences incurring a lot of cost and are possibly lethal. CVCs have a higher infection risk than other indwelling vascular access lines, being the significant reason for morbidity and mortality of critically ill patients (Frasca et al., 2010; Raad et al., 2007). CRBSIs are considered among the first and most "preventable" classes of nosocomial infections (Mermel et al., 2009). Patients with CVCs are at risk of developing local as well as systemic infectious complications such as local insertion site infection, CRBSI, septic thrombophlebitis, endocarditis, and other metastatic infections. The most serious complications are bacteremia, sepsis, and death (Pronovost et al., 2006). The definitive diagnosis of catheter infection can be made by using a combination of clinical signs and symptoms together with the quantitative culture techniques (Richet et al., 1990).

7. Model of biofilm infection

Costerton et al. (1999) proposed a simple model of biofilm infection. The evolving biofilm infection paradigm was a significant departure from the then widely held view of infection, which envisioned single-species bacteria in a planktonic mode of growth utilizing virulence factors to cause infection (Donati et al., 2010; Rayner et al., 1998). This planktonic view of infection could explain most acute infections but was wholly inadequate for understanding chronic infections. However, Costerton and Stewart's early innovative biofilm model of infection demonstrated, at the biochemical and cellular levels, a new bacterial strategy by which communities of bacteria produce infection. Their biofilm model of infection explained the ineffectiveness of antibodies (Lam et al., 1987) and white blood cells (Leid et al., 2005) in combating biofilms. Other work showed the ineffectiveness of antibiotics for clearing a biofilm infection (Stewart & Costerton, 2001). Many new studies utilizing an emerging and sophisticated science have generated a wealth of fresh insights into the nature of biofilm infection.

8. Identification of genes and proteins

Biofilm-forming cells are distinctive from the well-investigated planktonic cells and exhibit a different type of gene expression. Several new *Escherichia coli* genes related to biofilm formation have been identified through genomic approaches such as DNA microarray analysis. However, many others involved in this process might have escaped detection due to poor expression, regulatory mechanism, or genetic backgrounds. Screening a collection of single-gene deletion mutants of *E. coli* named "Keio collection" to identify genes (Tabe Eko NIBA et al., 2008) required for biofilm formation is another method. However, this study requires literature survey as the method to form a list of proteins (Eyko et al., 2005; Jeske et al., 2003; Murga et al., 2001; Niba et al., 2008) associated with biofilm formation in bacteria.

9. Docking and combinatorial library

The identification of interactions between drugs/compounds and their targets is crucial for the development of new drugs. In vitro screening experiments (i.e., bioassays) are frequently used for this purpose; however, experimental approaches are insufficient to explore novel drug–target interactions, mainly because of feasibility problems, as they are labor-intensive, costly, and time-consuming. A computational field known as "virtual screening" (VS) has emerged in the past decades to aid experimental drug discovery studies by statistically estimating unknown biointeractions between compounds and biological targets.

10. Advanced strategies to combat biofilms

Conventional methods use the physicochemical and structural properties of compounds and/or target proteins along with the experimentally verified biointeraction information to generate predictive models. Lately, sophisticated deep learning and machine intelligence (niche fields of artificial intelligence) techniques are applied in VS to elevate the predictive performance (Rifaioglu et al., 2018), which may be the futuristic application of these.

The increased expression of resistance markers within the biofilm and the diffusion limitations of the extracellular matrix have made biofilm bacteria recalcitrant to treatment with antibiotics. In UPECs, antibiotic resistance mechanisms have evolved as it has been observed that there is a decrease in susceptibility to first-line agents such as nitrofurantoin, ampicillin, fluoroquinolones, and sulfamethoxazole/trimethoprim. Therefore, there is an urgent need for the development of new therapeutic strategies to eradicate biofilm infections by *E. coli* (Chibeu et al., 2012). The recent advances in strategies designed to treat biofilms are antiadhesion agents, phage therapy, phytochemicals, antimicrobial peptides, and nanoparticles.

The inhibition of curli and pili formation can help in the treatment of *E. coli* biofilm. Structural knowledge of the target protein has led to design and screening of small molecule-based inhibitors. These molecules can traverse through cell membrane and thus target various components of cellular machinery. Type 1 pilus in *E. coli* was inhibited by N-(4-chlorophenyl)-2-{5-[4-(pyrrolidine-1-sulfonyl)-phenyl]-[1,3,4] oxadiazol-2-yl sulfanyl}-acetamide (AL1) (Lo et al., 2014). Mannosides were shown to act synergistically with trimethoprim–sulfamethoxazole in treatment of infection due to UPEC (Guiton et al., 2012).

In another study, curli subunit CsgA was targeted by rationally designed 2-pyridone compounds, which prevented *E. coli* biofilm formation. Curli is abundant in matrix of bacterial biofilm, and its inhibition prevents colonization and invasion and reduces the biofilm biomass (Andersson et al., 2013; Cegelski et al., 2009). The curlicides BibC6 and FN075 have a common chemical lineage to ring-fused 2-pyridones known as pilicides. They inhibit major curli subunit protein CsgA and hence curtail the formation of curli in UPEC. The curlicides also retain pilicide activities and inhibit both curli-dependent and type 1-dependent biofilms (Cegelski et al., 2009). Fluorescent pilicides and curlicides have been synthesized using coumarin and 4,4-difluoro-4-bora-3a,4a-diaza-s-indacene (BODIPY) fluorophores, and it was found to have improved antibiofilm activity (Chorell et al., 2012).

Phages have been effective in eradicating biofilm of single or mixed bacterial species and could lyse biofilm grown on medical devices and filtration membranes. Certain phages possess virion-associated polysaccharide depolymerases, which help them to degrade capsules. It was found that mature biofilms are more effectively eradicated in depolymerase-producing phages than the nonsecreting ones; therefore, they have been used along with antimicrobials to facilitate deeper penetration by degrading the EPS. The phage depolymerases have an important part in the degradation of the EPS matrix of the biofilm, which facilitate the permeation of the phages inside the biofilm layers resulting in bacterial cell lysis. However, phages that are not capable of producing EPS depolymerases are also used in biofilm degradation. These include phages such as T7 engineered to express recombinant dispersin B (DsbB) and naturally occurring phages such as T4 (Chibeu et al., 2012).

A cocktail of bacteriophages has a greater impact on the biofilm bacteria as their combination causes lysis of the bacterial cells of the biofilm much more than that caused by a single bacteriophage. One phage could facilitate the infection by the other phage by degrading the polysaccharides in the bacterial biofilm matrix with depolymerase. These enzymes are highly species-specific and hence target different bacteria especially for mixed biofilm. T7 phage has been engineered to express lactonase, which can degrade AHLs from many bacteria. They were found to be effective against mixed biofilm of *Pseudomonas aeruginosa* and *E. coli* (Pei & Lamas-Samanamud, 2014). The combination of phage with antibiotic has resulted in remarkable decrease in antibiotic-resistant *E. coli* (Kim et al., 2015). Recently, it has been shown that phage-resistant bacteria overproduce colanic acid. Second infection with phage-carrying colanic acid–degrading enzyme can restrain the development of phage-resistant bacteria (Coulter et al., 2014).

Plants are being increasingly explored as the possible antitherapeutic agent, as they can kill the microorganism with diverse mechanisms of action with minimal chance for bacteria to develop resistance to it. The phytochemicals such as 7-hydroxycoumarin (7-HC), indole-3-carbinol (I3C), salicylic acid, and saponin have shown inhibitory activity against the planktonic culture of *E. coli* and *Staphylococcus aureus* and were also able to restrict the growth of biofilm partially. The phytochemicals I3C and 7-HC had a more pronounced effect on QS inhibition and bacterial motility for both *E. coli* and *S. aureus*. I3C exhibited synergistic activity with antibiotics against resistant strains of *S. aureus* (Monte et al., 2014). Ginkgolic acid and *Ginkgo biloba* extract have shown significant inhibition of enterohemorrhagic *E. coli* O157:H7 biofilm formation by downregulating curli and prophage genes (Lee et al., 2014). The β-sitosterol glucoside isolated from citrus fruit inhibited *E. coli* O157:H7 biofilm formation and motility by suppressing the levels of RssAB and HNS of flagellar master operon flhDC (Vikram et al., 2013).

Phenolic acids (gallic acid and ferulic acid) inhibit bacterial motility of *E. coli*. Both gallic acid and ferulic acid caused total inhibition of swarming in *E. coli* and thus reduced the biofilm mass considerably (Borges et al., 2012). In another study, phenolic-rich maple syrup extract (PRMSE) was tested for its antibiofilm activity on pathogenic bacteria including *E. coli*. The transcriptome analysis revealed that PMRSE effectively repressed multiple drug resistance genes and genes associated with motility, adhesion, and biofilm formation (Maisuria et al., 2015).

Antimicrobial peptides (AMPs) are the integral part of innate immunity and are produced by varied living organisms to fight against infection. It has been seen that a chance for the development of bacterial resistance is very limited when the bacterial growth is restricted with AMP. Based on the ability of amino acid residues to form a helix, adhere to the surface, and possess antimicrobial activity, KABT-AMP was designed, which showed antimicrobial activity against *E. coli* (Thankappan et al., 2013).

Two antibacterial peptides containing tryptophan (KT2 and RT2) were designed and found to be highly effective against multidrug-resistant, enterohemorrhagic *E. coli* O157:H7 biofilm at 1 $\mu mol\ L^{-1}$ concentration. It was proposed that these peptides could traverse inside the cell and eventually bind the DNA for its antimicrobial action (Anunthawan et al., 2015). Bacteriocin isolated from *Citrobacter freundii* showed antimicrobial activity against a wide range of bacteria including *E. coli* in both planktonic and biofilm formation (Shanks et al., 2012).

Nanoparticles, being more stable and having high bioavailability, can be delivered efficiently as antimicrobial agents. Silver nanoparticles are known to be flexible and stable and can restrict the infection and biofilm formation of *E. coli*. The silver nanoparticles due to their small size and enhanced surface to volume ratio can be incorporated in medical devices and wound dressings. The mechanism of silver toxicity has been attributed to thiol group that renders many enzymes inactive, inhibiting DNA replication and translation of crucial proteins. Silver nanoparticles have been synthesized from the aqueous extract of *Calotropis procera* flower and were found to be effective against enterotoxic *E. coli* biofilm and significantly decreased the colonization in the small intestine of infant mouse model (Salem et al., 2015).

In another study, the silver nanoparticles were embedded in orthorhombic nanotubes of lithium vanadium oxide (Li2O5 /Ag), and they could restrict the growth of *E. coli* biofilm at the concentration of 60–120 μg mL^{-1}. The images obtained by scanning electron microscopy revealed the action of the nanocomposites on the surface of the bacteria in the form of surface perturbation, thereby facilitating its use as a candidate for biofabrication of medical devices to prevent infectious diseases (Diggikar et al., 2013).

Wei He et al. tested a nanotechnology antimicrobial spray, JUC, against the *E. coli* biofilm formed in CAUTI. It formed an invisible, protective positively charged film on the surface after it was sprayed, and prevented bacterial growth. In the clinical study, catheter sprayed with JUC therapy group, 4·52% of patients were diagnosed with CAUTI as compared with 13·04% in the control group (catheter sprayed with distilled water) (He et al., 2012). In a recent study, selenium and tellurium nanoparticles obtained from the strains *Stenotrophomonas maltophilia* and *Ochrobactrum* sp. MPV1, respectively, were found to be effective against both planktonic and biofilm forms of *E. coli* JM109, *P. aeruginosa* PAO1, and *S. aureus* ATCC 25923 (Zonaro et al., 2015). AMP LL-37 coated on magnetic nickel nanoparticles with the aid of polyacrylic acid, created as adhesion layer on nanoparticles, was found to be effective in killing *E. coli* (Chen et al., 2009).

Though many studies on experimental methods in *E. coli* biofilms are common, researches using in silico techniques in proteomics are few. The advantage of using in silico methods is the ease and speed at which a drug can be brought to the market to combat biofilm-related infections. This study is one such method where biofilm-associated proteins and phytochemicals, which may inhibit them, may be researched in detail, and the same may be handed over as a model for wet-lab scientists. Thereby, they can directly target the proteins associated by synthesizing the effective components of phytochemicals experimentally prepared and also report the bioactivity.

References

Anderl, J. N., Franklin, M. J., & Stewart, P. S. (2000). Role of antibiotic penetration limitation in *Klebsiella pneumoniae* biofilm resistance to ampicillin and ciprofloxacin. *Antimicrobial Agents and Chemotherapy, 44*(7), 1818–1824.

An, Y. H., Dickinson, R. B., & Doyle, R. J. (2020). Mechanisms of bacterial adhesion and pathogenesis of implant and tissue infections. In N. J. Towota (Ed.), *Handbook of bacterial adhesion: Principles, methods, and applications*. Humana Press.

Andersson, E. K., Bengtsson, C., Evans, M. L., Chorell, E., Sellstedt, M., Lindgren, A. E., Hufnagel, D. A., Bhattacharya, M., Tessier, P. M., Wittung-Stafshede, P., Almqvist, F., & Chapman, M. R. (2013). Modulation of curli assembly and pellicle biofilm formation by chemical and protein chaperones. *Chemistry and Biology, 20*(10), 1245–1254.

Anunthawan, T., Fuente-Núñez, C. D. L., Hancock, R. E., & Klaynongsruang, S. (2015). Cationic amphipathic peptides KT2 and RT2 are taken up into bacterial cells and kill planktonic and biofilm bacteria. *Biochimica et Biophysica Acta (BBA)—Biomembranes, 1848*(6), 1352–1358.

Bassler, B. L. (2002). *Small Talk. Cell., 109*(4), 421–424.

Beloin, C., Valle, J., Latour-Lambert, P., Faure, P., Kzreminski, M., Balestrino, D., Haagensen, J. A. J., Molin, S., Prensier, G., Arbeille, B., & Ghigo, J. M. (2004). Global impact of mature biofilm lifestyle on *Escherichia coli* K-12 gene expression. *Molecular Microbiology, 51*, 659–674.

Borges, A., Saavedra, M. J., & Simões, M. (2012). The activity of ferulic and gallic acids in biofilm prevention and control of pathogenic bacteria. *Biofouling, 28*(7), 755–767.

Budzik, J. M., & Schneewind, O. (2006). Pili prove pertinent to enterococcal endocarditis. *Journal of Clinical Investigation, 116*(10), 2582–2584.

Cegelski, L., Pinkner, J. S., Hammer, N. D., Cusumano, C. K., Hung, C. S., Chorell, E., Aberg, V., Walker, J. N., Seed, P. C., Almqvist, F., Chapman, M. R., & Hultgren, S. J. (2009). Small-molecule inhibitors target *Escherichia coli* amyloid biogenesis and biofilm formation. *Nature Chemical Biology, 5*(12), 913–919.

Chen, G., Zhou, M., Chen, S., Lv, G., & Yao, J. (2009). Nanolayer biofilm coated on magnetic nanoparticles by using a dielectric barrier discharge glow plasma fluidized bed for immobilizing an antimicrobial peptide. *Nanotechnology, 20*(46), 465706.

Chibeu, A., Lingohr, E. J., Masson, L., Manges, A., Harel, J., Ackermann, H.-W., Kropinski, A. M., & Boerlin, P. (2012). Bacteriophages with the ability to degrade uropathogenic *Escherichia coli* biofilms. *Viruses, 4*(4), 471–487.

Chorell, E., Pinkner, J. S., Bengtsson, C., Edvinsson, S., Cusumano, C. K., Rosenbaum, E., Johansson, L. B. A., Hultgren, S. J., & Almqvist, F. (2012). Design and synthesis of fluorescent pilicides and curlicides: Bioactive tools to study bacterial virulence mechanisms. *Chemistry, 18*(15), 4522–4532. https://doi.org/10.1002/chem.201103936

Costerton, J. W., Geesey, G. G., & Cheng, K.-J. (1978). How bacteria stick. *Scientific American, 238*(1), 86–95.

Costerton, J. W., Stewart, P. S., & Greenberg, E. P. (1999). Bacterial biofilms: A common cause of persistent infections. *Science, 284*, 1318–1322.

Coulter, L., Mclean, R., Rohde, R., & Aron, G. (2014). Effect of bacteriophage infection in combination with tobramycin on the emergence of resistance in *Escherichia coli* and *Pseudomonas aeruginosa* biofilms. *Viruses, 6*(10), 3778–3786.

Diggikar, R. S., Patil, R. H., Kale, S. B., Thombre, D. K., Gade, W. N., Kulkarni, M. V., & Kale, B. B. (2013). Silver-decorated orthorhombic nanotubes of lithium vanadium oxide: An impeder of bacterial growth and biofilm. *Applied Microbiology and Biotechnology, 97*(18), 8283–8290.

Domka, J., Lee, J., Bansal, T., & Wood, T. K. (2007). Temporal gene-expression in *Escherichia coli* K-12 biofilms. *Environmental Microbiology, 9*, 332–346.

Donati, C., Hiller, N. L., Tettelin, H., Muzzi, A., Croucher, N. J., Angiuoli, S. V., Oggioni, M., Dunning Hotopp, J. C., Hu, F. Z., Riley, D. R., Covacci, A., Mitchell, T. J., Bentley, S. D., Kilian, M., Ehrlich, G. D., Rappuoli, R., Richard Moxon, E., & Masignani, V. (2010). Structure and dynamics of the pan-genome of *Streptococcus pneumoniae* and closely related species. *Genome Biology, 11*, R107–R126.

Dunne, W. M. M. (1993). Diffusion of rifampin and vancomycin through a Staphylococcus epidermidis biofilm. *Antimicrobial Agents and Chemotherapy, 37*, 2522–2526.

Eyko, Aoki Elisabeth, Carlos, Pizzolitto Antonio, Botelho, Garcia Lourdes, & Elisabeth Loshchagin, Pizzolitto (December 2005). *Staphylococcus aureus* biofilms on central venous haemodialysis catheters. *Brazilian Journal of Microbiology, 36*(4), 342–346.

Flemming, H. C., & Wingender, J. (2002). *Extracellular polymeric substances: Structure, ecological functions, technical relevance*. NY: Wiley.

Frasca, D., Dahyot-Fizelier, C., & Mimoz, O. (2010). Prevention of central venous catheter-related infection in the intensive care unit. *Critical Care, 14*, 212.

Fuqua, C., Parsek, M. R., & Greenberg, E. P. (2001). Regulation of gene expression by cell-to-cell communication: Acyl-homoserine lactone quorum sensing. *Annual Review of Genetics, 35*(1), 439–468.

Guiton, P. S., Cusumano, C. K., Kline, K. A., Dodson, K. W., Han, Z., Janetka, J. W., Henderson, J. P., Caparon, M. G., & Hultgren, S. J. (2012). Combinatorial small-molecule therapy prevents uropathogenic *Escherichia coli* catheter-associated urinary tract infections in mice. *Antimicrobial Agents and Chemotherapy, 56*(9), 4738–4745.

He, W., Wang, D., Ye, Z., et al. (2012). Application of a nanotechnology antimicrobial spray to prevent lower urinary tract infection: A multicenter urology trial. *Journal of Translational Medicine, 10*(Suppl. 1), S14. https://doi.org/10.1186/1479-5876-10-S1-S14. Suppl 1.

Jeske, Christian, Raedler, Claus, von Goedecke, Achim, Mayr, Andreas, Guido, Hinterberger, Aspoeck, Ch, Lass-Floerl, Cornelia, & Benzer, Arnulf (October 2003). Early identification of bacteria leading to central venous catheter contamination. *Anesthesia and Analgesia, 97*(4), 940–943.

Junker, L. M., Toba, F. A., & Hay, A. G. (2007). Transcription in *Escherichia coli* PHL628 biofilms. *FEMS Microbiology Letters, 268*, 237–243.

Keren, I., Shah, D., Spoering, A., Kaldalu, N., & Lewis, K. (2004). Specialized persister cells and the mechanism of multidrug tolerance in *Escherichia coli*. *Journal of Bacteriology, 186*(24), 8172–8180.

Kim, M. S., Kim, Y. D., Hong, S. S., Park, K., Ko, K. S., & Myung, H. (2015). Phage-encoded colanic acid-degrading enzyme permits lytic phage infection of a capsule-forming resistant mutant *Escherichia coli* strain. *Applied and Environmental Microbiology, 81*(3), 900–909. https://doi.org/10.1128/AEM.02606-14

Lam, J. S., MacDonald, L. A., Lam, M. Y., Duchesne, L. G., & Southam, G. G. (1987). Production and characterization of monoclonal antibodies against serotype strains of *Pseudomonas aeruginosa*. *Infection and Immunity, 55*, 1051–1057.

Lazazzera, B. A. (2005). Lessons from DNA microarray analysis: The gene expression profile of biofilms. *Current Opinion in Microbiology, 8*, 222–227.

Lee, J.-H., Kim, Y.-G., Ryu, S. Y., Cho, M. H., & Lee, J. (2014). Ginkgolic acids and *Ginkgo biloba* extract inhibit *Escherichia coli* O157:H7 and *Staphylococcus aureus* biofilm formation. *International Journal of Food Microbiology, 174*, 47–55.

Leid, J. G., Willson, C. J., Shirtliff, M. E., Hassett, D. J., Parsek, M. R., & Jeffers, A. K. (2005). The exopolysaccharide alginate protects *Pseudomonas aeruginosa* biofilm bacteria from IFN-gamma-mediated macrophage killing. *The Journal of Immunology, 175*, 7512–7518.

Lo, A. W., Van de Water, K., Gane, P. J., Edith Chan, A. W., Steadman, D., Stevens, K., Selwood, D. L., Waksman, G., & Remaut, H. (2014). Suppression of type 1 pilus assembly in uropathogenic *Escherichia coli* by chemical inhibition of subunit polymerization. *Journal of Antimicrobial Chemotherapy, 69*(4), 1017–1026. https://doi.org/10.1093/jac/dkt467

Maisuria, V. B., Hosseinidoust, Z., & Tufenkji, N. (2015). Polyphenolic extract from maple syrup potentiates antibiotic susceptibility and reduces biofilm formation of pathogenic bacteria. *Applied and Environmental Microbiology, 81*(11), 3782–3792. https://doi.org/10.1128/AEM.00239-15

Minutes of the national advisory dental and craniofacial research council, 153rd meeting. (September 9, 1997). National Institutes of Health. http://www.nidcr.nih.gov/AboutNIDCR/CouncilAndCommittees/NADCRC/Minutes/Minutes153.htm.

Mermel, L. A., Allon, M., Bouza, E., Craven, D. E., Flynn, P., O'Grady, N. P., Raad, I. I., Rijnders, B. J. A., Sherertz, R. J., & Warren, D. K. (2009). Clinical practice guidelines for the diagnosis and management of intravascular catheter-related infection: 2009 update by the infectious diseases society of America. *Clinical Infectious Diseases, 49*, 1.

Monte, J., Abreu, A. C., Borges, A., Simões, L. C., & Simões, M. (2014). Antimicrobial activity of selected phytochemicals against *Escherichia coli* and *Staphylococcus aureus* and their biofilms. *Pathogens, 3*(2), 473–498. https://doi.org/10.3390/pathogens3020473. Published 2014 Jun 18.

Murga, R., Miller, J. M., & Donlan, R. M. (2001). Biofilm Formation by gram-negative bacteria on central venous catheter connectors: Effect of conditioning films in a laboratory model. *Journal of Clinical Microbiology, 39*(6), 2294–2297.

Niba, E. T. E., Naka, Y., Nagase, M., Mori, H., & Kitakawa, M. (2008). A genome-wide approach to identify the genes involved in biofilm formation in *E. coli*. *DNA Research, 14*(6), 237–246.

Nielsen, P. H., & Jahn, A. (1999). Extraction of EPS. In *Microbial extracellular polymeric substances: characterization, structure and function* (pp. 49–72). Berlin, Germany: Springer.

Parsek, M. R., & Greenberg, E. (2005). Sociomicrobiology: The connections between quorum sensing and biofilms. *Trends in Microbiology, 13*(1), 27–33.

Pei, R., & Lamas-Samanamud, G. R. (2014). Inhibition of biofilm formation by T7 bacteriophages producing quorum-quenching enzymes. *Applied and Environmental Microbiology, 80*(17), 5340–5348. https://doi.org/10.1128/AEM.01434-14

Prigent-Combaret, C., Prensier, G., Le Thi, T. T., Vidal, O., Lejeune, P., & Dorel, C. (2000). Developmental pathway for biofilm formation in curli-producing *Escherichia coli* strains: Role of flagella, curli and colanic acid. *Environmental Microbiology, 2*, 450–464.

Pronovost, P., Needham, D., Berenholtz, S., Sinopoli, D., Chu, H., Cosgrove, S., Sexton, B., Hyzy, R., Welsh, R., Roth, G., Bander, J., Kepros, J., & Goeschel, C. (2006). An intervention to decrease catheter-related bloodstream infections in the ICU. *New England Journal of Medicine, 355*, 2725–2732.

Raad, I., Hanna, H., & Maki, D. (2007). Intravascular catheter-related infections: Advances in diagnosis, prevention, and management. *The Lancet Infectious Diseases, 7*, 645–657.

Rayner, M. G., Zhang, Y., Gorry, M. C., Chen, Y., Post, J. C., & Ehrlich, G. D. (1998). Evidence of bacterial metabolic activity in culture-negative otitis media with effusion. *JAMA, 279*, 296–299.

Reisner, A., Haagensen, J. A., Schembri, M. A., Zechner, E. L., & Molin, S. (2003). Development and maturation of *Escherichia coli* K-12 biofilms. *Molecular Microbiology, 48*, 933–946.

Ren, D., Bedzyk, L. A., Thomas, S. M., Ye, R. W., & Wood, T. K. (2004). Gene expression in *Escherichia coli* biofilms. *Applied and Environmental Microbiology, 64*, 515–524.

Richet, H., Hubert, B., Nitemberg, G., Andremont, A., Buu-Hoi, A., Ourbak, P., Galicier, C., Veron, M., Boisivon, A., & Bouvier, A. M. (1990). Prospective multicenter study of vascular-catheter–related complications and risk factors for positive central catheter cultures in intensive care unit patients. *Journal of Clinical Microbiology, 28*, 2520–2525.

Rifaioglu, A. S., Atas, H., Martin, M. J., Cetin-Atalay, R., Atalay, V., & Doğan, T. (2018). Recent applications of deep learning and machine intelligence in *in silico* drug discovery: Methods, tools and databases. *Briefings in Bioinformatics, 20*.

Salem, W., Leitner, D. R., Zingl, F. G., Schratter, G., Prassl, R., Goessler, W., Reidl, J., & Schild, S. (2015). Antibacterial activity of silver and zinc nanoparticles against *Vibrio cholerae* and enterotoxic *Escherichia coli*. *International Journal of Medicine Microbiology, 305*(1), 85–95. https://doi.org/10.1016/j.ijmm.2014.11.005

Schembri, M. A., Kjærgaard, K., & Klemm, P. (2003). Global gene expression in *Escherichia coli* biofilms. *Molecular Microbiology, 48*, 253–267.

Shanks, R. M., Dashiff, A., Alster, J. S., & Kadouri, D. E. (2012). Isolation and identification of a bacteriocin with antibacterial and antibiofilm activity from *Citrobacter freundii*. *Archieves of Microbiology, 194*(7), 575–587. https://doi.org/10.1007/s00203-012-0793-2

Stewart, P. S., & Costerton, J. W. (2001). Antibiotic resistance of bacteria in biofilms. *Lancet, 358*, 135–138.

Sturme, M. H. J., Kleerebezem, M., Nakayama, J., Akkermans, A. D. L., Vaughan, E. E., & de Vos, W. M. (December 2002). Cell to cell communication by autoinducing peptides in gram-positive bacteria. *Antonie Van Leeuwenhoek, 81*(1–4), 233–243.

Genome-wide approach to identify the genes involved in biofilm formation. (January 7, 2008). In Tabe Eko Niba. E. coli Emma, Naka, Yoshiaki, Nagase, Megumi, Mori, Hirotada, & Kitakawa, Madoka (Eds.), (January 7, 2008).

Thankappan, B., Jeyarajan, S., Hiroaki, S., Anbarasu, K., Natarajaseenivasan, K., & Fujii, N. (2013). Antimicrobial and antibiofilm activity of designed and synthesized antimicrobial peptide, KABT-AMP. *Applied Biochemistry and Biotechnology, 170*(5), 1184–1193.

Valle, J., Latasa, C., Gil, C., Toledo-Arana, A., Solano, C., Penadés, J. R., & Lasa, I. (2012). Bap, a biofilm matrix protein of *Staphylococcus aureus* prevents cellular internalization through binding to GP96 host receptor. *PLoS Pathogens, 8*(8).

Vikram, A., Jayaprakasha, G., Uckoo, R. M., & Patil, B. S. (2013). Inhibition of *Escherichia coli* O157:H7 motility and biofilm by β-Sitosterol glucoside. *Biochimica et Biophysica Acta (BBA)—General Subjects, 1830*(11), 5219–5228.

Watnick, P., & Kolter, R. (2000). Biofilm, city of microbes. *Journal of Bacteriology, 182*(10), 2675–2679.

Zonaro, E., Lampis, S., Turner, R. J., Qazi, S. J., & Vallini, G. (2015). Biogenic selenium and tellurium nanoparticles synthesized by environmental microbial isolates efficaciously inhibit bacterial planktonic cultures and biofilms. *Frontiers in Microbiology, 6*, 584. https://doi.org/10.3389/fmicb.2015.00584. Published 2015 Jun 16.

CHAPTER

4

Metabolomic study of biofilm-forming natural microbiota of oral biofilm

Sylvia Denis, Suhana Sultana, Ankita Banerjee and Arup Kumar Mitra

Department of Microbiology, St. Xavier's College (Autonomous), Kolkata, West Bengal, India

1. Introduction

In accordance with the "insurance hypothesis," a single isolated species is considered to be more vulnerable by the environment than a collective of multiple species. Bacteria, which were previously considered to live rather asocial and reclusive lives, are now known to have elaborate chemical signaling systems to communicate within and between other species and form biofilms to colonize different biotic and abiotic substrates.

A biofilm is best described as a diverse and sessile microbial community that is characterized by microbial cells, which are attached irreversibly to an interface or a substratum or to each other. They are typically implanted in EPS—an extracellular polymeric substance matrix, secreted by the members of the biofilm. The microbial communities constitute microcolonies and interact with each other through methods such as quorum sensing and often exhibit altered phenotypes with respect to their gene transcription and growth rate (Donlan & William Costerton, 2002). Biofilms are complex systems with high cell densities that ranges from 10^8 to 10^{11} cells per g wet-weight (Wingender et al., 2022).

Biofilms confer certain survival advantage to the members of the bacterial community and protect them from various host defense mechanisms or applied antibiotics. The biofilm matrix also reduces the chances of mechanical injury to the individual members and additionally makes cell-to-cell communication frequent and effective. The antigenic epitopes of the bacterial members are hidden in the biofilm matrix, thus reducing their susceptibility to the host immune system. As more and more species take part in the development of biofilm, the structural complexity of the biofilm increases, and this is also linked to greater antibiotic resistance

Microbial Biofilms
https://doi.org/10.1016/B978-0-323-95715-1.00014-5

by the bacterial cells, in comparison with planktonic cells. It has been reported that biofilms can resist antimicrobial concentrations 100 to 1000 times of that needed to kill the same planktonic cells (Lewis, 2001). Plasmids carrying genes conferring antibiotic resistance can be transferred among bacteria through conjugation, which results in an antibiotic resistant property throughout the biofilm.

A common example of a natural polymicrobial biofilm is the dental plaque found in humans. It is known that it plays an important role in the occurrence and pathogenesis of many oral diseases such as gingivitis, caries, and periodontitis. *Streptococcus mutans, Streptococcus gordonii, Streptococcus oralis, Porphyromonas gingivalis, Aggregatibacter actinomycetemcomitans,* and *Fusobacterium* sp. are some bacteria that are known to form oral biofilms. The oral microflora usually colonizes the teeth, tongue, periodontal pockets, palate, etc. where they exist either as planktonic cells or in the form of biofilms (Berger et al., 2018a). Mature oral biofilms are made of towering microcolonies where the resident bacterial cells interact with each other and exchange messages via signaling molecules and metabolites. The primary colonizers that include *Streptococcus, Actinomyces* etc., lay the foundations of dental plaques. They provide a binding site for coadhesion with other bacteria (Watnick & Kolter, 2000). During the formation of dental plaque, the bacteria adherent to the teeth can sense their neighbors and generate responses appropriately. Some of their interactions involve signaling molecules that seem to have evolved specifically to elicit responses in neighboring bacteria.

2. Formation of biofilm in the buccal cavity

Biofilm formation within the buccal cavity can lead to the formation of dental plaques, resulting in dental cavities and severe periodontal diseases (Berger et al., 2018b). Biofilm development is a dynamic process. Formation of biofilm in the oral cavity is unique compared with any other type of biofilm because it generally requires salivary glycoproteins of the host for attachment (Huang et al., 2011). Oral biofilm formation usually starts with attachment of an acquired pellicle, which is a thin protein-containing film derived from salivary glycoproteins, to a clean tooth surface that is followed by a reversible attachment of a few bacterial cells on dental pellicles, implants, or dentures and then an irreversible attachment facilitated by their adhesive structures such as pilli, flagella, some proteins, and polysaccharide adhesins (Ramburrun et al., 2021). Biofilm formation by the free-living planktonic cells involves five major stages—(1) reversible attachment, (2) irreversible attachment, (3) growth, (4) maturation, and (5) dispersal (Muhammad et al., 2020). The initial attachment of the few free-living planktonic cells to the substratum is a critical step and depends upon the nature of the surface and the type of interaction between the cells and the substratum (Donlan, 2002; Petrova & Sauer, 2012). Studies have shown that bacteria tend to form a more robust biofilm on irregular surfaces compared with smoother ones owing to their large surface area and reduced effect of the shearing force (Donlan, 2002; Yu et al., 2016; Zheng et al., 2021). The surface charge also plays an important role in promoting cell adhesion to the substratum. For instance, the presence of phosphate, amino, and carboxyl groups in the cell wall components imparts a negative charge to bacteria that promotes its attachment to a positively charged surface (Hallab et al., 1995; Sousa et al., 2011). However, depending upon the growth medium, pH, ionic strength, bacterial age, etc. the bacterial surface charge may also vary (Katsikogianni et al., 2004). The initial reversible attachment of the free-living microbial cells is found to be facilitated by weak noncovalent interactions such as Lifshitz−van der Waals interactions, hydrophobic interactions, and electrostatic interactions, which enables the cells to get easily detached from the surface either utilizing their locomotory appendages or by the effect of the shearing force of fluid flowing over them (Muhammad et al., 2020). For an effective biofilm formation, the cells need to irreversibly attach to the surface. The presence of bacterial projecting adhesive structures such as pilli and flagella helps in irreversible attachment through strong covalent interaction, ionic, H-bonding, and dipole−dipole interactions (Muhammad et al., 2020). This leads to the establishment of cell-to-cell communication that harmonizes each and every cell to initiate biofilm formation through a mechanism known as quorum sensing (Preda & Săndulescu, 2019). Irreversible adhesion is also facilitated by the production of EPS (exopolysaccharides) that serve as an essential component of the biofilm. The extracellular secreted EPS comprises polysaccharides, lipids, proteins, DNAs, and other polymeric compounds that help in the cohesion of bacteria and also the adhesion of the bacteria to the substratum (Jain et al., 2022; di Martino, 2018). The bacteria remain embedded within the EPS matrix functioning as a biological "glue" to hold the cells together (di Martino, 2018). The bacteria multiply and mature within the embedded polysaccharide matrix rewiring their essential metabolic pathways required for survival inside the EPS layer (Karygianni et al., 2020). Previous studies tell us that there exists a significant difference in the gene expression profile in the case of sessile cells from biofilm compared with that of free-living cells (Hall-Stoodley & Stoodley, 2002; Shemesh et al., 2007). The EPS layer formation is succeeded by the formation of small water channels that serve as the circulatory system transporting nutrients to individual cells and removing metabolic waste in the biofilm microenvironment (Garnett & Matthews, 2013). The terminal phase of biofilm involves the dispersal of the cells from the biofilm enabling them to colonize over new surfaces, thereby starting a new biofilm life cycle (Rumbaugh & Sauer, 2020). The detachment of cells from the biofilm occurs under specific environmental conditions and is regulated by specific environmental cues and complex signaling cascades (Rumbaugh & Sauer, 2020).

3. Defining metabolome

The entire collection of all the metabolites contained in a biological cell compartment, cell, tissue, organ, or organism that has been studied as a cellular extract is referred as the metabolome. Numerous metabolites, which are products of diverse metabolic processes, are found in living cells. These metabolites are the by-products of several biochemical cellular activities, such as protein synthesis, mRNA translation, gene transcription, and metabolic enzymatic reactions. Extensively characterizing and quantifying these metabolites is known as "metabolomics." The genome, proteome, and metabolome are all components of biology's hierarchical structure, with the metabolome serving as the product of all biological function (as shown in Fig. 4.1). Thus, we can say that a thorough analysis of the metabolome is known as metabolome analysis.

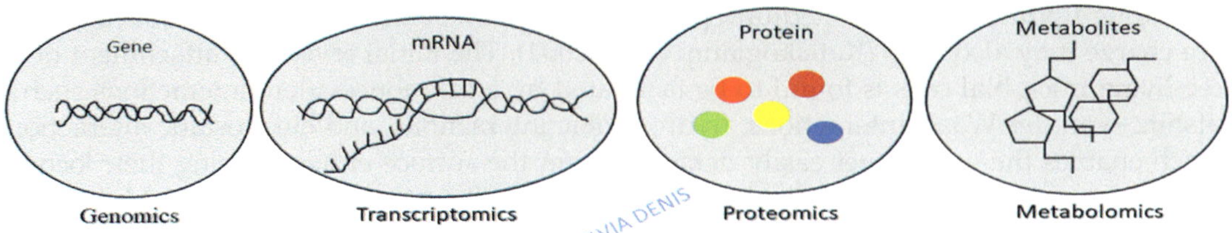

FIGURE 4.1 Pathway leading to metabolomics.

It has been suggested that there is a biological hierarchy with the genome at the top and the metabolome at the bottom. The genome is a collection of genes that serves as a repository for data on gene function. The transcriptome is made up of a group of messenger RNAs that can be produced from this group of genes. The proteome, or collection of proteins, is created after the transcriptome. Through the transcriptome and proteome, the metabolic processes can be performed, which are catalyzed by enzymes. As a result, a so-called metabolome, a collection of metabolites, is then created. The transcriptome or proteome resembles a series of microbial enzymes The genome is analogous to the microbiome (microbial composition).

4. Metabolomics of biofilms

When studying oral biofilms, researchers mainly analyze three aspects—the microbiome, the microenvironments, and the functions of the microbial community, which includes their metabolic activity (Takahashi et al., 2012). The oral biofilm's metabolome, which is closely related to the pathogenicity hierarchy, is the main product of the oral biofilm's metabolism. Not only the presence of the microbes but their metabolic activity is crucial in disease initiation and progress. A variety of metabolites are present in our saliva, and they might be associated with oral disease so metabolomic analysis of saliva can help us identify disease-specific biomarkers.

It has been reported in previous studies that organic acids, such as lactic acid, that are produced in the supragingival plaque biofilm by bacterial sugar metabolism through the EMP pathway are involved in initiating dental caries. Bacteria responsible for periodontitis, which are a part of oral biofilm, utilize various amino acids along with the sugars as their energy and carbon substrates. Short-chain fatty acids, sulfur compounds, and ammonia that are made through protein and amino acid metabolism are involved in initiating periodontal disease (Keyes, 1968).

Previously, technical difficulties hindered the study of metabolic systems that operate in oral biofilm. Advancement of techniques such as CE (capillary electrophoresis) and TOF-MS (time of flight—mass spectrometry) has been revolutionary in these studies. We can now accurately separate minute ionic molecules and determine their masses with CE-TOF-MS enabling us to recognize metabolites present even in very little amounts of samples.

It is known that oral biofilm utilizes a diverse range of substrates for its metabolism other than glucose and accordingly produces a wide variety of metabolites. However, with the use of CE-TOF-MS for the analysis, it is difficult to quantify some smaller (MW $<$ 50), nonionic (neutral), or volatile metabolites, such as hydrogen sulfide, acetic acid, and ammonia, which are commonly produced in the oral biofilms. For carrying out greater comprehensive metabolomic evaluation, we would require a combination of CE-TOFMS along with techniques such as gas and liquid chromatography, which enable us to measure the levels of those small and volatile metabolites.

Thus, metabolomic study of oral biofilms gives us valuable insights regarding the metabolic pathways and their regulatory mechanisms. This knowledge can help develop strategies to prevent diseases.

Our objective here is to discuss biofilm formation by *Staphylococcus aureus, Fusobacterium nucleatum,* and *S. mutans* using a metabolomic approach based on previous studies.

5. Studying biofilm production by *Staphylococcus aureus, Fusobacterium nucleatum,* and *Streptococcus mutans* using a specific metabolomic approach

S. aureus colonizes the teeth of humans in the form of biofilms. Antibiotic resistance genes and virulence factors can be efficiently transferred between the bacterial members of the biofilm. *S. aureus* is an opportunistic pathogen and can adapt well to the environmental conditions it is exposed to. The close aggregation among the bacterial members of the biofilm matrix can inhibit penetration of the antibiotics applied to kill the organisms.

In an experiment conducted (Baqai et al., 2011), 114 *S. aureus* strains were isolated from 250 bacterial isolates of blood cultures of different patients who underwent surgical procedures. It was found that 56% of the population was made of *S. aureus* followed by *E. coli* (25%), *Pseudomonas* spp. (13%), *S. typhi* (4%), and *Shigella* spp. (2%). Biofilm assay performed using Tryptone soy broth followed by staining with crystal violet and taking optical densities at 570 nm using a spectrophotometer revealed that biofilm was detected in 11 (16.17%) of the 68 *S. aureus* strains used, while ATCC culture of *S. aureus* (2523) showed weak biofilm production.

Metabolomics provides novel methods for studying various processes involved in biofilm formation and its response to certain antimicrobial therapies. LC-MS metabolomics has been used to demonstrate differences in metabolism between planktonic and biofilm cells of *S. aureus* grown under identical conditions. Data obtained from *S. aureus* cell extractions of planktonic shaking cultures and static biofilm cultures show significant differences in their expression of certain metabolites. While 151 metabolites were significantly upregulated in their expression in planktonic cells in comparison with that in biofilms (with $\log_2$ fold changes of ≥ 1), on the contrary, 177 metabolites showed significant upregulation in biofilms in comparison with planktonic cells (with $\log_2$ fold changes of ≥ 1) (Stipetic et al., 2016). Arginine metabolism shows notable changes between planktonic and biofilm cell samples.

KEGG is a widely used database system in metabolomics, for understanding high-level functions of the biological systems, especially for molecular level information, generated by genome sequencing. In this study, the results showed around 129 pathways with at least two identified ($HRMS^1{}_a$, R_{ta}) or annotated ($HRMS^1{}_{PL}$) metabolites, which exhibited significant ($P < .05$) variation in expression between planktonic cells and biofilm cell samples. Table 4.1 enlists 20 of these pathways annotated with the most frequently detected metabolites.

TABLE 4.1 Top 20 metabolic pathways that have various metabolites displaying significant changes in intensity or expression between planktonic cells and biofilm samples.

Pathway name	KEGG map ID	Number of metabolites	Annotated (HRMS1PL)	Identified (HRMS1a Rta)	Coverage (%)	P versus B (Planktonic versus Biofilm)
Arginine and proline metabolism/arginine biosynthesis	00330/ 00220	90	49	5	60	33
Protein digestion and absorption	04974	47	22	7	61.7	19
Tyrosine metabolism	00350	76	35	2	48.7	18
Histidine metabolism	00340	45	25	3	62.2	17
Galactose metabolism	00052	41	22	0	53.7	17
Aminoacyl-tRNA biosynthesis	00970	52	13	7	38.5	17
Cyanoamino acid metabolism	00460	46	31	1	69.6	16
Linoleic acid metabolism	00591	28	26	0	92.9	15
Limonene and pinene degradation	00903	64	53	0	82.8	15
C5-branched dibasic acid metabolism	00660	32	20	4	75	14
Mineral absorption	04978	29	11	4	51.7	14
Phosphotransferase system	02060	48	20	1	43.8	14
Two-component system	02020	41	10	6	39	14
Fructose and mannose metabolism	00051	51	16	0	31.4	14
Alanine, aspartate, and glutamate metabolism	00250	24	15	2	70.8	13
Lysine degradation	00310	47	26	1	57.4	13
Glycine, serine, and threonine metabolism	00260	51	27	2	56.9	13
Phenylalanine metabolism	00360	72	33	1	47.2	13
Pyrimidine metabolism	00240	66	28	5	50	12
Aminobenzoate degradation	00627	84	28	1	34.5	12
Purine metabolism	00230	92	16	9	27.2	12
Citrate cycle (TCA cycle)	00020	20	5	3	40	4

Most importantly, mapping data to arginine biosynthesis (Kegg pathway 00220) led to the observation that four identified ($HRMS^1{}_a$, R_{ta}) and five putatively annotated ($HRMS^1{}_{PL}$) metabolites participate in this pathway of *S. aureus* (Table 4.2).

The kind of changes shown here provides evidence that bacteria are dynamic entities, which are able to respond to their changing environment and thus display significantly altered phenotypes with respect to their metabolic processes.

Fusobacterium nucleatum, a gram-negative bacterium, is a member of the buccal microflora and is known to cause periodontitis. *Streptococcus gordonii*, a gram-positive bacterium, is another commensal bacterium found in the periodontal environment. A novel metabolic interaction has been identified between *S. gordonii* and *F. nucleatum* with respect to the metabolism of arginine by *S. gordonii* as a substrate where arginine is converted to ornithine. Along with it, ammonia and ATP are also produced. The arginine–ornithine antiporter, ArcD, present in *S. gordonii*, excretes ornithine as its by-product. This in turn accelerates the growth and development of biofilm by *F. nucleatum*. Cross-feeding behavior also helps *F. nucleatum* to acquire alanine, glutamate, and alanyl-alanine, released from *S. gordonii*. Coculturing of *F. nucleatum* with *S. gordonii* also enhances the production of butyrate and putrescine, a polyamine produced through ornithine decarboxylation. It has been scientifically proven that the ArcD-dependent ornithine that is secreted by *S. gordonii* induces synergistic putrescine production by *F. nucleatum*, and this metabolism creates a putrescine-rich microenvironment in the *F. nucleatum* biofilms. Putrescine is an essential polyamine required for the growth of many bacteria. It accelerates biofilm life cycle of maturation and helps in thickening of biofilms (Sakanaka et al., 2022). This provides evidence of interspecies interactions within the microbial communities of the biofilm in a buccal cavity.

TABLE 4.2 Arginine Biosynthesis metabolites identified from planktonic and biofilm data sets Stipetic et al. (2016).

Peak number	Metabolite	Elemental formula	KEGG ID	Metabolite code	Mass [M−H] (Da)	$R^c{}_t$ (sec)	Log_2 fold change
2782	Aspartate	C4H7NO4	C00049	HRMS1PL	132.0302	749.1	1.577
2873	Glutamate	C5H9NO4	C00025	$HRMS^1{}_a$ R_{ta}	146.0458	725.7	1.0159
2981	Glutamine	C5H10N2O3	C00064	$HRMS^1{}_a$ R_{ta}	145.0618	781.3	n/a(e)
3171	Citrulline	C6H13N3O3	C00327	$HRMS^1{}_a$ R_{ta}	174.0884	813.1	3.5413
3252	*N*-acetyl-L-glutamate	C7H11NO5	C00624	$HRMS^1{}_a$ R_{ta}	188.0564	691.1	4.3654
3505	L-arginosuccinate	C10H18N4O6	C03406	HRMS1PL	289.1156	800.5	n/a[f]
3608	*N*-acetyl-L-citrulline	C8H15N3O4	C15532	HRMS1PL	216.099	594.1	5.049
3636	*N*-acetyl-ornithine	C7H14N2O3	C00437	HRMS1a Rta	173.0931	791.2	1.068
3838	Arginine	C6H14N4O2	C00062	HRMS1PL	173.1044	1304.1	1.3781

- J. Clarke discovered a species from carious lesions in 1924 and named it *Streptococcus mutans* because he believed the oval-shaped cells he had seen were mutant strains of streptococci (Clarke, 1924). The scientific world began to pay serious attention to *S. mutans* in the late 1950s, and by the middle of the 1960s, clinical and animal-based laboratory research had established *S. mutans* as a significant causative agent in dental caries. The human oral cavity, and more especially dental plaque, a multispecies biofilm generated on hard surfaces of the tooth, is *S. mutans'* native environment. The ability of *S. mutans* to synthesize significant amounts of extracellular polymers of glucan from sucrose, which facilitates the long-term colonization of hard surfaces and the formation of the extracellular polymeric matrix in situ, and its capacity to survive in an unfavorable environment, particularly low pH (aciduricity), are the key characteristics that contribute to the organism's cariogenic potential.

S. mutans, a significant etiological factor in human dental caries, typically resides in biofilms on tooth surfaces, or "dental plaque." Three glucosyltransferases, GtfB, C, and D, are produced by *S. mutans* strains. They also exploit the glucose moiety of sucrose as their substrate. This results in the synthesis of glucose polymers of glucans (mutans) (Loesche, 1986). Water-insoluble glucans rich in α-linkages (Donlan & William Costerton, 2002; Lewis, 2001; Wingender et al., 2022) are produced by GtfB. Insoluble and soluble glucans with α-linkages (Berger et al., 2018a, 2018b; Donlan & William Costerton, 2002; Lewis, 2001; Watnick & Kolter, 2000; Wingender et al., 2022) are produced by GtfC and GtfD, respectively (often called dextran). These polymers are vital components of plaque biofilm matrices, especially the α-linked (Donlan & William Costerton, 2002; Lewis, 2001) water insoluble glucans (Bowen & Koo, 2011). Furthermore, Gtfs interacts to oral microbes, especially those that do not typically express Gtfs. This transforms them into unofficial glucan manufacturers. Additionally, *S. mutans* produces GbpA, GbpB, GbpC, and GbpD, which are surface-associated glucan-binding proteins. Acid production is a virulence factor in *S. mutans* because acid demineralizes the tooth surface, leading to tooth decay. Lactic acid is a metabolite in the Embden–Meyerhof carbon metabolic pathway, which is vital in ATP production which promotes dental plaque. When the pH falls below 5.5, demineralization of the tooth surface occurs, resulting in tooth decay. In order for such an organism to construct the sucrose-dependent biofilm that is crucial for its carcinogenicity, adhesive glucans, Gtfs, and Gbps behave as an organized scaffold (Lemos & Burne, 2008). To achieve this, they promote the local aggregation of microbial cells. Generation of a diffusion-limiting polymeric matrix that protects the implanted bacteria is also observed.

S. mutans secretes glycosyltransferases (GFTs), which adhere to bacteria or enamel pellicle surfaces. In presence of sucrose, its catabolism by GFTs forms the extracellular polysaccharide matrix (EPS) and its insoluble components (as shown in Fig. 4.2). This process generates significant quantities of both soluble and insoluble glucans. Since glucans offer binding sites for *S. mutans,* glucan-binding proteins (GBPs), and other organisms, the EPS matric operates as an architectural framework for the biofilm structure. This mediates strong adhesion to the tooth enamel and bacteria.

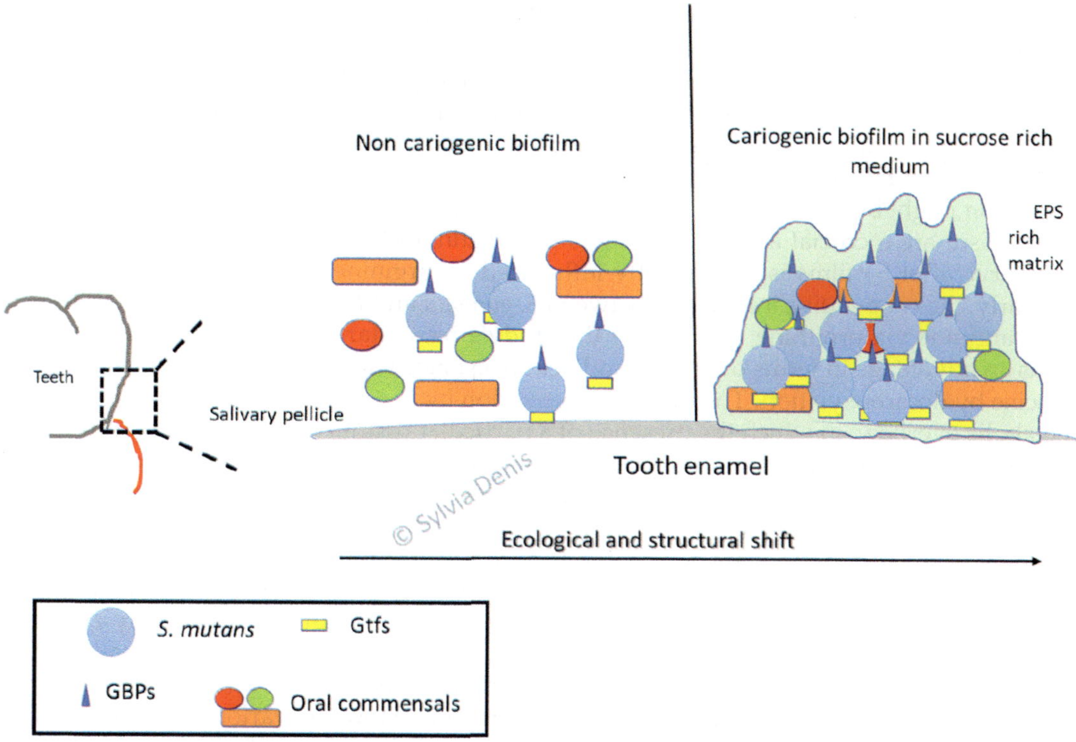

FIGURE 4.2 *Streptococcus mutans* biofilm development and host interactions.

6. Conclusion

Modern molecular biology methods have found roughly 1000 distinct bacterial species in the dental biofilm. This is twice as many as can be grown. The perks of a biofilm include a greater habitat range for development, increased resistance to microbiological medications and host immune, and enhanced potential to spread illness. Biofilms are done to boost the flow of nutrients and various by-products. Bacteria that form biofilms have a physiological makeup that differs from planktonic cells and live under nutrient limitation and are frequently in a dormant state.

The bulk of oral disorders caused by bacteria, including dental caries, periodontal disease, and oral malodor, are sparked. They are spread by microbial metabolic activity rather than merely by the presence of specific microbes. Controlling these microbes' metabolic activity, and therefore their development and pathogenicity, can prevent certain illnesses without necessarily killing or destroying the germs. Questions such as "What the microbiota are doing?" and "How they are controlled?" may be answered by metabolome analysis, which may examine metabolic features of oral biofilm in vivo, including as metabolic pathways and metabolic regulatory mechanisms.

The arginine cycle in *S. aureus*, especially, is one metabolic route that experiences major alterations. Strong upregulation of intermediates in planktonic samples or their downregulation in biofilm samples indicated their depletion in response to rapid fluxes. According to studies, *S. aureus* biofilm development results in an increase of genes related to the urea cycle. Although the amino acid glutamine was recognized (HRMS1a Rta), it was shown that its concentration did not differ between samples taken from planktonic and biofilm organisms. Only planktonic samples contained the metabolite L-arginosuccinate, according to the annotation (HRMS1PL). The metabolite might not be present at all in biofilms or its concentration might be below the instrumentation's detection thresholds. Changes in amino acid metabolism are a crucial aspect of biofilms that distinguish them from planktonic samples. The survival of biofilms is significantly influenced by arginine metabolism and catabolism. Similarly, the cooperative interaction between *S. gordonii* and *F. nucleatum* to create putrescine from arginine via ornithine was one of the study's most startling discoveries. *F. nucleatum* gains an advantage in terms of energy and acid resistance when ornithine is converted to putrescine by decarboxylation. The putrescine secretion by *F. nucleatum* is probably the result of metabolic overflow, in which the presence of a particular quantity of ornithine in the medium causes an excess generation of putrescine that leads to its secretion. It revealed a great detail of cooperative metabolism across oral bacteria. This would not have been possible without the cooperation of metabolic pathways in multiple taxa. Therefore, it provided an understanding of the metabolic characteristics of *F. nucleatum* from the perspective of the pathogenicity of microbial communities. Given its ability to regulate modifications in the plaque microbiota via EPS, evidence amassed over many years has made abundantly clear that *S. mutans* is a crucial factor in dental caries. The pursuit of new targets for caries treatment and prevention can thus be aided by continuous attempts to clarify how S. mutans recognizes and reacts to external stimuli through biofilm formation. How the various *S. mutans* isolates interact with the microbiome in synergistic or antagonistic methods to impact the different levels of disease is a topic that has received limited research.

7. Future prospects

Since the concept of the oral ecosystem has gained popularity and oral bacterial interactions have been better understood, it is possible that in the future, the development of the drug that take advantage of interbacterial associations to treat and prevent these ailments can be considered. The interactions within biofilm, which are currently understudied, must be the focus of future study (ten Cate, 2006). This objective can only be achieved with more research on the oral ecosystems due to the complex nature of biofilm. There are metabolic techniques to treat dental caries, but they have not yet been developed for treating periodontitis and oral malodor (Takahashi, 2005). This is likely because it has been difficult to identify metabolites that are important for metabolic pathways, including amino acids and amino sugars. We need to do fundamental research on the microbial metabolic pathway as well as further improve analytical methods and metabolite databases. The complete oral biofilm may be examined using metabolome analysis, which is a potent approach that can reveal the biofilm's metabolic features and reveal new information on the effects of oral biofilm on health and various diseases. Additionally, metabolomics may aid us to characterize the metabolic characteristics of oral biofilm in vivo, as well as understand the efficacy of various drugs and novel biomarkers associated with oral diseases. Humans are a mix of the human body and the microbiota, a so-called "superorganism" in which host tissue and symbiotic microbes interact and affect one another. This idea suggests that the study of symbiotic microorganisms, particularly dental biofilm, may begin a new phase because of the metabolomic technology. Metabolomics is only the start of this new era.

References

Baqai, R., Memon, M., & Khan, M. K. (2011). *Biofilm formation and isolation of staphylococcus aureus from blood cultures of dental patients undergoing oral surgical procedures*. undefined.

Berger, D., Rakhamimova, A., Pollack, A., & Loewy, Z. (August 31, 2018). Oral biofilms: Development, control, and analysis. *High-Throughput, 7*(3). Available from: http://www.ncbi.nlm.nih.gov/pubmed/30200379.

Berger, D., Rakhamimova, A., Pollack, A., & Loewy, Z. (August 31, 2018). Oral biofilms: Development, control, and analysis [cited September 3, 2022] *High-Throughput [Internet], 7*(3), 24. Available from: https://www.mdpi.com/2571-5135/7/3/24/htm.

Bowen, W. H., & Koo, H. (April 2011). Biology of streptococcus mutans-derived glucosyltransferases: Role in extracellular matrix formation of cariogenic biofilms [cited September 17, 2022] *Caries Research [Internet], 45*(1), 69–86. Available from: https://www.karger.com/Article/FullText/324598.

ten Cate, J. M. (September 2006). Biofilms, a new approach to the microbiology of dental plaque [cited September 17, 2022] *Odontology [Internet], 94*(1), 1–9. Available from: https://link.springer.com/article/10.1007/s10266-006-0063-3.

Clarke, J. K. (1924). On the bacterial factor in the aetiology of dental caries [cited September 17, 2022] *British Journal of Experimental Pathology [Internet], 5*(3), 141. Available from: https://www.ncbi.nlm.nih.gov/pmc/articles/PMC2047899/.

Donlan, R. M. (2002). Biofilms: Microbial life on surfaces [cited September 3, 2022] *Emerging Infectious Diseases [Internet], 8*(9), 881. Available from: /pmc/articles/PMC2732559/.

Donlan, R. M., & William Costerton, J. (2002). Biofilms: Survival mechanisms of clinically relevant microorganisms. *Clinical Microbiology Reviews [Internet], 15*(2), 167–193. Available from: https://journals.asm.org/journal/cmr.

Garnett, J. A., & Matthews, S. (January 21, 2013). Interactions in bacterial biofilm development: A structural perspective. *Current Protein & Peptide Science, 13*(8), 739–755.

Hall-Stoodley, L., & Stoodley, P. (2002). Developmental regulation of microbial biofilms Article in current opinion in biotechnology influence of high speed water jets and sprays on biofilm removal view project biofilm therapeutics view project [cited September 3, 2022] *Current Opinion in Biotechnology [Internet], 13*, 228–233. Available from: www.microbelibrary.org/Visual/page1.htm.

Hallab, N. J., Bundy, K. J., O'Connor, K., Clark, R., & Moses, R. L. (January 1, 1995). Cell adhesion to biomaterials: Correlations between surface charge, surface roughness, adsorbed protein, and cell morphology [cited September 3, 2022] *Journal of Long-Term Effects of Medical Implants [Internet], 5*(3), 209–231. Available from: https://europepmc.org/article/med/10172729.

Huang, R., Li, M., & Gregory, R. L. (2011). Bacterial interactions in dental biofilm SpeciaL FOcuS reView: BiOFiLmS SpeciaL FOcuS: BacteriaL BiOFiLm. *Virulence [Internet], 2*(5), 435–444. Available from: https://www.tandfonline.com/action/journalInformation?journalCode=kvir20.

Jain, S., Bhowmick, A., Jeong, B., Bae, T., & Ghosh, A. (December 1, 2022). Unravelling the physiological roles of mazEF toxin–antitoxin system on clinical MRSA strain by CRISPR RNA-guided cytidine deaminase [cited September 3, 2022] *Journal of Biomedical Science [Internet], 29*(1), 1–16. Available from: https://link.springer.com/articles/10.1186/s12929-022-00810-5.

Karygianni, L., Ren, Z., Koo, H., & Thurnheer, T. (August 1, 2020). Biofilm matrixome: Extracellular components in structured microbial communities. *Trends in Microbiology, 28*(8), 668–681.

Katsikogianni, M., Missirlis, Y. F., Harris, L., & Douglas, J. (2004). Concise review of mechanisms of bacterial adhesion to biomaterials and of techniques used in estimating bacteria-material interactions [cited September 3, 2022] *European Cells & Materials [Internet], 8*, 37–57. Available from: https://pubmed.ncbi.nlm.nih.gov/15593018/.

Keyes, P. H. (1968). Research in dental caries [cited September 4, 2022] *Journal of the American Dental Association [Internet], 76*(6), 1357–1373. Available from: https://pubmed.ncbi.nlm.nih.gov/4870825/.

Lemos, J. A., & Burne, R. A. (November 1, 2008). A model of efficiency: Stress tolerance by Streptococcus mutans [cited September 17, 2022] *Microbiology (N Y) [Internet], 154*(11), 3247–3255. Available from: https://www.microbiologyresearch.org/content/journal/micro/10.1099/mic.0.2008/023770-0.

Lewis, K. (2001). Riddle of biofilm resistance [cited September 3, 2022] *Antimicrobial Agents and Chemotherapy [Internet], 45*(4), 999–1007. Available from: https://pubmed.ncbi.nlm.nih.gov/11257008/.

Loesche, W. J. (1986). Role of Streptococcus mutans in human dental decay [cited September 17, 2022] *Microbiological Reviews [Internet], 50*(4), 353–380. Available from: https://journals.asm.org/journal/mr.

di Martino, P. (2018). Extracellular polymeric substances, a key element in understanding biofilm phenotype [cited September 3, 2022] *AIMS Microbiology [Internet], 4*(2), 274. Available from:/pmc/articles/PMC6604936/.

Muhammad, M. H., Idris, A. L., Fan, X., Guo, Y., Yu, Y., Jin, X., et al. (May 21, 2020). Beyond risk: Bacterial biofilms and their regulating approaches. *Frontiers in Microbiology, 11*, 928.

Petrova, O. E., & Sauer, K. (May 2012). Sticky situations: Key components that control bacterial surface attachment [cited September 3, 2022] *Journal of Bacteriology [Internet], 194*(10), 2413–2425. Available from: https://journals.asm.org/journal/jb.

Preda, V. G., & Săndulescu, O. (September 30, 2019). Communication is the key: Biofilms, quorum sensing, formation and prevention. *Discoveries, 7*(3).

Ramburrun, P., Pringle, N. A., Dube, A., Adam, R. Z., D'souza, S., & Aucamp, M. (June 9, 2021). Recent advances in the development of antimicrobial and antifouling biocompatible materials for dental applications [cited September 3, 2022] *Materials [Internet], 14*(12), 3167. Available from: https://www.mdpi.com/1996-1944/14/12/3167/htm.

Rumbaugh, K. P., & Sauer, K. (June 12, 2020). Biofilm dispersion [cited September 3, 2022] *Nature Reviews Microbiology [Internet], 18*(10), 571–586. Available from: https://www.nature.com/articles/s41579-020-0385-0.

Sakanaka, A., Kuboniwa, M., Shimma, S., Alghamdi, S. A., Mayumi, S., Lamont, R. J., et al. (August 30, 2022). Fusobacterium nucleatum metabolically integrates commensals and pathogens in oral biofilms [cited September 16, 2022] *mSystems [Internet], 7*(4). Available from: https://journals.asm.org/doi/10.1128/msystems.00170-22.

Shemesh, M., Tam, A., & Steinberg, D. (May 1, 2007). Differential gene expression profiling of Streptococcus mutans cultured under biofilm and planktonic conditions [cited September 3, 2022] *Microbiology (N Y) [Internet], 153*(5), 1307–1317. Available from: https://www.microbiologyresearch.org/content/journal/micro/10.1099/mic.0.2006/002030-0.

Sousa, C., Rodrigues, D., Oliveira, R., Song, W., Mano, J. F., & Azeredo, J. (October 22, 2011). Superhydrophobic poly(l-lactic acid) surface as potential bacterial colonization substrate [cited September 3, 2022] *AMB Express [Internet], 1*(1), 1–9. Available from: https://amb-express.springeropen.com/articles/10.1186/2191-0855-1-34.

Stipetic, L. H., Dalby, M. J., Davies, R. L., Morton, F. R., Ramage, G., & Burgess, K. E. V. (April 1, 2016). A novel metabolomic approach used for the comparison of *Staphylococcus aureus* planktonic cells and biofilm samples [cited September 16, 2022] *Metabolomics [Internet], 12*(4). Available from: https://pubmed.ncbi.nlm.nih.gov/27013931/.

Takahashi, N. (September 1, 2005). Microbial ecosystem in the oral cavity: Metabolic diversity in an ecological niche and its relationship with oral diseases. *International Congress Series, 1284*, 103–112.

Takahashi, N., Washio, J., & Mayanagi, G. (August 1, 2012). Metabolomic approach to oral biofilm characterization—a future direction of biofilm research. *Journal of Oral Biosciences, 54*(3), 138–143.

Watnick, P., & Kolter, R. (May 2000). Biofilm, city of microbes [cited September 3, 2022] *Journal of Bacteriology [Internet], 182*(10), 2675–2679. Available from: https://journals.asm.org/doi/10.1128/JB.182.10.2675-2679.2000.

Wingender, J., Szewzyk, U., & Steinberg, P. *Biofilms: An emergent form of bacterial life*. (cited September 1, 2022). Available from: www.nature.com/nrmicro.

Yu, P., Wang, C., Zhou, J., Jiang, L., Xue, J., & Li, W. (2016). Influence of surface properties on adhesion forces and attachment of Streptococcus mutans to zirconia in vitro. *BioMed Research International, 2016*.

Zheng, S., Bawazir, M., Dhall, A., Kim, H. E., He, L., Heo, J., et al. (February 12, 2021). Implication of surface properties, bacterial motility, and hydrodynamic conditions on bacterial surface sensing and their initial adhesion. *Frontiers in Bioengineering and Biotechnology, 9*, 82.

CHAPTER

5

From symbiosis to dysbiosis in gut-consequence includes metabolic syndrome

Nithyakalyani Mohan and Anusha Sunder

Apollo Hospital and Research Foundation, Apollo Hospitals, Chennai, Tamil Nadu, India

1. Introduction/keyhole view of gut dysbiosis and metabolic syndrome

Diet decides our well-being, and the same holds true for the symbiotic microbes that colonize our digestive tract. Our gut microorganisms impact many health aspects, notably the digestion and the tendency for developing disorders such as metabolic syndrome. We humans lack appropriate enzymes to digest the nonstarch polysaccharide (NSP), mainly dietary fiber. And hence the dietary fiber transits the proximal intestine undigested, with its residues reaching the distal gut for symbiotic microbial action. Symbiotic bacteria of our gut hydrolyze and then ferment them to produce short-chain fatty acids (SCFAs), which can be absorbed by the host. Thus the dietary fiber's microbial fermentation results in SCFA production, which crucially mediates the gut microbiome's beneficial health effects. And this list of benefits includes glucose homeostasis, favoring lipid utilization rather than synthesis, appetite regulation, and obesity control. Hence, it is quite clear as to why a disturbance in the gut microbiome causes a clustering of interrelated risk factors such as abdominal obesity, hyperglycemia, hypertension, and hypercholesterolemia, which together are named as "metabolic syndrome" (Makki et al., 2018).

2. The gut microbiome—our body's largest endocrine organ

Our gut health is primarily influenced by certain friendly microorganisms that habituate our lower digestive tract. The gut microflora or microbiome refers to nearly 1 trillion microorganisms, of which ideally about 85% should be friendly to our health. An overgrowth of

Microbial Biofilms
https://doi.org/10.1016/B978-0-323-95715-1.00005-4

unfriendly, disease-causing microorganisms can result from an unhealthy lifestyle, which mostly includes processed food intake, inadequate physical activity, insufficient sleep, and stress. Such a happening can disrupt the balance between unhealthy and healthy microbes, unfavorably moving the pendulum toward the unhealthy ones that erode the intestinal mucosal layer. This in turn can cause undue inflammation and a poor absorption of nutrients such as B vitamins, making us prone to health disorders (Myhrstad et al., 2020).

3. What is symbiosis? How and why does it become dysbiosis

The gut microbiota's relationship with the host is symbiotic in humans. Their innumerable benefits on host health include maintenance of intestinal barrier's integrity, defense against pathogens, valuable participation in nutrient (especially B vitamins) production, and synthesis of SCFA (Barko et al., 2018; Sekirov et al., 2010). The gut microbiota comprises bacteria, archaea, and eukarya. Among the gut microorganisms, bacteria show abundance, with its species/phyla, namely, Bacteroidetes, Firmicutes and Actinobacteria dominating numerically (Guarner & Malagelada, 2003; Rinninella et al., 2019). A number of bacteria from these phyla have a link with metabolic diseases (Makki et al., 2018). And therefore, a balance in bacterial count and composition is vital for maintaining intestinal homeostasis and sustaining its immunity. "Dysbiosis" referring to an imbalance in gut microbiota has unhealthy metabolic consequences (Underwood, 2014). Particularly, gut dysbiosis underlies metabolic disorders such as obesity, type 2 diabetes, and cardiovascular diseases (Ley et al., 2006; Lin & Zhang, 2017).

Our gut microbiota are contoured right from early years as their composition relies on infant transition factors such as gestational nutrition, type of delivery, breastfeeding, and also on early-age antibiotic use. The core native microbiota, which are healthy, remain relatively stable in adulthood but show interindividual variability owing to enterotypes, body mass index (BMI), lifestyle factors (such as dietary habits, exercise regularity, sleep quality), and cultural practices. As a result, optimal gut microbiota composition varies for each individual and abides by a healthy host–gut microbe balance to favorably perform metabolic and immune functions and also to prevent diseases (Rinninella et al., 2019). Dietary intake has the potential to modify colonic microbiota and thus can impact health. Nature of diet, whether predominantly plant-based or animal-based, modifies gut microbial community and triggers interindividual differences in microbial gene expression. Animal-based diet amplifies the count of bile-tolerant microorganisms (*Alistipes, Bilophila*, and *Bacteroides*) while diminishing the levels of Firmicutes species (*Roseburia, Eubacterium rectale*, and *Ruminococcus bromii*), which metabolizes dietary plant polysaccharides. This, for instance, impacts the synthesis of health-loaded, SCFAs (Barber et al., 2020).

4. Current scenario—gut health imbalance and metabolic syndrome

We all wish to be happy, and the secret of happiness lies in good health. Our lifestyle choices, including diet, exercise, sleep, and stress management, decide our health and well-being. Chronic lifestyle-related health disorders including obesity, hypertension, dyslipidemia, and insulin resistance are health challenges that are not only common but also cumulative in prevalence. An obvious cause of this chronic disease burden points toward pathological pathways implicating inflammation and metabolic dysfunction (such as insulin resistance). The metabolic syndrome (MetS), characterized by obesity, insulin resistance, hypertension, and hyperlipidemia, is a growing epidemic globally and necessitates lifestyle-based prevention strategies (Barber et al., 2020).

A broad group of microorganisms inhabit the gastrointestinal tract (GIT) of humans, and they exert structural, protective, and metabolic functions on the intestinal mucosa. As mutualists, we coexist with our gut microbiota. Still, at times, this relationship can turn out to be pathological; exemplars are inflammatory bowel diseases (IBDs), obesity, atherosclerosis, and diabetes. Metabolic syndrome correlates with structural modifications in the gut microbiota resulting in a low-grade inflammatory response. This inflammation is induced by an intense penetration of bacterial components that impair gut membrane and possibly involves epigenetic alteration of inflammatory molecules such as Toll-like receptors (TLRs) (Remely et al., 2014).

Confounding factors such as age, genetics, and lifestyle effect gut microbiome composition. Among them, the easiest to modify is diet as it paves the simplest path for therapeutic intervention. The recent lifestyle adaptation to a high-sugar/high-fat, dietary fiber-deprived diet is worrisome as it alters gut-resident microbes' genetic composition and metabolic activity. Diet-induced changes of this sort alter the gut microbial communities in such a way that it contributes to metabolic disorders (Arumugam et al., 2011; Hooper & Gordon, 2001; Wu et al., 2011). Adding on to the faulty diet, desk-bound nature of work, sleep deprivation due to gadget overuse, and the current pandemic-induced stress are also contributors to the gut dysbiosis and metabolic syndrome.

So let us explore more about the lifestyle components and their impact on gut health, as this will help us in formulating prevention/management strategies for metabolic syndrome.

5. What is the role of diet in gut symbiosis?

The gut microbiome is hugely impacted by our diet as the macronutrients, especially non-digestible carbohydrates or dietary fiber enables them to produce SCFAs such as acetate, propionate, and butyrate in the large intestine. The SCFAs effect the hepatic glucose and lipid metabolism beneficially (Blaut, 2015). Butyrate and propionate favorably regulate intestinal gluconeogenesis, and this benefit is dysregulated in metabolic syndrome (De Vadder et al., 2014; Ríos-Covián et al., 2016). Individuals with type 2 diabetes have suppressed butyrate biosynthesis owing to a compromise in the butyrate-producing gut microbial species (Qin et al., 2012). Dietary changes (short- and long-term) have been shown to alter the gut microbiome's abundance and functional capacity to process various nutrients. Hence, such changes in humans are detectable within days of dietary interventions. The complex interaction between gut microbiome and dietary factors regulates nutrient availability. Thus, the gut microbiome composition controls metabolic responses labeling diet as a prime pathophysiological and treatment component for lifestyle-associated disorders, including the metabolic syndrome (Chan et al., 2009; Kovatcheva-Datchary et al., 2015; Muegge et al., 2011; Turnbaugh et al., 2009; Zeevi et al., 2015).

Unfavorable changes in gut microbiome can lead to insulin resistance and obesity. A Dutch, human clinical trial, spanning for 6 weeks, studied the effects of infusing intestinal microbiome from lean donors to male recipients with metabolic syndrome. Allogenic or autologous microbiota infused in small amounts increased the recipients' insulin sensitivity (median rate of glucose disappearance improved from 26.2 to 45.3 μmol/kg/min; $P < 0.05$) with an added benefit of escalation in butyrate-producing intestinal microbiota. This implies the potential of developing intestinal microbiota as therapeutic agents, in humans, to improve insulin sensitivity (Vrieze et al., 2012).

Prebiotics denotes a selective stimulation of our gut microbiome's growth and/or activity, which proves health beneficial for the host. A deep understanding of the gut microbiomes' composition and their health-advantageous metabolites will aid in choosing right food sources of prebiotic fibers. Adequate inclusion of dietary prebiotic fibers will thus facilitate prevention/management of metabolic syndrome (Jakobsdottir et al., 2014). Gut microbiome depicts variability among individuals, and this can probably impact the degree of response to prebiotic interventions.

5.1 Symbiotic gut microbes relish on dietary carbohydrates!

Carbohydrates are our major energy providers. Just like us, the friendly microbes of our gut also rely on a carbohydrate for their survival. Hence, a sneak peak of carbohydrates is needed for us to understand about gut health. Among carbohydrates, the NSPs, mainly dietary fiber, nourish the friendly microbes of our gut, and they in turn gratify us with health benefits.

Carbohydrates are synonymous with the term saccharides, and *saccharide* is derived from "*sákkharon*," the Greek synonym for "sugar." A carbohydrate being an organic compound consists of three prime molecules, namely carbon, hydrogen, and oxygen mostly with a proportion of hydrogen to oxygen molecules (2:1), in par with water. Carbohydrates with an empirical formula of $C_m(H_2O)_n$ are polyhydric alcohols with an active aldehyde or ketone group. Carbohydrates comprise of simple sugars, maltodextrins, and starches that are digestible and fuel sources of glucose within our body. Additionally, partially digestible and nondigestible carbohydrates such as dietary fiber provide bulk for laxation and fuel our gut microbiome (Gourineni et al., 2017).

Carbohydrates are chemically classified based on their molecular size. Simple sugars (1–2 monomers) and most oligosaccharides (3–9 monomers) are digestible, while polysaccharides (≥10 monomers) are generally nondigestible (rapidly digestible starch being an exception in this category).

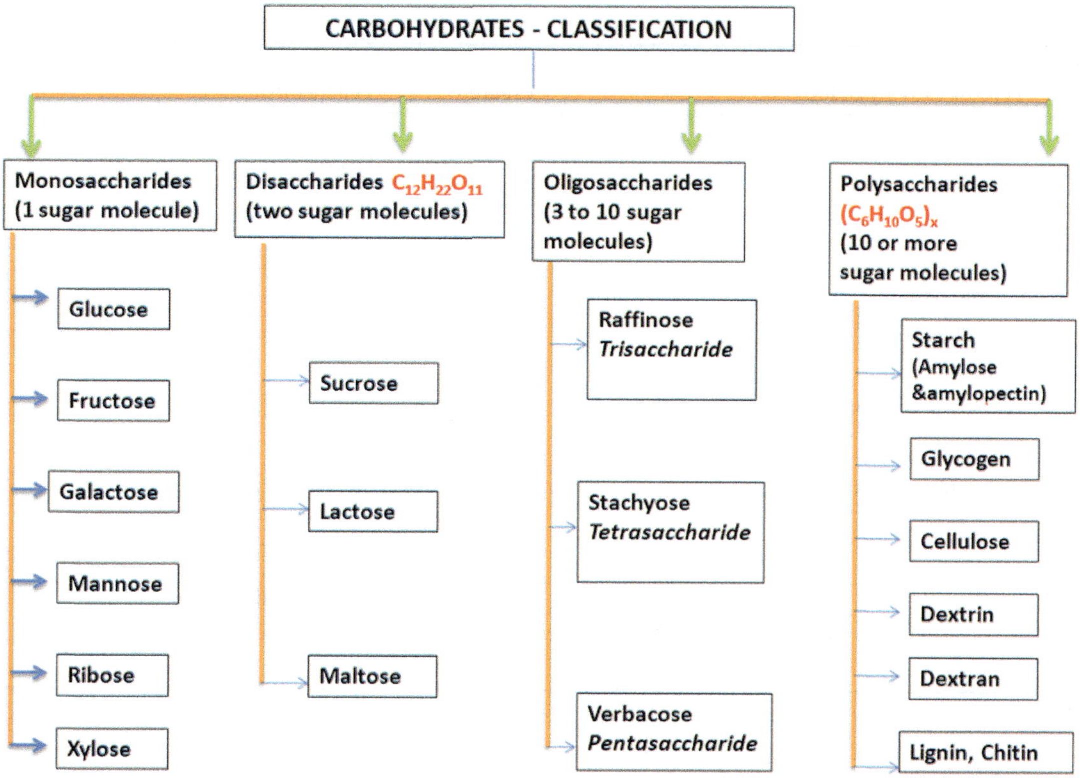

Food sources of various types of carbohydrates:

S. No.	Type of carbohydrate	Example	Food source
1	Monosaccharides	Glucose	Fruits, vegetables, honey, corn syrup
		Fructose	Fruits, vegetables, honey
		Galactose	Milk (not in free form, in combination with glucose)
		Xylose	Fruits, vegetables, cereals, mushrooms
2	Disaccharides	Sucrose	Cane, beet, molasses, fruits, vegetables
		Lactose	Milk
		Maltose	Malt products, sprouted grains

(*Continued*)

—cont'd

S. No.	Type of carbohydrate	Example	Food source
3	Oligosaccharides	Raffinose	Sugar beets, kidney beans, lentils, navy beans
		Stachyose	Beans
		Verbascose	Legumes
4	Polysaccharides		
4a	Digestible	Starch	Grain products, legumes, roots, and tubers
4b	Partially digestible indigestable (nonstarch polysaccharides/NSPs)	Inulin	Onion, garlic, mushroom
		Mannosans	Legumes
4c	Indigestible (NSPs)	Cellulose	Structural part of fruits, vegetables, whole grain
		Hemicellulose	Cereals, seeds, nuts
		Pectin	Fruits (mainly apple, guava)
		Gums and mucilages	Plant secretion and seeds; guar gum in cluster beans

5.2 Among dietary carbohydrates there are good and bad ones for gut health

Starch, the major storage carbohydrate of plants, has a mentionable impact on the texture of cooked food. Starch digestion triggers an insulinemic response by releasing glucose into the blood stream. Starch, therefore is the chief dietary glycemic carbohydrate. The health value of starch is decided by its rate and extent of digestion as well as small intestinal absorption. Accordingly, starch classification into rapidly digestible starch (RDS), slowly digestible starch (SDS), and resistant starch (RS) specifies its nutritional quality, which in turn determines the value of this food component in good health (Gourineni et al., 2017; Zhang & Hamaker, 2009).

RDS, going by its name, gets digested rapidly, say within 20 min after ingestion. Its rapidity in digestion follows a quick absorption in the upper small intestine resulting in a swift blood glucose rise initially, followed by an abrupt drop in the levels, owing to pancreatic insulin secretion and release into the bloodstream to lower blood glucose levels promptly. The commonest RDS food sources are cooked white rice, boiled potato, white bread made with refined flour. Thus, frequent and enormous ingestion of the rapidly digestible starch can potentiate hyper- and hypoglycemic episodes repetitively, consequently precipitating as obesity, insulin resistance, and type 2 diabetes (Lu et al., 2016; Zhang & Hamaker, 2009).

Digestibility of starch relates with glycemic index (GI). SDS stays intermediary to RDS and RS, as it gets slowly digested in small intestine to sustain glucose release. The SDS generates an initial rise in blood glucose levels, though low, and then subsequently causes a slow and prolonged release of glucose and insulin. Nuts, oilseeds, and legumes are the commonest food sources of SDS (Gourineni et al., 2017).

Starches that resist digestive enzymes are gaining focus as "**gut health savior**." In healthy humans, the starch that escapes small intestinal digestion is functionally termed as "resistant starch." The endosperm of cereal grains and seeds produces resistant starch, which gets embedded in protein matrix and cell wall material. Such physical structuring renders it non-digestibility and a low glycemic response. While cooking the whole kernels or coarsely ground seeds, the thick cell wall of legume seeds and the cereal grains' protein matrix prevent water penetration into the starch in the matrix. Therefore, the starch lacks moisture adequacy to readily gelatinize and swell. Without proper swelling to expose the starch molecules, the starch is not readily susceptible to enzymatic hydrolysis. Thus, the cell wall material and the protein matrix act as a physical barrier, defending the starch against hydrolysis by digestive enzymes. Consequently, ingestion of whole grains rich in resistant starch benefits us with a low postprandial glycemic response compared with white bread, which is sumptuous with easily digestible starch.

Enzyme-resistant starch traverses the upper digestive tract, undigested, and gets fermented by the colonic microflora, facilitating SCFA production. These metabolites, such as SCFAs, improve host's physical and mental health by effects including reduction of colon cancer precursors, systemic regulation of macronutrient metabolism, and favorable alteration in the secretion of certain hormones. ***The estimated daily intake of resistant starch is at least 6 g per meal recommended for health benefits*** (Birt et al., 2013).

Diets containing adequate amounts of resistant starch and nonstarch polysaccharides/NSPs have potential benefits in prevention of colorectal cancer through the delivery of fermentation acids, in particular butyrate, to the distal colon (Duncan et al., 2007; McIntyre et al., 1993). Microbial breakdown of NSP also releases bound phytochemicals into the colon (Gill & Rowland, 2002). These health benefits may be particularly important in obese and overweight subjects who are at increased risk of developing colorectal cancer and diabetes (Cani et al., 2007; David et al., 2014; Ley et al., 2006; Polednak, 2003; Walker et al., 2011).

Type of carbohydrate determines the digestive benefit (Groff & Gropper, 2000; McGrane, 2006; Shanmugam, 1998; Shils et al., 2005; Wardlaw et al., 1992; Williams, 1994)

- Carbohydrate digestion begins in the mouth. The salivary glands secrete saliva, which helps to moisten the food. The food is then chewed, while the salivary glands also release the enzyme **salivary amylase**, which begins the process of breaking down the polysaccharides in the food.
- After the food is chewed into smaller pieces and mixed with salivary amylase and other salivary juices, it is swallowed and passed through the esophagus. The mixture enters the stomach where it is known as **chyme.** There is no further digestion of carbohydrates in the stomach, as the gastric acidity inactivates the action of the salivary amylase.
- The chyme enters the beginning portion of the small intestine, the duodenum. In response to chyme being in the duodenum, the pancreas releases the enzyme **pancreatic amylase**, which hydrolyzes the polysaccharide into a disaccharide. The small intestine

then produces enzymes called lactase, sucrase, and maltase, which break down the disaccharides into monosaccharides. The monosaccharides are then absorbed in the small intestine.

- Carbohydrates that were not digested and absorbed by the small intestine reach the colon where they are partly broken down by intestinal bacteria. Fiber, which cannot be digested like other carbohydrates, adds bulk to the feces or gets fermented by the intestinal bacteria.

Starches are important as energy sources for humans and also for their interactions with the gut microflora throughout the digestive tract. Largely, those interactions promote human health. In the mouth, less gelatinized starches may lower risk of cariogenesis. In the large bowel, starches that have escaped small intestinal digestion (resistant starch), together with proteins, other undigested carbohydrates, and endogenous secretions, are fermented by the resident microflora. The resulting SCFAs contribute substantially to the normal physiological functions of the viscera. In the upper gut, these starches may assist in the transport of probiotic organisms, thus promoting the immune response and suppressing potential pathogens. SCFAs may modulate tissue levels and effects of growth factors in the gut and thus modify gut integrity. Disrupted gut integrity is a vital precipitating factor for diseases including colorectal cancer (Bird et al., 2000).

5.3 Is dietary fiber a gut-friendly carbohydrate?

Dietary fiber comprises the nondigestible carbohydrate forms such as NSPs, which are derived from plant-based foods. The American Association of Cereal Chemists defines dietary fiber as "the edible part of plant foods or analogous carbohydrates that are resistant to digestion and absorption in the human small intestine with complete or partial fermentation in the small intestine." Dietary fiber includes cellulose, hemicelluloses, lignin, pectin, and also the indigestible plant materials that are not cell wall components, for example, gums (such as guar and locust bean gums) and mucilages (National Research Council (US) Committee on Diet and Health, 1989; Julie, 2001).

Dietary fiber inclusion in the diet is linked with overall metabolic health through key aspects such as insulin sensitivity, colonic health, and gut motility. It also has a crucial role in certain pathologies including cardiovascular disease and colorectal carcinoma. The health benefits of dietary fiber are vitally mediated by gut microflora, which participate in appetite regulation, metabolic processes, and chronic inflammatory pathways.

Broadly, dietary fiber can be classified as soluble or insoluble based on how well it gets dispersed in our gut, when mixed with water. Soluble fibers such as hemicelluloses (e.g., xyloglucans, galactomannans mixed-linkage glucans), pectins, gums, and mucilages are mainly present in fruits and vegetables. Among the soluble fiber, there are variations in fermentability; mostly they support the proliferation of health-promoting bacterial species such as *Bifidobacterium*, *Lactobacillus*, and *Eubacterium*. Resistant starch, cellulose, and lignin are good examples of insoluble fiber, and their storehouses are cereals and whole-grain products. Interestingly, a majority of high-fiber natural foods contain both soluble and insoluble fiber in varying amounts. Our gastrointestinal tract lodges the gut microbiota, which ferments soluble fiber more readily than the insoluble one (Barber et al., 2020; Weickert & Pfeiffer, 2008).

Globally, cereal grains are an important energy source and hence are consumed widely. The DF components of cereal grains are cellulose and hemicelluloses (such as arabinoxylan and mixed-linkage glucans). Legumes contain dietary fiber in the form of cellulose and hemicellulose such as oligosaccharides including those of the raffinose family.

5.4 How much of dietary fiber is beneficial to our gut, and why so?

The Indian Council of Medical Research insists that 25–40 g of dietary fiber should be present in an adult's daily diet (based on a 2000 kilocalorie diet) (Position of the Indian Dietetic Association: Dietary Fibre and Health, 2018). In most of the European countries and in the United States, daily recommendation for dietary fiber ranges between 30 and 35 g in men and between 25 and 32 g in women (Stephen et al., 2017).

Dietary fiber's essentiality in human health is justified by its characteristics such as water-holding capacity, absorptive capacity, fecal bulking capacity, viscosity, binding ability, and fermentability (Williams et al., 2019). Three physiological mechanisms predominantly influence its health benefits, and they include the following:

- Physical structuring of digesta to favor gastric distension, which strongly relates with satiated feeling and food intake control (Capuano, 2017).
- Modulation of digestive processes that impact transit time, and thereby normalize circulating levels of blood glucose and lipids. This slow gastric emptying blunts the postprandial glycemic response (Gidley, 2013).
- Serve as energy sources of microbial fermentation, especially in the large intestine. Bacterial utilization of fermentable carbohydrates results predominantly in SCFA production (acetic, propionic, and butyric acids). Additionally, carboxylic acids, including lactic acid, are also produced (Williams et al., 2017). In our large intestine, SCFAs enhance colonic health by favorably influencing its mobility and blood flow and also by maintaining the GIT pH at an optimal level. These beneficial effects improve nutrient and electrolyte absorption (Tan et al., 2014; Topping & Clifton, 2001).

5.5 Hidden angel in dietary fiber, the "short-chain fatty acids"

The NSPs are the major components of dietary fiber and are fermented by the colon microbiota producing SCFAs. SCFAs produced primarily from the microbial fermentation of dietary fiber appear to be key mediators of the beneficial effects elicited by the gut microbiome. Not only does dietary fiber fermentation regulate microbial activity in the gut, but SCFAs also directly modulate host health through a range of tissue-specific mechanisms related to gut barrier function, glucose homeostasis, immunomodulation, appetite regulation, and obesity (Jakobsdottir et al., 2014; Lovegrove et al., 2017; Myhrstad et al., 2020; Pudlo et al., 2015).

Undue inflammation has a causative role in metabolic syndrome. Gut microbiome balances our inflammatory markers at an optimal level, and thus, gut symbiosis is vital in preventing metabolic syndrome. SCFAs produced by fermenting bacteria can suppress proinflammatory cytokines and interact with regulatory T cells to attenuate colitis (Smith et al., 2013). Endogenous tryptophan (Trp) metabolites from resident microbiota have an important role in gut immune homeostasis. The bacterial metabolite indole stimulates the production of interleukin-22 (IL-22), which stimulates the production of antimicrobial peptides, thus serving a protective role against pathogens. The IL-22-dependent balanced mucosal response allows for survival of mixed microbial communities yet provides colonization resistance to the fungus *Candida albicans* and mucosal protection from inflammation (Zelante et al., 2013). Polysaccharide A downregulates the production of the proinflammatory IL-17, while upregulating the production of IL-10, which together serve to protect against colitis (Mazmanian et al., 2008). The production of IL-6 and IL-1β can be stimulated by the gut microbiome, which can lead to regulatory B cell differentiation (Rosser et al., 2014). Overall, there are well-established links between the immune system and the gut microbiome in humans.

- The predominant SCFA in venous blood is acetic acid (Wong et al., 2006). Large intestine is its site of production, and from there, it is absorbed across the epithelial wall of GIT into the portal vein, diffusing in the peripheral venous system. Gut microbial fermentation of pectin and xylan principally yields acetic acid (Macfarlane & Macfarlane, 2006). Acetic acid beneficially impacts immune response and inflammation through its interaction with the G protein–coupled FFAR2 receptor (Maslowski et al., 2009).
- Propionic acid after absorption into the portal vein gets metabolized by hepatocytes. Almost 90% of propionic acid is metabolized by the liver after absorption into the portal vein and is substantially used for gluconeogenesis and interaction with the immune system via FFAR2 receptor (Brown et al., 2003; Padayachee et al., 2017). Propionate has the potential to alter cholesterol synthesis (Macfarlane & Macfarlane, 2012). Additionally, it can stimulate satiety and influence food intake (Al-Lahham et al., 2010).
- Butyric acid is the colonocytes' choicest oxidative fuel, furnishing around 60%–70% of their energy requirements (Robles Alonso & Guarner, 2013). In association with this, butyrate influences gut metabolic pathways by enhancing cellular growth and metabolism (Macfarlane & Macfarlane, 2012). Thus butyric acid has a pivotal role in colon cancer prevention (Andriamihaja et al., 2009).

6. Move a little to get closer to gut symbiosis!

Exercise is a bodily movement that requires energy expenditure. Exercising for a minimum of five times in a week with a daily duration of 30 min is ideal. But in case of time constraint, go for two to three segments of 10–15 min exercise per day (Oja & Titze, 2011).

Regular exercise is scientifically proven to improve the count and enrich the diversity of gut microflora. It also enhances the development of commensal bacteria, which are health beneficial for the host (Monda et al., 2017).

6.1 Exercise improves gut health—how?

Low-intensity exercise can influence the GIT reducing the transient stool time and thus the contact time between the pathogens and the gastrointestinal mucus layer (Bermon et al., 2015). As a consequence, it seems that exercise has protective effects, reducing the risk of colon cancer, diverticulosis, and IBD (Peters et al., 2001).

Exercise protects intestinal integrity by minimizing the inflammatory infiltrate (Campbell et al., 2016). Rise in inflammatory plasmacytoid and lymphocytic infiltrates increase the villi width in response to a combo of high-fat diet and sedentary behavior. Exercise averts such morphological changes by reducing cyclooxygenase 2 (Cox-2) expression in proximal as well as distal gut.

Endurance exercise determines a variation in the GIT due to the reduction of the splanchnic blood flow, as much as 80% of basal levels, resulting in toxicity effects (Peters et al., 2001; Rehrer et al., 2001). This reduction depends on the increase of arterial resistance in the splanchnic vascular bed, secondary to augmentation of sympathetic nervous system input. Prolonged exercise also determines an increase of intestinal permeability, compromising gut–barrier function and resulting in bacterial translocation from the colon (Gisolfi, 2000; Peters et al., 2001). Regular exercise increases n-butyrate concentration, and this is beneficial as n-butyrate protects against colon cancer and IBD by affecting cellular nuclear factor-kappa B (NF-κB) activation (Matsumoto et al., 2008; Monda et al., 2017).

A study conducted on elite rugby players demonstrated that exercise enriched the diversity of gut microflora and positively correlated with protein intake and creatine kinase levels. In particular, there was a greater diversity among the Firmicutes phylum (such as *Faecalibacterium prausnitzii*) that helped to maintain a healthier intestinal environment. Moreover, subjects with low BMI and athletes showed higher *Akkermansia muciniphila* levels in their gut microflora. *A. muciniphila* are mucin-degrading bacteria, which reside in the mucus layer, and they are inversely correlated with BMI, obesity, and metabolic disorders probably because they improve barrier function. These results indicated that both diet and exercise determined the microbial biodiversity of the gut (Clarke et al., 2014; Everard et al., 2013). Another study supports this finding, wherein fit individuals showed a microbiome enriched in butyrate-producing taxa, such as Clostridiales, *Roseburia*, Lachnospiraceae, and Erysipelotrichaceae, resulting in increased butyrate production, an indicator of favorable gut health. Exercise could be used as a therapeutic support in the treatment of dysbiosis-associated diseases, including metabolic syndrome (Estaki et al., 2016).

Exercise enriches the microflora diversity, and it balances the Bacteroidetes–Firmicutes ratio. This ratio is important for the prevention of gastrointestinal disorders and obesity-associated pathologies as it stimulates bacterial proliferation, which beneficially modulates mucosal immunity and improves barrier functions. Exercise, therefore, can be used as an effective and doable strategy to optimally maintain the gut microbial balance or to reverse an eventual dysbiosis into symbiosis, thus ascertaining health betterment.

6.2 Diet and exercise support each other in weight control and gut symbiosis

Probiotics are live microorganism (generally lactic acid bacteria)-containing food supplements, which are host beneficial. Lactic acid is a fuel for the exercising muscle to produce adenosine triphosphate (ATP) through anaerobic glycolysis. *Lactobacillus* spp. are valuable for exercise performance owing to their lactic acid production. This lactic acid serves as fuel for lactate-utilizing bacteria to synthesize butyrate. Thus, probiotics that supply lactic acid are gut-friendly as they favor butyrate production. Excess lactate in our blood can lead to muscle fatigue, and this can be offset by our gut microbiome. The abundance of *Eubacterium hallii* in the colonic ecosystem suggests that these bacteria play important roles in preventing lactate accumulation. To reap the benefits of regular exercise, our gut should be optimally colonized with *E. hallii*. Thus, probiotic supplementation fuels exercise, ensures gut symbiosis, and thus prevents metabolic syndrome (Duncan et al., 2004; Frémont et al., 2013).

Our symbiotic, gut microbe *F. prausnitzii* protects the digestive tract by producing butyrate and by lowering the oxygen tension in the lumen through flavin/thiol electron shuttle (Campbell et al., 2016; Monda et al., 2017). Nutritional status and physical activity alter gut microbiota composition. Optimal weight maintenance is hugely dependent on our leptin and ghrelin levels. Why so, because, leptin is secreted by adipocytes and provides the central nervous system with a signal of the state of energy balance, aiding in control of appetite and food intake, and thus favoring maintenance of a stable body weight. Contrary to this, ghrelin is mainly produced by the stomach, and it stimulates appetite and food intake, enabling fat mass deposition and weight gain.

Obesity is a health issue by itself, and it also leads to metabolic syndrome. So weight control is mandatory, but it should happen in a healthy way through proper diet and regular exercise. Food restriction is not an option for weight control as it leads to gut dysbiosis. In restricted eaters, a significant increase in the number of harmful pathogens such as *Proteobacteria, Bacteroides, Clostridium, Enterococcus, Prevotella, and Methanobrevibacter smithii* and a significant decrease in the quantities of gut-friendly microbes such as *Actinobacteria, Firmicutes, Bacteroidetes, B. coccoides-E. rectale group, Lactobacillus,* and *Bifidobacterium* is seen. Moreover, a significant increase in the number of gut-friendly microbes such as *Lactobacillus, Bifidobacterium* and *B. coccoides—E. rectale* group is observed in people who exercise regularly. Gut-friendly microbes such as *Bifidobacterium* and *Lactobacillus* correlate positively with serum leptin levels, while harmful pathogens such as *Clostridium, Bacteroides,* and *Prevotella* correlate negatively with serum leptin levels. Conversely, ghrelin's association with the gut-friendly microbes was negative (Queipo-Ortuño et al., 2013).

Regular exercise is a healthy habit, and it has to be inculcated right from the early years to prevent health issues such as metabolic syndrome in the adulthood. Exercise or physical activity practiced in juvenile period modifies our gut microbiome in such a way that there is an increase in the gut-friendly Bacteroidetes and a decrease in harmful pathogens such as Firmicutes. Furthermore, juveniles exercise, compared with adult exercise, modified more genera and led to an increase in lean body mass. These data suggest that early-life exercise can influence the gut microbiome composition in such a way that it favorably accommodates adaptive changes in host metabolism. Furthermore, exercise initiated in early life may favor optimal development of brain function, promoting health-enhancing microbial species in the gut (Mika et al., 2015; Stilling et al., 2015).

7. Sleep well to reap the benefits of gut symbiosis

Sleep is the best and easiest meditation, and we all love it. Still, do we value it these days? Absolutely not! Why so? Because we are busy staying awake with our electronic gadgets. Just before bedtime, our do's include checking emails, watching recent releases on Netflix, catching up on social media, even reading from a tablet.

The screen light from electronic devices such as smartphones, laptops, video games, televisions, and tablets lowers melatonin causing **sleep problems.** Going against nature, during bedtime, the light emitted from the electronic device screen combined with changing visuals and sound effects keeps the mind alert, averting the feel to relax. Electronic devices tempt us to stay awake with their high entertainment quotient. As a consequence, short sleep duration and poor sleep quality make us feel tired and less attentive throughout the day. Inadequate sleep can lead to eye disorders, lack of concentration, and digestion issues, and it can even be the hidden culprit behind weight fluctuations. Partial sleep deprivation can alter the gut microbiome composition in as little as 48 h (Benedict et al., 2016).

6–8 h of sound and refreshing sleep is sufficient to keep us healthy and energized throughout the day. Our body has a natural sleep–wake cycle, called the circadian rhythm, which is in sync with the rising and setting of the sun. Light exposure keeps us awake, while darkness makes us sleep. Our sleep–wake cycle happens because of a hormone called melatonin. During the day, melatonin levels are low. and that is why we are awake. While at night, the darkness around increases melatonin and makes us sleep.

The gut microbiome exhibits a circadian rhythm, and there is substantial evidence to say that this may be perturbed following circadian misalignment. Dysregulated gut microbiome has been found to be increased in people suffering from chronic sleep loss, which has become increasingly common in modern stressful and desk-bound lifestyle. Merely curtailing sleep to half the recommended amount for a single night acutely impairs fasting insulin sensitivity. When prolonged, sleep restriction can promote weight gain, possibly by altering energy expenditure as well as food choices and the behavioral response to especially hedonic food stimuli. Adverse metabolic changes that may increase the risk of impaired insulin sensitivity, type 2 diabetes, obesity, and dyslipidemia are associated with changes in the gut microbiome (their SCFA production), and this could result from recurrent sleep loss (Cedernaes et al., 2015, 2016; Ford et al., 2015; Spaeth et al., 2013).

Cytokines connect the dots between sleep physiology and gut microbiome composition. The acute phase pathway cytokines, namely, interleukin one beta (IL-1β) and interleukin 6 (IL-6) in particular are strongly associated with sleep physiology. IL-1β is a major somnogenic factor. IL-1β administration increases spontaneous sleep and fatigue, and IL-1β increases with ongoing sleep loss. Unlike IL-1β, IL-6 is not a direct somnogenic factor, but sleep loss results in increased IL-6 levels. In the gut, IL-6- and IL-1β-mediated inflammation fluctuates in response to the stressor, sleep deprivation (Frey et al., 2007; Kamada et al., 2013; Kapsimalis et al., 2005; Krueger et al., 2001; Liu et al., 2017; Obal & Krueger, 2003; Ranjbaran et al., 2007; Smith et al., 2019).

Obesity, through inflammation, insulin resistance, and visceral adiposity, is also considered a major cause of several sleep disorders, such as obstructive sleep apnea (OSA) sleepiness, and the associated cardiovascular comorbidities. In people with obesity-related sleep disorders, diet and exercise modification can improve sleep quality and bring in favorable changes in the gut microbiome composition (Tan et al., 2013; Vgontzas, 2008).

Excessive use of electronic devices during bedtime can compromise your sleep quality. So prioritize your sleep above your electronic device usage. Power up your day with an adequate night's sleep, and remember, to do this, you should power off your electronic devices. At bedtime, relax your mind rather than keeping it alert and entertained. A more recent study showed that high sleep quality was associated with a gut microbiome containing a high proportion of bacteria from the Verrucomicrobia and Lentisphaerae phyla and that this was associated with improved performance on cognitive tasks (Anderson et al., 2017).

7.1 How to better your sleep patterns?

Having known about the ill effects of using electronic devices at bedtime, it is high time for you to resist yourself.

✔ Power off screens to relax yourself

Turn off your smart phones, laptops, video games, TV, and tablets at least an hour before going to bed. If your work situations make it difficult, try powering down at least 15–30 min before bedtime.

✔ Reduce your screen light's brightness

Minimal light exposure from electronic devices lowers melatonin suppression of melatonin, thus promoting sleep. During the dusky hours, switch on the "night mode" in your smartphone and probably invert your display color as white text against a black background to curtail the blue light emission.

✔ Shield your eyes from the harmful screen light

Protect your eyes with blue light blocking glasses while using smart devices.

✔ Do your favorites before bedtime

At the end of the day, your mind needs its share of relaxation. This can be in any form, say listening to soothing music, practicing light yoga or a few minutes meditation, reading a nice book, or any other calming activity.

✔ Your bed has a role to play

Among the many reasons for lack of quality sleep at night, a significant one could be the bed used is not comfortable enough to help your body relax.

8. Stress is harmful even to our gut microbiome!

The brain and the gut have a lively ongoing dialogue through the gut–brain axis. And that is exactly the reason why negative emotions and stress can perturb gut motility. Psychological stress and depression can reshape the gut bacteria's composition through stress hormones, inflammation, and autonomic alterations. In turn, the gut bacteria release metabolites, toxins, and neurohormones that can alter eating behavior and mood. Stress can affect health through its impact on gut bacteria. The autonomic and circulatory systems carry distress signals to the gut. Additionally, a new bone marrow–mediated pathway was recently discovered, highlighting the role that immune cells play as messengers that convey psychological stress to the gut. The heightened inflammation that frequently accompanies stress and depression triggers blooms of pathogenic bacteria that encourage dysbiosis and a leaky gut (Madison & Kiecolt-Glaser, 2019).

Stress and depression can increase gut barrier permeability. The result, a "leaky gut," allows bacteria to seep into circulation, producing an inflammatory response. Both chronic and acute stressors can shift the gut bacteria in multiple regions and habitats—both the inside (lumen) and border (mucosal lining) of the gut. In depressed individuals, proinflammatory species can dominate at the expense of health-promoting species. Comparison of depressed and nondepressed individuals revealed 279 different bacteria-synthesized proteins, primarily related to glucose and amino acid metabolism, possible inflammatory correlates. Some bacterial species may encourage dysregulated eating. The gut bacteria may also upregulate stress responsiveness and heighten the risk for depression, which probiotic supplementation may attenuate. Digestive disorders such as irritable bowel syndrome commonly coincide with mood disorders and may reflect a dysfunctional composition of gut microbiome and related chronic inflammation. And inflammation, as we know, underlies metabolic syndrome. Henceforth, manipulating the gut microbiome and their functions with probiotics and relaxing health behaviors is a promising therapeutic strategy (Fond et al., 2014).

Stress and depression not only influence food choices, but they can also alter metabolic responses to food. Following a fast-food type meal, women who reported prior day stressors had lower fat oxidation, higher insulin, and lower resting energy expenditure than those reporting no prior day stressors with lower caloric expenditure that could potentially fuel 7–11 pounds of weight gain per year. Similarly, women with a history of depression had higher postprandial cortisol and fat oxidation, compared with women without a depression history. These kinds of metabolic changes could have downstream effects on the gut microbiota, and the reverse is also possible with a healthy diet. Diet determines which bacteria will thrive in the gut, and the gut bacteria in turn aid digestion. Healthier diets can reduce the risk of depression (Kiecolt-Glaser et al., 2015).

Gut bacteria influence food choices. The gut bacteria produce molecules that mimic or interfere with human appetite-regulating peptides and hormones (Aarts et al., 2017; Tennoune et al., 2014). Also, the gut bacteria can modify reward pathways (Bravo et al., 2011), communicate with the appetite-modulating vagus nerve (Swartz et al., 2012), and may even influence the expression of taste receptors (Cryan & Dinan, 2012). Lastly, through their release of neurotransmitters, such as serotonin, acetylcholine, and norepinephrine, the gut bacteria may indirectly influence eating behavior through mood changes (Chambers et al., 2015; Kaczmarek et al., 2017).

9. What is my take-home message?

An extensive list of microorganisms inhabits our gastrointestinal tract and offers structural, metabolic, and protective functions to the intestinal mucosa. Gut homeostasis and its integrity are vital responsibilities taken up by the inhabitant microbiome. Among the vital responsibilities of the gut microbiome, noteworthy are regulation of epithelial development, participation in signaling network, contribution to nutrient (B vitamin), and production and enhancement of immune system. Owing to its responsiveness to physiological and homeostatic changes, the gut microbiome has impressed researchers to crown it as an endocrine organ. Health-impacting factors, both built-in and external, such as host physiology and lifestyle (including diet, exercise, sleep, and stress), influence the development and stability of gut ecosystem. For instance, a diminished microbiome diversity and an unfavorably high ratio of harmful pathogens versus symbiotic microbes (for example, higher Firmicutes vs. Bacteroidetes ratio) are proven causative in insulin resistance, altered blood glucose and cholesterol levels, type 2 diabetes, and obesity. To the contrary are exercise regularity and adequacy in dietary fiber (through whole grain products, fruits, vegetables, and legumes), which has the potential to improve the gut microbiome diversity and promote its symbiosis (Clarke et al., 2014; Flint, 2012; Flint et al., 2012; Mackie et al., 1999; Payne et al., 2012).

✔ Eat adequate amounts of prebiotic, fermentable fiber. To achieve this, set your goal with a daily intake of minimum five servings of fruits and vegetables.

Replacing refined grains (nearly 20% of energy) with brown rice, whole wheat or barley, for 5 weeks each, can result in an approximate 1 kg weight loss. 75 g of oats contain 3 g of beta-glucan, a soluble fiber that reduces LDL/bad cholesterol levels. Half a cup of quinoa can surprise you with 2.6 g of heart-healthy soluble fiber, which guarantees bad cholesterol reduction (McRae, 2017).

✔ For exercise regularity, go with an easy goal of daily 30 min and habituate it for at least five times in a week. In case of time constraint, fragment your day's exercise into two to three segments each comprising of 10–15 min.

1 pound/0.45 kg weight loss requires 3500 calorie expenditure

100 calorie expenditure requires 2000 steps

1 pound/0.45 kg of weight loss in a week requires 500 calorie expenditure in a day or 10,000 steps daily

Sawyer, B. J., Bhammar, D. M., Angadi, S. S., Ryan, D. M., Ryder, J. R., Sussman, E. J., Bertmann, F. M., & Gaesser, G. A. (2015). Predictors of fat mass changes in response to aerobic exercise training in women. *Journal of Strength and Conditioning Research*, *29*(2), 297–304. https://doi.org/10.1519/JSC.0000000000000726.

✔ 6–8 h of sound and refreshing sleep is sufficient to keep us healthy and energized throughout the day.

Poor sleep is linked with risk factors for heart disease and stroke.

- **Poor sleep quality and short sleep duration are linked to higher levels of inflammation (fibrinogen, IL-6, and CRP). High levels of inflammation increase the risk of heart disease and stroke.**
- **Inadequate sleep significantly increases arterial stiffness. Arterial stiffness or hardening of the arterial wall (probably due to calcium deposition) renders lack of flexibility in the vessel wall causing high blood pressure and making the heart work harder**

Irwin, M. R., Olmstead, R., & Carroll, J. E. (2016). Sleep disturbance, sleep duration, and inflammation: A systematic review and meta-analysis of cohort studies and experimental sleep deprivation. *Biological Psychiatry*, *80*(1), 40–52. https://doi.org/10.1016/j.biopsych.2015.05.014; Kadoya, M., Kurajoh, M., Kakutani-Hatayama, M., Morimoto, A., Miyoshi, A., Kosaka-Hamamoto, K., Shoji, T., Moriwaki, Y., Inaba, M., & Koyama, H. (2018). Low sleep quality is associated with progression of arterial stiffness in patients with cardiovascular risk factors: HSCAA study. *Atherosclerosis*, *270*, 95–101. https://doi.org/10.1016/j.atherosclerosis.2018.01.039.

✔ Stress is inevitable, yet harmful! So have some suitable coping strategies.

It takes a bunch of risk factors such as insulin resistance, obesity, dyslipidemia, and hypertension to evolve into metabolic syndrome. ***As is the problem, so is the result***; meaning, again a lifestyle cluster including "dietary prebiotic fiber, exercise regularity, sound sleep, and stress relievers" is needed to prevent/manage metabolic syndrome. Given a health disorder or disease, universal choice would be prevention over cure! ***Gut health is the heart of this problem, as well as its solution!***

References

Aarts, E., Ederveen, T., Naaijen, J., Zwiers, M. P., Boekhorst, J., Timmerman, H. M., Smeekens, S. P., Netea, M. G., Buitelaar, J. K., Franke, B., van Hijum, S., & Arias Vasquez, A. (2017). Gut microbiome in ADHD and its relation to neural reward anticipation. *PloS One, 12*(9), e0183509. https://doi.org/10.1371/journal.pone.0183509

Al-Lahham, S. H., Peppelenbosch, M. P., Roelofsen, H., Vonk, R. J., & Venema, K. (2010). Biological effects of propionic acid in humans; metabolism, potential applications and underlying mechanisms. *Biochimica et biophysica acta, 1801*(11), 1175–1183. https://doi.org/10.1016/j.bbalip.2010.07.007

Anderson, J. R., Carroll, I., Azcarate-Peril, M. A., Rochette, A. D., Heinberg, L. J., Peat, C., Steffen, K., Manderino, L. M., Mitchell, J., & Gunstad, J. (2017). A preliminary examination of gut microbiota, sleep, and cognitive flexibility in healthy older adults. *Sleep Medicine, 38*, 104–107. https://doi.org/10.1016/j.sleep.2017.07.018

Andriamihaja, M., Chaumontet, C., Tome, D., & Blachier, F. (2009). Butyrate metabolism in human colon carcinoma cells: Implications concerning its growth-inhibitory effect. *Journal of Cellular Physiology, 218*(1), 58–65. https://doi.org/10.1002/jcp.21556

Arumugam, M., Raes, J., Pelletier, E., Le Paslier, D., Yamada, T., Mende, D. R., Fernandes, G. R., Tap, J., Bruls, T., Batto, J. M., Bertalan, M., Borruel, N., Casellas, F., Fernandez, L., Gautier, L., Hansen, T., Hattori, M., Hayashi, T., Kleerebezem, M., Kurokawa, K., ... Bork, P. (2011). Enterotypes of the human gut microbiome. *Nature, 473*(7346), 174–180. https://doi.org/10.1038/nature09944

Barber, T. M., Kabisch, S., Pfeiffer, A., & Weickert, M. O. (2020). The health benefits of dietary fibre. *Nutrients, 12*(10), 3209. https://doi.org/10.3390/nu12103209

Barko, P. C., McMichael, M. A., Swanson, K. S., & Williams, D. A. (2018). The gastrointestinal microbiome: A review. *Journal of Veterinary Internal Medicine, 32*(1), 9–25. https://doi.org/10.1111/jvim.14875

Benedict, C., Vogel, H., Jonas, W., Woting, A., Blaut, M., Schürmann, A., & Cedernaes, J. (2016). Gut microbiota and glucometabolic alterations in response to recurrent partial sleep deprivation in normal-weight young individuals. *Molecular Metabolism, 5*(12), 1175–1186. https://doi.org/10.1016/j.molmet.2016.10.003

Bermon, S., Petriz, B., Kajėnienė, A., Prestes, J., Castell, L., & Franco, O. L. (2015). The microbiota: An exercise immunology perspective. *Exercise Immunology Review, 21*, 70–79.

Bird, A. R., Brown, I. L., & Topping, D. L. (2000). Starches, resistant starches, the gut microflora and human health. *Current Issues in Intestinal Microbiology, 1*(1), 25–37. https://pubmed.ncbi.nlm.nih.gov/11709851/.

Birt, D. F., Boylston, T., Hendrich, S., Jane, J. L., Hollis, J., Li, L., McClelland, J., Moore, S., Phillips, G. J., Rowling, M., Schalinske, K., Scott, M. P., & Whitley, E. M. (2013). Resistant starch: Promise for improving human health. *Advances in Nutrition (Bethesda, Md.), 4*(6), 587–601. https://doi.org/10.3945/an.113.004325

Blaut, M. (2015). Gut microbiota and energy balance: Role in obesity. *The Proceedings of the Nutrition Society, 74*(3), 227–234. https://doi.org/10.1017/S0029665114001700

Bravo, J. A., Forsythe, P., Chew, M. V., Escaravage, E., Savignac, H. M., Dinan, T. G., Bienenstock, J., & Cryan, J. F. (2011). Ingestion of Lactobacillus strain regulates emotional behavior and central GABA receptor expression in a mouse via the vagus nerve. *Proceedings of the National Academy of Sciences of the United States of America, 108*(38), 16050–16055. https://doi.org/10.1073/pnas.1102999108

Brown, A. J., Goldsworthy, S. M., Barnes, A. A., Eilert, M. M., Tcheang, L., Daniels, D., Muir, A. I., Wigglesworth, M. J., Kinghorn, I., Fraser, N. J., Pike, N. B., Strum, J. C., Steplewski, K. M., Murdock, P. R., Holder, J. C., Marshall, F. H., Szekeres, P. G., Wilson, S., Ignar, D. M., Foord, S. M., ... Dowell, S. J. (2003). The Orphan G protein-coupled receptors GPR41 and GPR43 are activated by propionate and other short chain carboxylic acids. *The Journal of Biological Chemistry, 278*(13), 11312–11319. https://doi.org/10.1074/jbc.M211609200

Campbell, S. C., Wisniewski, P. J., Noji, M., McGuinness, L. R., Häggblom, M. M., Lightfoot, S. A., Joseph, L. B., & Kerkhof, L. J. (2016). The effect of diet and exercise on intestinal integrity and microbial diversity in mice. *PloS One, 11*(3), e0150502. https://doi.org/10.1371/journal.pone.0150502

Cani, P. D., Amar, J., Iglesias, M. A., Poggi, M., Knauf, C., Bastelica, D., Neyrinck, A. M., Fava, F., Tuohy, K. M., Chabo, C., Waget, A., Delmée, E., Cousin, B., Sulpice, T., Chamontin, B., Ferrières, J., Tanti, J. F., Gibson, G. R., Casteilla, L., Delzenne, N. M., … Burcelin, R. (2007). Metabolic endotoxemia initiates obesity and insulin resistance. *Diabetes, 56*(7), 1761–1772. https://doi.org/10.2337/db06-1491

Capuano, E. (2017). The behavior of dietary fiber in the gastrointestinal tract determines its physiological effect. *Critical Reviews in Food Science and Nutrition, 57*(16), 3543–3564. https://doi.org/10.1080/10408398.2016.1180501

Cedernaes, J., Lampola, L., Axelsson, E. K., Liethof, L., Hassanzadeh, S., Yeganeh, A., Broman, J. E., Schiöth, H. B., & Benedict, C. (2016). A single night of partial sleep loss impairs fasting insulin sensitivity but does not affect cephalic phase insulin release in young men. *Journal of Sleep Research, 25*(1), 5–10. https://doi.org/10.1111/jsr.12340

Cedernaes, J., Schiöth, H. B., & Benedict, C. (2015). Determinants of shortened, disrupted, and mistimed sleep and associated metabolic health consequences in healthy humans. *Diabetes, 64*(4), 1073–1080. https://doi.org/10.2337/db14-1475

Chambers, E. S., Viardot, A., Psichas, A., Morrison, D. J., Murphy, K. G., Zac-Varghese, S. E., MacDougall, K., Preston, T., Tedford, C., Finlayson, G. S., Blundell, J. E., Bell, J. D., Thomas, E. L., Mt-Isa, S., Ashby, D., Gibson, G. R., Kolida, S., Dhillo, W. S., Bloom, S. R., Morley, W., … Frost, G. (2015). Effects of targeted delivery of propionate to the human colon on appetite regulation, body weight maintenance and adiposity in overweight adults. *Gut, 64*(11), 1744–1754. https://doi.org/10.1136/gutjnl-2014-307913

Chan, J. C., Malik, V., Jia, W., Kadowaki, T., Yajnik, C. S., Yoon, K. H., & Hu, F. B. (2009). Diabetes in Asia: Epidemiology, risk factors, and pathophysiology. *JAMA, 301*(20), 2129–2140. https://doi.org/10.1001/jama.2009.726

Clarke, S. F., Murphy, E. F., O'Sullivan, O., Lucey, A. J., Humphreys, M., Hogan, A., Hayes, P., O'Reilly, M., Jeffery, I. B., Wood-Martin, R., Kerins, D. M., Quigley, E., Ross, R. P., O'Toole, P. W., Molloy, M. G., Falvey, E., Shanahan, F., & Cotter, P. D. (2014). Exercise and associated dietary extremes impact on gut microbial diversity. *Gut, 63*(12), 1913–1920. https://doi.org/10.1136/gutjnl-2013-306541

Cryan, J. F., & Dinan, T. G. (2012). Mind-altering microorganisms: The impact of the gut microbiota on brain and behaviour. *Nature Reviews. Neuroscience, 13*(10), 701–712. https://doi.org/10.1038/nrn3346

David, L. A., Maurice, C. F., Carmody, R. N., Gootenberg, D. B., Button, J. E., Wolfe, B. E., Ling, A. V., Devlin, A. S., Varma, Y., Fischbach, M. A., Biddinger, S. B., Dutton, R. J., & Turnbaugh, P. J. (2014). Diet rapidly and reproducibly alters the human gut microbiome. *Nature, 505*(7484), 559–563. https://doi.org/10.1038/nature12820

De Vadder, F., Kovatcheva-Datchary, P., Goncalves, D., Vinera, J., Zitoun, C., Duchampt, A., Bäckhed, F., & Mithieux, G. (2014). Microbiota-generated metabolites promote metabolic benefits via gut-brain neural circuits. *Cell, 156*(1–2), 84–96. https://doi.org/10.1016/j.cell.2013.12.016

Duncan, S. H., Belenguer, A., Holtrop, G., Johnstone, A. M., Flint, H. J., & Lobley, G. E. (2007). Reduced dietary intake of carbohydrates by obese subjects results in decreased concentrations of butyrate and butyrate-producing bacteria in feces. *Applied and Environmental Microbiology, 73*(4), 1073–1078. https://doi.org/10.1128/AEM.02340-06

Duncan, S. H., Louis, P., & Flint, H. J. (2004). Lactate-utilizing bacteria, isolated from human feces, that produce butyrate as a major fermentation product. *Applied and Environmental Microbiology, 70*(10), 5810–5817. https://doi.org/10.1128/AEM.70.10.5810-5817.2004

Estaki, M., Pither, J., Baumeister, P., Little, J. P., Gill, S. K., Ghosh, S., Ahmadi-Vand, Z., Marsden, K. R., & Gibson, D. L. (2016). Cardiorespiratory fitness as a predictor of intestinal microbial diversity and distinct metagenomic functions. *Microbiome, 4*(1), 42. https://doi.org/10.1186/s40168-016-0189-7

Everard, A., Belzer, C., Geurts, L., Ouwerkerk, J. P., Druart, C., Bindels, L. B., Guiot, Y., Derrien, M., Muccioli, G. G., Delzenne, N. M., de Vos, W. M., & Cani, P. D. (2013). Cross-talk between Akkermansia muciniphila and intestinal epithelium controls diet-induced obesity. *Proceedings of the National Academy of Sciences of the United States of America, 110*(22), 9066–9071. https://doi.org/10.1073/pnas.1219451110

Flint, H. J. (2012). The impact of nutrition on the human microbiome. *Nutrition Reviews, 70*(1), S10–S13. https://doi.org/10.1111/j.1753-4887.2012.00499.x

Flint, H. J., Scott, K. P., Louis, P., & Duncan, S. H. (2012). The role of the gut microbiota in nutrition and health. *Nature Reviews. Gastroenterology & Hepatology, 9*(10), 577–589. https://doi.org/10.1038/nrgastro.2012.156

Fond, G., Loundou, A., Hamdani, N., Boukouaci, W., Dargel, A., Oliveira, J., Roger, M., Tamouza, R., Leboyer, M., & Boyer, L. (2014). Anxiety and depression comorbidities in irritable bowel syndrome (IBS): A systematic review and meta-analysis. *European Archives of Psychiatry and Clinical Neuroscience, 264*(8), 651–660. https://doi.org/10.1007/s00406-014-0502-z

Ford, E. S., Cunningham, T. J., & Croft, J. B. (2015). Trends in self-reported sleep duration among US adults from 1985 to 2012. *Sleep, 38*(5), 829–832. https://doi.org/10.5665/sleep.4684

Frémont, M., Coomans, D., Massart, S., & De Meirleir, K. (2013). High-throughput 16S rRNA gene sequencing reveals alterations of intestinal microbiota in myalgic encephalomyelitis/chronic fatigue syndrome patients. *Anaerobe, 22*, 50–56. https://doi.org/10.1016/j.anaerobe.2013.06.002

Frey, D. J., Fleshner, M., & Wright, K. P., Jr. (2007). The effects of 40 hours of total sleep deprivation on inflammatory markers in healthy young adults. *Brain, Behavior, and Immunity, 21*(8), 1050–1057. https://doi.org/10.1016/j.bbi.2007.04.003

Gidley, M. J. (2013). Hydrocolloids in the digestive tract and related health implications. *Current Opinion in Colloid & Interface Science, 18*(4), 371–378. https://doi.org/10.1016/j.cocis.2013.04.003

Gill, C. I., & Rowland, I. R. (2002). Diet and cancer: Assessing the risk. *The British Journal of Nutrition, 88*(1), S73–S87. https://doi.org/10.1079/BJN2002632

Gisolfi, C. V. (2000). Is the GI system built for exercise? *News in Physiological Sciences: An International Journal of Physiology Produced Jointly by the International Union of Physiological Sciences and the American Physiological Society, 15*, 114–119. https://doi.org/10.1152/physiologyonline.2000.15.3.114

Gourineni, V., Stewart, M. L., Skorge, R., & Sekula, B. C. (2017). Slowly digestible carbohydrate for balanced energy: In vitro and in vivo evidence. *Nutrients, 9*(11), 1230. https://doi.org/10.3390/nu9111230

Groff, J. L., & Gropper, S. S. (2000). *Advanced nutrition and human metabolism* (3rd ed., pp. 144–146). Wadsworth, USA.

Guarner, F., & Malagelada, J. R. (2003). Gut flora in health and disease. *Lancet (London, England), 361*(9356), 512–519. https://doi.org/10.1016/S0140-6736(03)12489-0

Hooper, L. V., & Gordon, J. I. (2001). Commensal host-bacterial relationships in the gut. *Science (New York, N.Y.), 292*(5519), 1115–1118. https://doi.org/10.1126/science.1058709. https://www.ncbi.nlm.nih.gov/pmc/articles/PMC7589116/

Irwin, M. R., Olmstead, R., & Carroll, J. E. (2016). Sleep disturbance, sleep duration, and inflammation: A systematic review and meta-analysis of cohort studies and experimental sleep deprivation. *Biological Psychiatry, 80*(1), 40–52. https://doi.org/10.1016/j.biopsych.2015.05.014

Jakobsdottir, G., Nyman, M., & Fåk, F. (2014). Designing future prebiotic fiber to target metabolic syndrome. *Nutrition (Burbank, Los Angeles County, Calif.), 30*(5), 497–502. https://doi.org/10.1016/j.nut.2013.08.013. https://pubmed.ncbi.nlm.nih.gov/24262515/

Julie, M.,J. (2001). The definition of dietary fibers. *Cereal Foods World, 46*(3), 112–129.

Kaczmarek, J. L., Musaad, S. M., & Holscher, H. D. (2017). Time of day and eating behaviors are associated with the composition and function of the human gastrointestinal microbiota. *The American Journal of Clinical Nutrition, 106*(5), 1220–1231. https://doi.org/10.3945/ajcn.117.156380

Kadoya, M., Kurajoh, M., Kakutani-Hatayama, M., Morimoto, A., Miyoshi, A., Kosaka-Hamamoto, K., Shoji, T., Moriwaki, Y., Inaba, M., & Koyama, H. (2018). Low sleep quality is associated with progression of arterial stiffness in patients with cardiovascular risk factors: HSCAA study. *Atherosclerosis, 270*, 95–101. https://doi.org/10.1016/j.atherosclerosis.2018.01.039

Kamada, N., Seo, S. U., Chen, G. Y., & Núñez, G. (2013). Role of the gut microbiota in immunity and inflammatory disease. *Nature Reviews. Immunology, 13*(5), 321–335. https://doi.org/10.1038/nri3430

Kapsimalis, F., Richardson, G., Opp, M. R., & Kryger, M. (2005). Cytokines and normal sleep. *Current Opinion in Pulmonary Medicine, 11*(6), 481–484. https://doi.org/10.1097/01.mcp.0000183062.98665.6b

Kiecolt-Glaser, J. K., Habash, D. L., Fagundes, C. P., Andridge, R., Peng, J., Malarkey, W. B., & Belury, M. A. (2015). Daily stressors, past depression, and metabolic responses to high-fat meals: A novel path to obesity. *Biological Psychiatry, 77*(7), 653–660. https://doi.org/10.1016/j.biopsych.2014.05.018

Kovatcheva-Datchary, P., Nilsson, A., Akrami, R., Lee, Y. S., De Vadder, F., Arora, T., Hallen, A., Martens, E., Björck, I., & Bäckhed, F. (2015). Dietary fiber-induced improvement in glucose metabolism is associated with increased abundance of Prevotella. *Cell Metabolism, 22*(6), 971–982. https://doi.org/10.1016/j.cmet.2015.10.001

Krueger, J. M., Obál, F. J., Fang, J., Kubota, T., & Taishi, P. (2001). The role of cytokines in physiological sleep regulation. *Annals of the New York Academy of Sciences, 933*, 211–221. https://doi.org/10.1111/j.1749-6632.2001.tb05826.x

Ley, R. E., Turnbaugh, P. J., Klein, S., & Gordon, J. I. (2006). Microbial ecology: Human gut microbes associated with obesity. *Nature, 444*(7122), 1022–1023. https://doi.org/10.1038/4441022a

Lin, L., & Zhang, J. (2017). Role of intestinal microbiota and metabolites on gut homeostasis and human diseases. *BMC Immunology, 18*(1), 2. https://doi.org/10.1186/s12865-016-0187-3

Liu, Y. Z., Wang, Y. X., & Jiang, C. L. (2017). Inflammation: The common pathway of stress-related diseases. *Frontiers in Human Neuroscience, 11*, 316. https://doi.org/10.3389/fnhum.2017.00316

Lovegrove, A., Edwards, C. H., De Noni, I., Patel, H., El, S. N., Grassby, T., Zielke, C., Ulmius, M., Nilsson, L., Butterworth, P. J., Ellis, P. R., & Shewry, P. R. (2017). Role of polysaccharides in food, digestion, and health. *Critical Reviews in Food Science and Nutrition, 57*(2), 237–253. https://doi.org/10.1080/10408398.2014.939263

Lu, L. W., Monro, J., Lu, J., & Rush, E. (2016). Effect of cold storage, reheating, and particle sizes on in vitro glucose release and starch digestibility among five rice products in Auckland, New Zealand. *J Rice Res, 4*, 171. https://doi.org/10.4172/2375-4338.1000171

Macfarlane, S., & Macfarlane, G. T. (2006). Composition and metabolic activities of bacterial biofilms colonizing food residues in the human gut. *Applied and Environmental Microbiology, 72*(9), 6204–6211. https://doi.org/10.1128/AEM.00754-06

Macfarlane, G. T., & Macfarlane, S. (2012). Bacteria, colonic fermentation, and gastrointestinal health. *Journal of AOAC International, 95*(1), 50–60. https://doi.org/10.5740/jaoacint.sge_macfarlane

Mackie, R. I., Sghir, A., & Gaskins, H. R. (1999). Developmental microbial ecology of the neonatal gastrointestinal tract. *The American Journal of Clinical Nutrition, 69*(5), 1035S–1045S. https://doi.org/10.1093/ajcn/69.5.1035s

Madison, A., & Kiecolt-Glaser, J. K. (2019). Stress, depression, diet, and the gut microbiota: Human-bacteria interactions at the core of psychoneuroimmunology and nutrition. *Current Opinion in Behavioral Sciences, 28*, 105–110. https://doi.org/10.1016/j.cobeha.2019.01.011

Makki, K., Deehan, E. C., Walter, J., & Bäckhed, F. (2018). The impact of dietary fiber on gut microbiota in host health and disease. *Cell Host & Microbe, 23*(6), 705–715. https://doi.org/10.1016/j.chom.2018.05.012

Maslowski, K. M., Vieira, A. T., Ng, A., Kranich, J., Sierro, F., Yu, D., Schilter, H. C., Rolph, M. S., Mackay, F., Artis, D., Xavier, R. J., Teixeira, M. M., & Mackay, C. R. (2009). Regulation of inflammatory responses by gut microbiota and chemoattractant receptor GPR43. *Nature, 461*(7268), 1282–1286. https://doi.org/10.1038/nature08530

Matsumoto, M., Inoue, R., Tsukahara, T., Ushida, K., Chiji, H., Matsubara, N., & Hara, H. (2008). Voluntary running exercise alters microbiota composition and increases n-butyrate concentration in the rat cecum. *Bioscience, Biotechnology, and Biochemistry, 72*(2), 572–576. https://doi.org/10.1271/bbb.70474

Mazmanian, S. K., Round, J. L., & Kasper, D. L. (2008). A microbial symbiosis factor prevents intestinal inflammatory disease. *Nature, 453*(7195), 620–625. https://doi.org/10.1038/nature07008

McGrane, M. M. (2006). *Carbohydrate metabolism: Synthesis and oxidation* (pp. 258–277). Missouri: Saunders: Elsevier, ISBN 1-4160-0209-X.

McIntyre, A., Gibson, P. R., & Young, G. P. (1993). Butyrate production from dietary fibre and protection against large bowel cancer in a rat model. *Gut, 34*(3), 386–391. https://doi.org/10.1136/gut.34.3.386

McRae, M. P. (2017). Dietary fiber is beneficial for the prevention of cardiovascular disease: An umbrella review of meta-analyses. *Journal of Chiropractic Medicine, 16*(4), 289–299. https://doi.org/10.1016/j.jcm.2017.05.005

Mika, A., Van Treuren, W., González, A., Herrera, J. J., Knight, R., & Fleshner, M. (2015). Exercise is more effective at altering gut microbial composition and producing stable changes in lean mass in juvenile versus adult male F344 rats. *PloS One, 10*(5), e0125889. https://doi.org/10.1371/journal.pone.0125889

Monda, V., Villano, I., Messina, A., Valenzano, A., Esposito, T., Moscatelli, F., Viggiano, A., Cibelli, G., Chieffi, S., Monda, M., & Messina, G. (2017). Exercise modifies the gut microbiota with positive health effects. *Oxidative Medicine and Cellular Longevity, 2017*. https://doi.org/10.1155/2017/3831972, 3831972.

Muegge, B. D., Kuczynski, J., Knights, D., Clemente, J. C., González, A., Fontana, L., Henrissat, B., Knight, R., & Gordon, J. I. (2011). Diet drives convergence in gut microbiome functions across mammalian phylogeny and within humans. *Science (New York, N.Y.), 332*(6032), 970–974. https://doi.org/10.1126/science.1198719

Myhrstad, M., Tunsjø, H., Charnock, C., & Telle-Hansen, V. H. (2020). Dietary fiber, gut microbiota, and metabolic regulation-current status in human randomized trials. *Nutrients, 12*(3), 859. https://doi.org/10.3390/nu12030859. https://www.ncbi.nlm.nih.gov/pmc/articles/PMC7146107/

National Research Council (US) Committee on Diet and Health. (1989). *Diet and health: Implications for reducing chronic disease risk.* Washington (DC): National Academies Press (US), 10, Dietary Fiber. Available from: https://www.ncbi.nlm.nih.gov/books/NBK218764/.

Obal, F., Jr., & Krueger, J. M. (2003). Biochemical regulation of non-rapid-eye-movement sleep. *Frontiers in Bioscience: A Journal and Virtual Library, 8*, d520–d550. https://doi.org/10.2741/1033

Oja, P., & Titze, S. (2011). Physical activity recommendations for public health: Development and policy context. *The EPMA Journal, 2*(3), 253–259. https://doi.org/10.1007/s13167-011-0090-1

Padayachee, A., Day, L., Howell, K., & Gidley, M. J. (2017). Complexity and health functionality of plant cell wall fibers from fruits and vegetables. *Critical Reviews in Food Science and Nutrition, 57*(1), 59–81. https://doi.org/10.1080/10408398.2013.850652

Payne, A. N., Chassard, C., & Lacroix, C. (2012). Gut microbial adaptation to dietary consumption of fructose, artificial sweeteners and sugar alcohols: Implications for host-microbe interactions contributing to obesity. *Obesity Reviews: An Official Journal of the International Association for the Study of Obesity, 13*(9), 799–809. https://doi.org/10.1111/j.1467-789X.2012.01009.x

Peters, H. P., De Vries, W. R., Vanberge-Henegouwen, G. P., & Akkermans, L. M. (2001). Potential benefits and hazards of physical activity and exercise on the gastrointestinal tract. *Gut, 48*(3), 435–439. https://doi.org/10.1136/gut.48.3.435

Polednak, A. P. (2003). Trends in incidence rates for obesity-associated cancers in the US. *Cancer Detection and Prevention, 27*(6), 415–421. https://doi.org/10.1016/j.cdp.2003.09.002

Position of the Indian Dietetic Association: Dietary fibre and health. (2018).

Pudlo, N. A., Urs, K., Kumar, S. S., German, J. B., Mills, D. A., & Martens, E. C. (2015). Symbiotic human gut bacteria with variable metabolic priorities for host mucosal glycans. *mBio, 6*(6). https://doi.org/10.1128/mBio.01282-15, 012822-15 https://www.ncbi.nlm.nih.gov/pmc/articles/PMC4659458/.

Qin, J., Li, Y., Cai, Z., Li, S., Zhu, J., Zhang, F., Liang, S., Zhang, W., Guan, Y., Shen, D., Peng, Y., Zhang, D., Jie, Z., Wu, W., Qin, Y., Xue, W., Li, J., Han, L., Lu, D., Wu, P., … Wang, J. (2012). A metagenome-wide association study of gut microbiota in type 2 diabetes. *Nature, 490*(7418), 55–60. https://doi.org/10.1038/nature11450

Queipo-Ortuño, M. I., Seoane, L. M., Murri, M., Pardo, M., Gomez-Zumaquero, J. M., Cardona, F., Casanueva, F., & Tinahones, F. J. (2013). Gut microbiota composition in male rat models under different nutritional status and physical activity and its association with serum leptin and ghrelin levels. *PloS One, 8*(5), e65465. https://doi.org/10.1371/journal.pone.0065465

Ranjbaran, Z., Keefer, L., Stepanski, E., Farhadi, A., & Keshavarzian, A. (2007). The relevance of sleep abnormalities to chronic inflammatory conditions. *Inflammation Research: Official Journal of the European Histamine Research Society ... [et al.], 56*(2), 51–57. https://doi.org/10.1007/s00011-006-6067-1

Rehrer, N. J., Smets, A., Reynaert, H., Goes, E., & De Meirleir, K. (2001). Effect of exercise on portal vein blood flow in man. *Medicine and Science in Sports and Exercise, 33*(9), 1533–1537. https://doi.org/10.1097/00005768-200109000-00017

Remely, M., Aumueller, E., Jahn, D., Hippe, B., Brath, H., & Haslberger, A. G. (2014). Microbiota and epigenetic regulation of inflammatory mediators in type 2 diabetes and obesity. *Beneficial Microbes, 5*(1), 33–43. https://doi.org/10.3920/BM2013.006

Rinninella, E., Raoul, P., Cintoni, M., Franceschi, F., Miggiano, G., Gasbarrini, A., & Mele, M. C. (2019). What is the healthy gut microbiota composition? A changing ecosystem across age, environment, diet, and diseases. *Microorganisms, 7*(1), 14. https://doi.org/10.3390/microorganisms7010014

Ríos-Covián, D., Ruas-Madiedo, P., Margolles, A., Gueimonde, M., de Los Reyes-Gavilán, C. G., & Salazar, N. (2016). Intestinal short chain fatty acids and their link with diet and human health. *Frontiers in Microbiology, 7*, 185. https://doi.org/10.3389/fmicb.2016.00185

Robles Alonso, V., & Guarner, F. (2013). Linking the gut microbiota to human health. *The British Journal of Nutrition, 109*(2), S21–S26. https://doi.org/10.1017/S0007114512005235

Rosser, E. C., Oleinika, K., Tonon, S., Doyle, R., Bosma, A., Carter, N. A., Harris, K. A., Jones, S. A., Klein, N., & Mauri, C. (2014). Regulatory B cells are induced by gut microbiota-driven interleukin-1β and interleukin-6 production. *Nature Medicine, 20*(11), 1334–1339. https://doi.org/10.1038/nm.3680

Sawyer, B. J., Bhammar, D. M., Angadi, S. S., Ryan, D. M., Ryder, J. R., Sussman, E. J., Bertmann, F. M., & Gaesser, G. A. (2015). Predictors of fat mass changes in response to aerobic exercise training in women. *Journal of Strength and Conditioning Research, 29*(2), 297–304. https://doi.org/10.1519/JSC.0000000000000726

Sekirov, I., Russell, S. L., Antunes, L. C., & Finlay, B. B. (2010). Gut microbiota in health and disease. *Physiological Reviews, 90*(3), 859–904. https://doi.org/10.1152/physrev.00045.2009

Shanmugam, A. (1998). *Fundamentals of biochemistry for medical students* (7th ed., pp. 403–477). Chennai: Karthik Offset Printers.

Shils, M. E., Olson, J. A., & Shike, M. (2005). *Modern nutrition in health and disease* (9th ed., pp. 46–52). Philadelphia: Lea and Febiger, 75-80, 387-392, 527-553, 1199.

Smith, R. P., Easson, C., Lyle, S. M., Kapoor, R., Donnelly, C. P., Davidson, E. J., Parikh, E., Lopez, J. V., & Tartar, J. L. (2019). Gut microbiome diversity is associated with sleep physiology in humans. *PloS One, 14*(10), e0222394. https://doi.org/10.1371/journal.pone.0222394

Smith, P. M., Howitt, M. R., Panikov, N., Michaud, M., Gallini, C. A., Bohlooly-Y, M., Glickman, J. N., & Garrett, W. S. (2013). The microbial metabolites, short-chain fatty acids, regulate colonic Treg cell homeostasis. *Science (New York, N.Y.), 341*(6145), 569–573. https://doi.org/10.1126/science.1241165

Spaeth, A. M., Dinges, D. F., & Goel, N. (2013). Effects of experimental sleep restriction on weight gain, caloric intake, and meal timing in healthy adults. *Sleep, 36*(7), 981–990. https://doi.org/10.5665/sleep.2792

Stephen, A. M., Champ, M. M., Cloran, S. J., Fleith, M., van Lieshout, L., Mejborn, H., & Burley, V. J. (2017). Dietary fibre in Europe: Current state of knowledge on definitions, sources, recommendations, intakes and relationships to health. *Nutrition Research Reviews, 30*(2), 149–190. https://doi.org/10.1017/S095442241700004X

Stilling, R. M., Ryan, F. J., Hoban, A. E., Shanahan, F., Clarke, G., Claesson, M. J., Dinan, T. G., & Cryan, J. F. (2015). Microbes & neurodevelopment–Absence of microbiota during early life increases activity-related transcriptional pathways in the amygdala. *Brain, Behavior, and Immunity, 50*, 209–220. https://doi.org/10.1016/j.bbi.2015.07.009

Swartz, T. D., Duca, F. A., de Wouters, T., Sakar, Y., & Covasa, M. (2012). Up-regulation of intestinal type 1 taste receptor 3 and sodium glucose luminal transporter-1 expression and increased sucrose intake in mice lacking gut microbiota. *The British Journal of Nutrition, 107*(5), 621–630. https://doi.org/10.1017/S0007114511003412

Tan, J., McKenzie, C., Potamitis, M., Thorburn, A. N., Mackay, C. R., & Macia, L. (2014). The role of short-chain fatty acids in health and disease. *Advances in Immunology, 121*, 91–119. https://doi.org/10.1016/B978-0-12-800100-4.00003-9

Tan, X., Saarinen, A., Mikkola, T. M., Tenhunen, J., Martinmäki, S., Rahikainen, A., Cheng, S., Eklund, N., Pekkala, S., Wiklund, P., Munukka, E., Wen, X., Cong, F., Wang, X., Zhang, Y., Tarkka, I., Sun, Y., Partinen, M., Alen, M., & Cheng, S. (2013). Effects of exercise and diet interventions on obesity-related sleep disorders in men: Study protocol for a randomized controlled trial. *Trials, 14*, 235. https://doi.org/10.1186/1745-6215-14-235

Tennoune, N., Chan, P., Breton, J., Legrand, R., Chabane, Y. N., Akkermann, K., Järv, A., Ouelaa, W., Takagi, K., Ghouzali, I., Francois, M., Lucas, N., Bole-Feysot, C., Pestel-Caron, M., do Rego, J. C., Vaudry, D., Harro, J., Dé, E., Déchelotte, P., & Fetissov, S. O. (2014). Bacterial ClpB heat-shock protein, an antigen-mimetic of the anorexigenic peptide α-MSH, at the origin of eating disorders. *Translational Psychiatry, 4*(10), e458. https://doi.org/10.1038/tp.2014.98

Topping, D. L., & Clifton, P. M. (2001). Short-chain fatty acids and human colonic function: Roles of resistant starch and nonstarch polysaccharides. *Physiological Reviews, 81*(3), 1031–1064. https://doi.org/10.1152/physrev.2001.81.3.1031

Turnbaugh, P. J., Ridaura, V. K., Faith, J. J., Rey, F. E., Knight, R., & Gordon, J. I. (2009). The effect of diet on the human gut microbiome: A metagenomic analysis in humanized gnotobiotic mice. *Science Translational Medicine, 1*(6), 6ra14. https://doi.org/10.1126/scitranslmed.3000322

Underwood, M. A. (2014). Intestinal dysbiosis: Novel mechanisms by which gut microbes trigger and prevent disease. *Preventive Medicine, 65*, 133–137. https://doi.org/10.1016/j.ypmed.2014.05.010

Vgontzas, A. N. (2008). Does obesity play a major role in the pathogenesis of sleep apnoea and its associated manifestations via inflammation, visceral adiposity, and insulin resistance? *Archives of Physiology and Biochemistry, 114*(4), 211–223. https://doi.org/10.1080/13813450802364627

Vrieze, A., Van Nood, E., Holleman, F., Salojärvi, J., Kootte, R. S., Bartelsman, J. F., Dallinga-Thie, G. M., Ackermans, M. T., Serlie, M. J., Oozeer, R., Derrien, M., Druesne, A., Van Hylckama Vlieg, J. E., Bloks, V. W., Groen, A. K., Heilig, H. G., Zoetendal, E. G., Stroes, E. S., de Vos, W. M., Hoekstra, J. B., … Nieuwdorp, M. (2012). Transfer of intestinal microbiota from lean donors increases insulin sensitivity in individuals with metabolic syndrome. *Gastroenterology, 143*(4). https://doi.org/10.1053/j.gastro.2012.06.031, 913–6.e7.

Walker, A. W., Ince, J., Duncan, S. H., Webster, L. M., Holtrop, G., Ze, X., Brown, D., Stares, M. D., Scott, P., Bergerat, A., Louis, P., McIntosh, F., Johnstone, A. M., Lobley, G. E., Parkhill, J., & Flint, H. J. (2011). Dominant and diet-responsive groups of bacteria within the human colonic microbiota. *The ISME Journal, 5*(2), 220–230. https://doi.org/10.1038/ismej.2010.118

Wardlaw, G. M., Insel, P. M., & Seyler, M. F. (1992). *Contemporary nutrition – issues and insights* (1st ed.). London: Mosby Year Book.

Weickert, M. O., & Pfeiffer, A. F. (2008). Metabolic effects of dietary fiber consumption and prevention of diabetes. *The Journal of Nutrition, 138*(3), 439–442. https://doi.org/10.1093/jn/138.3.439

Williams, S. R. (1994). *Essentials of nutrition and diet therapy* (6th ed., pp. 402–406). St. Louis: Times Mirror/Mosby College Publishing, 43-46, 55, 391-399, 414-428, 565-581.

Williams, B. A., Grant, L. J., Gidley, M. J., & Mikkelsen, D. (2017). Gut fermentation of dietary fibres: Physico-chemistry of plant cell walls and implications for health. *International Journal of Molecular Sciences, 18*(10), 2203. https://doi.org/10.3390/ijms18102203

Williams, B. A., Mikkelsen, D., Flanagan, B. M., & Gidley, M. J. (2019). "Dietary fibre": Moving beyond the "soluble/insoluble" classification for monogastric nutrition, with an emphasis on humans and pigs. *Journal of Animal Science and Biotechnology, 10*, 45. https://doi.org/10.1186/s40104-019-0350-9

Wong, J. M., de Souza, R., Kendall, C. W., Emam, A., & Jenkins, D. J. (2006). Colonic health: Fermentation and short chain fatty acids. *Journal of Clinical Gastroenterology, 40*(3), 235–243. https://doi.org/10.1097/00004836-200603000-00015

Wu, G. D., Chen, J., Hoffmann, C., Bittinger, K., Chen, Y. Y., Keilbaugh, S. A., Bewtra, M., Knights, D., Walters, W. A., Knight, R., Sinha, R., Gilroy, E., Gupta, K., Baldassano, R., Nessel, L., Li, H., Bushman, F. D., & Lewis, J. D. (2011). Linking long-term dietary patterns with gut microbial enterotypes. *Science (New York, N.Y.), 334*(6052), 105–108. https://doi.org/10.1126/science.1208344

Zeevi, D., Korem, T., Zmora, N., Israeli, D., Rothschild, D., Weinberger, A., Ben-Yacov, O., Lador, D., Avnit-Sagi, T., Lotan-Pompan, M., Suez, J., Mahdi, J. A., Matot, E., Malka, G., Kosower, N., Rein, M., Zilberman-Schapira, G., Dohnalová, L., Pevsner-Fischer, M., Bikovsky, R., … Segal, E. (2015). Personalized nutrition by prediction of glycemic responses. *Cell, 163*(5), 1079–1094. https://doi.org/10.1016/j.cell.2015.11.001

Zelante, T., Iannitti, R. G., Cunha, C., De Luca, A., Giovannini, G., Pieraccini, G., Zecchi, R., D'Angelo, C., Massi-Benedetti, C., Fallarino, F., Carvalho, A., Puccetti, P., & Romani, L. (2013). Tryptophan catabolites from microbiota engage aryl hydrocarbon receptor and balance mucosal reactivity via interleukin-22. *Immunity, 39*(2), 372–385. https://doi.org/10.1016/j.immuni.2013.08.003

Zhang, G., & Hamaker, B. R. (2009). Slowly digestible starch: Concept, mechanism, and proposed extended glycemic index. *Critical Reviews in Food Science and Nutrition, 49*(10), 852–867. https://doi.org/10.1080/10408390903372466

CHAPTER

6

Metabolomic study of biofilm-forming natural microbiota of skin biofilm

Gokul Shankar Sabesan[1] *and Annie Jeyachristy Sam*[2]

[1]Faculty of Medicine, Manipal University College Malaysia, Melaka, Malaysia; [2]Faculty of Medicine, Universiti Kuala Lumpur Royal College of Medicine Perak, Ipoh, Perak, Malaysia

1. Introduction

Skin, the largest and prominent organ of the human body, is an essential site of interaction between the immune system and the microbes. Skin essentially connects the host with the environment. The immune system of the skin and its relationship with the skin microbiota have been well established (Belkaid & Segre, 2014). The adaptation of bacteria to inhabit and establish on the skin surfaces is made possible by the metabolic changes that produce substances extracellularly to retain the bacterial communities. The regulation of certain genes also contributes to this bacterial adaptation. This bacterial community forms networks with multicellular functions, thereby leading to establishment and differentiation of the community resulting in biofilm formation. These biofilms are advantageous to the community members in terms of virulence, pathogenesis, and antibiotic resistance (Hall & Mah, 2017; Santos et al., 2018; Vestby et al., 2020).

Anton Von Leeuwenhoek, who invented the microscope, first identified the microbial aggregates on the dental plaque scrapings from his teeth was later termed as "biofilm" by Bill Costerton in the year 1978.

Studies on skin-associated pathogens such as *Staphylococcus aureus* (*S. aureus*), *Staphylococcus epidermidis* (*S. epidermidis*), *Propionibacterium acnes* (*P. acnes*), *Malassezia* spp., and others, and their function in the complex mechanisms associated with skin diseases are still ongoing (Byrd et al., 2018; Carmona-Cruz et al., 2022; Claudel et al., 2019; Flowers & Grice, 2020). Associations between *P. acnes* and acne vulgaris, *S. aureus* and atopic dermatitis (AD), and *Malassezia* species and dandruff have been reported for decades (Brandwein et al., 2016; Dréno et al., 2018; Geoghegan et al., 2018; McLaughlin et al., 2019; Ogonowska et al., 2021;

Microbial Biofilms
https://doi.org/10.1016/B978-0-323-95715-1.00013-3

Rudramurthy et al., 2014; Saxena et al., 2018; Tomczak et al., 2019). Many skin-associated microbes, their capabilities to form biofilm and its relationship to disease states were documented in the past decade (Jordana-Lluch et al., 2020; Vaishnavi et al., 2019).

There are several factors that render an effect on the biofilm formation including the surface environment, host receptors, nutrition, metabolites, and the immune system.

In any cell regulatory process, metabolites are formed. They are classified as primary metabolites, required for cell maintenance, survival, development and growth, and secondary metabolites, also called as specialized metabolites or nonessential, help the organism by providing nutrients, protection against external stressors, and aiding their interaction and coexistence/antagonism with other organisms. The specialized metabolite production is influenced by pH, light, carbon, and nitrogen sources or by organisms living in the same habitat. The set of metabolites, called the metabolome, is responsible for the physiological state of an organism. Any changes in the metabolome are indicative of diseases (Tounta et al., 2021).

Metabolomics is a systems biology–driven omics method that helps in the profiling of all the metabolites. Alterations in any of the metabolite levels are correlated to the different biological processes and their contexts in cells. Metabolomics also enables the understanding of changes in spatial arrangement and chronological changes of different metabolic activities during the formation of biofilm.

This chapter focuses on the probable roles of skin inhibiting pathogens in biofilm formation, production of metabolites, metabolomics and pathogenesis of dermatological disorders, and the therapeutic approach of skin biofilm.

2. Skin and biofilm production

The bacterial communities are immobilized to the surfaces, either biotic or abiotic, as individual free-floating planktonic forms or as multicellular complexes to form the biofilm (Fig. 6.1). They range in size from small aggregates of cells to large biofilms of bacteria. The organisms are safely implanted/embedded in a glycocalyx, a pericellular matrix (extracellular) polymeric substance (EPS) that is produced within the biofilm. EPS such as polysaccharides, DNA, proteins, and lipids enhances the attachment of bacteria to the surface. Mineral scaffolds are also involved in the assembly of the extracellular matrix (Verderosa et al., 2019). Biofilms of consortiums are more resistant to antibiotics than their planktonic forms. Biofilms are formed by either one of the two mechanisms: reduction in the diffusion rate of antibiotics through the extracellular polymeric substances, or reduction in the metabolic changes that lead to altered phenotype of the bacteria in the biofilm (Muhammad et al., 2020; Srinivasan et al., 2021).

Multispecies biofilms found in nature are known as the polymicrobial biofilms. The factors that influence biofilm formation with polymicrobial composition are the physiochemical surface environment, increased mutation, phenotype switching, altered growth and metabolism, production of enzymes, host receptors, availability of the nutrients, coenzymes and cofactors, aggregation pattern, and local immune system activity. The co-occurrence of multispecies within a habitat involves various modes of interspecies communication such as quorum sensing. Biofilms can be formed on varied biological surfaces such as teeth (Neelakantan et al., 2017; Simon-Soro et al., 2022), heart valves (Lauten et al., 2021; Lerche et al.,

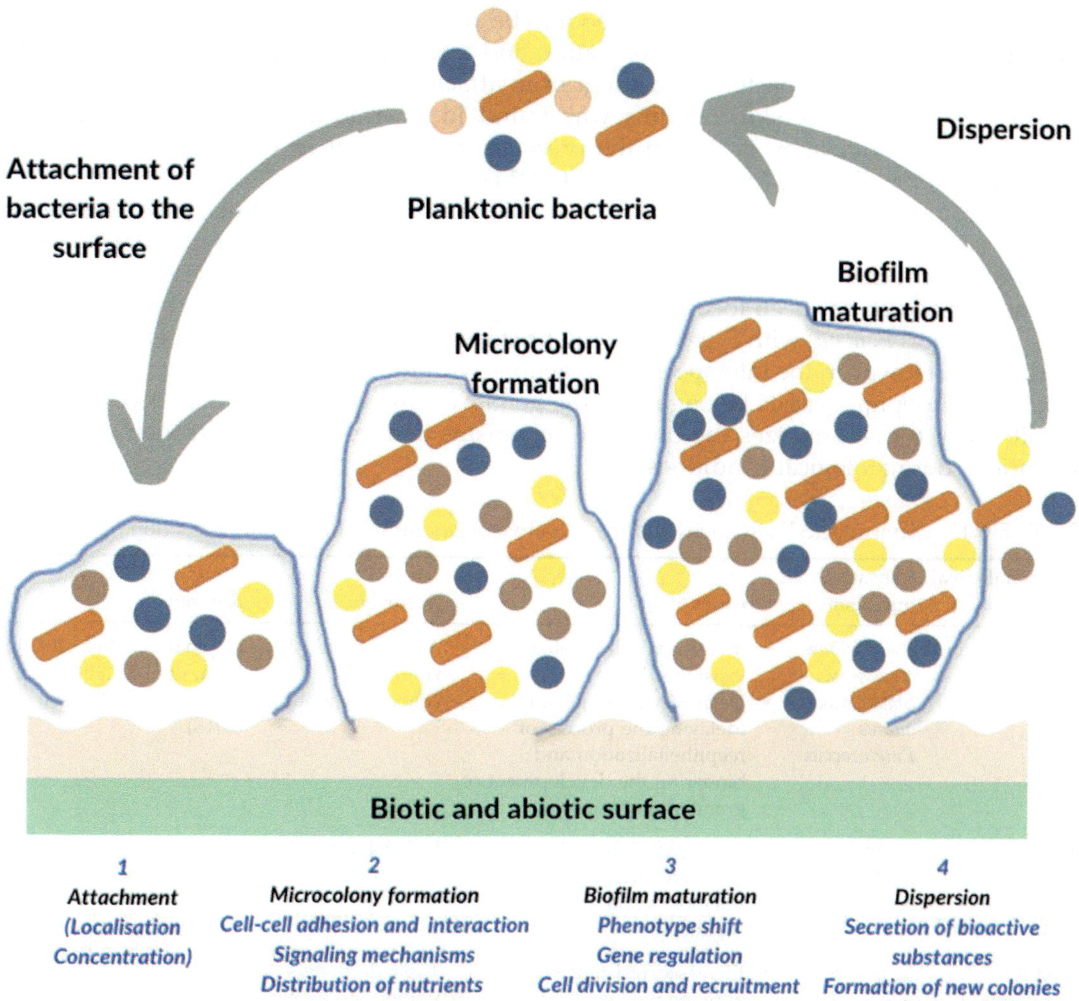

FIGURE 6.1 Biofilm formation.

2021), ear mucosa (Akyıldız et al., 2013), prosthetic valves, dental (Dhir, 2013) and orthopedic implants (Koseki et al., 2014), contact lenses (Dosler et al., 2020), intravenous catheters (Gominet et al., 2017), and invariably the skin (Brandwein et al., 2016).

Biofilm cells show an altered phenotype, a different growth rate, and varied process of gene transcription when compared with their free-living counterparts (Guzmán-Soto et al., 2021; Penesyan et al., 2019). Biofilms protect the polymicrobial consortiums from environmental perturbations leaving them undisturbed from external factors, allow communication between the microbial communities, and provide enhanced virulence and optimal utilization of nutrients for development (Kravvas et al., 2018). Exposure to a biofilm at the fetal stage helps in planned "conditioning" of the skin's surface and thereby enhances the colonization of "beneficial" bacteria that forms the protection to the host. Formation of biofilm by the skin's own aboriginal normal flora is known to be significant in the prevention of skin infection. The resistance conferred by colonization and a remarkable skin defensive mechanism prevents the exogenous bacteria and fungi from attaching to the skin surface.

The first evidence of skin biofilms was found in *S. epidermidis*. Studies revealed that delayed reepithelialization is due to the presence of an *S. aureus* or *S. epidermidis* biofilm (Brown & Horswill, 2020; Lister & Horswill, 2014).

Numerous studies on in vitro single-species biofilms of skin microbiota, such as *S. aureus*, *S. epidermidis*, and *P. acnes*, are reported (Gannesen et al., 2019; Kranjec et al., 2021; Neopane et al., 2018; Schilcher & Horswill, 2020). The studies on *Malassezia* spp. biofilms remain scarce. Extensive research has been focused on the interspecies interactions among the skin biofilms of *S. aureus* and *P. acnes* with *S. epidermidis* (Christensen et al., 2016; Coenye et al., 2021; Wang et al., 2014). However, the interspecies interactions with fungal commensals of the skin, principally the lipophilic yeasts belonging to the genus, *Malassezia*, are yet to be investigated in detail. Table 6.1 summarizes the causative organisms and its features and their occurrence in various dermatological conditions.

TABLE 6.1 Dermatological conditions and the causative organisms for the biofilm production.

Dermatological conditions	Causative organisms	Features	Occurrence	Reference
Chronic wounds	*Pseudomonas aeruginosa* *Staphylococcus aureus* *Enterococcus*	• Increased antimicrobial resistance • Delayed wound healing (delaying the process of reepithelialization and hindering the development of granulation tissue)	Acute, chronic wounds Diabetic wounds	Gopinathan et al. (2017); Kolar et al. (2020); Roux et al., 2022; Wilkinson et al. (2018)
Hidradenitis suppurativa	*Staphylococcus lugdunensis*	• Synergistic interaction between commensal microbials and atypical innate immunity	Hair follicle and sinus tracts	Nikolakis et al. (2015)
Atopic dermatitis	*S. aureus*	• Increased resistance to antibiotic • Biofilm-induced occlusion of sweat ducts	Topic dermatitis lesions and atopic skin	Allen et al. (2014); Jung et al. (2015)
Candidiasis	*Candida albicans*	• Alteration in the host immunity or local ecology • Involvement of BCR-1 transcription factor • Altered mucosal barrier	Mucosal surfaces	Dwivedi et al. (2011); Naglik & Moyes (2011); Tsui et al. (2016)
Acne	*Cutibacterium acnes*	• Follicular plugging and cohesiveness	Skin, oral cavity, large intestine, conjunctiva, external ear canal	Fournière et al. (2020); Mongaret et al. (2021); Jahns et al. (2012)
Onychomycosis	*Trichophyton* spp. *Candida* spp.		Skin, nail, and lungs	Ellabib et al. (2002)

3. Metabolites of skin biofilms

Transcriptome and proteome analysis helps to identify the upregulation and downregulation of genes and protein in bacterial biofilms, metabolite analysis helps to assess the cellular phenotype. Metabolomics, study of the metabolites, is the quantitative analysis technique that monitors variations in the metabolites produced by the metabolism of bacteria in response to particular environmental conditions. The regulation of gene and its protein activity takes place through transmembrane receptors, transcription factors, enzyme cofactors and co-substrates in the biochemical reactions, via posttranscriptional and posttranslational modifications. Metabolites function as signaling molecules of gene expression and epigenetic regulation through the transcription factors. The metabolites also function as co-substrates and/or cofactors in their role as chromatin-modifying enzymes (Fig. 6.2).

The molecules with low molecular weight that are involved in the metabolism are the metabolomes. Metabolomes are identified as biomarkers in disease diagnosis and prediction. It is estimated that more than 19,000 metabolomes are found in the body including those produced endogenous and exogenous (Wishart et al., 2013). Endogenous metabolites are usually linked to enzymatic activities that are encoded by specific genome, and exogenous metabolites are derived from medications, food, microbiota, and the environment.

The abundance of metabolites is related to the pathogenic mechanisms of the diseases. The analysis of the metabolites has a potential role in precision medicine (Clish, 2015). The identification of specific markers, drug pharmacodynamics, metabolic profile associated with external and internal environments, and the pathogenesis of the disease are vital for precision

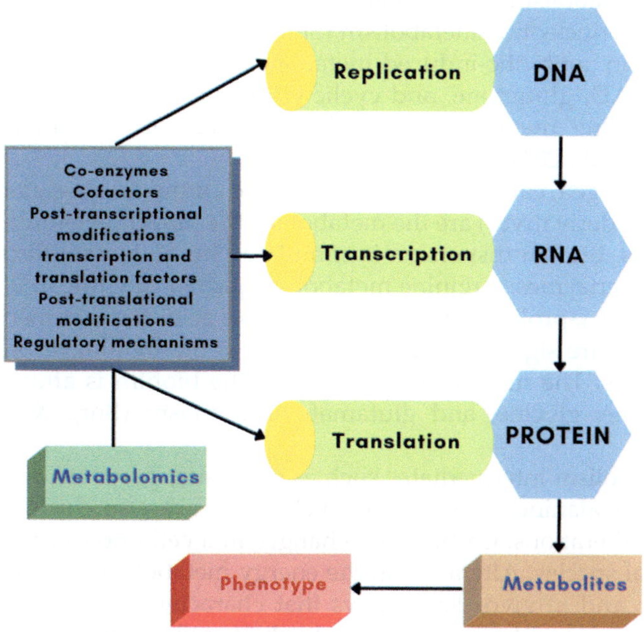

FIGURE 6.2 The sequence of events in the formation of metabolites includes the genome, proteome, and metabolome, the final outcome.

medicine. The recent development in metabolomics, the study of metabolic profile, has enabled the discovery of metabolic indicators related to the diseases.

Mass spectrometry (MS)–based analysis is usually the approach followed for the determination of more than 200 metabolites in a single experiment. This approach renders sensitive and specific results within a limited time frame. Liquid chromatography–mass spectrometry (LC-MS) and gas chromatography–mass spectrometry (GC-MS) are the techniques utilized to study the metabolites with excellent precision and are suitable for studies in human cohorts. The study of metabolites has proved to be a useful indicator in pancreatic cancer (Mehta et al., 2017), type 2 diabetes (Ahola-Olli et al., 2019; Gu et al., 2020; Yang et al., 2018; Long et al., 2020; Wang et al., 2011), memory impairment (Jiang et al., 2019; Tondo et al., 2020), and many other conditions (dos Santos et al., 2021; Kaushik & DeBerardinis, 2018; Qi et al., 2021). The metabolomic studies on the relationship that exists between the diet and disease is inspiring and has thrown highlights linking the elevated amino acid metabolites such as branched-chain amino acids to obesity and insulin resistance (Lee et al., 2021; Newgard, 2012; Newgard et al., 2009). Similarly, increased consumption of red meat results in high levels of trimethylamine-*N*-oxide that influences the composition of the gut microbiome and thereby related to the risk for cardiac-associated diseases (Koeth et al., 2013). This indicates the role of metabolites as efficacy markers of clinical conditions and in drug development. Table 6.2 depicts the distribution of microbiomes and its correlation to the metabolomes.

Alterations in the metabolism protects the bacteria by reducing the formation of metabolites that are toxic to the cell, inhibiting antibiotic uptake, and delaying the cell growth (Dwyer et al., 2014, 2015). Metabolites from the tricarboxylic acid cycle (TCA cycle) (Peng et al., 2015), NAD metabolism (Groth et al., 2021; Tan & Doig, 2021), amino acid metabolism (Lee et al., 2021), and nucleotide metabolism (Stokes et al., 2019; Xi et al., 2000) usually show significant changes in antibiotic-induced metabolic alteration. Additionally, changes in metabolites such as NAD, glutamine, and cyclic GMP due to the dysregulated purine metabolism and pyrimidine metabolism are also observed in antibiotic-induced metabolic alterations (Schelli et al., 2017).

The sources of nitrogen, carbon, and energy are the amino acids, glutamic acid, glutamine, and arginine (and its derivatives) are the metabolites that affect the biofilm formation through various mechanisms. Biofilm dispersion is stimulated by glutamate through the production of matrix-degrading enzymes. Arginine metabolism plays a key role in biofilm survival. Glutamic acid, L-ornithine, L-proline, and 4-aminobutanoic acid, the intermediates/derivatives of arginine metabolism, are significant markers to assess the biofilm response and survival to environmental factors. The nutrient supply within the biofilm is affected by the resistance to osmolytes, proline, glycine, and glutamate metabolism (Burg & Ferraris, 2008; Shen et al., 2020).

The carbon metabolism intermediates such as succinate, pyruvate, and fructose-6-p, some amino acids such as alanine, and several nucleotides are also dysregulated in antibiotic-induced metabolic alterations. Carbon flux changes in a cell impacts antibiotic susceptibility in multiple bacterial species. Alteration in the energy metabolism modulates antibiotic sensitivity both in vitro and in vivo. This shows that characterization of the antibiotic-induced metabolic changes would address the existing hitch of antibiotic resistance. In *S. aureus*, bactericidal antibiotics are reported to induce metabolic changes in an antibiotic sensitivity–dependent manner (Schelli et al., 2017; Shen et al., 2020).

TABLE 6.2 Microbiome distribution and its correlation to metabolome.

Phylum	Skin flora	Metabolome
Firmicutes	*Anaerococcus octavius* *Clostridium sphenoides* *Finegoldia magna* *Gemella haemolysans* *Gemella sanguinis* *Granulicatella adiacens* *Granulicatella elegans* *Lactobacillus rhamnosus* *Oribacterium sinus* *Parageobacillus caldoxylosilyticus* *Peptoniphilus grossensis* *Peptoniphilus* sp. *Ruminococcus gnavus* *Ruminococcus* sp. *Staphylococcus capitis* *Staphylococcus epidermidis* *Staphylococcus haemolyticus* *Staphylococcus hominis* *Staphylococcus warneri* *Streptococcus equinus* *Streptococcus gordonii* *Streptococcus mitis* *Streptococcus parasanguinis* *Streptococcus pneumoniae* *Streptococcus salivarius* *Streptococcus sanguinis* *Streptococcus thermophilus* *Veillonella atypica* *Veillonella rodentium* *Veillonella* sp. *Veillonella tobetsuensis*	Amino acids, lipids, and xenobiotics
Actinobacteria	*Actinomyces* sp. *Atopobium parvulum* *Corynebacterium diphtheriae* *Cutibacterium acnes* *Dermacoccus nishinomiyaensis* *Paenarthrobacter nicotinovorans* *Propionibacterium humerusii* *Trueperella pyogenes*	Xenobiotics, fatty acids, cofactors, and vitamins
Proteobacteria	*Acinetobacter baumannii* *Agrobacterium tumefaciens* *Brevundimonas diminuta* *Escherichia coli* *Helicobacter ganmani* *Moraxella osloensis* *Pseudomonas asiatica* *Salinivibrio kushneri*	Amino acids, lipids, and xenobiotics

(*Continued*)

TABLE 6.2 Microbiome distribution and its correlation to metabolome.—cont'd

Phylum	Skin flora	Metabolome
Bacteroidetes	*Bergeyella cardium* *Porphyromonas pasteri* *Prevotella histicola* *Prevotella melaninogenica* Prevotellaceae bacterium	Carbohydrates (glycoconjugates and glycosaminoglycans such as hyaluronic acid, cellulose, mucins)
Fusobacteria	*Fusobacterium massiliense* *Fusobacterium nucleatum* *Fusobacterium periodonticum* *Leptotrichia buccalis* *Leptotrichia goodfellowii* *Leptotrichia wadei*	Fatty acids, xenobiotics

The mechanism of the interactions of the microbial communities in a multispecies biofilms are

(i) Gene transfer (plasmid conjugation), from one species to another
(ii) Quorum sensing through interspecific and intraspecific communication by molecules
(iii) Metabolic cooperativity, the metabolite of one species serves as nutrient for another species
(iv) Cell-to-cell interactions between cells of different species through specific cell surface receptors (Burmølle et al., 2014)

These mechanisms are vital to understand the therapeutic strategies for various diseases associated with microbiome-induced metabolic alterations.

4. Natural microbiota of skin biofilms

The skin surface microbiomes plays a major role in skin health (Roux et al., 2022). Irrespective of the age group, the microbiota is present naturally in all age groups, from the infant stage to the adults. Table 6.3 shows the skin flora at different age groups.

4.1 *Staphylococcus epidermidis*

S. epidermidis, the most commonly found aerobic resident of skin, is often considered as skin contaminant during infection (Grice & Segre, 2011). *S. epidermidis* is a normal flora of skin that helps to maintain an unharmed relationship with its host. It prevents the growth and proliferation of pathogens (Otto, 2009). *S. epidermidis* is vital for maintaining the balance between the skin microflora. *S. epidermidis* biofilms are innocuous or, rarely opportunistic, exhibit antibiotic resistance and responsible for hospital-acquired infections (Brescó et al., 2017). *S. epidermidis* is suspected and implicated in enhancing the development of methicillin-resistant *S. aureus* (MRSA) and hence not just considered as a contaminant of infections (Lee et al., 2018).

TABLE 6.3 Distribution of skin flora of different age groups.

Age group	Criteria	Skin microflora	Significant parameter
Infant	Cesarean delivery	*Streptococcus, Staphylococcus, Propionibacterium*	Normal flora
	Normal vaginal delivery	*Lactobacillus, Prevotella, Candida*	Normal flora of female urogenital tract
Preteen/ teen	Puberty	*Corynebacterium* and *Propionibacterium*	Lipophilic organism in response to sebaceous gland secretion triggered by hormones
Adult	Acidic skin and hydrated skin	*Atopobium, Lactobacillus, Pseudomonas, Ruminococcus, and Schaalia*	Influenced by pH and moisture content
	Basic and slightly dry skin	*Cutibacterium was more abundant*	
	Slightly acid and slightly dry skin	*Moraxella, Agrobacterium, and Acinetobacter*	

4.2 *Staphylococcus aureus*

About 35%–60% of the population carry *Staphylococcus aureus. Though S. aureus* is commonly found in nasal flora, high prevalence rate is reported in diabetics, drug users, and immunocompromised individuals (Abu-Ashour et al., 2018; Quagliarello et al., 2002; Sasson et al., 2017).

S. aureus causes skin infections such as folliculitis, impetigo, furuncles, and subcutaneous abscesses and, also seen in toxic shock syndrome, pneumonia, osteomyelitis, and endocarditis (Tong et al., 2015).

The biofilm formation is rapid and develops within 8 hours in open wounds due to lack of cutaneous protection. Wound healing is delayed in biofilms of *S. aureus* due to the delayed re-epithelialization and inhibiting the development of granulation tissue. Around 6 percent of acute wounds and 60% of chronic wounds are found to be due to the biofilms (Malone et al., 2017). The organisms, *Pseudomonas aeruginosa, S. aureus,* and *Enterococcus* are involved in the biofilm formation on chronic or acute wounds (Attinger & Wolcott, 2012; Rajkumari et al., 2014; Tong et al., 2015). In diabetic wounds, the increased level of glucose contributes to the thickest biofilm.

S. aureus, one of the major Gram-positive opportunistic pathogens, is associated with nosocomial infections and food-borne illnesses. The most common sites are the skin (superficial layers) and the subcutaneous soft tissue that allows *S. aureus* to enter the human host and cause infection of which 75% is due to MRSA (Lee et al., 2018). The challenges in the control of *S. aureus* are due to its biofilm formation and the emergence and MRSA strains. There is alteration in the amino acid metabolic pathway when the biofilms are subjected to ampicillin, kanamycin, and norfloxacin (Peng et al., 2015). No significant changes were observed in arginine biosynthesis and aminoacyl-tRNA biosynthesis when subjected to conventional antibiotics. *S. aureus* takes up proline, glycine, and glutamic acid as osmoprotectants against antibiotic stress (Burg & Ferraris, 2008; Shen et al., 2020).

The comparative metabolomic analyses on MSSA and MRSA on different classes of antibiotics, β-lactams, aminoglycosides, and quinolones induce metabolic shift at different levels (Schelli et al., 2017). Important biochemical pathways such as nucleotides and amino acid metabolisms were among the most significantly altered metabolic pathways and in a dose-dependent manner (Liang et al., 2021). *S. aureus* is also a profuse biofilm producer that is facilitated through the expression of intracellular adhesion molecules followed by infection that poses a threat to both health and morbidity.

S. aureus produces a thick multilayered biofilm embedded within a glycocalyx, primarily composed of 80% teichoic acids. Besides teichoic acids, it is composed of staphylococcal and host proteins. A polysaccharide antigen named polysaccharide intercellular antigen (PIA) was also isolated from *S. aureus*. PIA constitutes 80%–85% β-1,6-linked *N*-acetylglucosamine residues and 15–20% of non-*N*-acetylated D-glucosaminyl residues with phosphate and ester-linked succinate (Archer et al., 2011).

4.3 *Staphylococcus lugdunensis*

Staphylococcus lugdunensis infections resemble those caused by *S. aureus* and are associated with skin abscesses, and wound infections (Parthasarathy et al., 2020), urinary tract infection (Chiu et al., 2020), and infection of intravascular catheters as well as other implanted medical devices (Frank et al., 2008). *S. lugdunensis* infections are largely due to them with the host cells to form biofilms. According to Tena et al. (2014), it is to be considered as a potential pathogen when isolated from patients with soft tissue and skin infection or after trauma or surgery. Though biofilms are known to be involved in the pathogenesis of many *S. lugdunensis* infections, studies are limited.

4.4 *Cutibacterium acnes*

The most commonly found skin-colonizing bacterium is *C. acnes*. It is predominantly found in facial skin and also can be found in almost all over the body. Acne vulgaris (pimples) is a chronic skin condition caused by *C. acnes* that affects almost more than half of adolescents. Acne vulgaris can also be a genetic disease that involves the inflammation of the sebaceous follicles (Mayslich et al., 2021).

C. acnes metabolizes free fatty acids within the sebaceous gland that leads to initiation and results in inflammation (Sanford et al., 2019). Infections due to use of foreign device are also associated with *C. acnes*.

The coexistence of *C. acnes* and *S. aureus* in many human diseases, that include acne lesions (Alkhawaja et al., 2020), infections due to implants (El-Mahdy et al., 2021), and sepsis (Lo et al., 2011; Abbott et al., 2022), is alarming. *C. acnes* forms a biofilm within the hair follicles of the skin. Follicular plugging and cohesiveness are the key mechanisms involved in the pathogenesis caused by *C. acnes* biofilm (Ellabib et al., 2002; Jahns et al., 2012).

In *C. acnes* biofilm, enzymes such as lipases, hyaluronidases, and chemotactic factors are secreted that increases the free fatty acid concentration, in turn provides nutrition for the bacterium, and enhances inflammation. The TLR2 and TLR4 receptors are activated, which in turn activates NF-κB, leading to proinflammatory response in acne (Bhat et al., 2017; Jappe et al., 2002).

4.5 *Acinetobacter* sp.

Acinetobacter baumannii is an aerobic, Gram-negative coccobacillus that causes skin infections particularly in wounds and burns. *A. baumannii* is associated with nosocomial and community-acquired infections (Antunes et al., 2014). They are associated with endocarditis (Ioannou et al., 2021), septicemia (Naghipour Erami et al., 2021), and respiratory tract infection (in particular ventilator-associated pneumonia) (Khalid et al., 2020; Lima et al., 2020) and meningitis (Cascio et al., 2010). The colonization of *Acinetobacter* was found in wounded military personnel. Though *Acinetobacter* is generally known as non-pathogenic, their colonization in immunocompromised individuals can be life-threatening. Nonhealing of wounds in *A. baumannii* infections is due to the biofilm formation that helps the organisms to resist immune and natural defenses of the host and antimicrobial interventions (Percival et al., 2015).

4.6 *Pseudomonas* sp.

P aeruginosa is a Gram-negative opportunistic pathogen found in human skin. This species survives on moist surfaces due to its affinity toward water. *P. aeruginosa* is implicated in cross-infection during the use of medical devices. They enter into the body through the colonized skin surfaces. The infections caused by *Pseudomonas* are opportunistic and are primarily associated with immunosuppression and immunocompromised patients.

P. aeruginosa causes dermatitis due to contaminated water while sharing hot tubs though less severe episodes are reported (Zichichi et al., 2000). *Pseudomonas* species forms good biofilms, and due to their antibacterial resistance the treatment becomes complicated (Masák et al., 2014).

4.7 *Corynebacterium* sp.

Corynebacteria, a group of bacteria, is known to colonize the animal and human skin. *Corynebacterium jeikeium*, an important opportunistic pathogen and the most prevalent species is often present as a part of the normal skin flora. The colonization rate is high in the skin of hospitalized patients. The clinical conditions attributed to *C. jeikeium* include skin and wound infections in adults, and a localized skin abscess in children (Percival et al., 2012).

5. Therapeutic strategies to combat biofilm formation

Research studies with the use of antibiotics and other compounds for biofilm disintegration via the disruption of the protective EPS matrix seem to be increasing in the past decades. The combination therapies with the enzymes or matrix-degrading agent or essential oils, natural products or nanoparticles or Quorum sensing (QS) inhibitors or inhibitors of metabolites and cellular signaling pathways along with the antimicrobial agent are the strategies used in the treatment to control biofilm formation (Table 6.4).

TABLE 6.4 Treatment approaches that affects the microbial biofilm.

Therapeutic strategies	Microbial biofilm
Alginase	*Pseudomonas aeruginosa* biofilm in cystic fibrosis
DNase1 and plant-based essential oils	Methicillin-resistant *Staphylococcus aureus*
Dispersin B and peptides	*Staphylococcus epidermidis* biofilm
Yeast *Saccharomyces cerevisiae*	*Listeria monocytogenes* biofilms
Honey	*P. aeruginosa* biofilm
Proteinase K-capped gold NPs	*Pseudomonas Fluorescens*
Iron chelators	Gram-positive and Gram-negative organisms
Ultrasonication ozone	*Listeria monocytogenes*
Photodynamic therapy, 5-aminolevulinic acid	*S. aureus and S. epidermidis*

6. Future of metabolomics in cosmetics and skin care

Skin biologists and cosmeceutical scientists opine that metabolomics shall be the major trend in cosmetics. It is important to furnish the skin ecosystem with the correct metabolites to ensure that the commensal microorganisms living on the skin are given adequate care. Major companies in the beauty care market such as L'Oreal, Johnson & Johnson Estee Lauder, L'Occitane, and P&G have taken a U-turn in their approach to skin microbiome. The new trend is to focus more on synergistic micronutrients to upkeep the beneficial skin bacteria than using the conventional preservatives that harm the skin flora. Research is already on to investigate the role of skin microbiome by providing the necessary metabolite, thereby regulating the skin conditions such as acne, dandruff, prickly heat, psoriasis, skin aging, and body odor.

DSM, an important ingredient maker, had taken an approach of formulating ingredients that "*interact with bacteria on the skin.*" DSM's research zeroed in on three prominent microorganisms living on human skin: *C. acnes, S. epidermidis, Corynebacterium kroppenstedtii* for its metabolome-based approach. Mibelle Biochemistry has introduced a new skin care ingredient called Black BeeOme. It is a nature-derived prebiotic made from the honey of bees residing in remote Swiss mountain valleys. This honey is unique and is obtained by the process of fermentation with a bacterium called *Zymomonas mobilis*. Black BeeOme and other such prebiotics are tested and shown to have a beneficial effect on several skin microbes. Future of the skin and beauty care market relies mainly on the do-good ingredients that promote skin health through symbiotic relationships with skin bacteria. This would be made possible by metabolomic research on skin bacteria.

In summary, the metabolic pathways and associated metabolites, of skin biofilm can be regarded as novel targets for the development of biofilm-based treatments and antibiotic discovery against the skin infection. The formation of biofilms is a major concern towards the efficacy of antimicrobial agents. It poses to be very vital in determining the clinical course of treatment for infections. This could provide an innovative insight into better understanding and elucidation on the mechanism and significance of biofilm formation caused by different microbial strains at metabolic level.

References

Abbott, C., Grout, E., Morris, T., & Brown, H. L. (2022). Cutibacterium acnes biofilm forming clinical isolates modify the formation and structure of Staphylococcus aureus biofilms, increasing their susceptibility to antibiotics. *Anaerobe, 76*, 102580. https://doi.org/10.1016/j.anaerobe.2022.102580.

Abu-Ashour, W., Twells, L. K., Valcour, J. E., & Gamble, J. M. (2018). Diabetes and the occurrence of infection in primary care: A matched cohort study. *BMC Infectious Diseases, 18*(1). https://doi.org/10.1186/S12879-018-2975-2

Ahola-Olli, A. V., Mustelin, L., Kalimeri, M., Kettunen, J., Jokelainen, J., Auvinen, J., Puukka, K., Havulinna, A. S., Lehtimäki, T., Kähönen, M., Juonala, M., Keinänen-Kiukaanniemi, S., Salomaa, V., Perola, M., Järvelin, M. R., Ala-Korpela, M., Raitakari, O., & Würtz, P. (2019). Circulating metabolites and the risk of type 2 diabetes: a prospective study of 11,896 young adults from four Finnish cohorts. *Diabetologia, 62*(12), 2298–2309. https://doi.org/10.1007/s00125-019-05001-w.

Akyıldız, İ., Take, G., Uygur, K., Kızıl, Y., & Aydil, U. (2013). Bacterial biofilm formation in the middle-ear mucosa of chronic otitis media patients. *Indian Journal of Otolaryngology and Head and Neck Surgery, 65*. https://doi.org/10.1007/s12070-012-0513-x

Alkhawaja, E., Hammadi, S., Abdelmalek, M., Mahasneh, N., Alkhawaja, B., & Abdelmalek, S. M. (2020). Antibiotic resistant Cutibacterium acnes among acne patients in Jordan: a cross sectional study. *BMC dermatology, 20*(1), 17. https://doi.org/10.1186/s12895-020-00108-9

Allen, H. B., Vaze, N. D., Choi, C., Hailu, T., Tulbert, B. H., Cusack, C. A., & Joshi, S. G. (2014). The presence and impact of biofilm-producing staphylococci in atopic dermatitis. *JAMA Dermatology, 150*(3), 260–265. https://doi.org/10.1001/JAMADERMATOL.2013.8627

Antunes, L. C., Visca, P., & Towner, K. J. (2014). Acinetobacter baumannii: evolution of a global pathogen. *Pathogens and disease, 71*(3), 292–301. https://doi.org/10.1111/2049-632X.12125

Archer, N. K., Mazaitis, M. J., William Costerton, J., Leid, J. G., Powers, M. E., & Shirtliff, M. E. (2011). *Staphylococcus aureus* biofilms: Properties, regulation and roles in human disease. *Virulence, 2*(5). https://doi.org/10.4161/viru.2.5.17724

Attinger, C., & Wolcott, R. (2012). Clinically addressing biofilm in chronic wounds. *Advances in Wound Care, 1*(3), 127. https://doi.org/10.1089/WOUND.2011.0333

Belkaid, Y., & Segre, J. A. (2014). Dialogue between skin microbiota and immunity. *Science, 346*(6212). https://doi.org/10.1126/science.1260144

Bhat, Y. J., Latief, I., & Hassan, I. (2017). Update on etiopathogenesis and treatment of Acne. *Indian Journal of Dermatology, Venereology and Leprology, 83*(3), 298–306. https://doi.org/10.4103/0378-6323.199581

Brandwein, M., Steinberg, D., & Meshner, S. (2016). Microbial biofilms and the human skin microbiome. *Npj Biofilms and Microbiomes, 2*(1). https://doi.org/10.1038/s41522-016-0004-z

Brescó, M. S., Harris, L. G., Thompson, K., Stanic, B., Morgenstern, M., O'Mahony, L., Richards, R. G., & Moriarty, T. F. (2017). Pathogenic mechanisms and host interactions in Staphylococcus epidermidis device-related infection. *Frontiers in Microbiology, 8*, 1401. https://doi.org/10.3389/FMICB.2017.01401/BIBTEX

Brown, M. M., & Horswill, A. R. (2020). Staphylococcus epidermidis-Skin friend or foe? *PLoS Pathogens, 16*(11). https://doi.org/10.1371/JOURNAL.PPAT.1009026

Burg, M. B., & Ferraris, J. D. (2008). Intracellular organic osmolytes: Function and regulation. *The Journal of Biological Chemistry, 283*(12), 7309–7313. https://doi.org/10.1074/JBC.R700042200

Burmølle, M., Ren, D., Bjarnsholt, T., & Sørensen, S. J. (2014). Interactions in multispecies biofilms: Do they actually matter? *Trends in Microbiology, 22*(2), 84–91. https://doi.org/10.1016/J.TIM.2013.12.004

Byrd, A. L., Belkaid, Y., & Segre, J. A. (2018). The human skin microbiome. *Nature reviews. Microbiology, 16*(3), 143–155. https://doi.org/10.1038/nrmicro.2017.157.

Carmona-Cruz, S., Orozco-Covarrubias, L., & Sáez-de-Ocariz, M. (2022). The Human Skin Microbiome in Selected Cutaneous Diseases. *Frontiers in cellular and infection microbiology, 12*, 834135. https://doi.org/10.3389/fcimb.2022.834135.

Cascio, A., Conti, A., Sinardi, L., Iaria, C., Angileri, F. F., Stassi, G., David, T., Versaci, A., Iaria, M., & David, A. (2010). Post-neurosurgical multidrug-resistant Acinetobacter baumannii meningitis successfully treated with intrathecal colistin. a new case and a systematic review of the literature. In , *14*. International journal of infectious diseases (pp. e572–e579). IJID : official publication of the International Society for Infectious Diseases. https://doi.org/10.1016/j.ijid.2009.06.032.

Chiu, K., Lam, R., Chan, E., Lau, S., & Woo, P. (2020). Emergence of *Staphylococcus lugdunensis* as a Cause of Urinary Tract Infection: Results of the Routine Use of MALDI-TOF MS. *Microorganisms, 8*(3), 381. https://doi.org/10.3390/microorganisms8030381.

Christensen, G. J., Scholz, C. F., Enghild, J., Rohde, H., Kilian, M., Thürmer, A., Brzuszkiewicz, E., Lomholt, H. B., & Brüggemann, H. (2016). Antagonism between Staphylococcus epidermidis and Propionibacterium acnes and its genomic basis. *BMC genomics, 17*, 152. https://doi.org/10.1186/s12864-016-2489-5.

Claudel, J. P., Auffret, N., Leccia, M. T., Poli, F., Corvec, S., & Dréno, B. (2019). Staphylococcus epidermidis: a potential new player in the physiopathology of acne? *Dermatology (Basel, Switzerland), 235*(4), 287–294. https://doi.org/10.1159/000499858.

Clish, C. B. (2015). Metabolomics: An emerging but powerful tool for precision medicine. *Cold Spring Harbor Molecular Case Studies, 1*(1), a000588. https://doi.org/10.1101/MCS.A000588

Coenye, T., Spittaels, K. J., & Achermann, Y. (2021). The role of biofilm formation in the pathogenesis and antimicrobial susceptibility of *Cutibacteriumacnes*. *Biofilm, 4*, 100063. https://doi.org/10.1016/j.bioflm.2021.100063.

Dhir, S. (2013). Biofilm and dental implant: The microbial link. *Journal of Indian Society of Periodontology, 17*(1). https://doi.org/10.4103/0972-124X.107466

Dosler, S., Hacioglu, M., Yilmaz, F. N., & Oyardi, O. (2020). Biofilm modelling on the contact lenses and comparison of the in vitro activities of multipurpose lens solutions and antibiotics. *PeerJ, 2020*(6). https://doi.org/10.7717/peerj.9419

Dréno, B., Pécastaings, S., Corvec, S., Veraldi, S., Khammari, A., & Roques, C. (2018). Cutibacterium acnes (Propionibacterium acnes) and acne vulgaris: a brief look at the latest updates. *Journal of the European Academy of Dermatology and Venereology : JEADV, 32*(Suppl 2), 5–14. https://doi.org/10.1111/jdv.15043.

Dwivedi, P., Thompson, A., Xie, Z., Kashleva, H., Ganguly, S., Mitchell, A. P., & Dongari-Bagtzoglou, A. (2011). Role of Bcr1-activated genes Hwp1 and Hyr1 in Candida albicans oral mucosal biofilms and neutrophil evasion. *PloS One, 6*(1). https://doi.org/10.1371/JOURNAL.PONE.0016218

Dwyer, D. J., Belenky, P. A., Yang, J. H., Cody MacDonald, I., Martell, J. D., Takahashi, N., Chan, C. T. Y., Lobritz, M. A., Braff, D., Schwarz, E. G., Ye, J. D., Pati, M., Vercruysse, M., Ralifo, P. S., Allison, K. R., Khalil, A. S., Ting, A. Y., Walker, G. C., & Collins, J. J. (2014). Antibiotics induce redox-related physi. *Proceedings of the National Academy of Sciences of the United States of America, 111*(20). https://doi.org/10.1073/PNAS.1401876111/-/DCSUPPLEMENTAL/PNAS.1401876111.SAPP.PDF

Dwyer, D. J., Collins, J. J., & Walker, G. C. (2015). Unraveling the physiological complexities of antibiotic lethality. *Annual Review of Pharmacology and Toxicology, 55*, 313–332. https://doi.org/10.1146/ANNUREV-PHARMTOX-010814-124712

Ellabib, M. S., Khalifa, Z., & Kavanagh, K. (2002). Dermatophytes and other fungi associated with skin mycoses in Tripoli, Libya. *Mycoses, 45*(3–4), 101–104. https://doi.org/10.1046/J.1439-0507.2002.00731.X

El-Mahdy, T. S., Mongaret, C., Varin-Simon, J., Lamret, F., Vernet-Garnier, V., Rammal, H., Mauprivez, C., Kerdjoudj, H., Gangloff, S. C., & Reffuveille, F. (2021). Interaction of implant infection-related commensal bacteria with mesenchymal stem cells: a comparison between Cutibacterium acnes and Staphylococcus aureus. *FEMS microbiology letters, 368*(4), fnab014. https://doi.org/10.1093/femsle/fnab014.

Flowers, L., & Grice, E. A. (2020). The skin microbiota: balancing risk and reward. *Cell host & microbe, 28*(2), 190–200. https://doi.org/10.1016/j.chom.2020.06.017.

Fournière, M., Latire, T., Souak, D., Feuilloley, M. G. J., & Bedoux, G. (2020). Staphylococcus epidermidis and Cutibacterium acnes: Two major sentinels of skin microbiota and the influence of cosmetics. *Microorganisms, 8*(11), 1–31. https://doi.org/10.3390/MICROORGANISMS8111752

Frank, K. L., Del Pozo, & Patel, R. (2008). From clinical microbiology to infection pathogenesis: how daring to be different works for Staphylococcus lugdunensis. *Clinical microbiology reviews, 21*(1), 111–133. https://doi.org/10.1128/CMR.00036-07.

Gannesen, A. V., Zdorovenko, E. L., Botchkova, E. A., Hardouin, J., Massier, S., Kopitsyn, D. S., Gorbachevskii, M. V., Kadykova, A. A., Shashkov, A. S., Zhurina, M. V., Netrusov, A. I., Knirel, Y. A., Plakunov, V. K., & Feuilloley, M.

(2019). Composition of the biofilm matrix of *Cutibacterium acnes* acneic strain RT5. *Frontiers in microbiology, 10*, 1284. https://doi.org/10.3389/fmicb.2019.01284.

Geoghegan, J. A., Irvine, A. D., & Foster, T. J. (2018). Staphylococcus aureus and atopic dermatitis: a complex and evolving relationship. *Trends in microbiology, 26*(6), 484–497. https://doi.org/10.1016/j.tim.2017.11.008.

Gominet, M., Compain, F., Beloin, C., & Lebeaux, D. (2017). Central venous catheters and biofilms: Where do we stand in 2017? *APMIS, 125*(4). https://doi.org/10.1111/apm.12665

Gopinathan, A., Mukhopadhyay, C., & K. E., V. (2017). Characterization of antimicrobial resistance mechanisms of multidrug resistant Gram negative bacterial wound infections and their clinical epidemiology from a tertiary care hospital in Karnataka, India. *International Journal of Research in Medical Sciences, 5*(3). https://doi.org/10.18203/2320-6012.ijrms20170535

Grice, E. A., & Segre, J. A. (2011). The skin microbiome. *Nature Reviews. Microbiology, 9*(4), 244. https://doi.org/10.1038/NRMICRO2537

Groth, B., Venkatakrishnan, P., & Lin, S. J. (2021). NAD+ metabolism, metabolic stress, and infection. *Frontiers in Molecular Biosciences, 8*, 446. https://doi.org/10.3389/FMOLB.2021.686412/BIBTEX

Gu, X., al Dubayee, M., Alshahrani, A., Masood, A., Benabdelkamel, H., Zahra, M., Li, L., Abdel Rahman, A. M., & Aljada, A. (2020). Distinctive metabolomics patterns associated with insulin resistance and type 2 diabetes mellitus. *Frontiers in Molecular Biosciences, 7*, 411. https://doi.org/10.3389/FMOLB.2020.609806/BIBTEX

Guzmán-Soto, I., McTiernan, C., Gonzalez-Gomez, M., Ross, A., Gupta, K., Suuronen, E. J., Mah, T. F., Griffith, M., & Alarcon, E. I. (2021). Mimicking biofilm formation and development: recent progress in *in vitro* and *in vivo* biofilm models. *iScience, 24*(5), 102443. https://doi.org/10.1016/j.isci.2021.102443.

Hall, C. W., & Mah, T. F. (2017). Molecular mechanisms of biofilm-based antibiotic resistance and tolerance in pathogenic bacteria. *FEMS Microbiology Reviews, 41*(3), 276–301. https://doi.org/10.1093/femsre/fux010

Ioannou, P., Mavrikaki, V., & Kofteridis, D. P. (2021). Infective endocarditis by *Acinetobacter* species: a systematic review. *Journal of chemotherapy (Florence, Italy), 33*(4), 203–215. https://doi.org/10.1080/1120009X.2020.1812804.

Jahns, A. C., Lundskog, B., Ganceviciene, R., Palmer, R. H., Golovleva, I., Zouboulis, C. C., McDowell, A., Patrick, S., & Alexeyev, O. A. (2012). An increased incidence of Propionibacterium acnes biofilms in acne vulgaris: A case-control study. *The British Journal of Dermatology, 167*(1), 50–58. https://doi.org/10.1111/J.1365-2133.2012.10897.X

Jappe, U., Ingham, E., Henwood, J., & Holland, K. T. (2002). Propionibacterium acnes and inflammation in acne; P. acnes has T-cell mitogenic activity. *The British Journal of Dermatology, 146*(2), 202–209. https://doi.org/10.1046/J.1365-2133.2002.04602.X

Jiang, Y., Zhu, Z., Shi, J., An, Y., Zhang, K., Wang, Y., Li, S., Jin, L., Ye, W., Cui, M., & Chen, X. (2019). Metabolomics in the development and progression of dementia: A systematic review. *Frontiers in Neuroscience, 13*. https://doi.org/10.3389/FNINS.2019.00343/FULL

Jordana-Lluch, E., Garcia, V., Kingdon, A., Singh, N., Alexander, C., Williams, P., & Hardie, K. R. (2020). A simple polymicrobial biofilm keratinocyte colonization model for exploring interactions between commensals, pathogens and antimicrobials. *Frontiers in microbiology, 11*, 291. https://doi.org/10.3389/fmicb.2020.00291.

Jung, M. Y., Chung, J. Y., Lee, H. Y., Park, J., Lee, D. Y., & Yang, J. M. (2015). Antibiotic susceptibility of *Staphylococcus aureus* in atopic dermatitis: Current prevalence of methicillin-resistant *Staphylococcus aureus* in Korea and treatment strategies. *Annals of Dermatology, 27*(4), 398. https://doi.org/10.5021/AD.2015.27.4.398

Kaushik, A. K., & DeBerardinis, R. J. (2018). Applications of metabolomics to study cancer metabolism. *Biochimica et Biophysica Acta (BBA) - Reviews on Cancer, 1870*(1), 2–14. https://doi.org/10.1016/J.BBCAN.2018.04.009

Khalid, F., Saleem, S., & Ahmad, I. (2020). High prevalence of carbapenem-resistant Acinetobacter baumannii associated respiratory tract infections in Pakistani hospitals. *JPMA. The Journal of the Pakistan Medical Association, 70*(9), 1630–1632. https://doi.org/10.5455/JPMA.35384.

Koeth, R. A., Wang, Z., Levison, B. S., Buffa, J. A., Org, E., Sheehy, B. T., Britt, E. B., Fu, X., Wu, Y., Li, L., Smith, J. D., Didonato, J. A., Chen, J., Li, H., Wu, G. D., Lewis, J. D., Warrier, M., Brown, J. M., Krauss, R. M., … Hazen, S. L. (2013). Intestinal microbiota metabolism of L-carnitine, a nutrient in red meat, promotes atherosclerosis. *Nature Medicine, 19*(5), 576. https://doi.org/10.1038/NM.3145

Kolar, M., Cermak, P., Hobzova, L., Bogdanova, K., Neradova, K., Mlynarcik, P., & Bostik, P. (2020). Antibiotic resistance in nosocomial bacteria isolated from infected wounds of hospitalized patients in Czech republic. *Antibiotics, 9*(6). https://doi.org/10.3390/antibiotics9060342

Koseki, H., Yonekura, A., Shida, T., Yoda, I., Horiuchi, H., Morinaga, Y., Yanagihara, K., Sakoda, H., Osaki, M., & Tomita, M. (2014). Early staphylococcal biofilm formation on solid orthopaedic implant materials: In vitro study. *PLoS ONE, 9*(10). https://doi.org/10.1371/journal.pone.0107588

Kranjec, C., Morales Angeles, D., Torrissen Mårli, M., Fernández, L., García, P., Kjos, M., & Diep, D. B. (2021). Staphylococcal biofilms: challenges and novel therapeutic perspectives. *Antibiotics (Basel, Switzerland), 10*(2), 131. https://doi.org/10.3390/antibiotics10020131.

Kravvas, G., Veitch, D., & Al-Niaimi, F. (2018). The increasing relevance of biofilms in common dermatological conditions. *Journal of Dermatological Treatment, 29*(2). https://doi.org/10.1080/09546634.2017.1360989

Lauten, A., Martinović, M., Kursawe, L., Kikhney, J., Affeld, K., Kertzscher, U., Falk, V., & Moter, A. (2021). Bacterial biofilms in infective endocarditis: an in vitro model to investigate emerging technologies of antimicrobial cardiovascular device coatings. *Clinical research in cardiology : official journal of the German Cardiac Society, 110*(3), 323–331. https://doi.org/10.1007/s00392-020-01669-y.

Lee, A. S., de Lencastre, H., Garau, J., Kluytmans, J., Malhotra-Kumar, S., Peschel, A., & Harbarth, S. (2018). Methicillin-resistant *Staphylococcus aureus*. *Nature Reviews Disease Primers, 4*(1), 1–23. https://doi.org/10.1038/nrdp.2018.33

Lee, J., Vijayakumar, A., White, P. J., Xu, Y., Ilkayeva, O., Lynch, C. J., Newgard, C. B., & Kahn, B. B. (2021). BCAA supplementation in mice with diet-induced obesity alters the metabolome without impairing glucose homeostasis. *Endocrinology (United States), 162*(7). https://doi.org/10.1210/endocr/bqab062

Lerche, C. J., Schwartz, F., Theut, M., Fosbøl, E. L., Iversen, K., Bundgaard, H., Høiby, N., & Moser, C. (2021). Antibiofilm approach in infective endocarditis exposes new treatment strategies for improved outcome. *Frontiers in Cell and Developmental Biology, 9*. https://doi.org/10.3389/fcell.2021.643335

Liang, C., Rios-Miguel, A. B., Jarick, M., Neurgaonkar, P., Girard, M., François, P., Schrenzel, J., Ibrahim, E. S., Ohlsen, K., & Dandekar, T. (2021). *Staphylococcusaureus* transcriptome data and metabolic modelling investigate the interplay of Ser/Thr Kinase PknB, lts phosphatase Stp, the *glmR/yvcK* regulon and the *cdaA* operon for metabolic adaptation. *Microorganisms, 9*(10), 2148. https://doi.org/10.3390/microorganisms9102148.

Lima, W. G., Brito, J., & da Cruz Nizer, W. S. (2020). Ventilator-associated pneumonia (VAP) caused by carbapenem-resistant Acinetobacter baumannii in patients with COVID-19: two problems, one solution? *Medical hypotheses, 144*, 110139. https://doi.org/10.1016/j.mehy.2020.110139.

Lister, J. L., & Horswill, A. R. (2014). *Staphylococcus aureus* biofilms: Recent developments in biofilm dispersal. *Frontiers in Cellular and Infection Microbiology, 4*. https://doi.org/10.3389/fcimb.2014.00178

Lo, C. W., Lai, Y. K., Liu, Y. T., Gallo, R. L., & Huang, C. M. (2011). *Staphylococcus aureus* hijacks a skin commensal to intensify its virulence: Immunization targeting β-hemolysin and CAMP factor. *The Journal of Investigative Dermatology, 131*(2), 401. https://doi.org/10.1038/JID.2010.319

Long, J., Yang, Z., Wang, L., Han, Y., Peng, C., Yan, C., & Yan, D. (2020). Metabolite biomarkers of type 2 diabetes mellitus and pre-diabetes: A systematic review and meta-analysis. *BMC Endocrine Disorders, 20*(1), 1–17. https://doi.org/10.1186/S12902-020-00653-X/FIGURES/5

Malone, M., Bjarnsholt, T., McBain, A. J., James, G. A., Stoodley, P., Leaper, D., Tachi, M., Schultz, G., Swanson, T., & Wolcott, R. D. (2017). The prevalence of biofilms in chronic wounds: A systematic review and meta-analysis of published data. *Journal of Wound Care, 26*(1), 20–25. https://doi.org/10.12968/JOWC.2017.26.1.20

Masák, J., Čejková, A., Schreiberová, O., & Řezanka, T. (2014). Pseudomonas biofilms: Possibilities of their control. *FEMS Microbiology Ecology, 89*(1), 1–14. https://doi.org/10.1111/1574-6941.12344

Mayslich C., Grange P.A., Dupin N. (2021). *Cutibacterium acnes* as an opportunistic pathogen: an update of its virulence-associated factors. *Microorganisms. 9(2),* 303. doi: 10.3390/microorganisms9020303. PMID: 33540667; PMCID: PMC7913060.

McLaughlin, J., Watterson, S., Layton, A. M., Bjourson, A. J., Barnard, E., & McDowell, A. (2019). *Propionibacterium acnes* and acne vulgaris: new insights from the integration of population genetic, multi-omic, biochemical and host-microbe studies. *Microorganisms, 7*(5), 128. https://doi.org/10.3390/microorganisms7050128.

Mehta, K. Y., Wu, H.-J., Menon, S. S., Fallah, Y., Zhong, X., Rizk, N., Unger, K., Mapstone, M., Fiandaca, M. S., Federoff, H. J., & Cheema, A. K. (2017). Metabolomic biomarkers of pancreatic cancer: A meta-analysis study. *Oncotarget, 8*(40), 68899. https://doi.org/10.18632/ONCOTARGET.20324

Mongaret, C., Velard, F., & Reffuveille, F. (2021). *Cutibacterium acnes*: the urgent need to identify diagnosis markers. *Infection and immunity, 89*(4), e00753–20. https://doi.org/10.1128/IAI.00753-20.

Muhammad, M. H., Idris, A. L., Fan, X., Guo, Y., Yu, Y., Jin, X., Qiu, J., Guan, X., & Huang, T. (2020). Beyond risk: bacterial biofilms and their regulating approaches. *Frontiers in microbiology, 11*, 928. https://doi.org/10.3389/fmicb.2020.00928.

Naglik, J. R., & Moyes, D. L. (2011). Mucosal immunity and Candida albicans infection. *Clinical and Developmental Immunology, 2011*. https://doi.org/10.1155/2011/346307

Naghipour Erami, A., Rasooli, I., Jahangiri, A., & Darvish Alipour Astaneh, S. (2021). Anti-Omp34 antibodies protect against Acinetobacter baumannii in a murine sepsis model. *Microbial pathogenesis, 161*(Pt B), 105291. https://doi.org/10.1016/j.micpath.2021.105291

Neelakantan, P., Romero, M., Vera, J., Daood, U., Khan, A. U., Yan, A., & Cheung, G. (2017). Biofilms in endodontics-current status and future directions. *International journal of molecular sciences, 18*(8), 1748. https://doi.org/10.3390/ijms18081748.

Neopane, P., Nepal, H. P., Shrestha, R., Uehara, O., & Abiko, Y. (2018). In vitro biofilm formation by *Staphylococcus aureus* isolated from wounds of hospital-admitted patients and their association with antimicrobial resistance. *International journal of general medicine, 11*, 25–32. https://doi.org/10.2147/IJGM.S153268.

Newgard, C. B. (2012). Interplay between lipids and branched-chain amino acids in development of insulin resistance. *Cell Metabolism, 15*(5), 606–614. https://doi.org/10.1016/J.CMET.2012.01.024

Newgard, C. B., An, J., Bain, J. R., Muehlbauer, M. J., Stevens, R. D., Lien, L. F., Haqq, A. M., Shah, S. H., Arlotto, M., Slentz, C. A., Rochon, J., Gallup, D., Ilkayeva, O., Wenner, B. R., Yancy, W. S., Eisenson, H., Musante, G., Surwit, R. S., Millington, D. S., … Svetkey, L. P. (2009). A branched-chain amino acid-related metabolic signature that differentiates obese and lean humans and contributes to insulin resistance. *Cell Metabolism, 9*(4), 311. https://doi.org/10.1016/J.CMET.2009.02.002

Nikolakis, G., Join-Lambert, O., Karagiannidis, I., Guet-Revillet, H., Zouboulis, C. C., & Nassif, A. (2015). Bacteriology of hidradenitis suppurativa/acne inversa: A review. *Journal of the American Academy of Dermatology, 73*(5), S12–S18. https://doi.org/10.1016/J.JAAD.2015.07.041

Ogonowska, P., Gilaberte, Y., Barańska-Rybak, W., & Nakonieczna, J. (2021). Colonization With *Staphylococcus aureus* in atopic dermatitis patients: attempts to reveal the unknown. *Frontiers in microbiology, 11*, 567090. https://doi.org/10.3389/fmicb.2020.567090.

Otto, M. (2009). Staphylococcus epidermidis - the "accidental" pathogen. *Nature Reviews Microbiology, 7*(8). https://doi.org/10.1038/nrmicro2182

Parthasarathy, S., Shah, S., Raja Sager, A., Rangan, A., & Durugu, S. (2020). Staphylococcus lugdunensis: review of epidemiology, complications, and treatment. *Cureus, 12*(6), Article e8801. https://doi.org/10.7759/cureus.8801.

Penesyan, A., Nagy, S. S., Kjelleberg, S., Gillings, M. R., & Paulsen, I. T. (2019). Rapid microevolution of biofilm cells in response to antibiotics. *NPJ biofilms and microbiomes, 5*(1), 34. https://doi.org/10.1038/s41522-019-0108-3.

Peng, B., Su, Y. B., Li, H., Han, Y., Guo, C., Tian, Y. M., & Peng, X. X. (2015). Exogenous alanine and/or glucose plus kanamycin kills antibiotic-resistant bacteria. *Cell Metabolism, 21*(2), 249–262. https://doi.org/10.1016/J.CMET.2015.01.008

Percival, S. L., Emanuel, C., Cutting, K. F., & Williams, D. W. (2012). Microbiology of the skin and the role of biofilms in infection. *International Wound Journal, 9*(1), 14. https://doi.org/10.1111/J.1742-481X.2011.00836.X

Percival, S. L., McCarty, S. M., & Lipsky, B. (2015). Biofilms and wounds: An overview of the evidence. *Advances in Wound Care, 4*(7), 373. https://doi.org/10.1089/WOUND.2014.0557

Qi, S., Wu, Q., Chen, Z., Zhang, W., Zhou, Y., Mao, K., Li, J., Li, Y., Chen, J., Huang, Y., & Huang, Y. (2021). High-resolution metabolomic biomarkers for lung cancer diagnosis and prognosis. *Scientific Reports, 11*(1), 1–10. https://doi.org/10.1038/s41598-021-91276-2

Quagliarello, B., Cespedes, C., Miller, M., Toro, A., Vavagiakis, P., Klein, R. S., & Lowy, F. D. (2002). Strains of *Staphylococcus aureus* obtained from drug-use networks are closely linked. *Clinical Infectious Diseases, 35*(6), 671–677. https://doi.org/10.1086/342196/2/35-6-671-FIG003

Rajkumari, N., Mathur, P., & Misra, M. C. (2014). Soft tissue and wound infections due to Enterococcus spp. among hospitalized trauma patients in a developing country. *Journal of Global Infectious Diseases, 6*(4), 189. https://doi.org/10.4103/0974-777X.145253

Roux, P. F., Oddos, T., & Stamatas, G. (2022). Deciphering the role of skin surface microbiome in skin health: An integrative multiomics approach reveals three distinct metabolite–microbe clusters. *Journal of Investigative Dermatology, 142*(2). https://doi.org/10.1016/j.jid.2021.07.159

Rudramurthy, S. M., Honnavar, P., Dogra, S., Yegneswaran, P. P., Handa, S., & Chakrabarti, A. (2014). Association of Malassezia species with dandruff. *The Indian journal of medical research, 139*(3), 431–437.

Sanford, J. A., O'Neill, A. M., Zouboulis, C. C., & Gallo, R. L. (2019). Short-chain fatty acids from Cutibacterium acnes activate both a canonical and epigenetic inflammatory response in human sebocytes. *Journal of Immunology (Baltimore, Md.: 1950), 202*(6), 1767. https://doi.org/10.4049/JIMMUNOL.1800893

dos Santos, G. C., Saldanha-Gama, R., de Brito, N. M., Renovato-Martins, M., & Barja-Fidalgo, C. (2021). Metabolomics in cancer and cancer-associated inflammatory cells. *Journal of Cancer Metastasis and Treatment, 7*, 1. https://doi.org/10.20517/2394-4722.2020.86

Santos, A., Galdino, A., Mello, T. P., Ramos, L. S., Branquinha, M. H., Bolognese, A. M., Columbano Neto, J., & Roudbary, M. (2018). What are the advantages of living in a community? a microbial biofilm perspective! *Memorias do Instituto Oswaldo Cruz, 113*(9), e180212. https://doi.org/10.1590/0074-02760180212

Sasson, G., Bai, A. D., Showler, A., Burry, L., Steinberg, M., Ricciuto, D. R., Fernandes, T., Chiu, A., Raybardhan, S., Science, M., Fernando, E., Morris, A. M., & Bell, C. M. (2017). *Staphylococcus aureus* bacteremia in immunosuppressed patients: A multicenter, retrospective cohort study. *European Journal of Clinical Microbiology & Infectious Diseases : Official Publication of the European Society of Clinical Microbiology, 36*(7), 1231–1241. https://doi.org/10.1007/S10096-017-2914-Y

Saxena, R., Mittal, P., Clavaud, C., Dhakan, D. B., Hegde, P., Veeranagaiah, M. M., Saha, S., Souverain, L., Roy, N., Breton, L., Misra, N., & Sharma, V. K. (2018). Comparison of healthy and dandruff scalp microbiome reveals the role of commensals in scalp health. *Frontiers in cellular and infection microbiology, 8*, 346. https://doi.org/10.3389/fcimb.2018.00346.

Schelli, K., Zhong, F., & Zhu, J. (2017). Comparative metabolomics revealing *Staphylococcus aureus* metabolic response to different antibiotics. *Microbial Biotechnology, 10*(6), 1764. https://doi.org/10.1111/1751-7915.12839

Schilcher, K., & Horswill, A. R. (2020). Staphylococcal biofilm development: structure, regulation, and treatment strategies. *Microbiology and molecular biology reviews : MMBR, 84*(3), e00026–19. https://doi.org/10.1128/MMBR.00026-19.

Shen, F., Ge, C., & Yuan, P. (2020). Metabolomics study reveals inhibition and metabolic dysregulation in *Staphylococcus aureus* planktonic cells and biofilms induced by carnosol. *Frontiers in Microbiology*, 11. https://doi.org/10.3389/FMICB.2020.538572

Simon-Soro, A., Ren, Z., Krom, B. P., Hoogenkamp, M. A., Cabello-Yeves, P. J., Daniel, S. G., Bittinger, K., Tomas, I., Koo, H., & Mira, A. (2022). Polymicrobial aggregates in human saliva build the oral. *Biofilm. mBio, 13*(1), Article e0013122. https://doi.org/10.1128/mbio.00131-22.

Srinivasan, R., Santhakumari, S., Poonguzhali, P., Geetha, M., Dyavaiah, M., & Xiangmin, L. (2021). Bacterial biofilm inhibition: a focused review on recent therapeutic strategies for combating the biofilm mediated infections. *Frontiers in microbiology, 12*, 676458. https://doi.org/10.3389/fmicb.2021.676458.

Stokes, J. M., Lopatkin, A. J., Lobritz, M. A., & Collins, J. J. (2019). Bacterial metabolism and antibiotic efficacy. *Cell metabolism, 30*(2), 251–259. https://doi.org/10.1016/j.cmet.2019.06.009.

Tan, A., & Doig, C. L. (2021). NAD+ degrading enzymes, evidence for roles during infection. *Frontiers in Molecular Biosciences, 8*, 803. https://doi.org/10.3389/FMOLB.2021.697359/BIBTEX

Tena, D., Martínez, N. M., Losa, C., & Solís, S. (2014). Skin and soft tissue infection caused by achromobacter xylosoxidans: Report of 14 cases. *Scandinavian Journal of Infectious Diseases, 46*(2), 130–135. https://doi.org/10.3109/00365548.2013.857043

Tomczak, H., Wróbel, J., Jenerowicz, D., Sadowska-Przytocka, A., Wachal, M., Adamski, Z., & Czarnecka-Operacz, M. M. (2019). The role of *Staphylococcus aureus* in atopic dermatitis: microbiological and immunological implications. *Postepy dermatologii i alergologii, 36*(4), 485–491. https://doi.org/10.5114/ada.2018.77056.

Tondo, M., Wasek, B., Escola-Gil, J. C., de Gonzalo-Calvo, D., Harmon, C., Arning, E., & Bottiglieri, T. (2020). Altered brain metabolome is associated with memory impairment in the rTg4510 mouse model of tauopathy. *Metabolites, 10*(2). https://doi.org/10.3390/METABO10020069

Tong, S. Y. C., Davis, J. S., Eichenberger, E., Holland, T. L., & Fowler, V. G. (2015). *Staphylococcus aureus* infections: Epidemiology, pathophysiology, clinical manifestations, and management. *Clinical Microbiology Reviews, 28*(3), 603. https://doi.org/10.1128/CMR.00134-14

Tounta, V., Liu, Y., Cheyne, A., & Larrouy-Maumus, G. (2021). Metabolomics in infectious diseases and drug discovery. *Molecular Omics, 17*(3). https://doi.org/10.1039/d1mo00017a

Tsui, C., Kong, E. F., & Jabra-Rizk, M. A. (2016). Pathogenesis of Candida albicans biofilm. *Pathogens and Disease, 74*(4), ftw018. https://doi.org/10.1093/FEMSPD/FTW018

Vaishnavi, K. V., Safar, L., & Devi, K. (2019). Biofilm in dermatology. *J Skin Sex Transm Dis, 1*, 3–7.

Verderosa, A. D., Totsika, M., & Fairfull-Smith, K. E. (2019). Bacterial biofilm eradication agents: A current review. *Frontiers in Chemistry, 7*. https://doi.org/10.3389/fchem.2019.00824

Vestby, L. K., Grønseth, T., Simm, R., & Nesse, L. L. (2020). Bacterial biofilm and its role in the pathogenesis of disease. *Antibiotics (Basel, Switzerland)*, *9*(2), 59. https://doi.org/10.3390/antibiotics9020059.

Wang, Y., Kuo, S., Shu, M., Yu, J., Huang, S., Dai, A., Two, A., Gallo, R. L., & Huang, C. M. (2014). Staphylococcus epidermidis in the human skin microbiome mediates fermentation to inhibit the growth of Propionibacterium acnes: Implications of probiotics in acne vulgaris. *Applied Microbiology and Biotechnology, 98*(1). https://doi.org/10.1007/s00253-013-5394-8

Wang, T. J., Larson, M. G., Vasan, R. S., Cheng, S., Rhee, E. P., McCabe, E., Lewis, G. D., Fox, C. S., Jacques, P. F., Fernandez, C., O'Donnell, C. J., Carr, S. A., Mootha, V. K., Florez, J. C., Souza, A., Melander, O., Clish, C. B., & Gerszten, R. E. (2011). Metabolite profiles and the risk of developing diabetes. *Nature Medicine, 17*(4), 448. https://doi.org/10.1038/NM.2307

Wilkinson, H. N., Iveson, S., Catherall, P., & Hardman, M. J. (2018). A novel silver bioactive glass elicits antimicrobial efficacy against *Pseudomonas aeruginosa* and *Staphylococcus aureus* in an ex vivo skin wound biofilm model. *Frontiers in Microbiology, 9*. https://doi.org/10.3389/fmicb.2018.01450

Wishart, D. S., Jewison, T., Guo, A. C., Wilson, M., Knox, C., Liu, Y., Djoumbou, Y., Mandal, R., Aziat, F., Dong, E., Bouatra, S., Sinelnikov, I., Arndt, D., Xia, J., Liu, P., Yallou, F., Bjorndahl, T., Perez-Pineiro, R., Eisner, R., … Scalbert, A. (2013). HMDB 3.0—the human metabolome database in 2013. *Nucleic Acids Research, 41*, D801. https://doi.org/10.1093/NAR/GKS1065

Xi, H., Schneider, B. L., & Reitzer, L. (2000). Purine catabolism in *Escherichia coli* and function of xanthine dehydrogenase in purine salvage. *Journal of Bacteriology, 182*(19), 5332–5341. https://doi.org/10.1128/JB.182.19.5332-5341.2000

Yang, S. J., Kwak, S. Y., Jo, G., Song, T. J., & Shin, M. J. (2018). Serum metabolite profile associated with incident type 2 diabetes in Koreans: findings from the Korean Genome and Epidemiology Study. *Scientific reports, 8*(1), 8207. https://doi.org/10.1038/s41598-018-26320-9.

Zichichi, L., Asta, G., & Noto, G. (2000). *Pseudomonas aeruginosa* folliculitis after shower/bath exposure. *International Journal of Dermatology, 39*(4), 270–273. https://doi.org/10.1046/J.1365-4362.2000.00931.X

CHAPTER

7

Metabolomic study of biofilm-forming natural microbiota of vaginal biofilm

Ashok Kumar Krishna Kumar[1], *Jayanthi Malaiyandi*[1], *Andrea Mary Francis*[2] *and Vidya Ramachandran*[3]

[1]Department of Biotechnology, School of Life Sciences, Vels Institute of Science Technology and Advanced Studies, Chennai, Tamil Nadu, India; [2]Department of Human Genetics, Sri Ramachandra Institute of Higher Education and Research, Chennai, Tamil Nadu, India; [3]Department of Microbiology, School of Life Sciences, Vels Institute of Science Technology and Advanced Studies, Chennai, Tamil Nadu, India

1. Introduction

Microorganisms typically survive in nature by adhering to living and inanimate surfaces. Generally, a biofilm of soil and the aquatic system is a spectrum of analysis for its benefits and defects. In recent years, the spectrum of biofilm analyses on medical devices, and in living tissues such as enamel, heart valves, the lung, and the middle ear, due to adverse medical effects after treatment. "Biofilm" is characteristic of adhering and growing on the surface, forming a structured aggregate by the microbes using an extracellular polymer matrix, which also covers the microorganism. It permanently adheres to the fetish or living surface; hence, removal is difficult, until, unless it is rapidly rinsed (Costerton et al., 1994; Hurlow et al., 2015). The attachment phase of a biofilm to the surface results in extracellular polymeric substances (EPS) formation. The production of an exopolysaccharide matrix provides the bacteria with interaction in the biofilm to determine whether a microbial biofilm will grow on a dead or solid surface. The thickness of the EPS matrix is typically 0.2–1.0 m, although the biofilm size usually does not exceed 10–30 nm (Sleytr, 1997). It is composed of 5%–35% biofilm volume, and the rest is made up of protein (Sun et al., 2005). The extracellular matrix creates a scavenging mechanism that traps some essential nutrients and minerals from the

Microbial Biofilms
https://doi.org/10.1016/B978-0-323-95715-1.00019-4

surrounding environment. Extracellular polymeric compounds contain a variety of constituents, including protein (>2%), polysaccharides (1%–12%), DNA molecules (1%), RNA (1%), ions (bound and free), and water (97%) in the majority of cases, which is responsible for the transport of vital nutrients inside it. Various factors influence the formation and progression of a biofilm, such as the bacterial strain, surface qualities, and environmental conditions, i.e., pH, nutrient content, and temperature (Donlan, 2002). Biofilm development occurs in five stages:

(i) Early adhesion
(ii) Irretrievable attachment
(iii) Initial development of biofilm architecture
(iv) Maturation
(v) Dissemination

Biofilm-formed microorganisms behave distinctly from free planktonic microbes. Bacteria in biofilms are more resistant to antibiotics and can facilely elude the immune system of the host generation. Probiotic bacteria, such as *Lactobacillus* spp., produce biofilms that prevent pathogenic bacteria from colonizing the host by promoting colonization and long-term persistence in the host's mucosa (Lecceseterraf et al., 2016). Some biofilm-producing *Lactobacillus* strains release extracellular polymeric substances that can inhibit harmful bacteria from developing biofilms (Lecceseterraf et al., 2016; Tahmourespour et al., 2019; Walencka et al., 2008). As a result, biofilms play an essential role in healthcare-associated infections (HAIs), particularly those involving implanted medical devices such as intravascular catheters, urinary catheters, and orthopedic implants.

Studies on the microbial flora of human bodies have long been on the oral cavity and the intestinal tract. Despite the importance of vaginal health for healthy reproduction and public health, it has garnered less attention (Chen et al., 2021). The vaginal cavity is home to a vast microecosystem with billions of bacteria. In recent years, more attention has been paid to female health, specifically the vaginal microbiota (Ravel et al., 2011).

2. Natural microbiota of vagina

Biofilm is an alternate survival strategy developed by various bacteria to aid in their growth and development. *Pseudomonas aeruginosa* is a gram-negative bacterium that forms biofilm efficiently, during virulence manifestation and disease progression (Lieberman, 2003). The vagina contains a diverse microbial population in a symbiotic relationship with the host. The human vaginal microbiota consists of many beneficial microbes and opportunistic pathogens that survive in the vaginal environment (Gajer et al., 2012; Smith & Ravel, 2017). Pathogens found in the niche include *Neisseria gonorrhoeao*, *Ureaplasma* species, *Mycoplasma genitalium*, *Streptococcus* species, *Escherichia coli*, *Chlamydia trachomatis*, and *Trichomonas vaginalis*.

The presence and absence of essential microbial community bound to the human vagina change over time by variables such as women's age and hormone shifts (e.g., menstrual cycle stage, contraception). It also varies during types and frequency of their sexual activities, urinary tract infection, health conditions (diabetes, urinary tract infection), medications intake, intravaginal washing, and hygiene). Hormonal changes, age, sexual practices, and antimicrobial treatment influence the microbiota and mycobiota (Hickey et al., 2015; Mulder et al., 2019; Plummer et al., 2019; Zapata & Quagliarello, 2015). *Lactobacillus* spp., beneficial bacteria often associated with healthy vaginal ecology, plays a vital role in defense of the female genital tract against infectious pathogens in healthful women (Filardo et al., 2017). Furthermore, pathogen growth is inhibited by competing for nutrients, liberating antimicrobial compounds, initiating immune system pathways, and sustaining a low vaginal pH through lactic acid production (Mastromarino et al., 2014; Sessa et al., 2017). Sustaining a harmonic balance of vaginal microbiota is essential for a healthy vaginal environment and a robust host–microbial relationship. In general, beneficial bacteria communities thrive in mutualism with human hosts, shielding the host vaginal environment from harmful microorganism colonization, while the host contributes resources for bacterial development.

The occurrence of fungal species in a human vaginal environment is dominated by lactobacilli (Drell et al., 2013). *Lactobacillus* account for 80.2% and 89.7% of the vaginal microbiota in Asian and white women, respectively, and is not the only genus that dominates the vaginal microbiota in black and Hispanic women (only 59.6% and 61.9%, respectively) (Ravel et al., 2011). Lactobacilli colonize and dominate the vaginal microbiota, which is usually seen in *Lactobacillus crispatus*, *Lactobacillus gasseri*, *Lactobacillus iners*, and *Lactobacillus jensenii* (Kroon et al., 2018; Ravel et al., 2011; Younes et al., 2018). *Lactobacillus* inhibits BVAB overproduction in the vaginal canal (Soper, 2019). *Gardnerella vaginalis* and *Candida* spp. also colonize the cervicovaginal microbiota. The microbial imbalance, including *G. vaginalis* overgrowth and *Lactobacillus* spp. Deficiencies, can contribute to genital infection. *Vulvovaginal candidiasis* (VVC) and *bacterial vaginosis* (BV) are common genital infections in women, accounting for 90% of all vaginitis cases (Eckert, 2006). Nearly 60% of BV vaginal infection occurs in gestation women (Machado & Cerca, 2015). Fungi, particularly *Candida* species, are thought to live as endosymbionts in the mucous layer of the vaginal mucosa, where they interact with diverse bacteria in a complex vaginal ecosystem (Gow & Hube, 2012; Hall & Noverr, 2017). *Candida albicans* is the most common vaginal colonizer and is the most prevalent cause of VVC, and 75% of women experience at least once, in their lifetime, 50% suffer twice, and 5%–10% of all women experience recurrent VVC (Bitew & Abebaw, 2018).

The growth of opportunistic pathogens in vaginal microbial dysbiosis contributes to disease onset (Bradford et al., 2017). Vaginal dysbiosis is a disorder of the microbial community in the vaginal that may lead to many gynecological ailments, in addition to various infections such as BV, VVC, and sexually transmitted infections (STIs) that are associated with vaginal dysbiosis (Eastment & McClelland, 2018; Mitra et al., 2016; Shannon et al., 2017; van de Wijgert, 2017; Ziklo et al., 2018). The change in vaginal pH is one of the most noticeable signs of vaginal dysbiosis. In a recent study, BV, CT, and VVC patients had a significantly higher vaginal pH than healthy women due to a drop in lactate concentration (Ceccarani et al., 2019). On the occurrence of vaginal dysbiosis, however, the *Lactobacillus*-dominated vaginal microbiota is supplanted by *Gardnerella* species (Muzny et al., 2019). *Lactobacillus* provides probiotic effects against pathogens in the GI tract, oral cavity, vaginal area, and epidermal layer (Chew et al., 2015; Humphreys & McBain, 2019; Prabhurajeshwar & Chandrakanth, 2017; Singh et al., 2017). *Lactobacillus acidophilus* KS400 has been shown to produce bacteriocin and inhibit the development of urogenital infections such as *G. vaginalis*, *Streptococcus agalactiae*, and *P. aeruginosa* by fermentation (Gaspar et al., 2018). By collapsing the pathogen's chemiosmotic potential, a bacteriocin from vaginal *Lactobacillus rhamnosus* was able to form transitory holes in the cytoplasmic membrane of *G. vaginalis* (Turovskiy et al., 2009). If left untreated, the change in vaginal microbial populations can cause gynecological problems such as pregnancy loss, premature labor, and poor conception rates (Amabebe & Anumba, 2018). Many "omics" techniques such as metabolomics, metagenomics, metatranscriptomics, and proteomics were used to study the microbiota of the vagina. Polymerase chain reaction—denaturing gradient gel electrophoresis, DNA pyrosequencing, fluorescence in situ hybridization, and microarrays are recently used molecular techniques for studying vaginal microbiota (Fredricks., 2011).

3. Metabolome of vagina

The muscular structure lined with epithelial cells is known as the vagina, which connects the uterus to the outer environment. The vaginal microbiota contains more than 50 different bacterial species, which happily coexist with the host and one another to develop complex interactions (Ma et al., 2012). The metabolome, which is a comprehensive collection of small molecules in a particular environment, has been researched in a variety of systems to uncover illness indicators. Few studies have demonstrated that certain metabolite concentrations correspond with the major bacteria prevalent in the human vaginal area. The following clinical parameters, including vaginal discharge, pH, and amine odor, may be impacted by the small molecule metabolites. Vaginal fluid is released by women daily at a rate of 6 g, with 0.5—0.075 g present at a regular interval (Owen & Katz, 1999). Fatty acids, proteins, salts, and carbohydrates form the major of the vaginal fluid; in this regard, it should be emphasized that there is significant inter- and intraindividual variance in these components.

Various factors have an impact on the vaginal metabolome directly or indirectly by changing the vaginal microbiome's composition, and this causes changes in the metabolic profile as well. In addition, polyamines such as putrescine and cadaverine, which are formed by the decarboxylation of amino acid—mediated bacteria and contribute to the fishy odor, are identified in the vaginal discharge (Table 7.1). The metabolites *N*-acetylputrescine, cadaverine, and tyramine are found in vaginal discharge when there is an elevation in pH. One of the most prevalent metabolites found in the vaginal fluid is succinate. Women's vaginal secretions showed metabolite changes linked to microbial invasion of the amniotic fluid. Leucine and isoleucine, branched-chain amino acids, and glycerophosphocholine concentrations in the vaginal discharge were decreased, while hypoxanthine, proline, choline, and acetylcholine concentrations were elevated (Vicente-Muñoz et al., 2020). Ion concentrations such as sodium, potassium, calcium, and chloride, as well as protein, have been found in vaginal fluid in other investigations (Owen & Katz, 1999).

The case of women affected by *VVC* revealed a high content of taurine. Similarly, BV showed the presence of isopropanol, acetylcholine, and glucose. In addition, women with VVC and BV exhibited decreased levels of lactate, 4-hydroxyphenylacetate, phenylalanine, pi-methylhistidine, and glycine. Extracellular enzymes produced by *Candida* spp., including

TABLE 7.1 Clinical criteria associated with metabolites in vaginal discharge.

S.No	Clinical criteria	Metabolites identified
1.	Bacterial vaginosis status	Phenylalanine Deoxycarnitine Pipecolate *N*-acetylputrescine Cadaverine
2.	Elevated pH	*N*-acetylputrescine Cadaverine Lactate Glutamate Sphingosine Tyrosine Tyramine
3.	Clue cells	Deoxycarnitine Glycylproline GSH Pipecolate
4.	Amine odor	Phenylalanine Lactate *N*-acetylputrescine
5.	Discharge from vagina	Agmatine Cadaverine

Srinivasan, S., Morgan, M. T, Fiedler, T. L, et al. (2015). Metabolic signatures of bacterial vaginosis. MBio, *6(2), e00204–e00215.*

proteinases, phospholipases, and hemolysins, are thought to be involved in the adhesion, invasion, and degeneration of vaginal epithelial cells during infection (Cauchie et al., 2017).

One of the most promising approaches in maternal and fetal medicine for analyzing the interrelations between the mother, placenta, and fetus is metabolomics. The result of all metabolic activities enables the identification and quantification of metabolome changes and more closely functionally simulates the phenotype. Furthermore, metabolomics improves knowledge of molecular pathways and their anomalies (Vicente-Muñoz et al., 2020). Women with microbial invasion of the amniotic cavity (MIAC) had higher vaginal metabolome levels of hypoxanthine, proline, choline, and acetylcholine than phenylalanine, glutamine, leucine, isoleucine, and glycerophosphocholine (Vicente-Muoz et al., 2020). Women infected with BV, VVC, and CT showed a decrease in lactate concentration, with a consequent increase in pH, which would be a biomarker for the diagnosis of dysbiosis. Furthermore, there was an increase in short-chain fatty acids such as butyrate, propionate, and acetate. Moreover, all three infections were characterized by a significant reduction of dimethylamine and a comparative increase in TMA, which resulted in vaginal dysbiosis and impaired balance (Ceccarani et al., 2019). As a result, more research is needed to determine whether microbial or metabolic changes aid in detecting disease onset or whether the pathogen itself disrupts the vaginal environment.

4. Bacterial vaginosis—pathogenesis

BV, also known as vaginal dysbiosis, is one of the most common vaginal illnesses and is associated with conditions such as preterm delivery, the transmission of STIs, and HIV. It is also one of the factors linked to abnormal vaginal microbiota alterations (Redelinghuys et al., 2020). Despite its significance in public health, the pathogenesis of BV is unknown, with some people calling it dysbiosis, which is a microbial imbalance in the vaginal flora that can cause changes in the vaginal environment (Hajishengallis, 2015; Nelson et al., 2015). A transition from a vaginal flora dominated by lactobacillus to one dominated by facultative and stringent anaerobes characterizes BV (Hillier, 1993).

The majority of epidemiological studies refer to BV as an STI. While sex with the same partner increases the chance of recurrent or chronic BV, a new sexual relationship increases the risk of incident BV. According to prospective research on vaginal flora, frequent condom use significantly reduced the incidence and prevalence of BV, especially in women with normal vaginal flora at baseline. More recently, it was demonstrated that hitherto unknown, noncultivable bacteria are also a part of the BV flora. Although studies have demonstrated that BV exhibits pathogenicity features and is therefore concordant between sexual partners, anaerobes are likely to be responsible for some of the symptoms. In their conceptual model for the pathogenesis of BV, Schwebke et al. (2014) identified *G. vaginalis* as the keystone pathogen. The role of G. vaginalis in BV pathogenesis is shown in a diagrammatic presentation (Fig. 7.1).

If we are to progress in the treatment and prevention of this common and deadly illness, elucidating the BV pathophysiology riddle is essential. Epidemiological data confirm that BV is transmitted sexually. Sexual networks may affect racial differences in BV-positive women similarly to other STIs. The striking similarities between periodontitis and BV pathogenesis may serve as a starting point for future studies on BV microbial interactions.

5. Metabolome alteration and biofilm formation in bacterial vaginosis

Several alterations in metabolome concentration have been linked to BV status. In BV, low levels of amino acids and high levels of amino acid catabolites such as amines putrescine, cadaverine, and tyramine were detected, which indicates increased amino acid use and reduced synthesis. *N*-acetylputrescine, a breakdown product of putrescine, was linked to an amine odor and an increased pH. In BV, lower arginine levels and greater putrescine levels favor arginine to putrescine conversion. However, the presence of alternative routes in BV is suggested by decreased spermine and greater succinate levels. The possible alteration may be that putrescine is transformed to succinate through γ-aminobutyrate (GABA), a novel alternative putrescine utilization pathway (Kurihara et al., 2010). Carnitine, a byproduct of lysine or methionine breakdown, was lower in BV, while its precursor deoxycarnitine was higher. In both the primary and validation trials, higher amounts of 12-HETE and lower levels of its precursor arachidonate were found in BV. 12-HETE is a biomarker for inflammation and a signaling eicosanoid that can modulate inflammatory response pathways. BV has decreased spermine and ascorbate metabolites and indicates oxidative stress and influences redox status (Henkel, 2011). Women with BV have increased levels of *N*-acetylneuraminate in their

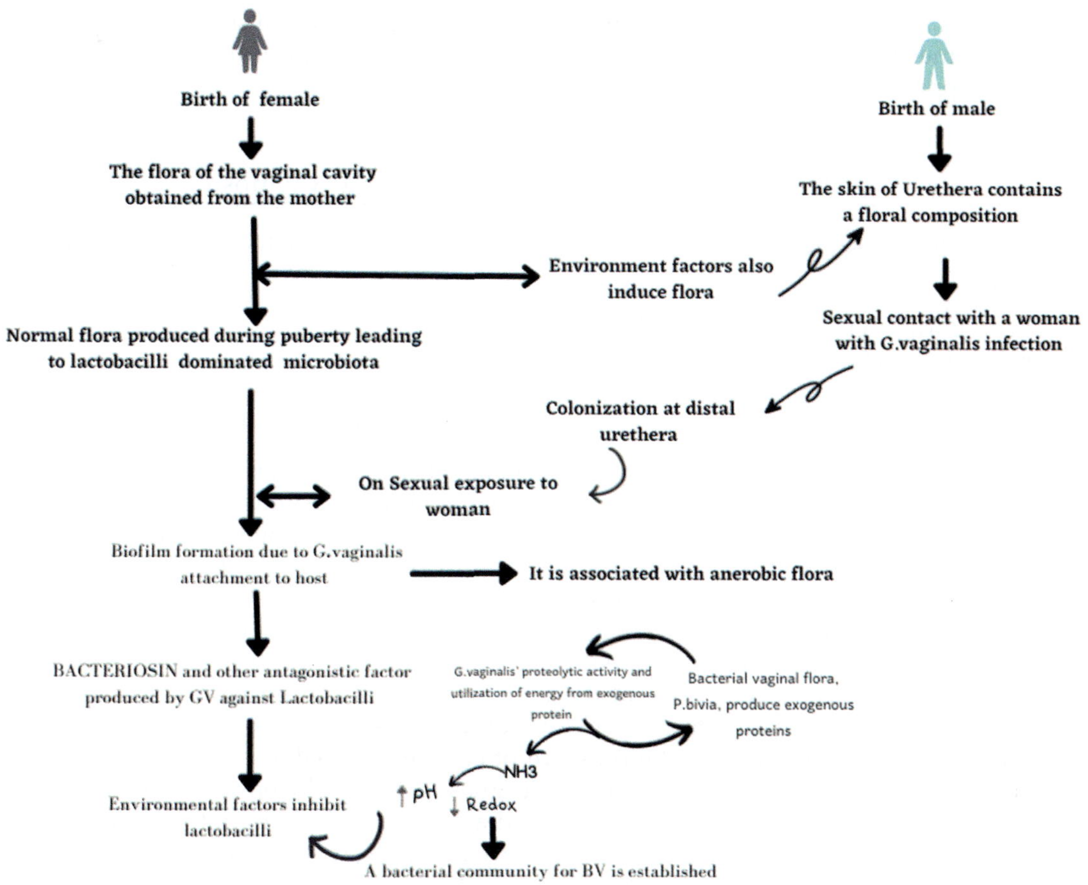

FIGURE 7.1 Diagrammatic representation of the role of G. vaginalis in the pathogenesis of BV.

vaginal fluid. The most common sialic acid is *N*-acetylneuraminate, and the measurement of sialidase activity in vaginal fluids is used in a current point-of-care diagnostic test for BV (Khatoon et al., 2013). In bacterial biofilms, *N*-acetylneuraminate appears to play an important function. Biofilms have been demonstrated to have a role in BV, with *G. vaginalis* being a significant member of these biofilms (Swidsinski et al., 2005). If *N*-acetylneuraminate plays a vital part in biofilm formation, interfering with its production or liberation in the vaginal canal could be one way to avoid BV. BV has been associated with an adhesive polymicrobial biofilm on vaginal epithelial cells comprising numerous *G. vaginalis* and smaller quantities of BVAB, including *Atopobium vaginae*. Vaginal infections have lately been attributed to biofilm formation. *G. vaginalis* biofilms, significant pathogens in BV, indicate potentially essential virulence features and resistance mechanisms (Muzny & Schwebke, 2015).

In microbiology, microbial metabolomics has gained a lot of interest. BV has a distinct metabolic fingerprint, which is consistent with substantial changes in vaginal bacterial assemblages. Metabolite profiles with and without BV were distinguished in various pathways, and metabolomics findings could be used to generate novel BV diagnostic biomarkers and new approaches in prevention and its therapy.

6. Diagnosis and treatment for bacterial vaginosis

If the patient satisfies three of the following four criteria, they may be subjected to the diagnosis of BV:

1. The uniform, thin, grayish-white discharge is present.
2. The vaginal pH of more than 4.5′
3. The existence of a positive whiff test.
4. The clue cells are present on microscopic inspection of vaginal fluid.

BV is differentiated from other vaginal infections such as candidiasis and trichomoniasis by examining vaginal discharge under a microscope. The appearance of clue cells under the microscope is a sign of BV. Clue cells are bacteria-covered vaginal cells that are the most reliable diagnosis of BV. Women infected with BV have clue cells and possess less normal vaginal bacteria, lactobacilli (Amsel et al., 1983).

Nugent scoring is another credible diagnostic technique that is frequently used in research investigations, although it is less frequently utilized in clinical practice since it necessitates Gram staining of the material. Three of the four Amsel's criteria must be necessary for the diagnosis of BV. These requirements include homogenous discharge, a pH of more than 4.5, a positive amine test, and/or the presence of clue cells on saline microscopy. Based on a Gram stain vaginal smear, Nugent and colleagues have developed a scoring system for the diagnosis of BV. The vaginal flora is assigned a score on a scale from 0 to 10 depending on the proportion of large gram-positive rods, tiny gram-negative or varied rods, or curved rod morphotypes. Currently, examination of molecular techniques is helpful to diagnose this condition as the discharge's culture does not help identify this ailment. Presently, a DNA probe for *G. vaginalis* detects large numbers of *G. vaginalis* organisms and can distinguish BV from *G. vaginalis* vaginal colonization (Nugent et al., 1991).

The Fem Exam card (Cooper Surgical, Shelton, CT) evaluates vaginal pH and identifies *G. vaginalis* metabolic products, such as amines and proline aminopeptidase activity. Two plastic cards make up the device; one measures pH and trimethylamine levels, while the other gauges proline aminopeptidase levels. Compared with the Nugent score, the combined sensitivity of cards 1 and 2 is 91%, and the specificity is 61%. It is rapid (2 min), precise, and simple to complete (West et al., 2003). The presence of species such as *A. vaginae*, *Ureaplasma* spp., and *Mycoplasma* spp., which are not detectable by Gram staining techniques or the Nugent score, lowers the Nugent score's sensitivity and necessitates the use of a confirmatory or additional molecular test to measure other etiological agents in the diagnosis of BV. The advantages of molecular technologies over microscopy-based testing include objectivity, detecting fastidious bacteria, quantification, and suitability for self-collected vaginal swabs. These technologies are more effective and rely on detecting certain bacterial nucleic acids (Kasper et al., 2005).

Many antimicrobial agents (e.g., ampicillin, penicillin, and metronidazole) have been used as a medicine of BV; with oral administration of metronidazole, nitroimidazole derivative is a widely recommended drug available in the form of tablets, gels, and suppositories. Metronidazole 0.75% gel twice weekly for 4–6 months is advised for treating recurrent BV. It is strongly recommended to use tinidazole instead of metronidazole since it has better pharmacokinetics and a longer half-life. Due to the development of ampicillin-resistant bacteria in BV patients, the use of ampicillin is discouraged (Mitchell et al., 2009). It has been observed that the lactic acid gel containing both lactic acid and metronidazole is more effective and secure in recolonizing the vaginal lumen with lactobacilli. Compared with individuals treated with metronidazole gels, recurrence of BV is less prevalent in people treated with lactic acid gel (Decena et al., 2006).

Before selecting a *Lactobacillus* species to treat BV, it is essential to consider the species' production of lactic acid, H_2O_2, adherence to the epithelial layer, and antimicrobials. Yogurt, acidophilus milk, and readily accessible *Lactobacillus* powder and pills are among the *Lactobacillus*-based items used to treat BV (Reid). It has been demonstrated that the most effective ways to raise the number of vaginal lactobacilli, and restore the vaginal microbiota to normal, are oral administration of *L. acidophilus* or intravaginal administration of *L. acidophilus* and *Lactobacillus* fermentum RC-14 (Reid et al., 2001).

Future-generation technologies, such as proteomics, genomics, immunomics, and metabolomics provide knowledge on the immune and functional processes during healthy, intermediate, and BV states, which can ultimately help with the creation of diagnostic procedures, therapeutic regimens, and preventative measures. The Ion Torrent 16STM Metagenomics kit (Thermo Fisher Scientific, Waltham, MA, United States) and Illumina 16S rRNA metagenomic sequencing (Illumina, San Diego, United States) are commercial assays that use affordable short-read 16S rRNA sequencing and are frequently used for research (Malla et al., 2019).

References

Amabebe, E., & Anumba, D. O. C. (2018). The vaginal microenvironment: The physiologic role of lactobacilli. *Frontiers in Medicine (Lausanne), 5*, 181.

Amsel, R., Totten, P. A., Spiegel, C. A., et al. (1983). Nonspecific vaginitis. Diagnostic criteria and microbial and epidemiological associations. *The American Journal of Medicine, 74*, 14–22.

Bitew, A., & Abebaw, Y. (2018). Vulvovaginal candidiasis: Species distribution of Candida and their antifungal susceptibility pattern. *BMC Womens Health, 18*(1), 94.

Bradford, L. L., Chibucos, M. C., Ma, B., et al. (2017). Vaginal Candida spp. genomes from women with vulvovaginal candidiasis. *Pathogens and Disease, 75*, 061.

Cauchie, M., Desmet, S., & Lagrou, K. (2017). Candida and its dual lifestyle as a commensal and a pathogen. *Research in Microbiology, 168*, 802–810.

Ceccarani, C., Foschi, C., Parolin, C., et al. (2019). Diversity of vaginal microbiome and metabolome during genital infections. *Scientific Reports, 9*, 14095.

Chen, X., Lu, Y., Chen, T., et al. (2021). The Female vagina Microbiome in health and Bacterial vaginosis. *Frontiers in Cellular and Infection Microbiology, 11*, 631972. https://doi.org/10.3389/fcimb.2021.631972

Chew, S. Y., Cheah, Y. K., & Seow, H. F. (2015). Probiotic Lactobacillus rhamnosus GR-1 and Lactobacillus reuteri RC-14 exhibit strong antifungal effects against vulvovaginal candidiasis-causing Candida glabrataisolates. *Journal of Applied Microbiology, 118*, 1180–1190.

Costerton, J. W., Lewandowski, Z., DeBeer, D., et al. (1994). Biofilms, the customized microniche. *Journal of Bacteriology, 176*, 2137–2142.

Decena, D. C., Co, J. T., Manalastas, R. M., et al. (2006). Metronidazole with Lactacyd vaginal gel in bacterial vaginosis. *Journal of Obstetrics and Gynaecology Research, 32*, 243–251.

Donlan, R. M. (2002). Biofilms: Microbial life on surfaces. *Emerging Infectious Diseases, 8*, 881–890.

Drell, T., Lillsaar, T., Tummeleht, L., et al. (2013). Characterization of the vaginal micro-and mycobiome in asymptomatic reproductive-age Estonian women. *PLoS One, 8*, e54379.

Eastment, M. C., & McClelland, R. S. (2018). Vaginal microbiota and susceptibility to HIV. *AIDS, 32*, 687–698.

Eckert, L. O. (2006). Acute vulvovaginitis. *New England Journal of Medicine, 355*(12), 1244–1252.

Filardo, S., Di Pietro, M., Porpora, M., et al. (2017). Diversity of cervical microbiota in asymptomatic *Chlamydia trachomatis* genital infection: A pilot study. *Frontiers in Cellular and Infection Microbiology, 7*, 321.

Fredricks, D. N. (2011). Molecular methods to describe the spectrum and dynamics of the vaginal microbiota. *Anaerobe, 17*, 191–195.

Gajer, P., Brotman, R. M., Bai, G., et al. (2012). Temporal dynamics of the human vaginal microbiota. *Science Translational Medicine, 4*, 132–152.

Gaspar, C., Donders, G. G., Palmeira-de-Oliveira, R., et al. (2018). Bacteriocin production of the probiotic Lactobacillus acidophilus KS400. *AMB Express, 8*, 153.

Gow, N. A. R., & Hube, B. (2012). Importance of the Candida albicans cell wall during commensalism and infection. *Current Opinion in Microbiology, 15*, 406–412.

Hajishengallis, G. (2015). Periodontitis: From microbial immune subversion to systemic inflammation. *Nature Reviews Immunology, 15*, 30–44.

Hall, R. A., & Noverr, M. C. (2017). Fungal interactions with the human host: Exploring the spectrum of symbiosis. *Current Opinion in Microbiology, 40*, 58–64.

Henkel, R. R. (2011). Leukocytes and oxidative stress: Dilemma for sperm function and male fertility. *Asian Journal of Andrology, 13*, 43–52.

Hickey, R. J., Zhou, X., Settles, M. L., et al. (2015). Aginal microbiota of adolescent girls prior to the onset of menarche resemble those of reproductive-age women. *MBio, 6*, 000977-15.

Hillier, S. L. (1993). Diagnostic microbiology of bacterial vaginosis. *American Journal of Obstetrics and Gynecology, 169*, 455–459.

Humphreys, G. J., & McBain, A. J. (2019). Antagonistic effects of Streptococcus and Lactobacillus probiotics in pharyngeal biofilms. *Letters in Applied Microbiology, 68*, 303–312.

Hurlow, J., Couch, K., Laforet, K., et al. (2015). Clinical biofilms: A challenging frontier in wound care. *Advances in Wound Care, 4*, 295–301.

Kasper, D. L., Braunwald, E., Hauser, S., Longo, D., Jameson, J. L., & Fauci, A. S. (2005). In *Harrison's principle of internal medicine* (16th ed.). New York, NY: McGraw-Hill.

Khatoon, R., Ahmad, S., Jahan, N., et al. (2013). Blue test: A new point-of-care test for diagnosing bacterial vaginosis and its comparison with gram staining. *African Journal of Microbiology Research, 7*(32), 4103–4106.

Kroon, S. J., Ravel, J., & Huston, W. M. (2018). Cervicovaginal microbiota, women's health, and reproductive outcomes. *Fertility and Sterility, 110*, 327–336.

Kurihara, S., Kato, K., Asada, K., et al. (2010). A putrescine-inducible pathway comprising PuuE-YneI in which γ-aminobutyrate is degraded into succinate in *Escherichia coli* K-12. *Journal of Bacteriology, 192*(18), 4582–4591.

LecceseTerraf, M. C., Juárez Tomás, M. S., & Rault, L. (2016). Biofilms of vaginal Lactobacillus reuteri CRL 1324 and Lactobacillus rhamnosus CRL 1332: Kinetics of formation and matrix characterization. *Archives of Microbiology, 198*, 689–700.

Lieberman, D. (2003). Pseudomonal infections in patients with COPD: Epidemiology and management. *American Journal of Respiratory and Critical Care Medicine, 2*(6), 459–468.

Machado, A., & Cerca, N. (2015). Influence of biofilm formation by Gardnerella vaginalis and other anaerobes on bacterial vaginosis. *Journal of Infectious Diseases, 212*(12), 1856–1861.

Ma, B., Forney, L. J., & Ravel, J. (2012). Vaginal microbiome: Rethinking health and disease. *Annual Review of Microbiology, 66*, 371–389.

Malla, M. A., Dubey, A., Kumar, A., et al. (2019). Exploring the human microbiome: The potential future role of next-generation sequencing in disease diagnosis and treatment. *Frontiers in Immunology, 9*, 2868. https://doi.org/10.3389/fimmu.2018.02868

Mastromarino, P., Di Pietro, M., Schiavoni, G., et al. (2014). Effects of vaginal lactobacilli in *Chlamydia trachomatis* infection. *International Journal of Medical Microbiology, 304*, 654–661.

Mitchell, C. M., Hitti, J. E., Agnew, K. J., et al. (2009). Comparison of oral and vaginal metronidazole for treatment of bacterial vaginosis in pregnancy: Impact on fastidious bacteria. *BMC Infectious Diseases, 9*, 89.

Mitra, A., MacIntyre, D. A., Marchesi, J. R., et al. (2016). The vaginal microbiota, human papillomavirus infection and cervical intraepithelial neoplasia: What do we know and where are we going next? *Microbiome, 4*, 58.

Mulder, M., Radjabzadeh, D., Hassing, R. J., et al. (2019). The effect of antimicrobial drug use on the composition of the genitourinary microbiota in an elderly population. *BMC Microbiology, 19*, 9.

Muzny, C. A., & Schwebke, J. R. (2015). Biofilms: An underappreciated mechanism of treatment failure and recurrence in vaginal infections. *Clinical Infectious Diseases, 61*(4), 601–606.

Muzny, C. A., Taylor, C. M., Swords, W. E., et al. (2019). An updated conceptual model on the pathogenesis of bacterial vaginosis. *The Journal of Infectious Diseases, 220*, 1399–1405.

Nelson, T. M., Borgogna, J. L., Brotman, R. M., et al. (2015). Vaginal biogenic amines: Biomarkers of bacterial vaginosis or precursors to vaginal dysbiosis? *Frontiers in Physiology, 6*, 253.

Nugent, R. P., Krohn, M. A., & Hillier, S. L. (1991). Reliability of diagnosing bacterial vaginosis is improved by a standardized method of gram stain interpretation. *Journal of Clinical Microbiology, 29*, 297–301. https://doi.org/10.1128/JCM.29.2.297-301.1991

Owen, D. H., & Katz, D. F. (1999). A vaginal fluid simulant. *Contraception, 59*(2), 91–95. https://doi.org/10.1016/s0010-7824(99)00010-4. PMID: 10361623.

Plummer, E. L., Vodstrcil, L. A., Fairley, C. K., et al. (2019). Sexual practices have a significant impact on the vaginal microbiota of women who have sex with women. *Scientific Reports, 9*, 19749.

Prabhurajeshwar, C., & Chandrakanth, R. K. (2017). Probiotic potential of lactobacilli with antagonistic activity against pathogenic strains: An in vitro validation for the production of inhibitory substances. *Biomedical Journal, 40*, 270–283.

Ravel, J., Gajer, P., Abdo, Z., et al. (2011). Vaginal microbiome of reproductive-age women. *Proceedings of the National Academy of Sciences, 108*, 4680–4687.

Redelinghuys, M. J., Geldenhuys, J., Jung, H., et al. (2020). Bacterial vaginosis: Current diagnostic avenues and future opportunities. *Frontiers in Cellular and Infection Microbiology, 10*, 354. https://doi.org/10.3389/fcimb.2020.00354

Reid, G., Bruce, A. W., Fraser, N., et al. (2001). Oral probiotics can resolve urogenital infections. *FEMS Immunology and Medical Microbiology, 30*(1), 49–52.

Schwebke, J. R., Muzny, C. A., & Josey, W. E. (2014). Role of Gardnerella vaginalis in the pathogenesis of bacterial vaginosis: A conceptual model. *The Journal of Infectious Diseases, 210*, 338–343. https://doi.org/10.1093/infdis/jiu089

Sessa, R., Di Pietro, M., Filardo, S., et al. (2017). Lactobacilli-lactoferrin interplay in *Chlamydia trachomatis* infection. *Pathogens and Disease, 75*(5), 054.

Shannon, B., Gajer, P., Yi, T. J., et al. (2017). Distinct effects of the cervicovaginal microbiota and herpes simplex type 2 infection on female genital tract immunology. *The Journal of Infectious Diseases, 215*, 1366–1375.

Singh, T. P., Kau, r G., Kapila, S., et al. (2017). Antagonistic activity of *Lactobacillus reuteri* strains on the adhesion characteristics of selected pathogens. *Frontiers in Microbiology, 8*, 486.

Sleytr, U. B. (1997). Basic and applied S-layer research: An overview. *FEMS Microbiology Reviews, 20*, 5–12.

Smith, S. B., & Ravel, J. (2017). The vaginal microbiota, host defence and reproductive physiology. *The Journal of Physiology, 595*, 451–463.

Soper, D. E. (2019). Bacterial vaginosis and surgical site infections. *American Journal of Obstetrics and Gynecology, 222*, 219–223.

Srinivasan, S., Morgan, M. T., Fiedler, T. L., et al. (2015). Metabolic signatures of bacterial vaginosis. *MBio, 6*(2), e00204–e00215.

Sun, D., Accavitti, M. A., & Bryers, J. D. (2005). Inhibition of biofilm formation by monoclonal antibodies against Staphylococcus epidermidis RP62A accumulation-associated protein. *Clinical and Diagnostic Laboratory Immunology, 12*, 93–100.

Swidsinski, A., Mendling, W., Loening-Baucke, V., et al. (2005). Adherent biofilms in bacterial vaginosis. *Obstetrics & Gynecology, 106*(5), 1013–1023.

Tahmourespour, A., Kasra-Kermanshahi, R., Salehi, R., et al. (2019). Lactobacillus rhamnosus biosurfactant inhibits biofilm formation and gene expression of caries-inducing Streptococcus mutans. *Dental Research Journal, 16*, 87.

Turovskiy, Y., Ludescher, R. D., Aroutcheva, A. A., et al. (2009). Lactocin 160, a bacteriocin produced by vaginal Lactobacillus rhamnosus, targets cytoplasmic membranes of the vaginal pathogen, *Gardnerella vaginalis*. *Probiotics and Antimicrobial Proteins, 1*, 67–74.

Vicente-Muñoz, S., Cobo, T., Puchades-Carrasco, L., et al. (2020). Vaginal metabolome: Towards a minimally invasive diagnosis of microbial invasion of the amniotic cavity in women with preterm labor. *Scientific Reports, 10*, 5465. https://doi.org/10.1038/s41598-020-62542-6

Walencka, E., Rózalska, S., & Sadowska, B. R. (2008). The influence of Lactobacillus acidophilus-derived surfactants on staphylococcal adhesion and biofilm formation. *Folia Microbiologica, 53*, 61–66.

West, B., Morison, L., Schim van der Loeff, M., et al. (2003). Evaluation of a new rapid diagnostic kit (Fem Exam) for bacterial vaginosis in patients with vaginal discharge syndrome in the Gambia. *Sexually Transmitted Diseases, 30*, 483–489. https://doi.org/10.1097/00007435-200306000-00003

van de Wijgert, J. H. H. M. (2017). The vaginal microbiome and sexually transmitted infections are interlinked: Consequences for treatment and prevention. *PLoS Med, 14*, e1002478.

Younes, J. A., Lievens, E., & Hummelen, R. (2018). Women and their microbes: The unexpected friendship. *Trends in Microbiology, 26*, 16–32.

Zapata, H. J., & Quagliarello, V. J. (2015). The microbiota and microbiome in aging: Potential implications in health and age-related diseases. *Journal of the American Geriatrics Society, 63*, 776–781.

Ziklo, N., Vidgen, M. E., Taing, K., et al. (2018). Dysbiosis of the vaginal microbiota and higher vaginal kynurenine/tryptophan ratio reveals an association with *Chlamydia trachomatis* genital infections. *Frontiers in Cellular and Infection Microbiology, 8*, 1.

CHAPTER

8

Metabolome analysis for host–microbiota interactions

Sreejita Ghosh[1], Dibyajit Lahiri[2], Moupriya Nag[2] and Rina Rani Ray[1]

[1]Department of Biotechnology, Maulana Abul Kalam Azad University of Technology, Haringhata, West Bengal, India; [2]Department of Biotechnology, University of Engineering and Management, Kolkata, West Bengal, India

1. Introduction

Guts in humans contain a huge number of symbiotic microorganisms, which are important for maintaining human health. On the other hand, communications inside the complicated community of microbiota and between the host and the microbiota are difficult to be explained, thereby restricting the fabrication of therapies for various diseases caused to dysbiosis of the microbiota. With the help of in silico methods of simulation on the basis of flux balance analysis, these communications can be studied better. Flux balance analysis makes use of an annotated genomic-scale restructuring of metabolic network for determining the distribution of the metabolic flux representing the entire metabolism of a single bacterium within a specific metabolic atmosphere like the gut. Simulation of a series of bacterial species within a shared metabolic ecology can facilitate the investigation of the effects of several disturbances such as dietary modifications or inclusion of a probiotic species in a customized way (Jansma & El Aidy, 2021). This chapter focuses on the application of the flux balance analysis of the host–microbiota interaction and how it can be utilized for improving the health of humans.

2. Use of metabolic networks of bacteria for studying metabolic communications

The microbiota of the gut is a microbial community present within the gut, and the community includes symbiotic, commensal, and pathogenic bacteria. Generally, the microbiota and its host remain in symbiosis (Li et al., 2008). Symbiosis disruption is adverse for the

Microbial Biofilms
https://doi.org/10.1016/B978-0-323-95715-1.00009-1

health of the host, resulting in diseases such as gastrointestinal problems (inflammatory bowel disorder), metabolic diseases (diabetes mellitus), and mental illnesses (major depressive disorder and autism spectrum disorder) (Cheung et al., 2019). To understand the symbiotic association, the various numbers of microbiota and the manner in which they interact with each other and with the host are needed to be studied. The microbiota interacts by metabolite production. Hence, it is chief in the study of host—microbe interactions for identifying which microbial species exist and what their metabolic production is. In contrast, this does not totally explain the dynamic communications inside the microbiota and between microbiota and the host because metabolite production by microbes depends on the surrounding environment (Martinet et al., 2019). Hence, metabolite production and also the host—microbiota interactions are different in different persons making successful therapies for the aforementioned diseases more difficult (Kolodziejczyk et al., 2019). Even though experimental strategies are critical for progression in the field of microbiota, they cannot completely grasp the mechanism, behavior, and interactions because of the enormous complexity of the microbiota ecology. Such disadvantages have given rise to a complementary strategy for complete understanding of the association between the microbes and the host or the bacterial metabolic network (Baldini et al., 2019). In this strategy, bacterial communications can be seen as a metabolic network. The metabolites consist of the nodes of metabolic network. Bioprocesses such as uptake, conversions, and secretions are denoted by the edges. By means of placing the series of metabolic reactions of a bacterium within a compartment, metabolic reactions of one bacterium can be distinguished from that of the other or from the host cells inside the metabolic network. By placing a single cellular compartment inside a common compartment, the manner in which the various cells can communicate with each other can be simulated (Embree et al., 2015).

Metabolic networks can be applied for the prediction of metabolic network modification in silico. A constraint-based reconstruction and analysis (COBRA) strategy is mostly utilized for simulating the metabolic network operation in presence of various external nutrient compositions. A COBRA approach is appropriate to investigate the metabolism of the microbiota by flax balance analysis (FBA) (Bauer & Thiele, 2018). Flux is defined as the turnover rate of the metabolites via a metabolic pathway. To carry out FBA, every flux within the network must be denoted by a series of linear equations. These equations are put in a stoichiometric matrix consisting of products, substrates, and reaction directionality. FBA also makes use of constraints for limiting metabolite flow via network and computes the metabolic flux distribution of a metabolic network for a specified objective function (OF), resulting in optimal flux distribution. For instance, an OF, which maximizes short-chain fatty acid (SCFA) acetate production, will lead to a separate distribution of flux in comparison with an OF, thereby maximizing production of butyrate. Other instances of OFs are minimizing particular metabolite production, maximization of cell growth in a bacterial community of single species, or growth maximization of the entire community (Bauer & Thiele, 2018).

FBA works in steady state, that is, the quantity of metabolites secreted is equal to the quantity of metabolite utilized. Hence, the set of linear equation can be written as follows:

$$S^{*}\ v = 0 \tag{8.1}$$

where S denotes stoichiometric matrix and v denotes distribution of flux. The formula that describes a biomass function comprises every metabolite within the system, which is needed for building a new cell. It can be also said that the biomass function stimulates the growth of cells.

Significantly, because Eq. (8.1) resembles a set of linear equations and usually, there occur more number of reactions than the compounds and so, greater than one flux distribution can take place. Constraints are included within a flux as lower and upper bounds for limiting minimum and maximum values, which each of the flux can hold. Constraints can represent conditions of the media, where rates of secretion and uptake are restricted or velocities of the reactions of interior enzymes are found from experimental results (Grobeholz et al., 2016). In accordance with this, each flux constraint can be written as follows:

$$V_i \min < V_i < V_i \max \tag{8.2}$$

In a similar way, the OF can be written as follows:

$$Z = c^{T*} v \tag{8.3}$$

where Z is solution for OF, C is vector of weights, which indicates how much of every reaction can contribute for OF, and T is the transpose of the matrix. For instance, when one flux is minimized or maximized, C is a vector of the zeroes with single one (Orth et al., 2010). On the other hand, fabricating an OF may become difficult and totally depends on the research problem.

FBA is flexible technique to be applied for many cases. Through adjustment of the lower and upper bounds of the metabolites, growth on various media or of the gut microbiota altered diet may be simulated. In this way, by means of setting flux of a particular metabolite to 0, a knockout of the gene or deletion of a microbe from the microbiota can also be simulated. In this manner, the viability of a community of microbes may be estimated in presence of various conditions and even the impact of introducing mew microbial species to a community of microbes on health of the host. For instance, FBA may be utilized for investigating the impact of lactic acid production over *Bifidobacterium adolescentis* showing reduction in secretion of ethanol, formic acid, and acetic acid and also reduction in the formation of biomass when production of lactic acid is increased manually. In this case, the OF is maximizing production of biomass indicating alteration in distribution of flux alters the environment (El-Semman et al., 2014). For investigating communication inside a community of microbes, researchers introduced a new bacterium, namely, *Faecalibacterium prausnitzii*, within the identical metabolic condition just like *B. adolescentis*. *F. prausnitzii* requires acetic acid for growing well in presence of glucose for butyrate production (Duncan et al., 2002). If acetic acid is not available, *F. prausnitzii* will utilize the acetic acid secreted by *B. adolescentis* for growing and subsequently producing butyrate. Because the reaction of biomass triggers growth of bacteria, flux alteration via reaction of biomass in one bacterium while maintaining the total system biomass as constant can trigger modifications in composition of the microbial community. Quantity of flux alteration via reaction of biomass will lead to modification of the distribution of flux in the entire system, thereby causing an elevation in the production of butyrate when flux via reaction of biomass in *F. prausnitzii* is augmented within the system in comparison with a condition, in which there is more flux via reaction of biomass in *B. adolescentis*. In this case, OF is reduction of glucose utilization for the 2 bacteria. In a similar manner, the addition of a compartment, which denotes a host cell, and then linking it with a common metabolic compartment, which further denotes lumen of the intestine, can be employed for studying the interactions between the microbiota and the host. By implementing this strategy, Heinken et al. (2013), aggregated the metabolic compartments of *Bacteroides thetaiotaomicron* and a generalized cell of mouse within a network of metabolites. The researchers studied metabolic reliance between the microbe and its host via parallel optimization of the rates of growth of the microbe inside the host cell. The researchers indicated that the existence of *B. thetaiotaomicron* may impact the generalized growth of the mouse cell by supplementing the cell of the host with nonessential as well as essential amino acids. Moreover, the researchers simulated knockouts of the gene of *B. thetaiotaomicron* and the cell of the mouse by adjusting the equivalent flux to zero. By optimization of the biomass functioning of the cell along with knockout, the researchers proved that *B. thetaiotaomicron* can protect the deadly knockouts within the cell of the host and vice versa (Heinken et al., 2013). Even though this metabolic network depicts just a communication between a type of host cell and a type of bacterium, which is far from confirming the microbiota interactions, scanning the network by introduction of more number of bacteria and the host compartments may provide newer details into systemic impacts of the host on the microbiota and vice versa. Metabolic networks showing metabolism of humans including host communications with the microbiota may be utilized for studying the casualties of microbiota-associated diseases, thereby formulating new hypotheses for treating these microbiota-associated diseases.

All the aforementioned examples of the metabolic networks indicate the utility of computational methodologies for understanding the complexities of microbiota. Luckily, the strategy of utilizing the methodologies of FBA within enclosed metabolic networks remains the same in case of larger metabolic networks. Hence, a detailed understanding of such principles will aid in formulating predictions further than what experimental researchers can deduce from logical reasoning to slowly facilitate inventing newer strategies for improving health of the humans.

Because FBA functions at steady state, it leads to an optimum flux distribution for a provided OF within small metabolic networks. In case of large metabolic networks, numerous solutions can be sought. To find such options, variance in analysis of flux may be utilized by means of using FBA for minimizing and maximizing each of the fluxes within the system (Orth et al., 2010). To find optimum solutions by metabolic network simulation, one must not take dynamic alterations in levels of metabolites and dynamic communications among the bacteria. Dynamic modifications play an important role for investigating communities of bacteria, since the cells can divide and die. In addition to this, in human host, there is cellular movement from upper portion of the intestine to the lower portion because of motility of the gut and the levels of metabolites keep on varying dynamically during day because of the cycles of food intake (Ji et al., 2020). Hence, capturing the dynamic modifications within metabolic networks is required for studying the microbiota. For capturing the dynamic alterations, dynamic FBA (DFBA) has been formulated. DFBA makes use of ordinary differential equations (ODEs) for coupling FBA with a kinetic model (Watanabe et al., 2017). This can be performed by means of three strategies: dynamic optimization approach (DOA), static optimization approach (SOA), or direct approach (DA) (Gomez et al., 2014). The mostly used strategy is SOA. Primarily, SOA fabricates a line of snapshots by the use of FBA. The initial conditions of a single snapshot are measured by the result of the initial snapshot. SOA needs small steps among the snapshots for exact capturing of dynamic alterations, thereby making it computationally costly (Hoffner et al., 2013). DOA extracts time profiles for levels of metabolites and fluxes by optimization of the entire period of time for simulation. The problem of dynamic optimization gets transformed into a nonlinear problem of programming that needs to be solved only once (Hoffner et al., 2013). In comparison with DOA, DA makes use of a solver of linear program on the right-hand side of the ODE. Just like SOA, DA is even computationally costly because the linear program has to be solved every time the right-hand side of ODE is evaluated (Gomez et al., 2014). SQA application can be used for the purpose of reframing and investigating *Escherichia coli* in the process of metabolic communication in diauxic growth. The researchers triggered a batch culture for 10 h, which got divided into time intervals of 10,000. The metabolite concentration at the initiation of every interval was directly measured from the initial interval. This strategy demonstrated that within the initial 4.6 h, the uptake rate of glucose and oxygen of the cells served as the limiting conditions for formation of biomass. Within the second interval between 4.6 and 6.9 h, the limiting factor for formation of biomass was the surrounding oxygen concentration. Within the last interval, where acetic acid gets utilized between 6.9 and 10 h, the coefficient of mass transfer acted as the limiting factor for formation of biomass. The computational results were identical to the experimental results, thereby proving the utility of DFBA for determining the limiting constraints of bacterial growth. DFBA can be utilized for studying microbial communities consisting of *Saccharomyces cerevisiae* and *E. coli* (Henson & Hanly, 2014). On the other hand, DFBA can be just used for small bacterial communities since addition of more species within a network dramatically raises reaction amount, which subsequently raises the costs and time taken for the simulations. In case of research on microbes, DFBA is extensively studied. For simulating communities of microbes by means of FBA, not only the reactions for secretions and uptakes are important, but also the molecular capacities of every bacterium within the metabolic network must be known. Hence, a metabolic network of each single bacterium is required. This can be performed by generation of metabolic models of every bacterium directly via their genomes, which have been already sequenced.

3. Construction of metabolic networks from annotated genomes

By creating ideas based on genome-scale metabolic models and validating them in the lab, computational approaches can direct the investigation of the microbiome. The COBRA methodology (Palsson, 2015), which has been effectively used in investigations on metabolic pathways, individual species metabolism, and interspecies metabolic interactions (Oberhardt et al., 2009, 2015; Heinken & Thiele, 2015), is one technique that can help elucidate such processes. Genome-scale metabolic reconstructions (GENREs), which reflect the whole known set of metabolic pathways that occur in that organism based on genomic and experimental information, are the foundation for COBRA research of the gut microbiome (Magnúsdóttir & Thiele, 2018).

The metabolic capacity of the bacteria in the network must be understood and converted into a model that can be utilized by FBA to construct and simulate a metabolic network. All of a bacterium's metabolic capabilities are included in genomic-scale metabolic models (GEMs), often referred to as GENREs (Magnúsdóttir et al., 2017), which are created automatically from annotated genomes (Zelezniak et al., 2015). The restrictions of these automatically created GEMs need to be manually revised because not all of an organism's genes are activated throughout each stage of growth or in every environment (Oberhardt et al., 2009, 2015;

Robador et al., 2018; Thiele and Palsson, 2010). AGORA, a semiautomated GEM database of 818 gut microbiota members, just became accessible (Magnúsdóttir et al., 2017). The AGORA GEMs, however, are distinct from other GEM databases such as BiGG, KBase, and CarveMe that are readily available (Arkin et al., 2018; King et al., 2015; Machado et al., 2018). It is feasible to utilize these GEMs in various ways since the AGORA database has a greater range of taxonomically categorized species and the GEMs are built with more generic construction. The AGORA database is also designed to model the whole gut microbial ecosystem. In comparison with the BiGG database, which has a smaller number of bacteria and is typically used to simulate the metabolic capabilities of a single bacterium and predict potential changes in this bacterium's metabolic capacity in the event that it harbors inactive genes, the AGORA GEMs, for example, have more carbon-uptake reactions (Lieven et al., 2020; Babaei et al., 2018). Since *E. coli* is a well-researched organism and the BiGG database contains many GEMs of it, it is possible to reliably identify the metabolic reactions and constraints from the literature. The AGORA database, on the other hand, comprises a large number of understudied species. This emphasizes the need for further research, an update of the metabolism of gut bacteria, and the appropriate refinement of the GEMs (Babaei et al., 2018). This provides a new outlook to the AGORA database as a repository for generic GEMs of gut bacteria, which need to be first converted into condition-specific GEMs by adding the appropriate restrictions and creating OFs based on the user's research topic. Before employing GEMs derived from databases that also contain condition-specific GEMs, this must be completed (Magnúsdóttir et al., 2018). Study done by Pryor et al. demonstrated that bacterially generated agmatine increased the longevity of *Caenorhabditis elegans* worms in vivo. Agmatine production is stimulated by metformin, a medication that is frequently prescribed to people with type 2 diabetes. The reported in vivo results in *C. elegans* were extrapolated to humans using GEMs from the AGORA database by the authors. They discovered that patients receiving metformin had a larger synthesis of agmatine, demonstrating a biochemical connection between the host and its microorganisms. The authors manually selected the GEMs and added agmatine uptake and secretion responses before they could utilize the GEMs from the AGORA database. The dietary information available for the patients in the data set used for the construction of metabolic communities required the authors to modify the AGORA GEM parameters, which were originally constructed using a western diet (Pryor Rosina et al., 2019). *Citrobacter sedlakii* is a nonpathogenic bacterium that is present in human stool. Kuang et al. coupled FBA with an untargeted mass spectroscopic analysis technique to discover compounds generated by *C. sedlakii* to enhance GEMs of bacteria. By collecting samples at various development stages and evaluating the extracts using two liquid chromatography—mass spectrometry (LCMS) techniques, reverse phase (RP) and hydrophilic interaction liquid chromatography, the scientists studied the metabolic output of *C. sedlakii* (HILIC). Employing a program named MS FBA, the collected data were compared with a forecasted list of metabolites produced using FBA. The comparison revealed that not all of the metabolites detected with LCMS were covered by the metabolic output of the *C. sedlakii* GEM, exemplifying the need to scrutinize the authenticity of bacterial GEMs and optimize genome annotation (Kuang et al., 2020). The metabolic microenvironment of the microbiota should be established when assessing flux distribution between diverse sorts of microbial profile, such as between patients and healthy controls. To infer the metabolic milieu of the microbial population, the metabolic environment from abundance distribution in individual samples is another strategy that should be altered with. To deduce the metabolome in the sequenced stool specimen, the Metabolic Analysis of Metagenomes using FBA and Optimization (MAMBO) technique integrates 16S sequencing data from individual stool specimens with GEMs collected from baseline genomes of the sequenced bacteria. This strategy is justified by the idea that the metabolic environment influences the abundance profile in a particular sample since the metabolic capability of the bacteria reveals which species will proliferate there (Rothschild et al., 2018). GEMs were created using the modelSEED pipeline using annotated reference genomes that were made accessible through the human microbiome project (Tumbaugh Peter et al., 2007) to provide an answer to this topic. The initial metabolite concentration in the metabolic environment was determined at random, and FBA was carried out. The OF is defined as the requirement that all GEMs' biomass functions be as near as possible to the discovered abundance distribution. The metabolites in the metabolic environment were slightly changed in a subsequent stage, and FBA using the identical OF was once more carried out. The metabolite alteration is accepted, and the procedure is continued until the Pearson correlation does not improve if the Pearson correlation between the biomass functions and the abundance profile is greater than it was in the prior metabolic environment. Which metabolic environment best fits the abundance distribution may be determined by comparing the biomass distribution in the FBA findings with the abundance profile obtained from the fecal sample (Gazra et al., 2018).

4. GEM application in gut microbiota modeling

For understanding the impact of the microbiota over the host, metabolic networks of the microbial may be extrapolated for including host metabolic networks. This allows the prediction of the impacts of gut microbiota over the host, thereby signifying probable interferences for promoting health of the host. Diener et al. (2020) fabricated metabolic networks of microbes from the metagenomic data of a cohort of the people of Sweden without and with diabetes mellitus and utilized FBA for analyzing production of SCFA. Here, abundance data of the 16S were incorporated along with metabolic flux for constructing tailored calculation of metabolic production and interferences such as dietary modifications and medicinal therapies. The researchers made use of GEMs from the AGORA database; in contrast, the AGORA database did not wrap every species existing in the cohort of Sweden. As an alternative, the researchers carried out FBA on the level of species and genus, wherein models of AGORA were united jointly into higher ranks of phylogeny. This gave rise to networks having 12–30 GEMs at the level of genus and 23–81 GEMs at the level of species. The GEMs were positioned within a communal metabolic surrounding for every sample of the cohort that represents the lumen of the gut. The relative biomass of each GEMs was predictable from the relative read distribution in every sample being sequenced. The researchers found a minimum overlie of resource consumption among microorganisms inside diverse niches, thereby implicating an upper boundary on alpha diversity of the gut. Following the ecological aspects, the researchers could conclude through their model that production of SCFA is extremely specific for each individual. Nevertheless, butyric acid and propionic acid production was decreased in diabetic individuals in comparison with the healthy individuals, and the general production of SCFA profile might get restored upon treatment with metformin

(Diener et al., 2020). The researchers did not calculate levels of metabolite in vitro, but the result is in line with other experimental research data (Zhao et al., 2018). Another investigation extended GEMs present in the AGORA database having reactions for metabolism of bile acid. In this case, corresponding bacteria were positioned in a common metabolic situation for comparing their capability of bile acid metabolization with that of a single bacterium. A standard European diet supplemented with bile acids such as glycocholate, taurocholate, glycochenodeoxycholate, and taurochenodeoxycholate was utilized as modeling factors, thereby maximizing the bile acid exchange used as the OF. The researchers demonstrated that bile acid metabolism is greater within a community of microbes in comparison with a single bacterium since individual bacteria cannot metabolize each bile acid. In a similar way, the researchers incorporated publicly accessible abundance data of microbial in healthy persons and patients suffering from inflammatory bowel disorders (IBD) into metabolic network for studying metabolism of bile acids in IBD. The experiment indicated that bile acid metabolism was lesser in patients having IBD, which is in line with in vivo data (Duboc et al., 2013). On the other hand, the researchers did not do any interference such as dietary modification or integration of other bacteria within the metabolic network for predicting promising therapy of patients suffering from IBD. Interventions in diet were further studied using 16S sequencing records for construction of metabolic networks of 28 patients suffering from Crohn's disease (CD) and 26 healthy subjects as control. The investigation positioned the GEMs, which were derived from the AGORA database and made use of BacArena, which is a technique to model communities of bacteria and placed the GEMs inside a grid for each patient (Bauer et al., 2017). The composition of bacteria derived from 16S sequencing results was utilized for determining the quantity of GEMs for every species, which are being placed inside the grid at the beginning of the simulation. The biomass of the microbes was utilized as OF, and all available metabolites, which can be utilized by the GEMs, were introduced at high concentrations in the surrounding. Growth of bacteria was simulated within 24 steps for a time period of 1 h. The final timepoint was used for comparing the abundance of microbes and production of metabolites between patients suffering from CD and healthy individuals. Fascinatingly, the results indicated a greater SCFA production in case of the control individuals. Following this, the researchers recognized metabolites that could result in higher SCFA production in every structured metabolic network of the patients suffering from CD. By addition of extra recognized metabolites within the surrounding in a tailored way, the researchers demonstrated that modifying the diet in silico may impose varying outcomes in every individual on the basis of their composition of the microbiota, which is in line with experimental data (Green et al., 2019).

Recently, numerous techniques have been made available to model communities of bacteria, such as BacArena, OptCom, COMETS, MiMoSa, MICOM, and FLYCOP. To totally confine the impact of the community of microbes on the host, metabolic networks in the microbial community requires to be expanded with GEMs of the host cells. Several GEMs for tissues of humans such as blood vessels, liver, and epithelial cells of the gut are already accessible for expansion of the present metabolic networks of the microbiota for including the cells of human (Fouladiha & Marashi, 2017). Combination of the metabolic networks on the basis of 16S sequencing results along with a representative human metabolism GEM could be utilized for investigating the impact of the metabolic output of the microbiota on health of the host. A representative human metabolism model is Recon3D that comprises more than 13,000 metabolic reactions and may be applied to integrating metabolism of microbiota and the host (Noronha et al., 2019). In contrast to this, combination of bacterial GEMs and host GEMs can be difficult because of the OF formulation and spatial arrangement. For instance, the intestine is spatially arranged, in which the cells of the intestine enclose the microbiota. In accordance with this, metabolites of the gut microbiota possess the maximum concentration inside the lumen and will be initially available for the microbiota and not to the cells of the host. While simulating the communications between the host, and the microbiota of the gut, metabolic networks must consider these gradients. In a similar manner, pH and metabolic gradients continue living along the length of the tract of the intestine, and the composition of the microbiota within the small intestine varies from that of the colon (Kuc'D et al., 2008). Moreover, cells of the bacteria pass through the upper portion of the intestine to the colon and finally get discarded through the feces (Kastl et al., 2020). One method for simulating such gradients and spatial arrangement is by adding blank compartments between compartments of the cells of host and the compartments of bacteria within a metabolic model. The compartments get structured within a 2D grid, and every bacterium has its individual compartment, from where it can swap nutrients. The compartments get linked via the swapping reactions; hence the bacteria can communicate with each other. Through addition of the compartments in absence of bacteria in between the compartments, which are occupied by bacteria, a new bacterium can fulfill the blank compartment, thereby simulating bacterial movement or replication. Furthermore, the blank compartments simulate the gradients, because all the nutrients, which enter the blank compartment, will not move to the following compartment, which is already filled with a bacterium (Bauer et al., 2017). Another method of simulating bacterial movement and gradient is by means of differential equations. A metabolic model was used, wherein the spatial arrangement is incorporated for investigation of cross-feeding. The researchers extended a *Lactobacillus plantarum* GEM along with the metabolic reactions, which are usually found within the gut-like butyric acid and propionic acid fermentation. Subsequently, the researchers positioned multiple extended *L. plantarum* GEMs at random within a tube-like grid surrounding and utilized SOA for simulating the network in 80,000 time steps. For each of the cells within the grid, OF maximizes the ATP production rate, which is a substitute to formation of biomass. The system was supplemented with glucose at regular intervals through the proximal end of the tube and metabolite concentration in every compartment was moved to the right end till they exit the system through the distal end of the tube. Collectively, this denotes metabolic gradients occurring inside the gut. The researchers subsequently introduced removal, growth, single microbial evolution, and movement within the network and demonstrated that versatility can be a sprouting characteristic of cross-feeding between communities of microbes. When diarrhea was being simulated through maximization of the speed of the flux within the system, the diversity of microbes was disrupted, which is in line with in vivo results (Vandeputte et al., 2016). Chan et al. (2019) suggested that aerobic and anaerobic bacteria can be found in various niche depending on the availability of oxygen. These instances indicate the significance of the ecology of the gut, and such factors must be considered. Persi et al. (2018) introduced enzymatic activity depending on the pH, and these enzymes were derived from experimental results from the BRENDA database in a cancer cell GEM. The researchers demonstrated that cancer cells divide in a different way in silico in accordance with the pH. In contrast to this, integration of pH-associated enzymatic activity within a GEM is strongly dependent on experimental research carried out in vitro. Because most of the enzymes of the bacteria residing within the gut are under study, introduction of pH-associated functions in GEMs utilized for metabolic modeling is at present not practical.

Another limitation in combining host GEMs and bacterial GEMs is the fabrication of an accurate OF. In case of most of the metabolic networks, the total growth of the biomass, or metabolite production, remains optimized. Inside the gut, the host takes most of the benefits from an equilibrium between the bacterial metabolite production that needs a steady state between the various microbial species distribution. For solving this issue, OptCom has been designed. This strategy makes use of 2 OF layers. The first layer increases the formation of biomass to the highest in every individual species. The second layer increases the entire communal growth, thereby giving rise to a more realistic distribution of growth within a community of bacteria (Zomorrodi et al., 2012). The expansion to d-OptCom may be applied for DFBA (Zomorrodi et al., 2012). On the other hand, OptCom is not easy be applied in communities comprising of a wide number of bacteria because of increased time required for computation with escalating complexity of the community. Thus, community- and systems-level interactive optimization (CASINO) was designed. The CASINO structure makes use of 2 OF layers; however, it varies from OptCom by optimization of both the OF layers iteratively. A limitation with CASINO and OptCom is the explanation of the biomass function. As already mentioned, the host gets benefits the maximum from an equilibrium of the metabolite production of every bacteria present. Hence, for a strong microbiota, the community requires to work at steady state. On the other hand, every single bacterium grows having a specific rate of growth (Ley et al., 2006). CASINO and OptCom consider the growth of bacteria, excluding the steady state of the bacterial community. Thus, optimization of formation of biomass leads to the predominance of the bacterium that grows very fast within the system, leading to a distribution of flux, which does not symbolize the distribution of flux in the community of the gut. For overcoming this disadvantage, SteadyCom was designed that considers steady state of the community (Chan et al., 2017).

5. Dysbiosis simulation and therapies by use of metabolic networks

Although the gut microbiota composition varies in due course of time, since, out of all others, the composition of the diet largely is relatively stable and can improve with perturbations for a short time namely, administration of antibiotics for a short time period, starvation periods, and fundamental dietary changes (Suez et al., 2018). On contrary, perturbations for a longer period of time like long-drawn-out use of antibiotics, other than dietary changes may lead to a modification in the composition of microbiota, thereby resulting in modification of its metabolites (Zinocker et al., 2018). Modifications in the composition of microbiota, also called dysbiosis, may impose negative effects on health of the host. Dysbiosis has been found among patients suffering from Huntington's disease, Alzheimer's disease, and Parkinson's disease (PD) (Sun et al., 2018). The microbiome modeling tool may utilize a metabolic network for construction of the microbiota from the relative abundance data derived from 16S sequencing studies of fecal samples and might be applied for studying dysbiosis (Baldini et al., 2019). Baldini et al. (2019) made use of the microbiome modeling tool for studying the variations in the metabolite production between the microbiota of healthy individuals (control) and patients suffering from PD. The researchers positioned GEMs obtained from the AGORA database within a common metabolic surrounding for a metabolic network construction. Variations in composition of bacteria among every person were performed by adjustment of the coefficient for the biomass function inside the stoichiometric matrix. Increasing the entire output to a maximum of the community of microbes in case of an already produced metabolite was utilized as OF. Overall, 129 metabolites were studied for every individual. The average diet of Europe was utilized as an input for simulating modifications in the output of the metabolites. The experiment reported nine metabolites to possess the capacity to be remarkably modified in patients suffering from PD in comparison with healthy individuals (controls) including cysteinylglycine and methionine that are part of the metabolism of sulfur. Moreover, the researchers indicated that an increased presence of *Bilophila wadsworthia*, and *Akkermansia muciniphila* in patients suffering from PD, recognized a novel research target in research of PD by using metabolic networks (Balindi et al., 2019). Hertel et al. (2019) proved such results by means of the similar strategy as used by Baldini et al. (2019) for construction and simulation of customized community networks of 31 early stages; patients suffering from PD and who have not been administered with any drug and age of 28 acted as matched controls from results derived by Bedarf et al. (2017). FBA simulation confirmed that four reactions of the microbes taking part in metabolism of homoserine get modified in patients with PD, which remains constant with calculated levels in the plasma of patients with PD, and homoserine, methionine precursor, is a part of the metabolism of sulfur. Moreover, the researchers pointed an elevation in two species of bacteria, both of which take part in metabolism of sulfur, *A. muciniphila* and *B. wadsworthia* (Hertel et al., 2019). Both of these species have been reported in advancement of PD (Sampson et al., 2016). The aforementioned investigations indicate the capacity of FBA and the metabolic networks in studying the underlying associations in understanding disease theory. In addition to this, dysbiosis may take part in treating neurodegenerative diseases. For instance, patients suffering from PD were administered levodopa (L-DOPA) as medicine. In contrast, the dosage changes extensively for different patients (Katzenschlager & Lees, 2002). Microbes of the microbiota in gut could alter L-DOPA. Hence, possessing more quantity of L-DOPA altering microbes within the microbiota leads to a greater L-DOPA dosage among patients with PD (Rekdal et al., 2019). However, therapies applied in the treatment of brain-associated disorders may impose negative effects by dysbiosis induction within the gut. For instance, fluoxetine, which is a drug applied in the form of an antidepressant, results in sporulation among the microbes of the microbiota in the gut. Therefore, modification of the metabolite production by the microbiota can affect health of the host (Fung et al., 2019). The aforementioned investigations demonstrate a connection among dysbiosis and medication, however, lacking the dysbiosis causality and substitutes for treating dysbiosis for improvement of the effectiveness of drug. Construction of metabolic networks of a dysbiosis may provide an insight in the communications of dysbiosis. Subsequently, treatment efficacy may be tested in silico on the microbiota of every patient separately, which would enhance the procedure to choose the best therapy, while prescribing medicines.

Over the recent years, dysbiosis may be cured in numerous manners such as fecal matter transplantation (FMT), use of antibiotics and applying pre-, pro-, or psychobiotics. FMT has been applied with great success in persistent infection caused by *Clostridium difficile*, while it was not much effective for the treatment of IBD. Fascinatingly, an experiment studied FMT in case of IBD, which confirmed that patients, who received treatment with fecal material from a specific donor, evoked greater response in comparison with patients, who were being treated with fecal matter from other donors . This indicates that for effective usage of FMT, the species of microbes and metabolites, which are accountable for the useful impacts of FMT, requires to be recognized. On contrary, for identification of the varieties among donors and comprehending the reason behind, certain donors are better than the others, not just the microbiota and composition of the metabolites is required, but also the bacterial interaction among the host and the bacteria, hinting at the necessity for constructing the metabolic networks of donors, is required

Additional promising therapy for dysbiosis is the application of probiotics. Probiotics can be defined as live microbes that, while administered in sufficient quantities, provide the host with a health benefit (Hill et al., 2014). Several studies have demonstrated positive impacts of an individual strain or a probiotic mixture on health of the host within humans or animals. Among humans, the impact of administration of probiotic is widely studied in clinical trials, wherein the impact on health is calculated (Markowiak & Ślizewska, 2017). On the other hand, these studies frequently result in unexpected or contradictory outcomes. For instance, Suez et al. (2018) found that probiotic administration followed by treatment with antibiotics reduced the recovery rate of the microbiota, while other investigations reported that probiotic administration prior to treatment with antibiotics does not impose any effects on the microbiota recovery (Korpela et al., 2018). Such conflicting outcomes may be described by variations in the strains of probiotics utilized, antibiotic types, treatment dosage, and also medical history, diet, patient genetics, and preliminary composition of the microbiota. To overcome these limitations, there is a requirement to understand the molecular mechanisms behind the observed impacts of probiotics. Besides, the probiotic viability in vivo, and the communication inside the microbiota and the metabolite production, requires to be studied. With respect to this, metabolic networks offer a helpful apparatus to consider the result of probiotics on the microbiota of gut. When a strain of probiotic is introduced in the form of a compartment within a metabolic network of a gut community, the distribution of the metabolic flux of the network will alter. Moreover, specific bacterial metabolite levels produced inside the network will be amplified or reduced. As a result, this could pose an impact on the composition of the microbiota presented inside the network. From these modifications observed in silico, the efficacy of treatment with probiotic can be assumed. On the other hand, for accurate prediction of the impact of adding a novel species in a community, the metabolic activities of the newly introduced species must be identified at the strain level, and the metabolic capacity subjected to various environmental conditions must be known. Species of *Bifidobacterium* are broadly applied as probiotics (Azad et al., 2018). As aforementioned, the *Bifidobacteria* GEMs in the AGORA database are not condition-sensitive and do not present the metabolic behavior in depth. For instance, the *Bifidobacteria* GEMs from the AGORA database did not grow in presence of starch, while a majority of *Bifidobacteria* could metabolize starch (Liu et al., 2015). Thus, cautious curation of the GEMs obtained from the AGORA database is required prior to the utilization of the FBA strategy in research on probiotics. Devika & Raman (2019), developed GEMs of 36 *Bifidobacteria* strains counting strains of probiotic utilized for commercialized goods. The researchers compared the metabolic capacities of the GEMs subjected to 30 various surrounding conditions, with maximized biomass function acting as OF. On the basis of metabolic capacities within various environments, the *Bifidobacteria* might be classified into three groups. According to the metabolic abilities of the GEMs, the researchers assumed that the shielding impact of the probiotic candidate *Bifidobacterium thermophilum* strain RBL67 on *Listeria* and *Salmonella* species comes from the SCFA production. Moreover, the researchers assume that *Bifidobacterium kashiwanohense* strain DSM21854 and *Bifidobacterium gallicum* strain DSM20093 may aid in relieving constipation through acetic acid production suggesting that FBA may be a helpful technique to identify novel species of probiotics (Devika & Raman, 2019).

6. Conclusion

Communications inside the complicated community of microbiota and among the host and the microbiota are difficult to clearly understand the vitality of such communications to successfully intervene within the microbiota. This chapter demonstrates the use of computational methodologies on the basis of FBA, thereby providing with a novel insight into the microbial communications and helping in fabricating newer hypotheses, which can be further tested and aid in the investigation of the efficacies of the administration of the probiotics in a customized way.

References

Arkin, A. P., Cottingham, R. W., Henry, C. S., Harris, N. L., Stevens, R. L., Maslov, S., et al. (2018). In *KBase: The United States department of energy systems biology knowledgebase* (Vol 36, pp. 566–569). Nature Publishing Group.

Azad, M. A. K., Sarker, M., Li, T., & Yin, J. (2018). Probiotic species in the modulation of gut microbiota: An overview. *BioMed Research International, 2018*.

Babaei, P., Shoaie, S., Ji, B., & Nielsen, J. (2018). Challenges in modeling the human gut microbiome. *Nature Biotechnology, 36*(8), 682–686.

Baldini, F., Heinken, A., Heirendt, L., Magnusdottir, S., Fleming, R. M. T., & Thiele, I. (2019). The microbiome modeling toolbox: From microbial interactions to personalized microbial communities. *Bioinformatics (Oxford, England), 35*(13), 2332–2334.

Bauer, E., & Thiele, I. (2018). From network analysis to functional metabolic modeling of the human gut microbiota. *mSystems, 3*(3), 27.

Bauer, E., Zimmermann, J., Baldini, F., Thiele, I., & Kaleta, C. (2017). BacArena: Individual based metabolic modeling of heterogeneous microbes in complex communities. *PLoS Computational Biology, 13*(5), e1005544.

Bedarf, J. R., Hildebrand, F., Coelho, L. P., Sunagawa, S., Bahram, M., Goeser, F., et al. (2017). Functional implications of microbial and viral gut metagenome changes in early stage L-DOPA-naïve Parkinson's disease patients. *Genome Medicine, 9*(1), 39.

Chan, S. H. J., Simons, M. N., & Maranas, C. D. (2017). SteadyCom: Predicting microbial abundances while ensuring community stability. *PLoS Computational Biology, 13*(5), e1005539.

Chan, P. C., Lu, Q., de Toledo, R. A., Gu, J. D., & Shim, H. (2019). *Improved anaerobic co-digestion of food waste and domestic wastewater by copper supplementation – Microbial community change and enhanced effluent quality. Science of The Total Environment, 670*, 337–344.

Cheung, S. G., Goldenthal, A. R., Uhlemann, A. C., Mann, J. J., Miller, J. M., & Sublette, M. E. (2019). In *Systematic review of gut microbiota and major depression* (Vol 10). Frontiers media S.A.

Devika, N. T., & Raman, K. (2019). Deciphering the metabolic capabilities of Bifidobacteria using genome-scale metabolic models. *Scientific Reports, 9*(1), 1–9.

Diener, C., Gibbons, S. M., & Resendis-Antonio, O. (2020). MICOM: Metagenome-scale modeling to infer metabolic interactions in the gut microbiota. *mSystems, 5*(1), 21.

Duboc, H., Rajca, S., Rainteau, D., Benarous, D., Maubert, M. A., Quervain, E., et al. (2013). Connecting dysbiosis, bile-acid dysmetabolism and gut inflammation in inflammatory bowel diseases. *Gut, 62*(4), 531–539.

Duncan, S. H., Hold, G. L., Harmsen, H. J. M., Stewart, C. S., & Flint, H. J. (2002). Growth requirements and fermentation products of Fusobacterium prausnitzii, and a proposal to reclassify it as Faecalibacterium prausnitzii gen. Nov., comb. nov. *International Journal of Systematic and Evolutionary Microbiology, 52*(6), 2141–2146.

El-Semman, I. E., Karlsson, F. H., Shoaie, S., Nookaew, I., Soliman, T. H., & Nielsen, J. (2014). Genome-scale metabolic reconstructions of Bifidobacterium adolescentis L2-32 and Faecalibacterium prausnitzii A2-165 and their interaction. *BMC Systems Biology, 8*(1), 41.

Embree, M., Liu, J. K., Al-Bassam, M. M., & Zengler, K. (2015). Networks of energetic and metabolic interactions define dynamics in microbial communities. *Proceedings of the National Academy of Sciences, 112*(50), 15450–15455.

Fouladiha, H., & Marashi, S. A. (2017). In *Biomedical applications of cell- and tissue-specific metabolic network models* (Vol 68, pp. 35–49). Academic Press Inc.

Fung, T. C., Vuong, H. E., Luna, C. D., Pronovost, G. N., Aleksandrova, A. A., Riley, N. G., et al. (2019). Intestinal serotonin and fluoxetine exposure modulate bacterial colonization in the gut. *Natural Resources Forum, 4*, 2064–2073.

Garza, D. R., Van Verk, M. C., Huynen, M. A., & Dutilh, B. E. (2018). Towards predicting the environmental metabolome from metagenomics with a mechanistic model. *Nature Microbiology, 3*(4), 456–460.

Gomez, J. A., Höffner, K., & Barton, P. I. (2014). DFBAlab: A fast and reliable MATLAB code for dynamic flux balance analysis. *BMC Bioinformatics, 15*(1), 18.

Green, N., Miller, T., Suskind, D., & Lee, D. (2019). A review of dietary therapy for IBD and a vision for the future. *Nutrients, 11*, 5.

Großeholz, R., Koh, C. C., Veith, N., Fiedler, T., Strauss, M., Olivier, B., et al. (2016). Integrating highly quantitative proteomics and genome-scale metabolic modeling to study pH adaptation in the human pathogen enterococcus faecalis. *NPJ Systems Biology and Applications, 2*(1), 1–9.

Heinken, A., Sahoo, S., Fleming, R. M. T., & Thiele, I. (2013). Systems-level characterization of a host-microbe metabolic symbiosis in the mammalian gut. *Gut Microbes, 4*(1), 28–40.

Heinken, A., & Thiele, I. (2015). Systems biology of host–microbe metabolomics. *Wiley Interdisciplinary Reviews: Systems Biology and Medicine, 7*, 195–219.

Henson, M. A., & Hanly, T. J. (2014). Dynamic flux balance analysis for synthetic microbial communities. *IET Systems Biology, 8*(5), 214–229.

Hertel, J., Harms, A. C., Heinken, A., Baldini, F., Thinnes, C. C., Glaab, E., et al. (2019). Integrated analyses of microbiome and longitudinal metabolome data reveal microbial-host interactions on sulfur metabolism in Parkinson's disease. *Cell Reports, 29*(7), 1767–1777.e8.

Hill, C., Guarner, F., Reid, G., Gibson, G. R., Merenstein, D. J., Pot, B., et al. (2014). Expert consensus document: The international scientific association for probiotics and prebiotics consensus statement on the scope and appropriate use of the term probiotic. *Nature Reviews Gastroenterology & Hepatology, 11*(8), 506–514.

Höffner, K., Harwood, S. M., & Barton, P. I. (2013). A reliable simulator for dynamic flux balance analysis. *Biotechnology and Bioengineering, 110*(3), 792–802.

Jansma, J., & El Aidy, S. (2021). Understanding the host-microbe interactions using metabolic modeling. *Microbiome, 9*(1), 16. https://doi.org/10.1186/s40168-020-00955-

Ji, B. W., Sheth, R. U., Dixit, P. D., Tchourine, K., & Vitkup, D. (2020). Macroecological dynamics of gut microbiota. *Nature Microbiology, 5*(5), 768–775.

Kastl, A. J., Terry, N. A., Wu, G. D., & Albenberg, L. G. (2020). In *The structure and function of the human small intestinal microbiota: Current understanding and future directions* (Vol. 9, pp. 33–45). Elsevier Inc.

Katzenschlager, R., & Lees, A. J. (2002). Treatment of Parkinson's disease: Levodopa as the first choice. *Journal of Neurology, 249*, ii19.

King, Z. A., Lu, J., Dräger, A., Dräger, D., Miller, P., Federowicz, S., et al. (2015). BiGG models: A platform for integrating, standardizing and sharing genome-scale models. *Nucleic Acids Research, 44*, 515–522.

Kolodziejczyk, A. A., Zheng, D., & Elinav, E. (2019). In *Diet–microbiota interactions and personalized nutrition* (Vol 17, pp. 742–753). Nature Publishing Group.

Korpela, K., Salonen, A., Vepsäläinen, O., Suomalainen, M., Kolmeder, C., Varjosalo, M., et al. (2018). Probiotic supplementation restores normal microbiota composition and function in antibiotic-treated and in caesarean-born infants. *Microbiome, 6*(1), 1.

Kuang, E., Marney, M., Cuevas, D., Edwards, R. A., & Forsberg, E. M. (2020). Towards predicting gut microbial metabolism: Integration of flux balance analysis and untargeted metabolomics. *Metabolites, 10*(4), 1.

Kuc, D., Zgrajka, W., Parada-Turska, J., Urbanik-Sypniewska, T., & Turski, W. A. (2008). Micromolar concentration of kynurenic acid in rat small intestine short communication. *Amino Acids, 35,* 503–505.

Ley, R. E., Peterson, D. A., & Gordon, J. I. (2006). Ecological and evolutionary forces shaping microbial diversity in the human intestine. *Cell, 124,* 837–848.

Lieven, C., Beber, M. E., Olivier, B. G., Bergmann, F. T., Ataman, M., Babaei, P., et al. (2020). MEMOTE for standardized genome-scale metabolic model testing. *Natural Resources Forum, 38,* 272–276.

Liu, S., Ren, F., Zhao, L., Jiang, L., Hao, Y., Jin, J., et al. (2015). Starch and starch hydrolysates are favorable carbon sources for Bifidobacteria in the human gut. *BMC Microbiology, 15*(1), 54.

Li, M., Wang, B., Zhang, M., Rantalainen, M., Wang, S., Zhou, H., et al. (2008). Symbiotic gut microbes modulate human metabolic phenotypes. *Proceedings of the National Academy of Sciences, 105*(6), 2117–2122.

Machado, D., Andrejev, S., Tramontano, M., & Patil, K. R. (2018). Fast automated reconstruction of genome-scale metabolic models for microbial species and communities. *Nucleic Acids Research, 46*(15), 7542–7553.

Magnúsdóttir, S., Heinken, A., Kutt, L., Ravcheev, D. A., Bauer, E., Noronha, A., et al. (2017). Generation of genome-scale metabolic reconstructions for 773 members of the human gut microbiota. *Nature Biotechnology, 35*(1), 81–89.

Magnúsdóttir, S., Heinken, A., Fleming, R. M. T., & Thiele, I. (2018). Reply to challenges in modeling the human gut microbiome. *Nature Biotechnology, 36*(8), 686–691.

Magnúsdóttir, S., & Thiele, I. (2018). In *Modeling metabolism of the human gut microbiome* (Vol 51, pp. 90–96). Elsevier Ltd.

Markowiak, P., & Ślizewska, K. (2017). Effects of probiotics, prebiotics, and synbiotics on human health. *MDPI AG, 9,* 1.

Martinet, L., Naômé, A., Deflandre, B., Maciejewska, M., Tellatin, D., Tenconi, E., et al. (2019). A single biosynthetic gene cluster is responsible for the production of bagremycin antibiotics and ferroverdin iron chelators. *mBio, 10*(4), 7.

Noronha, A., Modamio, J., Jarosz, Y., Guerard, E., Sompairac, N., Preciat, G., et al. (2019). The virtual metabolic human database: Integrating human and gut microbiome metabolism with nutrition and disease. *Nucleic Acids Research, 47*(D1), D614–D624.

Oberhardt, M. A., Palsson, B. O., & Papin, J. A. (2009). Applications of genomescale metabolic reconstructions. *Molecular Systems Biology, 5,* 320.

Oberhardt, M. A., Zarecki, R., Gronow, S., Lang, E., Klenk, H. P., Gophna, U., et al. (2015). Harnessing the landscape of microbial culture media to predict new organism-media pairings. *Nature Communications, 6.*

Orth, J. D., Thiele, I., & Palsson, B. O. (2010). In *What is flux balance analysis?* (Vol 28, pp. 245–248). Nature Publishing Group.

Palsson, B. (2015). *Systems biology: Constraint-based reconstruction and analysis.* Cambridge University Press.

Persi, E., Duran-Frigola, M., Damaghi, M., Roush, W. R., Aloy, P., Cleveland, J. L., et al. (2018). Systems analysis of intracellular pH vulnerabilities for cancer therapy. *Nature Communications, 9*(1), 1–11.

Pryor Rosina, R., Norvaisas, P., Marinos, G., Best, L., & Thingholm, L. B. (2019). Host-microbe-drug-nutrient screen identifies bacterial effectors of metformin therapy. *Cell, 178*(6), 1299–1312.

Rekdal, V. M., Bess, E. N., Bisanz, J. E., Turnbaugh, P. J., & Balskus, E. P. (2019). Discovery and inhibition of an interspecies gut bacterial pathway for levodopa metabolism. *Science, 364,* 6445.

Robador, A., LaRowe, D. E., Finkel, S. E., Amend, J. P., & Nealson, K. H. (2018). Changes in microbial energy metabolism measured by nanocalorimetry during growth phase transitions. *Frontiers in Microbiology, 9,* 1.

Rothschild, D., Weissbrod, O., Barkan, E., Kurilshikov, A., Korem, T., Zeevi, D., et al. (2018). Environment dominates over host genetics in shaping human gut microbiota. *Nature, 555*(7695), 210–215.

Sampson, T. R., Debelius, J. W., Thron, T., Janssen, S., Shastri, G. G., Ilhan, Z. E., et al. (2016). Gut microbiota regulate motor deficits and neuroinflammation in a model of Parkinson's disease. *Cell, 167*(6), 1469–1480.e12.

Suez, J., Zmora, N., Zilberman-Schapira, G., Mor, U., Dori-Bachash, M., Bashiardes, S., et al. (2018). Post-antibiotic gut mucosal microbiome reconstitution is impaired by probiotics and improved by autologous FMT. *Cell, 174*(6), 1406–1423.e16.

Sun, M. F., & Shen, Y. Q. (2018). Dysbiosis of gut microbiota and microbial metabolites in Parkinson's disease. In *, Vol 45. Ageing Res Rev* (pp. 53–61). Elsevier Ireland Ltd.

Thiele, I., & Palsson, B. (2010). A protocol for generating a high-quality genome-scale metabolic reconstruction. *Nature Protocols, 5*(1), 93–121.

Turnbaugh Peter, J. P., Ley, R. E., Hamady, M., Fraser-Liggett, C. M., Knight, R., et al. (2007). *Nature, 449*(7164), 804–810.

Vandeputte Doris, D., Falony, G., Vieira-Silva, S., Tito, R. Y., & Joossens, M. (2016). Stool consistency is strongly associated with gut microbiota richness and composition, enterotypes and bacterial growth rates. *Gut, 65*(1), 57–62.

Watanabe, L. H., König, M., & Myers, C. J. (2017). *Dynamic flux balance analysis models in SBML.*

Zelezniak, A., Andrejev, S., Ponomarova, O., Mende, D. R., Bork, P., & Patil, K. R. (2015). Metabolic dependencies drive species co-occurrence in diverse microbial communities. *Proceedings of the National Academy of Sciences, 112*(20), 6449–6454.

Zhao, L., Zhang, F., Ding, X., Wu, G., Lam, Y. Y., Wang, X., et al. (2018). Gut bacteria selectively promoted by dietary fibers alleviate type 2 diabetes. *Science, 359*(6380), 1151.

Zinöcker, M. K., & Lindseth, I. A. (2018). In *The western diet—microbiome-host interaction and its role in metabolic disease* (Vol 10). MDPI AG.

Zomorrodi, A. R., & Maranas, C. D. (2012). OptCom: A multi-level optimization framework for the metabolic modeling and analysis of microbial communities. *PLoS Computational Biology, 8*(2), e1002363.

CHAPTER

9

Multispecies metabolomics interactions resulting in the development of resistance

Bhavdip Anavadiya[1], *Dhaval Acharya*[2], *Meenu Saraf*[1] *and Dweipayan Goswami*[1]

[1]Department of Microbiology & Biotechnology, University School of Sciences, Gujarat University, Ahmedabad, Gujarat, India; [2]Department of Microbiology, B N Patel Institute of Paramedical and Sciences, Anand, Gujarat, India

1. Introduction

Biofilm is a cluster of microorganisms attached with each other that are able to produce a complex structure of extracellular matrix that consists of protein, polysaccharide, and extracellular DNA on various surfaces including living tissue, aquatic ecosystem, terrestrial ecosystem, and nonliving things (Donlan, 2002). Microorganisms (bacteria or fungi) in biofilm can survive or inhabit in the natural environment by formation of macro- and microcolonies. Biofilm is a result of survival strategies of microorganisms against stress condition of environment. Bacteria are predominant group among microorganisms that can attach to surfaces and survive together by self-synthesized extracellular polymeric matrix (Jamal et al., 2015). These are also known as biofilms, of which bacteria interact with other microbial species on the basis of microbial competition, communication, and cooperation. The bacteria within biofilm show a modified phenotype as compared with corresponding planktonic cells, especially in the growth rate, synthesis of protein, gene transcription, gene expression, and intracellular metabolic activity. Bacteria may able to grow on a wide variety of living as well as nonliving surfaces and attach to alive surfaces, including medical devices, artificial devices, industrial instrument surface, etc. (Costerton et al., 1999). The biofilm formation is a naturally occurring process, which have a complex of microorganisms and surfaces—including slightly wet or slightly moist or whether possess a sufficient or nonsufficient level of nutrients. This phenomenon shows why

Microbial Biofilms
https://doi.org/10.1016/B978-0-323-95715-1.00016-9

TABLE 9.1 Distribution of multispecies biofilm.

Localization	Microbial species	References
Industrial bioreactor	*Desulfosarcina variabilis, Desulfovibrio aminophilus, Desulfotomaculum nigrificans, Flavobacterium, Desulfobulbus propionicus, Chryseobacterium* sp., *Desulfovibrio fructosivorans*	Urakawa et al. (1999)
Urinary catheter	*Pseudomonas aeruginosa, Escherichia coli, Staphylococcus epidermidis, Klebsiella pneumonia, Enterococcus faecalis*	Stickler (1996)
Marine sedimentary material	*Nitrospina gracilis, Verrucomicrobium* sp., *Pseudoalteromonas* sp., *Vibrio splendidus, Desulfosarcina variabilis, Anodontia phillippiana, Riftia pachyptila, Arhodomonas aquaeolei, Alvinella pompejana, Desulfocapsa sulfexigens, Lucina pectinata*	Dar et al. (2005)
Dental plaque	*Neisseria, Selenomonas, Actinomyces, Peptostreptococcus, Bifidobacterium, Streptococcus, Eubacterium, Veillonella, Propionibacterium, Campylobacter, Fusobacterium, Haemophilus, Treponema, Lactobacillus, Leptotrichia*	Marsh & Bradshaw (1995)
Chronic wounds	*Peptoniphilus, Bacteroides, Anaerococcus, Streptococcus, Staphylococcus, Peptostreptococcus* sp. *Enterococcus* sp., *Corynebacterium* sp., *Finegoldia*	Dowd et al. (2008)
Central venous catheter	*Klebsiella pneumoniae, Pseudomonas aeruginosa, Staphylococcus aureus, Staphylococcus epidermidis, Enterococcus faecalis, Candida albicans*	Donlan (2001)

biofilms can produce in laboratory easily (Brigette Carpentier., 1999). Biofilms have assembly of multiple bacterial species that shows resistance against antibacterial substances including antibiotics and detergents (Burmølle et al., 2014). Biofilms are involved in bioremediation of toxic compounds produced from human activities. They are also involved in removing unwanted organic compounds from wastewater treatment and supply clean water to the society (Chew & Yang, 2015). Microorganisms involved in multispecies biofilm have high antibiotic resistance and tolerance that play a major role in the causes and spreading of many chronic microbial diseases. Formation of biofilm confirms several benefits to bacteria including resistance against antibiotic exposure, metal toxicity, and osmotic stress (Dincer et al., 2001) (Table 9.1).

2. Formation of multispecies biofilm

2.1 Component involved in biofilm formation

The biofilm comprises extracellular components mainly one or more polysaccharides, DNA, and proteins (Flemming et al., 2007). Presence of certain membrane transport channels in the multispecies biofilm that allows passage for nutrients, oxygen, water, and other necessary biomolecules for the growth of microorganisms and development of biofilm (Zhang et al., 1998) (Fig. 9.1).

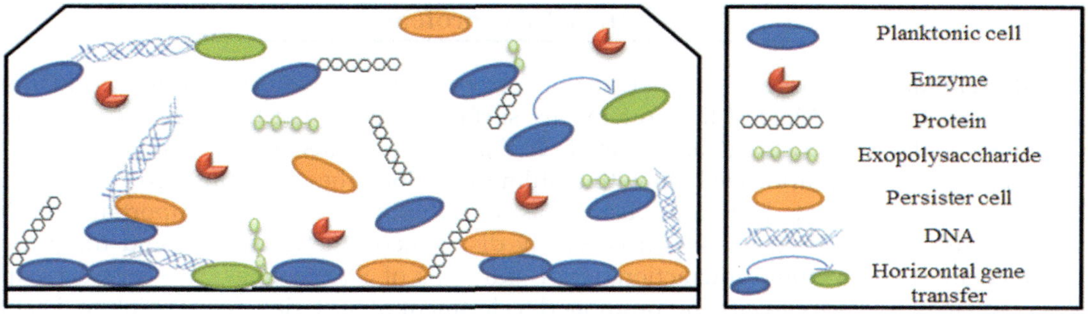

FIGURE 9.1 Components of biofilm.

3. Exopolysaccharides

Exopolysaccharides are either extracellular or intracellular product of a cell that is secreted into the outside environment (Nwodo et al., 2012). Exopolysaccharides are important in initiation and maintaining cell to cell interaction during biofilm development (Cugini et al., 2019). Basically, they appear as linear form or branched linear chain under electron microscope and are attached to outermost surface of cell that creates a large networks. Exopolysaccharides work as a ligament to other carbohydrates, lipids, eDNA, and proteins to adhere with each other in the development of biofilm. The structures, components, and properties of the exopolysaccharides are varied from one biofilm to another biofilm. The composition and linkages of exopolysaccharides matrix from *Klebsiella pneumoniae, Staphylococcus aureus, Acinetobacter baumannii, Pseudomonas aeruginosa, Enterococcus faecalis,* and *Escherichia coli* spp. biofilm were analyzed by the Nelson and his coworkers (Bales et al., 2013). Sugars are the most abundant carbohydrate group present in biofilm including glucose, mannose, galactose, arabinose, rhamnose, xylose, *N*-acetyl glucosamine, etc. (Bales et al., 2013). Most of the polysaccharides are not involved in biofilm formation, but their excess production leads to stress response such as alginic acid production in *P. aeruginosa* (Davies & Geesey, 1995) and colonic acid production in *E .coli* (Prigent-Combaret et al., 1999). Several gene clusters present in *P. aeruginosa,* which encode the proteins that are responsible for exopolysaccharide production, such as polysaccharide synthesis locus (*psl*) (Friedman & Kolter, 2004; Jackson et al., 2004), contribute in glucose, mannose, and rhamnose containing carbohydrate production, the pellicle (*pel*) (Friedman & Kolter, 2004a) gene involved in biofilm production and also responsible for pellicle organization in standing liquid cultures, and alginate synthesis gene for alginic acid production. PelC protein encoded by PelC gene is an outer membrane transport protein that allows transportation of polysaccharide to the outermost surfaces of bacterial cell. Mack and his coworkers characterized the polysaccharide intracellular adhesive (PIA) that is commonly known as poly *N*-acetyl glucosamine as exopolysaccharide of biofilm in *Staphylococcus.* PIA is a linear polymer form of β-1,6-linked glucosamine residues. EPS or poly-δ-glutamate is example of exopolysaccharide present in biofilm of *Bacillus subtilis* (Branda et al., 2006; Stanley & Lazazzera, 2005).

4. Extracellular proteins

Extracellular proteins are another major component involved in biofilm formation (Frølund et al., 1996). Most of the proteins are surface proteins that have several structural and functional characteristics (Chagnot et al., 2013). Some proteins are attached with cell surfaces and exopolysaccharides to help with biofilm formation. These proteins also provide stabilization to the biofilm. They are working as an adhesin and play an important role in multispecies biofilm. Protein adhesin is widely distributed among bacteria and fungi and can mediate bacteria–fungi interactions. SspB adhesins of *Streptococcus oralis* interact with Als3 protein of *Candida albicans* and promote development of bacterial–fungal biofilm. Lectin is one of the extracellular proteins found in *Azospirillum brasiliense* that links the bacterial cell to the exopolysaccharide and stabilizes the extracellular polymeric substance/matrix (EPS). Bap (biofilm-associated protein) is another group of high-molecular-mass surface protein, which provides primary attachment of cells to abiotic surfaces in biofilm of *S. aureus* (Agrobiotecnologı, 2001). In *S. mutants*, the glucan-binding proteins (gbps) link bacteria and exopolysaccharide in biofilm (Lynch et al., 2007). Amyloid proteins are water-insoluble fibrous in structure that provide support in biofilm architecture. Fap amyloid proteins are present in biofilm of *P. aeruginosa*, whereas TasA amyloid proteins form strong fibers that can hold biofilm of *B. subtilis* and protect against from harsh destructive forces (Romero et al., 2010).

5. Extracellular DNA

DNA can able to survive outside the cell; "eDNA" is defined as DNA not packed in the cell. DNA contains genetic information of living cells. The chromosomal DNA of microbial cell also known as genomic library (gDNA). Bacteria also contain small extracircular autonomously replicated DNA called plasmids. eDNA is an important element because it provides biofilm-regulating genes during biofilm formation. eDNA plays an important role in biofilm of gram-negative as well as gram-positive bacteria (Jamal et al., 2015). gDNA and eDNA may originate from the same cell, but they are not always similar on the basis of structure or composition. eDNA not only comes from the cell lysis, but sometimes it can be secreted by the cell that has an important role in formation of biofilm. It can also be secreted outside through membrane vesicles. Spontaneous phage induction can also help in secretion of eDNA. The negative charge of eDNA helps in the initial attachment to surfaces and also it interacts with substrate receptor of surface for facilitating adhesion (Das et al., 2010). eDNA also involved in twitching motility-mediated *P. aeruginosa* biofilm expansion (Gloag et al., 2013). eDNA- and DNA-binding protein are incorporated with each other and form biofilm in *E.coli.* eDNA can able to chelate metal cations from mixture due to its negative charge (Rabin et al., 2015). eDNA is also involved in biofilm of gram-positive *Staphylococcus;* strains of *S. aureus* have varying levels of eDNA in their biofilms (Rabin et al., 2015). *A. baumannii* (Sahu et al., 2012) and *Listeria monocytogenes* (Šuláková et al., 2019) are another examples of gram-positive bacteria that have eDNA in their structure of biofilms. Research studies that eDNA serves as nutrient source and spreads genetic material among microbial cells during biofilm formation (Panlilio & Rice, 2021).

6. Mechanisms for multispecies biofilm formation

Many bacteria have ability to alter between planktonic or free living forms to complex biofilm form. The growth rate and reproduction rate are relatively high in planktonic bacteria as compared with biofilm bacteria. The biofilm form is predominant state of bacteria in nature. When a bacteria approach at surface, attractive and repulsive forces come into play (Palmer et al., 2007). Formation of multispecies biofilm is a multistep process that consists of bacterial adhesion or attachment, maturation, and separation (dispersion or detachment) processes. Biofilm formation is also a dependent mechanism on the expression of biofilm-regulating genes that regulate the mechanism of biofilm formation (Okada et al., 2005). Successful adhesion of bacterial cell to solid surfaces can be carried out by an organic or inorganic molecules secreted from bacterial cells, and adhesion can be grouped into two stages. In the first stage, planktonic cells are moved toward to the suitable surface (Silva et al., 2014). Bacteria involved in biofilm formation possess flagella, which enables swimming motility of bacteria and fimbriae or pili, which enables active attachment of bacteria to an appropriate surface for multispecies biofilm formation. Bacterial attachment with appropriate surfaces depends on numbers of polysaccharides so surface attachment in parallel proportion with number of polysaccharides (Karatan & Watnick, 2009). The primary stage of bacterial adhesion is also referred to as reversible attachment stage because the primary stage of adhesion is detachable. Once attached, the bacteria start to colonize on the surface and form multiple colonies known as microcolonies. With increasing in the cell numbers, microcolonies form two-dimensional structures under favorable environmental condition. In which, the first step is to structural organization on inhabit surface. These organized structures further enter into maturation stage. In maturation stage, biofilm grows from a thin layer to 3D structures or mushroom-like assemblies with low surface coverage (Jahid & Ha, 2012; Karatan & Watnick, 2009). During maturation stage, cells of biofilm are trapped into exopolymeric substance secreted by the cells themselves (Costerton & Irvin, 1981). The final stage is dispersal or separation, which involves a number of cells leaving out from biofilm into surrounding environment. Dispersal can be done through active or passive detachment. In active detachment, dormant cell converts back to their planktonic form and promotes the initiation of new biofilms at other sites, whereas passive detachment occurs as a result of lack of nutrient availability, intense competition between species, movement of surrounding liquid, excessive growth of microbial communities, and erosion of parts of the biofilm by chemical substances or forces (Kaplan, 2010) (Fig. 9.2).

A preliminary mechanism for the formation of multispecies biofilms involves.

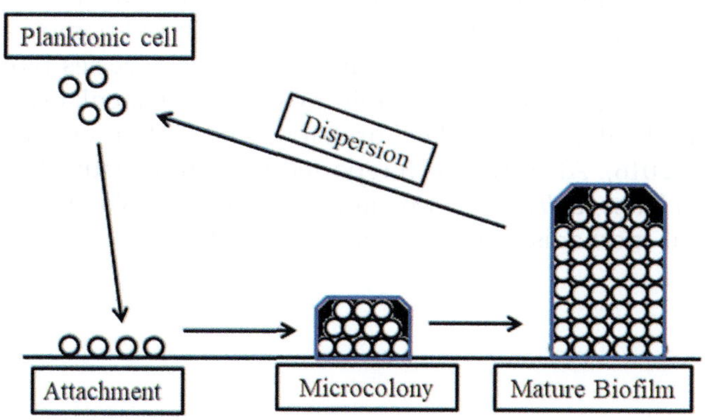

FIGURE 9.2 Biofilm formation process.

7. Quorum sensing signal molecules

Quorum sensing (QS) is a signal system in which species can sense cell density in microbial community and is also involved in regulation of gene expression, which takes part in multispecies biofilm formation (Bassler, 2002). QS signal is defined as a system that allows microorganisms to sense their population in narrower region in the environment and also participates in activation of biofilm-regulating genes for biofilm formation. Researches show that one or more microorganisms can recognize specific QS signal molecules, which enable them to communicate with other microorganisms nearby through chemical signal molecules called autoinducers (AIs) (Grandclément et al., 2015). The AIs can be used in multiway by cells to communicate with other cells in biofilm (Papenfort & Bassler, 2016). The concentration of signal molecule reaches the threshold value that leads to QS system activation. The activated QS system is responsible for expression of genes, which are required for essential cellular activity, production of virulence factors, signal molecules, and regulation of biofilm production (Waters & Bassler, 2005). Signal molecule–mediated QS is most found in bacteria during biofilm formation (Singh et al., 2017). The *N*- acyl homoserine lactone (AHL) or AI-I, AI-II, and oligopeptides are three most observed QS molecules in bacterial biofilm. Among them, acyl homoserine lactone is the most well-characterized QS molecule in gram-negative bacteria (Bjarnsholt & Givskov, 2007). However, fungi have farnesol, phenylethanol, tyrosol, and tryptophol as QS molecules in their biofilm (Wongsuk et al., 2016). Acyl homoserine lactone-mediated QS plays a vital role in formation and regulation of gram-negative bacterial biofilm; it also involves in maturation of biofilms (Landini et al., 2010). Different types of AHLs are dependent on corresponding bacterial strain from which they are produced. AHL-mediated QS is well studied and described by Waters and Bassler (2005). Acyl homoserine lactone is product of Luxl gene (Chew & Yang, 2015). All AHLs have the same homoserine lactone molecule but differ in the length and structure of acyl groups. A huge variety of AHLs are produced by bacterial species including *N*-3-oxohexanoyl HSL from *Hafnia alvei* (Bruhn et al., 2004), 3-oxo-C8-HSL, and *N*-butyryl-L-homoserine lactone signaling molecules from *Pseudomonas fluorescens* (Liu et al., 2007) (Fig. 9.3).

AI-II, another QS molecule, is nonspecific in nature so it can be used in both interspecies and intraspecies bacterial communication. AI-II synthesis is done in two major enzymatic steps including S-ribosyl-homocysteine (SRH) is produced from S-adenosyl-homocysteine (SAH) through cleavage of adenine by methylthioadenosine nucleosidase. SRH is subsequently converted into 4, 5-dihydroxy-2, 3-pentanedione (DPD) by LuxS protein. This DPD is referred to as QS molecule, which involves in biofilm formation. In biofilm of *L. monocytogenes*, the major important QS molecule is S-ribosyl-homocysteine (SRH) (Belval et al., 2006). AI-II signal also involves in formation of multispecies biofilm between *E. coli* and *E. faecalis* is done by accumulation of LSR proteins (Laganenka & Sourjik, 2018). AI-II-mediated QS signal activated through *luxS* protein is reported in some food borne pathogens (Duanis-Assaf et al., 2016). AI-III has been reported in gram-negative bacteria *E.coli, K. pneumoniae, Enterobacter cloacae, Salmonella* spp., and *Shigella* spp. (Walters et al., 2006; Sperandio et al., 2003). General mechanism of AI-III is not yet clear (reviewed by Bai & Rai, 2011) (Fig. 9.4).

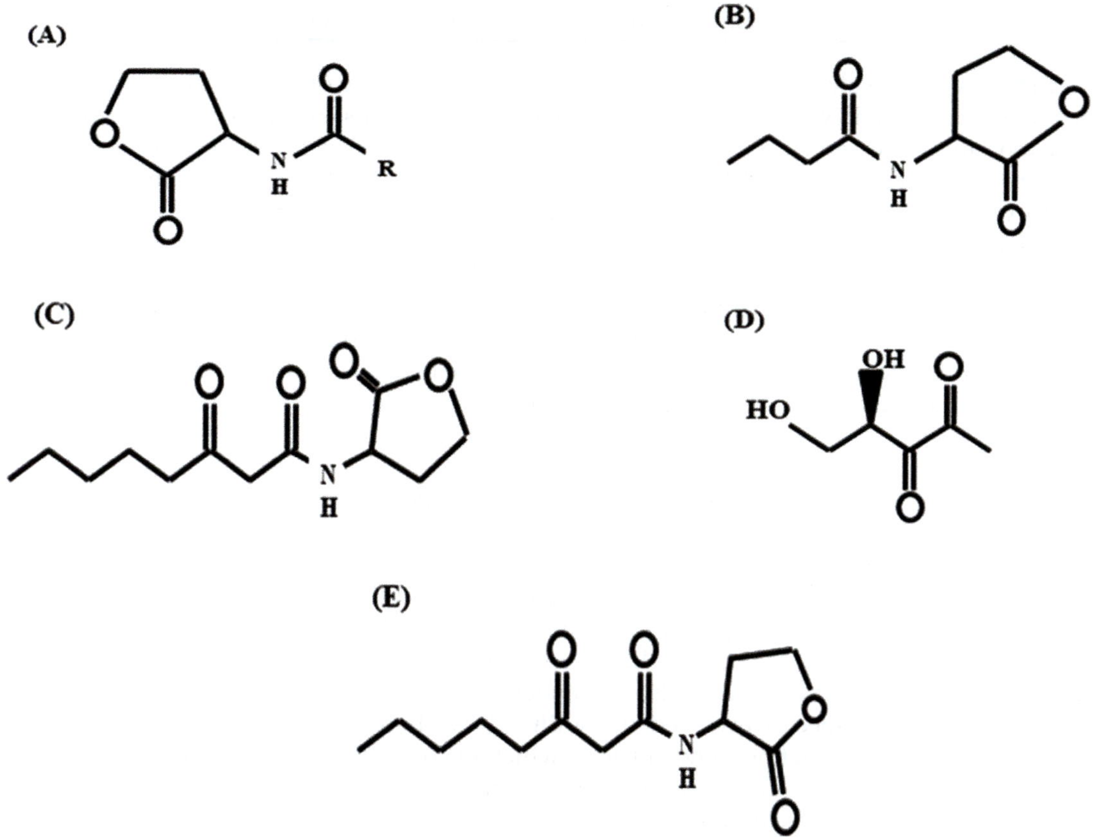

FIGURE 9.3 Key structural autoinducers used in quorum sensing during multispecies biofilm formation: (A) *N*-acyl homoserine lactone, (B) *N*-butyryl-L-homoserine lactone, (C) 3-oxo-C8-HSL, (D) 4,5-dihydroxy-2, 3-pentanedione (DPD), (E) *N*-3-oxohexanoyl HSL.

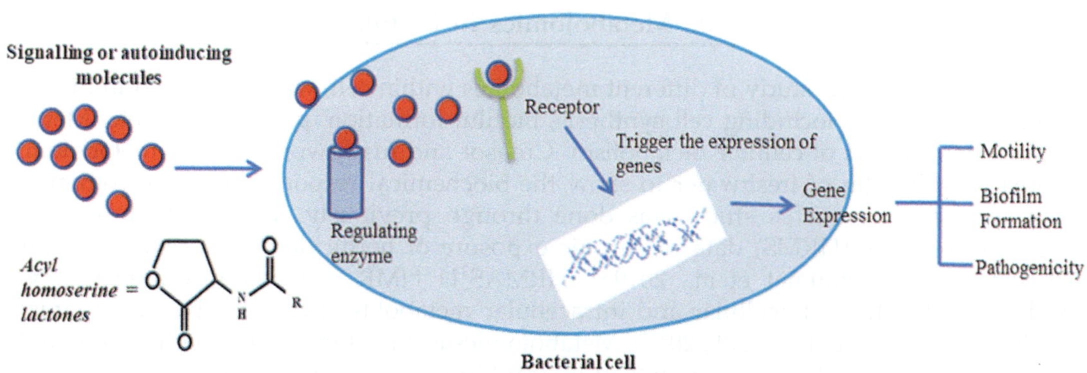

FIGURE 9.4 General mechanism of quorum sensing in biofilm formation.

8. Extracellular polymeric substances

EPS is a complex structure of biomolecules mainly comprising of carbohydrates, lipids, proteins, extracellular DNA, and extracellular RNA, which helps in microbial interaction and transportation of essential biomolecules including nutrients, QS signals, and oxygen to form biofilm matrix (Rice et al., 2016). EPS also promotes the growth of cells (Kim et al., 2014) and expansion of cell clusters in biofilm (Knecht et al., 2016). EPS provides nutrient-rich area for EPS-producing strains, while EPS-nonproducing strains will compete for nutrients and spaces in multispecies biofilms (Wang et al., 2013). EPS components are most of responsible for cell surface attachment components. These include flagella, fimbriae, and amyloids, which are responsible for cell motility, cell to cell attachment, stability, and autoimmune response (Karygianni et al., 2020). Cellulose and capsular polysaccharides and lipopolysaccharides promote the viability and resistance of multispecies biofilms (Wang, 2019).

9. Biofilm-regulating genes

Most of the cellular activity of cell depends on regulation of gene expression. Biofilm formation is one of the most stable and dormant processes, which constitute through the regulation of hundreds of biofilm-specific genes that are required for QS signal, immune response, stress response, cellular metabolism, EPS, and transportation of biomolecules (Domka et al., 2007). Development of biofilm is also dependent on genetic characteristics and interaction among of multiple species involved in biofilm (Rosenberg et al., 2016). Transcriptome is one of the biofilm-regulating gene clusters, which encodes the collection of all RNA molecules, especially mRNA. This transcriptome involves in formation of dual-species biofilm of *Azotobacter chroococcum and Trichoderma viride* (Velmourougane et al., 2019). The expression of biofilm regulatory genes involved in multispecies biofilms of microorganisms shows great significance and importance for understanding the evolution and adaptation of microbial species on their physiological characteristics (LI et al., 2021).

10. Metabolomics in biofilm

Metabolomics is the study of different metabolites within cell, which involved in different cellular mechanisms including cell synthesis, biofilm formation, pathogenesis, etc. Metabolites are the product of cellular metabolism. Creusot and his coworkers used metabolomics approach on biofilm of freshwater to show the biochemical response of whole community level against diuron. This study was done through previously achieved high-resolution mass spectrometry (HRMS) data of chronic exposure of herbicide diuron against biofilms (Chaumet, 2018; Chaumet et al., 2019a). HRMAS^{1}H NMR metabolomics approach was used to identify the extracellular and intracellular metabolites produced by the organisms involved in biofilm (Zabek et al., 2017). Metabolomics is one of the useful methods for studying the whole metabolic system involved in oral biofilm and oral cancer. A complete set of metabolites within cell is known as metabolome (Fig. 9.5).

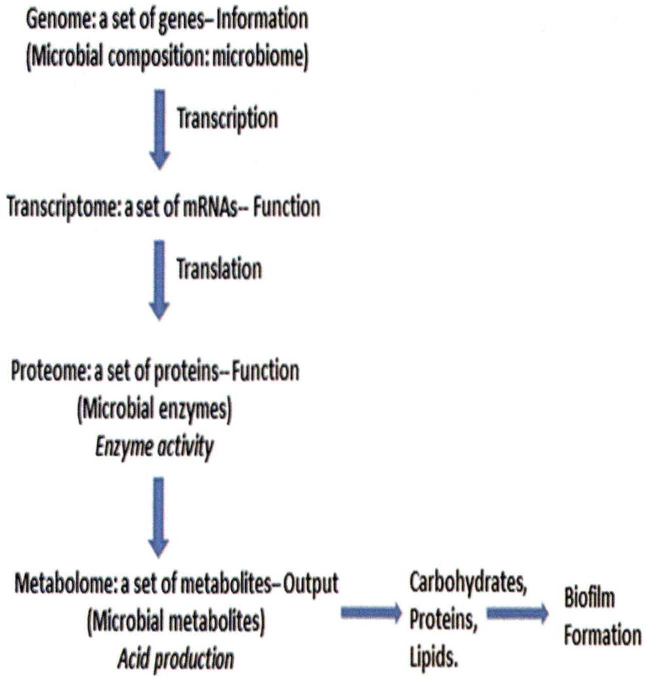

FIGURE 9.5 Biological hierarchy from genome to metabolome.

11. Microbial interaction in multispecies biofilms

Multispecies biofilms are result of communication and interaction among the microbial cells present on the biotic or abiotic surfaces in nature. Biofilm formation involves the exchange of metabolic products and intracellular signaling between interspecies and intraspecies of microbial communities. Interspecies interactions lead to the changes in phenotypic characteristics of bacteria involved in biofilm. The interaction established among multiple species is generally species competition, cooperative, or neutral (Dincer et al., 2001) (Fig. 9.6).

12. Competitive interaction

In competitive interaction, microbial individuals or population may strive for survival resources including space, nutrient availability, and energy with help of rapid growth rate, direct competition with other species by secreting bacteriostatic substance, and superior position in the habitat or through contact-dependent inhibition (Kim et al., 2014; Pfeiffer et al., 2001; Riley, 1999; Scholz & Greenberg, 2015). Generally, in dual-species or multispecies competition, one population show dominance, while other population show persistence in the biofilm. Invasion of *Pseudomonas putida* on biofilm of *Hyphomicrobium* spp. shows dominancy by *P. putida*, but the population of *Hyphomicrobium* spp. cells remained constant in

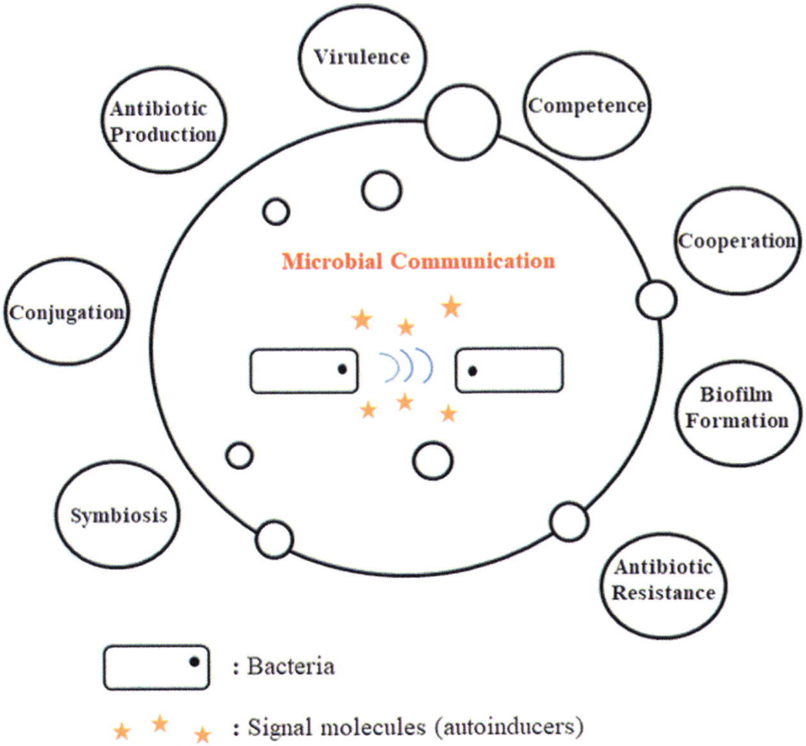

FIGURE 9.6 Microbial interaction in multispecies biofilm.

biofilm (Banks & Bryers, 1991). *K. pneumoniae* invades on the biofilm of *P. aeruginosa* in which population of *P. aeruginosa* remained in increasing number, which shows competition among them (Sturman et al., 1994). Moreover, competitive interaction may increase the resistance of dominant bacteria against antibiotic in multispecies biofilms (Kocot & Olszewska, 2020).

13. Cooperative interaction

Microbes can form multispecies biofilm for protection against stress and unfavorable living condition through cooperative interaction with each other. Bacteria in biofilms are close to each other, which can help in exchange of metabolites between cells. The increasing resistance of biofilm against antibiotic is a result of cooperative interaction. In multispecies biofilm, bacteria can also inhibit contamination through production of antimicrobial substances. In cooperative interaction, dual-species biofilm containing *Pseudomonas putida* SB5 and *Chryseobacterium* spp. SB9 with *P. putida* SB5 could initiate and promote the formation of bacterial species biofilm. Microbial cells may able to secret or release metabolites through cell breakage or by the end product of their cellular metabolic activity, which can be absorbed by other microbial cell in biofilm (Lilja & Johnson, 2016; Morris, 2015).

14. Horizontal gene transfer

Some bacteria can get resistance against antibiotic through gene mutation, through direct uptake of antibiotic-resistant genes from plasmids, and through mutualistic association with other antibiotic producers. In multispecies biofilms, exchange of extra chromosomal DNA (plasmid) is most common weapon used in horizontal gene transfer among multiple species. Conjugation is one of the gene transfer mechanisms, which shows greater rate in cell of biofilm as compared with planktonic cells or free living cells (Roberts et al., 1999). In conjugation, generally, F pilus (consists of pilin proteins) provides adhesin for cell to surface attachment and cell to cell interaction in formation of biofilm (Ghigo, 2001). Biofilms are group of living microorganisms that can help in detection of horizontal gene transfer through conjugation and transformation.

15. Other interaction

Despite competition and cooperative interaction, multiple species can coexist in neutral state, which shows noninterference with each other in multispecies biofilm (West et al., 2007). In dual-species biofilm of *L. monocytogenes*, the presence of metabolic by product of *E.coli* did not show any harmness in biofilm formation mechanism (de Grandi et al., 2018). The microbial interaction in multispecies biofilm can also be affected by environmental condition and their cellular properties.

16. General mechanism of multispecies biofilms for antibiotic resistance

Researches show that various microbial interactions between interspecies and intraspecies may increase the resistance of multispecies biofilms against antimicrobial agents (Sanchez-Vizuete et al., 2015). Multispecies biofilms have much more resistance against antibiotic as compared with single-species biofilms.

17. Penetration power of antibiotic

Antibiotic molecules can penetrate through the complex structure of biofilm matrix to inhibit microbial growth, but the extracellular polymeric substance reduces the antibiotic penetration through interaction with antibiotic so EPS provides an antispread block for antimicrobial agents (Dincer et al., 2001). EPS can bind with toxic compounds, which leads to the formation of barrier that can help reduce penetration of antibiotics in deeper parts of the biofilm (Chew & Yang, 2015). EPS provides numerous ionic molecules including anionic and cationic molecules. Most of these ionic molecules are carbohydrates, amino acids, proteins, glycolipids, and glycoproteins that react with charged antimicrobial agents and protect the microbial communities against antibiotic (Nadell et al., 2015). Pel exopolysaccharides bind with aminoglycosides cationic antibiotics and provide resistance against these molecules in *P. aeruginosa* biofilms (Colvin et al., 2013).

18. Persisters

When a planktonic cells convert into dormant stage of their life in stressful condition, they are known as persisters during biofilm formation. Most of the persisters are highly resistant to antibiotics. Presence of persisters in the multispecies biofilms may account for showing resistance to antimicrobial agents because they have lower metabolism rate as compared with planktonic bacteria (Dincer et al., 2001). Persisters may present relatively in small numbers, but these few cells are highly resistant against antibiotics in biofilm. The presence and efficiency of persisters is much higher in biofilm communities as compared with planktonic communities. Genes are responsible for persister state and may also include encoding regulatory system that regulates the entry and exit from persister state. *Hip* genes are responsible for persister in *E. coli.*

19. Efflux pumps

Efflux pumps work as membrane proteins that release harmful compounds from the bacterial cell into outside environment. Genes for efflux pumps can be found in bacterial genome. Efflux pumps can export the substances such as dyes, antibiotics, toxins, and waste metabolites from cell (Alav et al., 2018). Antimicrobial efflux pumps confer resistance to cells through excluding the antimicrobial agents from intracellular target sites (Nehme & Poole, 2007). MexAB-OprM and PA1875-1877 are examples of biofilm-mediated multidrug efflux pumps that are responsible for antibiotic resistance to planktonic cells of *P. aeruginosa* (Nehme & Poole, 2007). The major facilitator superfamily (MF), the small multidrug resistance family (SMR), the resistance nodulation division family (RND), the ATP-binding cassette family (ABC), and the multidrug and toxic compound extrusion family (MATE) are five known major families of the efflux pumps (Singh et al., 2017).

20. Antibiotic-degrading enzymes

There are certain enzymes responsible for degradation of antibiotics in biofilm matrix. The microorganisms involved in multispecies biofilm are able to accumulate β-lactamase in the biofilm complex for defense purpose. Accumulation of β-lactamase in biofilm of *P. aeruginosa* leads to increase in hydrolysis of ceftazidime and imipenem (Bagge et al., 2004). Aminoglycoside-modifying enzyme or chloramphenicol acetyltransferases are another examples involved in antibiotic degradation (Anderl et al., 2000; Darouiche et al., 1994; Dunne et al., 1993) (Fig. 9.7).

21. Adaptive response

In multispecies biofilm, bacteria can interact with other host microorganisms and provide strength to host for environmental stress responses including in temperature changes, DNA damage, oxidative stress, starvation, and other stresses. Stress responses may be induced in

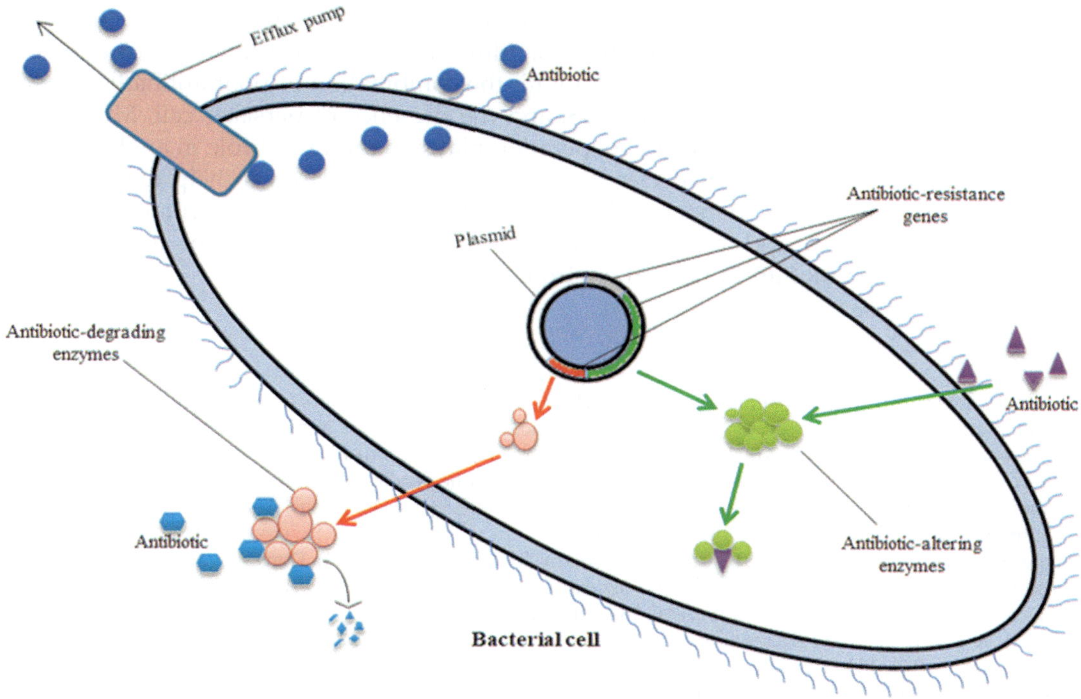

FIGURE 9.7 Summary of antibiotic resistance mechanisms.

biofilm because of changes in environmental condition. Many of these stress responses have been characterized at molecular and genetic levels by using planktonic cells. These protective responses may bring effective action in biofilms (Matin, 1991; Poolman & Glaasker, 1998).

22. Multispecies interactions

Different interactions among different microorganisms in a multispecies biofilm may help in altering general mechanism of antimicrobial resistance of microorganisms. Multispecies biofilm shows much more antibiotic resistance as compared with monospecies biofilms (Dincer et al., 2001). *P. aeruginosa* is more sensitive to gentamicin antibiotic in monospecies biofilm but less sensitive in growing with *S. aureus, E. faecalis, and Finegoldia magna* in multispecies biofilm (Dalton et al., 2011). *Candida albicans* and *S. aureus* show a high resistance against vancomycin antibiotic in dual-species biofilm (Langsrud et al., 2016).

23. Conclusion and future prospects

Interactions among different microorganisms result into the formation of strong, dense, and complex structure of multispecies biofilms. Metabolites study (metabolomics) within cell is also a useful method for understanding general mechanism of multispecies biofilm formation. Metabolomics helps to identify which type of extracellular and intracellular metabolites takes part in formation of biofilm. Biofilms show high resistant against antibiotic. Antibiotic resistance is increased in biofilm with the combination of poor antibiotic penetration, efflux pumps, adaptive responses, QS, and the presence of persister cell. Multispecies interaction including competition and cooperation plays an important role in antibiotic resistance during formation of multispecies biofilm. Research database can help the biofilm researchers to study and understand the general process of biofilm formation between microbial communities and also help in development of some control techniques for pathogen-associated biofilm in natural environment.

References

Agrobiotecnologı, I. D. (2001). Bap, a. *Microbiology, 183*(9), 2888–2896. https://doi.org/10.1128/JB.183.9.2888

Alav, I., Sutton, J. M., & Rahman, K. M. (2018). Role of bacterial efflux pumps in biofilm formation. *Journal of Antimicrobial Chemotherapy, 73*(8), 2003–2020. https://doi.org/10.1093/jac/dky042

Anderl, J. N., Franklin, M. J., & Stewart, P. S. (2000). Role of antibiotic penetration limitation in Klebsiella pneumoniae biofilm resistance to ampicillin and ciprofloxacin. *Antimicrobial Agents and Chemotherapy 44, 7*(July 1), 1818–1824. https://doi.org/10.1128/aac.44.7.1818-1824.2000

Bagge, N., Hentzer, M., Andersen, J. B., Ciofu, O., Givskov, M., & Høiby, N. (2004). Dynamics and spatial distribution of β-lactamase expression in *Pseudomonas aeruginosa* biofilms. *Antimicrobial Agents and Chemotherapy, 48*(4), 1168–1174. https://doi.org/10.1128/AAC.48.4.1168-1174.2004

Bai, A. J., & Rai, V. R. (2011). Bacterial quorum sensing and food industry. *Comprehensive Reviews in Food Science and Food Safety, 10*(3), 183–193. https://doi.org/10.1111/j.1541-4337.2011.00150.x

Bales, P. M., Renke, E. M., May, S. L., Shen, Y., & Nelson, D. C. (2013). Purification and characterization of biofilm-associated EPS exopolysaccharides from ESKAPE organisms and other pathogens. *PLoS One, 8*(6). https://doi.org/10.1371/journal.pone.0067950

Banks, M. K., & Bryers, J. D. (1991). Bacterial species dominance within a binary culture biofilm. *Applied and Environmental Microbiology, 57*(7), 1974–1979. https://doi.org/10.1128/aem.57.7.1974-1979.1991

Bassler, B. L. (2002). Small talk: Cell-to-cell communication in bacteria. *Cell, 109*(4), 421–424. https://doi.org/10.1016/S0092-8674(02)00749-3

Belval, S. C., Gal, L., Margiewes, S., Garmyn, D., Piveteau, P., & Guzzo, J. (2006). Assessment of the roles of LuxS, S-ribosyl homocysteine, and autoinducer 2 in cell attachment during biofilm formation by Listeria monocytogenes EGD-e. *Applied and Environmental Microbiology, 72*(4), 2644–2650. https://doi.org/10.1128/AEM.72.4.2644-2650.2006

Bjarnsholt, T., & Givskov, M. (2007). The role of quorum sensing in the pathogenicity of the cunning aggressor *Pseudomonas aeruginosa. Analytical and Bioanalytical Chemistry, 387*(2), 409–414. https://doi.org/10.1007/s00216-006-0774-x

Branda, S. S., Chu, F., Kearns, D. B., Losick, R., & Kolter, R. (2006). A major protein component of the Bacillus subtilis biofilm matrix. *Molecular Microbiology, 59*(4), 1229–1238. https://doi.org/10.1111/j.1365-2958.2005.05020.x

Brigette Carpentier. (1999). *Biofilm.Review.2.Pdf.*

Bruhn, J. B., Christensen, A. B., Flodgaard, L. R., Nielsen, K. F., Larsen, T. O., Givskov, M., & Gram, L. (2004). Presence of acylated homoserine lactones (AHLs) and AHL-producing bacteria in meat and potential role of AHL in spoilage of meat. *Applied and Environmental Microbiology, 70*(7), 4293–4302. https://doi.org/10.1128/AEM.70.7.4293-4302.2004

Burmølle, M., Ren, D., Bjarnsholt, T., & Sørensen, S. J. (2014). Interactions in multispecies biofilms: Do they actually matter? *Trends in Microbiology, 22*(2), 84–91. https://doi.org/10.1016/j.tim.2013.12.004

Chagnot, C., Zorgani, M. A., Astruc, T., & Desvaux, M. (2013). Proteinaceous determinants of surface colonization in bacteria: Bacterial adhesion and biofilm formation from a protein secretion perspective. *Frontiers in Microbiology, 4,* 1–26. https://doi.org/10.3389/fmicb.2013.00303

Chaumet, B. (2018). *Transfert et distribution des pesticides dans les bioflms en lien avec les efets toxiques associés.* PhD Thesis. University Bordeaux.

Chaumet, B., Morin, S., Boutry, S., & Mazzella, N. (2019). Diuron sorption isotherms in freshwater bioflms. *Science of the Total Environment, 651*, 1219–1225. https://doi.org/10.1016/j.scitotenv.2018.09.286

Chew, S. C., & Yang, L. (2015). Biofilms. *Encyclopedia of Food and Health*, 407–415. https://doi.org/10.1016/B978-0-12-384947-2.00069-6

Colvin, K. M., Alnabelseya, N., Baker, P., Whitney, J. C., Lynne Howell, P., & Parsek, M. R. (2013). PelA deacetylase activity is required for pel polysaccharide synthesis in pseudomonas aeruginosa. *Journal of Bacteriology, 195*(10), 2329–2339. https://doi.org/10.1128/JB.02150-12

Costerton, J. W., & Irvin, R. T. (1981). *The Bacterial Glycocalyx in Nature and Disease*, 299–324.

Costerton, J. W., Stewart, P. S., & Greenberg, E. P. (1999). Bacterial biofilms: A common cause of persistent infections. *Science, 284*(5418), 1318–1322. https://doi.org/10.1126/science.284.5418.1318

Cugini, C., Shanmugam, M., Landge, N., & Ramasubbu, N. (2019). The role of exopolysaccharides in oral biofilms. *Journal of Dental Research, 98*(7), 739–745. https://doi.org/10.1177/0022034519845001

Dalton, T., Dowd, S. E., Wolcott, R. D., Sun, Y., Watters, C., Griswold, J. A., & Rumbaugh, K. P. (2011). An in vivo polymicrobial biofilm wound infection model to study interspecies interactions. *PLoS One, 6*(11). https://doi.org/10.1371/journal.pone.0027317

Donlan, R. M. (2002). *Biofilms: Microbial life on surfaces.*

Dar, S. A., Kuenen, J. G., & Muyzer, G. (2005). Nested PCR-denaturing gradient gel electrophoresis approach to determine the diversity of sulfate-reducing bacteria in complex microbial communities. *Applied and Environmental Microbiology, 71*(5), 2325–2330. https://doi.org/10.1128/AEM.71.5.2325-2330.2005

Darouiche, R. O., Dhir, A., Miller, A. J., Landon, G. C., Raad, I. I., & Musher, D. M. (1994). Vancomycin penetration into biofilm covering infected prostheses and effect on bacteria. *Journal of Infectious Diseases, 170*(3), 720–723. https://doi.org/10.1093/infdis/170.3.720

Das, T., Sharma, P. K., Busscher, H. J., Van Der Mei, H. C., & Krom, B. P. (2010). Role of extracellular DNA in initial bacterial adhesion and surface aggregation. *Applied and Environmental Microbiology, 76*(10), 3405–3408. https://doi.org/10.1128/AEM.03119-09

Davies, D. G., & Geesey, G. G. (1995). Regulation of the alginate biosynthesis gene algC in *Pseudomonas aeruginosa* during biofilm development in continuous culture. *Applied and Environmental Microbiology, 61*(3), 860–867. https://doi.org/10.1128/aem.61.3.860-867.1995

Dincer, S., Uslu, F. M., & Delik, A. (2001). *Antibiotic resistance in biofilm.*

Domka, J., Lee, J., Bansal, T., & Wood, T. K. (2007). Temporal gene-expression in *Escherichia coli* K-12 biofilms. *Environmental Microbiology, 9*(2), 332–346. https://doi.org/10.1111/j.1462-2920.2006.01143.x

Donlan, R. M. (2001). Biofilms and device-associated infections. *Emerging Infectious Diseases, 7*(2), 277–281. https://doi.org/10.3201/eid0702.010226

Dowd, S. E., Wolcott, R. D., Sun, Y., McKeehan, T., Smith, E., & Rhoads, D. (2008). Polymicrobial nature of chronic diabetic foot ulcer biofilm infections determined using bacterial tag encoded FLX amplicon pyrosequencing (bTEFAP). *PLoS One, 3*(10). https://doi.org/10.1371/journal.pone.0003326

Duanis-Assaf, D., Steinberg, D., Chai, Y., & Shemesh, M. (2016). The LuxS based quorum sensing governs lactose induced biofilm formation by Bacillus subtilis. *Frontiers in Microbiology, 6*, 1–10. https://doi.org/10.3389/fmicb.2015.01517

Dunne, W. M., Mason, E. O., & Kaplan, S. L. (1993). Diffusion of rifampin and vancomycin through a Staphylococcus epidermidis biofilm. *Antimicrobial Agents and Chemotherapy, 37*(12), 2522–2526. https://doi.org/10.1128/AAC.37.12.2522

Flemming, H. C., Neu, T. R., & Wozniak, D. J. (2007). The EPS matrix: The "house of biofilm cells". *Journal of Bacteriology, 189*(22), 7945–7947. https://doi.org/10.1128/JB.00858-07

Friedman, L., & Kolter, R. (2004). Two genetic loci produce distinct carbohydrate-rich structural components of the *Pseudomonas aeruginosa* biofilm matrix. *Journal of Bacteriology, 186*(14), 4457–4465. https://doi.org/10.1128/JB.186.14.4457-4465.2004

Friedman, L., & Kolter, R. (2004a). Genes involved in matrix formation in *Pseudomonas aeruginosa* PA14 biofilms. *Molecular Microbiology, 51*(3), 675–690. https://doi.org/10.1046/j.1365-2958.2003.03877.x

Frølund, B., Palmgren, R., Keiding, K., & Nielsen, P. H. (1996). Extraction of extracellular polymers from activated sludge using a cation exchange resin. *Water Research, 30*(8), 1749–1758. https://doi.org/10.1016/0043-1354(95)00323-1

Ghigo, J.-M. (2001). Natural conjugative plasmids induce bacterial bioflm development horizontal gene transfer is a principal source of evolution leading to change in the ecological character of bacterial species. *Nature, 412*(July), 442–445. https://www.nature.com/nature/journal/v412/n6845/pdf/412442a0.pdf.

Gloag, E. S., Turnbull, L., Huang, A., Vallotton, P., Wang, H., Nolan, L. M., Mililli, L., Hunt, C., Lu, J., Osvath, S. R., Monahan, L. G., Cavaliere, R., Charles, I. G., Wand, M. P., Gee, M. L., Prabhakar, R., & Whitchurch, C. B. (2013). Self-organization of bacterial biofilms is facilitated by extracellular DNA. *Proceedings of the National Academy of Sciences of the United States of America, 110*(28), 11541–11546. https://doi.org/10.1073/pnas.1218898110

Grandclément, C., Tannières, M., Moréra, S., Dessaux, Y., & Faure, D. (2015). Quorum quenching: Role in nature and applied developments. *FEMS Microbiology Reviews, 40*(1), 86–116. https://doi.org/10.1093/femsre/fuv038

de Grandi, A. Z., Pinto, U. M., & Destro, M. T. (2018). Dual-species biofilm of Listeria monocytogenes and *Escherichia coli* on stainless steel surface. *World Journal of Microbiology and Biotechnology, 34*(4). https://doi.org/10.1007/s11274-018-2445-4

Jackson, K. D., Starkey, M., Kremer, S., Parsek, M. R., & Wozniak, D. J. (2004). Identification of psl, a locus encoding a potential exopolysaccharide that is essential for *Pseudomonas aeruginosa* PAO1 biofilm formation. *Journal of Bacteriology, 186*(14), 4466–4475. https://doi.org/10.1128/JB.186.14.4466-4475.2004

Jahid, I. K., & Ha, S. Do (2012). A review of microbial biofilms of produce: Future challenge to food safety. *Food Science and Biotechnology, 21*(2), 299–316. https://doi.org/10.1007/s10068-012-0041-1

Jamal, M., et al. (2015). Bacterial biofilm: Its composition, formation and role in human infections. *Research & Reviews: Journal of Microbiology and Biotechnology, 4*(3), 1–14. Retrieved from http://www.rr.

Kaplan, J. B. (2010). Biofilm dispersal: Mechanisms, clinical implications, and potential therapeutic uses. *Journal of Dental Research, 89*(3), 205–218. https://doi.org/10.1177/0022034509359403

Karatan, E., & Watnick, P. (2009). Signals, regulatory networks, and materials that build and break bacterial biofilms. *Microbiology and Molecular Biology Reviews, 73*(2), 310–347. https://doi.org/10.1128/mmbr.00041-08

Karygianni, L., Ren, Z., Koo, H., & Thurnheer, T. (2020). Biofilm matrixome: Extracellular components in structured microbial communities. *Trends in Microbiology, 28*(8), 668–681. https://doi.org/10.1016/j.tim.2020.03.016

Kim, W., Racimo, F., Schluter, J., Levy, S. B., & Foster, K. R. (2014). Importance of positioning for microbial evolution. *Proceedings of the National Academy of Sciences of the United States of America, 111*(16). https://doi.org/10.1073/pnas.1323632111

Knecht, L. D., O'Connor, G., Mittal, R., Liu, X. Z., Daftarian, P., Deo, S. K., & Daunert, S. (2016). Serotonin activates bacterial quorum sensing and enhances the virulence of *Pseudomonas aeruginosa* in the host. *EBioMedicine, 9*, 161–169. https://doi.org/10.1016/j.ebiom.2016.05.037

Kocot, A. M., & Olszewska, M. A. (2020). Interaction of *Pseudomonas aeruginosa* and *Staphylococcus aureus* with Listeria innocua in dual species biofilms and inactivation following disinfectant treatments. *Lwt, 118*, 108736. https://doi.org/10.1016/j.lwt.2019.108736

Laganenka, L., & Sourjik, V. (2018). Autoinducer 2-dependent *Escherichia coli* biofilm formation is enhanced in a dual-species coculture. *Applied and Environmental Microbiology, 84*(5). https://doi.org/10.1128/AEM.02638-17

Landini, P., Antoniani, D., Burgess, J. G., & Nijland, R. (2010). Molecular mechanisms of compounds affecting bacterial biofilm formation and dispersal. *Applied Microbiology and Biotechnology, 86*(3), 813–823. https://doi.org/10.1007/s00253-010-2468-8

Langsrud, S., Moen, B., Møretrø, T., Løype, M., & Heir, E. (2016). Microbial dynamics in mixed culture biofilms of bacteria surviving sanitation of conveyor belts in salmon-processing plants. *Journal of Applied Microbiology, 120*(2), 366–378. https://doi.org/10.1111/jam.13013

Li, Q., Liu, L., Guo, A., Zhang, X., Liu, W., & Ruan, Y. (2021). formation of multispecies biofilms and their resistance to disinfectants in food processing environments: A review. *Journal of Food Protection, 84*(12), 2071–2083. https://doi.org/10.4315/jfp-21-071

Lilja, E. E., & Johnson, D. R. (2016). Segregating metabolic processes into different microbial cells accelerates the consumption of inhibitory substrates. *ISME Journal, 10*(7), 1568–1578. https://doi.org/10.1038/ismej.2015.243

Liu, M., Wang, H., & Griffiths, M. W. (2007). Regulation of alkaline metalloprotease promoter by N-acyl homoserine lactone quorum sensing in Pseudomonas fluorescens. *Journal of Applied Microbiology, 103*(6), 2174–2184. https://doi.org/10.1111/j.1365-2672.2007.03488.x

Lynch, D. J., Fountain, T. L., Mazurkiewicz, J. E., & Banas, J. A. (2007). Glucan-binding proteins are essential for shaping Streptococcus mutans biofilm architecture. *FEMS Microbiology Letters, 268*(2), 158–165. https://doi.org/10.1111/j.1574-6968.2006.00576.x

Marsh, P. D., & Bradshaw, D. J. (1995). Dental plaque as a biofilm. *Journal of Industrial Microbiology, 15*(3), 169–175. https://doi.org/10.1007/BF01569822

Matin, A. (1991). The molecular basis of carbon-starvation-induced general resistance in *Escherichia coli. Molecular Microbiology, 5*(1), 3–10. https://doi.org/10.1111/j.1365-2958.1991.tb01819.x

Morris, J. J. (2015). Black Queen evolution: The role of leakiness in structuring microbial communities. *Trends in Genetics, 31*(8), 475–482. https://doi.org/10.1016/j.tig.2015.05.004

Nadell, C. D., Drescher, K., Wingreen, N. S., & Bassler, B. L. (2015). Extracellular matrix structure governs invasion resistance in bacterial biofilms. *ISME Journal, 9*(8), 1700–1709. https://doi.org/10.1038/ismej.2014.246

Nehme, D., & Poole, K. (2007). Assembly of the MexAB-OprM multidrug pump of *Pseudomonas aeruginosa*: Component interactions defined by the study of pump mutant suppressors. *Journal of Bacteriology, 189*(17), 6118–6127. https://doi.org/10.1128/JB.00718-07

Nwodo, U. U., Green, E., & Okoh, A. I. (2012). Bacterial exopolysaccharides: Functionality and prospects. *International Journal of Molecular Sciences, 13*(11), 14002–14015. https://doi.org/10.3390/ijms131114002

Okada, M., Sato, I., Cho, S. J., Iwata, H., Nishio, T., Dubnau, D., & Sakagami, Y. (2005). Structure of the Bacillus subtilis quorum-sensing peptide pheromone comx. *Nature Chemical Biology, 1*(1), 23–24. https://doi.org/10.1038/nchembio709

Palmer, J., Flint, S., & Brooks, J. (2007). Bacterial cell attachment, the beginning of a biofilm. *Journal of Industrial Microbiology and Biotechnology, 34*(9), 577–588. https://doi.org/10.1007/s10295-007-0234-4

Panlilio, H., & Rice, C. V. (2021). The role of extracellular DNA in the formation, architecture, stability, and treatment of bacterial biofilms. *Biotechnology and Bioengineering, 118*(6), 2129–2141. https://doi.org/10.1002/bit.27760

Papenfort, K., & Bassler, B. L. (2016). Quorum sensing signal-response systems in Gram-negative bacteria. *Nature Reviews Microbiology, 14*(9), 576–588. https://doi.org/10.1038/nrmicro.2016.89

Pfeiffer, T., Schuster, S., & Bonhoeffer, S. (2001). Cooperation and competition in the evolution of ATP-producing pathways. *Science, 292*(5516), 504–507. https://doi.org/10.1126/science.1058079

Poolman, B., & Glaasker, E. (1998). Regulation of compatible solute accumulation in bacteria. *Molecular Microbiology, 29*(2), 397–407. https://doi.org/10.1046/j.1365-2958.1998.00875.x

Prigent-Combaret, C., Vidal, O., Dorel, C., & Lejeune, P. (1999). Abiotic surface sensing and biofilm-dependent regulation of gene expression in *Escherichia coli. Journal of Bacteriology, 181*(19), 5993–6002. https://doi.org/10.1128/jb.181.19.5993-6002.1999

Rabin N, Zheng Y, Opoku-Temeng C, Du Y, Bonsu E, Sintim HO. Biofilm formation mechanisms and targets for developing antibiofilm agents. Future Med Chem. 2015;7(4):493-512. doi: 10.4155/fmc.15.6. Erratum in: 2015. *Future Med Chem. 7*(10),1362. PMID: 25875875.

Rice, S. A., Wuertz, S., & Kjelleberg, S. (2016). Next-generation studies of microbial biofilm communities. *Microbial Biotechnology, 9*(5), 677–680. https://doi.org/10.1111/1751-7915.12390

Riley, M. A. (1999). The ecological role of bacteriocins in. *Trends in Microbiology, 99*, 129–133.

Roberts, A. P., Pratten, J., Wilson, M., & Mullany, P. (1999). Transfer of a conjugative transposon, Tn5397 in a model oral biofilm. *FEMS Microbiology Letters, 177*(1), 63–66. https://doi.org/10.1016/S0378-1097(99)00288-8

Romero, D., Aguilar, C., Losick, R., & Kolter, R. (2010). Amyloid fibers provide structural integrity to Bacillus subtilis biofilms. *Proceedings of the National Academy of Sciences of the United States of America, 107*(5), 2230–2234. https://doi.org/10.1073/pnas.0910560107

Rosenberg, G., Steinberg, N., Oppenheimer-Shaanan, Y., Olender, T., Doron, S., Ben-Ari, J., Sirota-Madi, A., Bloom-Ackermann, Z., & Kolodkin-Gal, I. (2016). Not so simple, not so subtle: The interspecies competition between Bacillus simplex and Bacillus subtilis and its impact on the evolution of biofilms. *Npj Biofilms and Microbiomes, 2*, 1–11. https://doi.org/10.1038/npjbiofilms.2015.27

Sahu, P. K., Iyer, P. S., Oak, A. M., Pardesi, K. R., & Chopade, B. A. (2012). Characterization of eDNA from the clinical strain acinetobacter baumannii AIIMS 7 and its role in biofilm formation. *The Scientific World Journal, 2012*. https://doi.org/10.1100/2012/973436

Sanchez-Vizuete, P., Orgaz, B., Aymerich, S., Le Coq, D., & Briandet, R. (2015). Pathogens protection against the action of disinfectants in multispecies biofilms. *Frontiers in Microbiology, 6*, 1–12. https://doi.org/10.3389/fmicb.2015.00705

Scholz, R. L., & Greenberg, E. P. (2015). Sociality in *Escherichia coli*: Enterochelin is a private good at low cell density and can be shared at high cell density. *Journal of Bacteriology, 197*(13), 2122–2128. https://doi.org/10.1128/JB.02596-14

Silva, V. O., Silva, A., & Chang, Y. (2014). Biofilm formation on biotic and abiotic surfaces in the presence of antimicrobials by *Escherichia coli* isolates from cases of bovine mastitis. *Applied and Environmental Microbiology, 80*(19), 6136–6145. https://doi.org/10.1128/AEM.01953-14

Singh, S., Singh, S. K., Chowdhury, I., & Singh, R. (2017). Understanding the mechanism of bacterial biofilms resistance to antimicrobial agents. *The Open Microbiology Journal, 11*(1), 53–62. https://doi.org/10.2174/1874285801711010053

Sperandio, V., Torres, A. G., Jarvis, B., Nataro, J. P., & Kaper, J. B. (2003). Bacteria-host communication: The language of hormones. *Proceedings of the National Academy of Sciences of the United States of America, 100*(15), 8951–8956. https://doi.org/10.1073/pnas.1537100100

Stanley, N. R., & Lazazzera, B. A. (2005). Defining the genetic differences between wild and domestic strains of Bacillus subtilis that affect poly-γ-DL-glutamic acid production and biofilm formation. *Molecular Microbiology, 57*(4), 1143–1158. https://doi.org/10.1111/j.1365-2958.2005.04746.x

Stickler, D. J. (1996). Bacterial biofilms and the encrustation of urethral catheters. *Biofouling, 9*(4), 293–305. https://doi.org/10.1080/08927019609378311

Sturman, P. J., Jones, W. L., & Characklis, W. G. (1994). Interspecies competition in colonized porous pellets. *Water Research, 28*(4), 831–839. https://doi.org/10.1016/0043-1354(94)90089-2

Šuláková, M., Pazlarová, J., Meyer, R. L., & Demnerová, K. (2019). Distribution of extracellular DNA in Listeria monocytogenes biofilm. *Czech Journal of Food Sciences, 37*(6), 409–416. https://doi.org/10.17221/9/2019-CJFS

Urakawa, H., Kita-Tsukamoto, K., & Ohwada, K. (1999). Microbial diversity in marine sediments from Sagami Bay and Tokyo Bay, Japan, as determined by 16S rRNA gene analysis. *Microbiology*, (Pt 11), 3305–3315. https://doi.org/10.1099/00221287-145-11-3305

Velmourougane, K., Prasanna, R., Supriya, P., Ramakrishnan, B., Thapa, S., & Saxena, A. K. (2019). Transcriptome profiling provides insights into regulatory factors involved in Trichoderma viride-Azotobacter chroococcum biofilm formation. *Microbiological Research, 227*, 126292. https://doi.org/10.1016/j.micres.2019.06.002

Walters, M., Sircili, M. P., & Sperandio, V. (2006). AI-3 synthesis is not dependent on luxS in *Escherichia coli*. *Journal of Bacteriology, 188*(16), 5668–5681. https://doi.org/10.1128/JB.00648-06

Wang, R. (2019). Biofilms and meat safety: A mini-review. *Journal of Food Protection, 82*(1), 120–127. https://doi.org/10.4315/0362-028X.JFP-18-311

Wang, R., Kalchayanand, N., Schmidt, J. W., & Harhay, D. M. (2013). Mixed biofilm formation by Shiga toxin-producing *Escherichia coli* and *Salmonella enterica* serovar typhimurium enhanced bacterial resistance to sanitization due to extracellular polymeric substances. *Journal of Food Protection, 76*(9), 1513–1522. https://doi.org/10.4315/0362-028X.JFP-13-077

Waters, C. M., & Bassler, B. L. (2005). Quorum sensing: Cell-to-cell communication in bacteria. *Annual Review of Cell and Developmental Biology, 21*, 319–346. https://doi.org/10.1146/annurev.cellbio.21.012704.131001

West, S. A., Diggle, S. P., Buckling, A., Gardner, A., & Griffin, A. S. (2007). The social lives of microbes. *Annual Review of Ecology, Evolution, and Systematics, 38*, 53–77. https://doi.org/10.1146/annurev.ecolsys.38.091206.095740

Wongsuk, T., Pumeesat, P., & Luplertlop, N. (2016). Fungal quorum sensing molecules: Role in fungal morphogenesis and pathogenicity. *Journal of Basic Microbiology, 56*(5), 440–447. https://doi.org/10.1002/jobm.201500759

Zabek, A., Junka, A., Szymczyk, P., Wojtowicz, W., Klimek-Ochab, M., & Młynarz, P. (2017). Metabolomics analysis of fungal biofilm development and of arachidonic acid-based quorum sensing mechanism. *Journal of Basic Microbiology, 57*(5), 428–439. https://doi.org/10.1002/jobm.201600636

Zhang, X., Bishop, P. L., & Kupferle, M. J. (1998). Measurement of polysaccharides and proteins in biofilm extracellular polymers. *Water Science and Technology, 37*(4–5), 345–348. https://doi.org/10.1016/S0273-1223(98)00127-9

CHAPTER

10

Natural phytochemicals: a potential alternative antibiofilm agent

Kanchanlata Tungare[1], *Ameyota De*[1], *Shreya Das*[1], *Vaishnavi Kesarkar*[1], *Mustansir Bhori*[1] *and Renitta Jobby*[2,3]

[1]School of Biotechnology and Bioinformatics, D Y Patil Deemed to be University, Belapur, Maharashtra, India; [2]Amity Institute of Biotechnology, Amity University Maharashtra, Mumbai, Maharashtra, India; [3]Amity Centre of Excellence in Astrobiology, Amity University Maharashtra, Mumbai, Maharashtra, India

1. Introduction to biofilms—formation and habitat

Biofilm, a heterogenous community of microorganisms, was recently identified on an array of surfaces, including living tissues, natural aquatic or drinkable water systems, medical equipment, and so on (Singh et al., 2017). Biofilm is composed of DNA, polysaccharides, and proteins in an extracellular polymeric matrix. Interstitial voids provide biofilm colonies in the wild, both macro- and microsized, and distinctive architectural features. Biofilms' porous structures allow nutrients, gases, and antimicrobial agents to permeate them, but the structures themselves change over time in response to environmental and microbial pressures. Microcolonies mature into fully formed, intricate three-dimensional biofilms in the later phases of biofilm development (Mishra et al., 2020). It has been estimated that bacteria make up only 35% of the dry weight of a biofilm, with the remaining consisting of water (Yaron & Römling, 2014). During the early phases of biofilm development, bacterial motility is crucial because it enables free movement of the bacteria to find an appropriate surface. In this way, the flagella serve as both adhesion factors that help the cell stick to a surface and motility organelles that help the cell move to more favorable surroundings. It is critical to control the rotation and functionality of flagella for efficient biofilm development. Biofilm proteins have both structural and physiological roles (Yaron & Römling, 2014). The extracellular polymeric substances (EPS), which are high-molecular-weight compounds, unite the cells in biofilms, which provide them the resilience to persist despite adverse growth conditions. The EPS, which is composed of polysaccharides, lipids, proteins,

Microbial Biofilms
https://doi.org/10.1016/B978-0-323-95715-1.00012-1

polysaccharides, and extracellular DNA, plays a crucial role in the pathogenesis of different microbial infections (Mishra et al., 2020). There are several potential benefits for the bacteria living in the biofilm as opposed to a planktonic existence. Morphological changes in colonies, increased expression of helpful genes, production of large amounts of EPS, acquisition of antibiotic resistance genes via plasmid transfer, enhanced nutrient accessibility, and closer cell proximity that enables for synergistic associations and defense are just a few examples (Percival et al., 2011).

When bacteria colonize and form biofilms on a surface, they link together to create an extracellular polysaccharide matrix that encloses the earliest colonizers and biofilm formers. Coaggregation is a sort of cell-to-cell adhesion that is engaged in the production of multispecies biofilms in a wide range of environments, involving aquatic biofilm-forming microbes, dental plaque, chicken feed, and urogenital system (Percival et al., 2011). The bacteria in a biofilm regularly engage with each other due to their close proximity, which can be either beneficial or detrimental to the microorganisms existing. Consumption of oxygen by aerobes increases the localized redox potential, permitting anaerobes to coexist, illustrating the importance of synergistic interactions in the development of mixed-species biofilms. Other impacts include metabolic pathway sharing to maximize the use of available resources, coordinated resistance against host immune defenses, and antimicrobial medicines (Percival et al., 2011). Biofilms can grow on abiotic and biotic surfaces, such as plant leaf and artificial organs. For the biofilm to be formed successfully, several genes need to be expressed in unison. The cell-cell communication mechanism QS regulates the change from planktonic to biofilm formation. QS is associated with bacterial autoinducers, which are signal molecules. In gram-positive bacteria, signaling is accomplished using autoinducing peptides (AIPs), whereas *N*-acyl-homoserine lactones (AHLs), which are lipid-based molecules, are used in gram-negative bacteria (Khare et al., 2021). Antibiotic resistance is more common in multispecies (polymicrobial) biofilms than in single-species biofilms (Mishra et al., 2020). Nosocomial transmission accounts for the vast majority of all cases of disease, and the unprecedented and alarming widespread of antibiotic-resistant bacteria is a substantial public health issue (Khare et al., 2021). For medical researchers in a wide range of healthcare settings, biofilm antimicrobial tolerance has emerged as a major issue. Antibiotic combinations, synthetic drugs, and other therapeutic modalities failed to produce the desired outcomes. As part of the broader effort to discover novel antimicrobials in response to the drug resistance challenge, the scientific community has been looking for novel natural antibiofilm agents (Mishra et al., 2020). Targeting QS constitutes one of the current methods for combating antibiotic resistance since it prevents the growth of new biofilms and slows the spread of existing ones (Paluch et al., 2020). To address the rising tide of antibiotic-resistant microorganisms, new antibacterial medications are considered vital. Plants may be the best choice for providing a broad range of chemotherapeutics in the form of secondary metabolites that can oppose bacterial infections. A substitute for traditional antibiotics used to treat infections brought on by bacteria resistant to them includes phytochemicals, or active substances obtained from plants. Included in this category are secondary metabolites also known as phytochemicals, such as alkaloids, flavonoids, quinones, and coumarins (Khare et al., 2021).

2. Phytochemicals, types, examples, and their varied uses

Owing to high potencies, plant extracts and powders have been used medicinally for centuries. More recently, purified products (pure active phytomolecules) have gained popularity. Scientists have taken an interest in phytochemicals because of their usefulness in a variety of contexts, such as dietary supplements, medicinal herbs, cosmetics, and antimicrobial properties. Several phytochemicals have already been confirmed for their antimicrobial potential (Enwa, 2014). When used as extracts, they have been shown to have a hindering impact on clinical isolates. Some phytochemicals have been demonstrated to have an effect on cell permeability, replication machinery, efflux pumps, and other events necessary for the pathogen's survival and tolerance. Phytochemicals have been used effectively to treat antibiotic-resistant microorganisms (Khare et al., 2021). There are five general types of naturally occurring compounds with potent antibiofilm properties. Common classes of these chemicals include terpenoids, alkaloids, lectins, phenolics, polypeptides, essential oils, and polyacetylenes (Mishra et al., 2020).

Their antibacterial properties and ability to inhibit biofilms can be attributed to the following:

i) Surface charge and hydrophobicity—Microbe's adhesion abilities are crucial for colonization. The mechanism by which microorganisms adhere to surfaces is intricate and is affected by a variety of circumstances. Both the rate and the extent of microbial adhesion are determined by a number of factors, including the hydrophobicity, charge of the cell surface, presence of bacterial adhesins (such as fimbriae, flagella, and pili), and the composition of produced EPS (Donlan, 2002). Surface charge is commonly evaluated using the zeta potential of cells, which is derived from cell mobility in the influence of an electrical field using a set of known parameters (pH and salt concentrations). The presence of anionic groups in their membranes, such as carboxyl and phosphate, causes most microorganisms to be negatively charged in physiological conditions. The surface charge of cells can be altered to be less negative through the interaction of bacteria and phenolic acids (such as gallic and ferulic acids). Thus, phytochemicals can lessen the ability to donate electrons. Interactions of phytochemicals with the surface components of bacteria can change their physicochemical characteristics (Monte et al., 2014).

ii) Cell motility—Swarming movement on the cell surface has been linked to biofilm development. The production of flagella and surface polysaccharides is required for both biofilm formation and swarming. Because the force-generating motion aids in overcoming electrostatic repulsive forces between the bacteria and the substratum, boosting first contacts between the two surfaces, swarming motility serves primarily to increase early attachment in biofilm growth. At minimal inhibitory concentration (MIC), phytochemicals are tested for their ability to impede swimming and swarming motilities movements. It was observed that some phytochemicals induced the swimming and swarming more intensely over time, while others had opposite effect. Loss of biofilm formation competence has been associated with alterations in bacterial cell motility (Monte et al., 2014).

iii) Quorum sensing—Inhibitors of QS can be advantageous in biofilm removal considering QS is involved in several steps in the formation and differentiation of bacteria in biofilms. Antibiotic-induced selective pressure can be minimized with the help of QS inhibitors. Different phytochemicals and compounds that affect QS can have various results such as signals degradation, AHL enzyme suppression, and blockage of signal production or associated activity (Monte et al., 2014).

iv) Biofilm control—Biofilms are extremely difficult to eliminate because they are resistant to a wide variety of treatment methods, including antibiotics, biocides, and the body's own defenses. The complexity of the processes that underlie biofilm resistance to antimicrobials makes it difficult to predict the behavior of biofilm cells. The capacity of phytochemicals to regulate (eliminate and inactivate) 24-hour-old biofilms was examined based on their effects on biomass and metabolic activity. Significant differences were revealed in the responses of gram-negative and gram-positive bacteria to the same concentration of the phytochemical. The biomass percentage removal and inactivation were higher for *Escherichia coli* than for *Staphylococcus aureus* for all tested phytochemicals and concentrations. It was revealed that phytochemicals were more effective against and in controlling biofilm development by gram-negative bacteria than by gram-positive bacteria (Monte et al., 2014).

3. Quorum sensing and biofilm inhibitors from phytochemicals

3.1 Phenolics

The effects of phenolics on QS and biofilm growth have been studied more than those of any other class of active compounds. The impact of these compounds may vary on various organisms.

3.1.1 Phenylpropenoids

i) Eugenol—At dosages ranging from 50 to 200 μM, the phytochemical reduces the levels of QS-mediated violacein in *Chromobacterium violaceum* and pathogenicity factors in *Pseudomonas aeruginosa* PAO1 by 32%–56% (Zhou et al., 2013). Additionally, *P. aeruginosa, Listeria monocytogenes,* and clinical isolates of *Klebsiella pneumoniae* were successfully combatted by this substance in biofilms. This study inhibited the development of new biofilms, neutralized preexisting biofilms in *L. monocytogenes*, suppressed biofilm-forming genes in *L. monocytogenes* (Upadhyay et al., 2013), and also prevented *K. pneumoniae* clinical isolates from forming biofilms (Kumar et al., 2013).

ii) Cinnamaldehyde—In *Vibrio* species, cinnamaldehyde phytochemicals specifically disrupt their autoinducer-II-mediated QS system. *Cronobacter sakazakii, L. monocytogenes,* and *Staphylococcus epidermidis* are just some of the gram-positive and gram-negative bacteria that cinnamaldehyde has proven to be effective against in terms of biofilm production. Biofilm production by *S. epidermidis* has been reported to get reduced (Sharma et al., 2014), while genes involved in biofilm formation by *L. monocytogenes* are suppressed by this bioactive molecule (Upadhyay et al., 2013).

3.1.2 *Benzoic acid derivatives-*

The growth of *C. violaceum* and *Aeromonas hydrophila* biofilms have been reported to be suppressed by vanillin as it inhibits QS. According to the literature, reduction of biofilm formation by *A. hydrophila* on membrane filters results due to pretreatment with vanillin at a concentration of 0.18 mg/mL (Kappachery et al., 2010). Gallic acid is also known to stimulate *S. epidermidis* biofilm development while inhibiting *Eikenella corrodens* (Matsunaga et al., 2010). However, ellagic acid reduces the growth of *Streptococcus dysgalactiae* biofilm and suppresses *Escherichia coli* QS by 40% (Huber et al., 2003).

3.1.3 *Tannins*

There is evidence that tannins such as proanthocyanidins and ellagitannins can prevent the growth of biofilms and QS. For both gram-positive and gram-negative bacteria, tannic acid has proven to be helpful (Huber et al., 2003). Biofilm formation by *Staphylococcus aureus* is limited by 1,2,3,4,6-penta-O-galloyl-ß-D-glucopyranose; violacein production by *C. violaceum* is curtailed by punicalagin (Li et al., 2014). Further the swarming and swimming movements are downregulated in *Salmonella typhimurium*. At a low concentration, tannic acid favors biofilm formation by *P. aeruginosa* but impedes QS in *Pseudomonas putida* at higher dosings (Plyuta et al., 2013).

3.1.4 *Flavonoids*

As a class of phenolic compounds, flavonoids have been demonstrated to reduce QS sensing and biofilm development. In *S. aureus*, quercetin decreases the gene expression related to QS and pathogenicity, and it inhibits biofilm formation in *E. coli* (>80%) and MRSA (>50%) strains at 1 μg/mL (Le et al., 2013). *E. coli* AHL synthesis was reported to boost by epicatechin, and *P. aeruginosa* elastase activity was decreased by catechin (Plyuta et al., 2013). Biofilm generation by *E. corrodens* is reduced by 60% when treated with gallocatechin gallate (Matsunaga et al., 2010). Epigallocatechin gallate diminishes biofilm formation by 30% and swarming motility in *Burkholderia cepacia* (Huber et al., 2003).

3.2 Terpenoids

Several terpenes, including triterpenes, limonoids, and monoterpenes, have been demonstrated to inhibit biofilm formation or disrupt QS. It has been documented that the carvacrol and monoterpenes thymol are effective against both novel and established biofilms of gram-positive and gram-negative bacteria (Qiu et al., 2010; Soumya et al., 2011). Thymol inhibited the development of new *L. monocytogenes* biofilms at 0.5 mM and inactivated existing biofilms at 5 mM (Upadhyay et al., 2013). At low concentrations, *L. monocytogenes* downregulated genes necessary for biofilm development. *L. monocytogenes* biofilm-associated genes were suppressed by carvacrol at 0.65 mM (Arnason & Ta, 2015). These monoterpenes also impede *P. aeruginosa* growth at 0.1% concentration (Soumya et al., 2011). According to other research, gymnemic acids reduced conidial germination and hyphal growth in *Aspergillus fumigates* by 74% at a concentration of 40 g/mL (Vediyappan et al., 2013). Furthermore, acanthospermolide is reported to prevent biofilm formation in *P. aeruginosa* by 70% at 2.5 g/mL. Other literature reveals ursolic acid to harbor potential of inhibiting biofilm formation in *P. aeruginosa* (Plyuta et al., 2013).

3.3 Sulfur-containing phytochemicals

Allicin, ajoene, and thiocyanates are only some of the sulfur-containing compounds shown to disrupt biofilm formation and QS in both gram-positive and gram-negative bacteria. In *P. aeruginosa* PAO1, allicin (garlic extract) prevents the synthesis of virulence components that are controlled by QS (Lin et al., 2013) and reduces biofilm adherence in *S. epidermidis* as well as *P. aeruginosa* PAO1. More than 90% of biofilm development is dramatically decreased by allicin at a dosage of 4 mg/mL in *S. epidermidis* strains (Perez et al., 2003). Two Brassicaceae thiocyanates that have been shown to block QS in *E. coli* and *C. violaceum* are sulforaphane and allyl isothiocyanate. Furthermore, it has demonstrated to lessen the development of biofilm in *P. aeruginosa* and *L. monocytogenes*, respectively(Mansuri et al., 2022). Sulforaphane completely inhibited QS in *E. coli* DH5 at 100 μM concentration. Other reports have revealed that biofilm generation and pyocyanin synthesis (QS-mediated) were both inhibited by 60% and 70%, respectively, in *P. aeruginosa* PAO1 when these bioactives were supplemented at 37 μM and 100 μM concentration. Additionally, at 5 μg/mL, allyl isothiocyanate reduced *C. violaceum's* ability to produce violacein by 70% (QS-controlled) (Ganin et al., 2012).

3.4 Coumarins

Several studies have found that coumarins can inhibit both QS and biofilms. When tested at 500 μM, aesculetin was viewed to inhibit QS in *C. violaceum*, *P. aeruginosa*, and *E. coli* by 30%–78% (Brackman et al., 2009). Aesculetin inhibited the synthesis of Shiga-like toxins in *E. coli* O157:H7 (Lee et al., 2014) and minimized pathogenicity in a model of *Caenorhabditis elegans* infection. At a concentration of 50 μg/mL, umbelliferone (another coumarin) considerably lowered the expression of the genes involved in motility and adhesion as well as biofilm formation in *E. coli* O157:H7 (Lee et al., 2014).

3.5 Quinones

Quinones have also been shown in studies to suppress the growth of fungal and bacterial biofilms. Treatment with quinones such as emodin, chrysophanol, and shikonin inhibited the growth of biofilms from *Stenotrophomonas maltophilia* and *P. aeruginosa* PAO1. *P. aeruginosa* PAO1 biofilm formation was lessened by 75% after treated with emodin at 20 μM (Plyuta et al., 2013) and by 43% after treatment with emodin against *S. maltophilia* (Ding et al., 2011). To achieve the same degree of action against both bacterial species, chysophanol and shikonin demanded greater concentration of 200 μM.

3.6 Alkaloids

Alkaloid like berberine is reported to inhibit biofilm formation in *S. epidermidis* and *K. pneumoniae*. Berberine reduced biofilm growth in a number of *K. pneumoniae* clinical isolates at a concentration of 63.5 μg/mL (Kumar et al., 2013). When applied at concentrations between 30 and 45 μg/mL, berberine was found to significantly reduce *S. epidermis* biofilm development (by 50%) (Artini et al., 2012). Both sanguinarine and chelerythrine were micromolar effective against gram-positive biofilms of *S. aureus* and *S. epidermidis*, with 50% inhibitory values of 15–25 μM for *S. aureus* and 5–9 μM for *S. epidermidis*, respectively (Wang et al., 2009). With a minimum inhibitory concentration of 15.6 μg/mL, reserpine, an alkaloid from *Rauwolfia* sp. (Apocynaceae), has likewise demonstrated inhibitory efficacy against *K. pneumoniae* biofilms (Arnason & Kim Ta, 2015).

4. Biofilm-imposed challenges

4.1 Limited antibiotic penetration

Bacteria structures are known for their outer cell matrix; those matrices outside cells of biofilms can act as a shield, effectively preventing the entry of antimicrobials into biofilms. As a result, antimicrobials are exposed throughout EC biofilms since the deep-layer cells are challenging for most antibiotics to penetrate. Cells in the biofilm's basal layer might well be subjected to inconceivable combinations of antimicrobial chemicals until they become fatal. Antimicrobial penetration can be enhanced if the biofilm matrix is disrupted, as described with delafloxacin (Kranjec et al., 2021).

Vancomycin, oxacillin, cefotaxime, and delafloxacin are examples of antibiotics with poor penetration into staphylococcal biofilms, but amikacin and ciprofloxacin are not (Singh et al., 2010). Furthermore, ECM functions as a binding enzyme, which has the ability to render antibiotics ineffective. It is unknown how limiting antibiotic penetration affects biofilm growth because even antibiotics that rapidly disseminate the biofilm do not induce considerable cell death. According to the theory, slowing antibiotic penetration gives time for a variety of phenotypic responses, which is likely to result in less exposure (Dincer et al., 2020).

4.2 Resistance detection and monitoring problems

Concerning studies of biofilm dynamics, the primary challenges are on identifying the unmistakable components of the dynamic biofilm at any given time. This problem can be vanquished with the use of other sensory and cognitive approaches, which reveal how biofilms adapt and alter in response to various stimuli, such as antibiotic therapy. But there are still some important questions that need answering, such as how to improve the spatial and temporal resolution of sensory approaches and how to minimize biofilm disturbance and interruption throughout the hearing process. One major obstacle is making a sensor requires piercing the biofilm, which damages the biofilm and makes it harder to reproduce the scale (Funari & Shen, 2022).

4.3 Bacterial pathogenicity in biofilms

The pathogenicity of biofilm bacterial forms can be indicated by a variety of criteria. With the use of a hearing aid, the external chemicals emitted by biofilms modify the genetic expression of many hazardous compounds. Furthermore, bacteria in a biofilm can raise the frequency of maturation, which helps to avoid infection, as well as exchange plasma for gene transfer in response to toxic compounds and antibiotic resistance, boost mutation frequency, and enhance pump efflux function (Khan et al., 2021). Bacterial film is linked to over 65% of all bacterial illnesses, according to some estimates. Both non–device-related and device-related infections are included here. Fewer than 5% of breast implants, 2% of synthetic organs, 4% of mechanical heart valves, 10% of ventricular shunts, 4% of pacemakers and defibrillators, and roughly 40% of ventricular assist devices have had data collected on device-related diseases. Inflammation of the pulmonic heart valves and the vascular endothelium is the result of microbial encounters, a condition known as native valve endocarditis (NVE) (Jamal et al., 2018).

Antibiotics with a minimum biofilm eradication concentration (MBEC) that is thousands of times higher than that of planktonic bacteria must be employed in biofilms (Wolcott & Ehrlich, 2008).

Biofilms have nutrient and oxygen gradients that result in metabolic substance depletion and the provision of a medium for cell doubling and dormancy. This may also contribute to biofilm's capacity to coexist with other bacteria. Certain antimicrobial drugs can cause cell dormancy, which spurs an antitoxin for cell downregulation and, when the antimicrobial stressor is excluded, can cause metabolic dormancy with dormancy reversal (Shoji & Chen, 2020).

Many antibiotics target rapidly dividing cells, metabolically active, which may aid in biofilm resistance to antibiotics following stressor exposure. Biofilms can persist under many conditions such as modest variations in pH (either rise or decrease), and common adjuvant therapies, such as hydrogen peroxide, povidone-iodine, or sodium hypochlorite, can only partially eradicate *S. aureus* biofilm. As a result, the potential of biofilms to endure in different conditions, such as extreme pH, poses a difficulty for the applicability of various chemical adjuvant therapies (Khan et al., 2021).

5. Phytochemical mechanism for inhibition

For many infections, biofilm formation is an important step in disease progression, and the first steps of surface attachment are an important target. The mechanism of pathogens attaching to host tissues overlaps with the mechanisms leading to biofilm formation. Inhibition of biofilms is facilitated by phytochemical mechanisms. Polyphenols, sulfur-containing phytochemicals, tannins, and steroids are all examples of phytochemicals (secondary metabolites) with well-established biological activity, including some antibacterial effects (Thakur et al., 2020). There are generally five types of naturally occurring chemicals that have strong antibiofilm abilities. That list includes phenolics, terpenoids, lectins, polypeptides, and polyacetylenes. There are many chemical substances known as phenolics. Stilbenes, alkaloids, alkyl phenols, tannins, coumarins, phenolic acids, flavonoids, and phenolic acids are some of the seven classes found here. Substrate deprivation, membrane rupture, adhesion complex binding, protein binding, DNA association, and preventing viral combination are just some of the core ways by which these bioactive chemicals affect biofilm.

5.1 Polyphenols

Epicatechin gallate and epigallocatechin gallate are both examples of catechin gallates, a class of phenolic compounds that act as mild inhibitors of efflux pumps. The efflux transporter contains binding sites for these compounds with different affinities. Low quantities of catechins occupy high-affinity binding sites, resulting in a prolonged efflux substrate. Toxicity concerns aside, epigallocatechin gallate improves the efficacy of tetracycline, erythromycin, and ciprofloxacin against overexpressing bacteria.

5.1.1 *Stilbene*

These chemicals can be identified in a wide variety of plant and tree species, such as those used to make wine, peanuts, and eucalyptus. They are created when harmful organisms invade plants and alter their DNA by suppressing cellular division, oxidizing membranes, or by inactivating metabolic enzymes. It has been shown that stilbenes can stimulate nuclear factor erythroid-2-related components (Nrf2) and the expression of Nrf2 target genes is involved in innate immunity and antioxidant defense. For cells to detoxify and defend themselves from oxidative stress, Nrf2 must be present in the cellular transcriptome. Nrf2 acts as an intermediary in the regulatory control of a large group of phase II metabolite conjugation and antioxidant genes, including several Nrf2 target genes such as heme oxygenase-1 (HO-1) and NADPH, which are found in the nucleus and can identify the antioxidant reaction component with a specific nucleotide binding in target genes' regulatory regions. By building a protein complex with Kelch-like ECH-associated protein 1, Nrf2 is secluded in the cytosol, where its activity in the Nrf2 pathway is inhibited (Keap1). Translocation of Nrf2 into the nucleus from the Keap1 complex occurs under stress or after exposure to electrophiles (Savinov and Eggler, 2013). Dissociation of Nrf2 occurs when electrophiles, oxidants, or enhanced supplements such as stilbenes modify particular Keap1 cysteine residues. The small Maf (sMaf) proteins, with which Nrf2 forms a heterodimer in the nucleus, appear to be essential partners for the subsequent transactivation of target genes. The transcriptional repressor BACH1 appears to play a vital role in the nucleus as an antagonist for Nrf2-mediated stimulation by binding ARE-like regions in Nrf2 target genes. Keap1 can be localized to the nucleus and then shuttle back to the cytoplasm, and this fact should be taken into account. Stilbenes are reported to regulate these pathways and exhibit its effect.

5.1.2 *Tannins*

Tannins cause protein binding, enzyme blocking, and substrate depletion. These compounds are found in tree bark, leaves, and other plant parts and directly operate on the metabolism of bacteria by blocking oxidative phosphorylation. Tannin's phenolic groups establish insoluble hydrogen bonds with the −NH groups of proteins and peptides, preventing digestive enzymes from breaking the bonds, receptor binding inhibition, and its impact on receptor activity. By binding to the cell receptor, it prevents the virus from attaching to those surfaces. It penetrates the bacterial cell wall and reaches the inner membrane, where it blocks the cell's metabolic processes and ultimately causes cell death. Variable effects of tannin concentration on growth, microbes, and morphology have been observed. The antibacterial effects of tannins make them an optimistic alternative to antibiotics for the treatment of intracellular infections. In the total fatty acids, condensed tannins reduce the amount of unsaturated fatty acids and disintegrate membranes. Tannins inhibit extracellular microbial enzymes and oxidative phosphorylation, which affects microbial metabolism, deprives microorganisms of growth substrates, and increases membrane permeability.

5.1.3 Coumarins

These bioactive molecules decrease the expression of biofilm-related genes (adhesion, virulence, motility). Inhibition by coumarins occurs via various mechanisms, some of which are unique, such as prodrug inhibitors. Carbonic anhydrase (CA) esterase actions cause the hydrolysis of the lactone ring. CAs are abundant metalloenzymes that can be encoded by eight different genes. The hydroxide species produced by the enzyme catalyze this chemical reaction. This is therefore crucial for antiglaucoma medicines and antiepileptics, as well as pH regulatory processes in all cells and organisms. In addition to the activation mechanism for CAs, which includes a catalytically important zinc ion at their active site, at least four ways for inhibiting CA activity have been described to date (Supuran, 2020). The CA modulators integrate various tails and scaffolds in their molecule, as listed in the following:

1. The zinc binders that are coupled to metal ion have a zinc-binding group; these chemotypes have a ZBG, but the rest of their molecule interact with either the hydrophilic or hydrophobic, or both portions of the active site. Some ZBGs also participate in hydrogen bond interactions found in all α-CAs with two amino acid residues, known as gatekeepers. They can have aromatic, heterocyclic, aliphatic, or carbohydrate scaffolds.
2. The molecules that have an anchoring group and bind to zinc-coordinated water form hydrogen bonds with water pulled together with the metal. These inhibitors contain an anchoring group (AG) as well as phenols, polyamines, sulfocoumarins, and other chemicals that block CAs by this mechanism. As a result, AG is typically composed of OH, NH_2, or SO_3H moieties.
3. Inhibitors that prevent access to the active site. Coumarins are the substances that attach in this manner at the cavity's entrance, quite far out from the metal center. AG is a phenolic OH or COOH moiety in this case.
4. Inhibitors that bind to the active site. Only one such derivative is known as 2-benzyl sulfonyl-benzoic acid, which binds to the entrance to the active site cavity in an adjacent hydrophobic pocket, obstructing the residue His64 in its out conformation, enzyme's proton shuttle, and causing the catalytic cycle to collapse.

5.1.4 Flavonoids

This molecule retards enzyme activity by binding to adhesion protein and forms a compound with the cell wall. Mechanisms of antioxidant action include scavenging reactive oxygen system and enhancing or protecting antioxidant defenses; inhibiting enzymes involved in free radical generation; and repressing creation of the reactive oxygen system. Some of the problems they mediate may be the combined effect of their interaction with enzyme functions and their activity as radical scavengers. Microsomal monooxygenase, mitochondrial succinoxidase, glutathione S-transferase, and NADH oxidase are all enzymes associated in the production of reactive oxygen species, and flavonoids suppress them. They are bactericidal and bacteriostatic, destroying the cytoplasmic membrane and halting the metabolism and synthesis of nucleic acids in targeted microbes, respectively. They also injure cells by increasing its osmotic pressure, moving the net charge of the cell, and as a result killing the cell. High-affinity interactions between flavonoid molecules and hydrophilic amino acid residues, as predicted by the theoretical affinity order, can also have a cumulative effect on the endogenous scavenging chemicals.

These phytochemicals are essential for QS regulation, which is necessary for virulence and biofilm formation, as they act as suppressors of the biofilm's efflux pump mechanism, which removes toxic compounds from the cells. Efflux pump use is crucial for biofilm management, and antibiotic tolerance reduction occurs during inhibition.

QS is a configuration of intercellular communication that can increase the pathogenicity of a biofilm. Disrupting the QS mechanism used by the diverse microbial species that shape the plaque biofilm has the potential to impede biofilm formation. Although the term "quorum quenching" (QQ) was initially intended to refer only to the prevention of QS initiated by enzymatic hydrolysis of AHL autoinducers, it has since come to refer more broadly to the suppression of QS by any means, whether enzymatic or nonenzymatic molecules are employed. Blocking QS, which synchronizes beneficial events crucial to biofilm growth and differentiation, requires interfering with either the generation or destruction of signal molecules or blocking the signal molecules from connecting to their receptors. By preventing the AHL molecule from attaching to its receptor, inhibitors can overcome the limitations of antibiotic use in treating biofilm infections. The AHL molecule may be subject to vigorous suppression by chemicals that attach to the preferred receptor. Molecules produced by a wide variety of plants, algae, and other animals can prevent the luxr-type receptor protein from binding and triggering signal production by making small changes to the acyl side chain of AHL, to the lactone ring, or to both. Likewise, structural analogs of S-adenosylmethionine and acyl carrier protein may be used to impede the production of AHL, which would be a significant advancement. Reducing the dosage of some macrolide antibiotics, such as erythromycin, can inhibit AHL production. Even today, the precise mechanism by which these drugs obstruct QS in bacteria remains unclear.

5.1.5 Alkaloids

Alkaloids are structurally distinct compounds with antimicrobial activity via inhibition of enzyme activity or other mechanisms. The mechanisms of action differ between different alkaloids such as follows:

i) Affecting cell division: Pergularinine and tylophorinidine, two phenanthroindolizidine plant alkaloids, block dihydrofolate reductase action and, by extension, nucleic acid synthesis. It is difficult to synthesize amino acids, RNA, or DNA without the enzyme dihydrofolate reductase. The FtsZ protein is the prokaryotic equivalent of the eukaryotic tubulin and is required for bacterial cell division. The alkaloid berberine inhibits cell division by binding with high affinity to the FtsZ protein and preventing its assembly and GTPase activity, which causes the cell to become elongated. Ungeremine alkaloid blocks the topoisomerases of bacteria (*E. coli*). All naturally occurring quinoline alkaloids are devoid of the 3-carboxy group that is required for binding as well as inhibiting DNA-type IIA topoisomerase complexes.
ii) Respiratory and enzyme inhibition in bacteria: Both bacterial respiration and enzyme activity can be stifled by alkaloids. Agelasine D, a diterpene alkaloid found in marine

sponges, has antimycobacterial impact by binding straight to the protein, ciphered through a dioxygenase gene, and inhibiting the function. This mechanism of action is unique to the alkyl methyl quinoline alkaloids and accounts for their potent antibacterial activity.

iii) Bacterial membrane disruption: Polyamine alkaloid squalamine depolarizes bacterial membranes and inhibits their ability to leak ions across their outer membranes in a detergent-like mechanism of action.

iv) Affecting virulence genes. Regulatory genes encode regulatory proteins, which are directly involved in the excitation of several virulence determinants, including the genes encoding the toxin and virulence factors; an isoquinoline alkaloid suppresses the virulence factors.

5.2 Biofilm inhibition

5.2.1 Reducing and preventing bacterial surface adherence

The process initiates with bacterial cells adhering to the surface, known as biofilm development. The antibacterial activity of 3-(trimethoxysilyl)-propyldimethyloctadecyl ammonium chloride (QAS) allows it to be covalently coupled with other molecules to prevent biofilm development (Wan et al., 2007). A surface treated with antiadhesion agents discourages bacteria from affixing themselves to hosts in the first place. Chemicals can be used to alter its surface's roughness, texture, and hydrophobicity/hydrophilicity (Ghosh et al., 2020).

5.2.2 Interference with the quorum sensing system

QS is based on density-based production and sensing of unique extracellular small molecules that may differ among bacterial species. Inhibiting QS systems using AHLs is common practice. AHLs are an unconventional signal for many different types of bacteria. Some pathogenic bacterial phenotypes, such as virulence and biofilm formation, are regulated by intercellular signaling (Tang & Zhang, 2014); species-specific AIPs play the important role of QS stimulation and act as a QS stimulator in microorganisms; and only a select few microorganisms are able to retort to many different QS stimulators of a complex regulatory network. QQ refers to the molecular mechanism by which microorganisms stop certain chemical messengers, and it includes both naturally occurring and synthetic (artificial) QS inhibitors (QSIs). The importance of methylthioadenosine nucleosidase (MTAN) inhibitors in limiting the selection of resistant mutants is highlighted by their antimicrobial action, in particular their ability to suppress QS-derived biofilm development without affecting the bacterial growth profile. Incorporating QSI into antibiotic treatment can shorten treatment time, lessen side effects, and boost the host's innate immune response, all while preventing biofilm generation and reducing virulence in harmful microbes (Ghosh et al., 2020; Lee & Lee, 2010)

5.2.3 Modulating nucleotide signaling molecules

Guanosine pentaphosphate (pppGpp) and guanosine tetraphosphate (ppGpp) play vital roles in a wide range of organic processes that ensure life under nutrient-restricted situations. Intracellular (p)ppGpp levels affect QS, biofilm dispersion in microorganisms, and biofilm formation with the exclusion of ppGpp effects (Liu et al., 2015). The bis(3′,5′)-cyclic diguanylate acid (c-di-GMP) is known to manage different physiological pathways, including biofilm development and QS. This nucleotide messenger is produced by diguanylate cyclases (DGCs) that change two GTP molecules to c-di-GMP, which is then broken down by phosphodiesterase (PDE) enzymes into 5′-phospho guanylyl-(3′,5′)-guanosine (pGpG) and/or GMP (Syal et al., 2017). The intracellular c-di-GMP level is decided/controlled by the cost of synthesis and breakdown, which plays a part in sluggish development, drug tolerance, and biofilm formation. All signaling pathways contribute directly to the development of antimicrobial resistance and serve as a promising curing strategy for reducing biofilm population, resulting in a greater susceptibility to standard medicines (Sambanthamoorthy et al., 2012). The second messenger synthetase enzymes could be considered as the most sensible therapeutic targets to lower the intracellular concentrations of pppGpp and c-di-GMP, respectively (Ghosh et al., 2020).

5.2.4 *Chemical inhibition of biofilm maturation*

Both synthetic and natural antibiofilm chemicals are used to influence biofilm formation and maturity, typically by disrupting microbial surface characteristics (Naresh et al., 2011). As biofilms form, deacylated lipopolysaccharide (dLPS) interacts with naturally existing LPS in the cell wall to inhibit their growth (Ghosh et al., 2020).

5.2.5 *Disruption of mature biofilms*

Many bacterial diseases are linked to the development of biofilms, which must be specially targeted with chemicals to be eradicated. Since exopolysaccharides are vital to the functioning of a biofilm, monoclonal antibodies (mAbs) are able to specifically target these exopolysaccharides that are polysaccharides composed of D-mannose, L-rhamnose, and D-glucose (Mathur et al., 2018). It is well documented that the glycoside hydrolase enzymes enhance the efficacy of antibiotics and facilitate neutrophil-mediated death, despite being noncytotoxic. Surface-associated and secreted proteins play critical role in biofilm growth as well, since extracellular molecules release staphopain A, cysteine, staphopain B, and proteases, all of which are considered to compromise the integrity of an already-established biofilm (Gawande et al., 2014). Specific proteins in biofilm can be degraded by the extracellular serine protease, leading to the breakdown of the biofilm's structure both in vivo and in vitro. The biofilm can be broken up by the release of endopeptidase and extracellular protease. Since extracellular DNA (eDNA) is a crucial substance of the matrix of majority biofilms, it is an alluring aim for biofilm deterioration. Treatment with DNase can break up an existing biofilm up to a certain microbial limit, but beyond that point, the biofilm becomes impenetrable to the enzyme. The enzyme DNase dornase alfa is widely employed in clinical methods to break up biofilm layers; eDNA, on the other hand, is expected to reduce interactions with other matrix components such as polysaccharide. Chemical suppression of biofilm formation and destruction of robust biofilms can be accomplished using certain precise pathway inhibitors. Small organic chemicals can decrease bacterial surface adherence and interact with the QS system (Ghosh et al., 2020).

6. Future perspective of novel phytochemical strategies

6.1 Phenolic compounds

Numerous mechanisms, such as decreased membrane fluidity, cell wall synthesis, associations with bacterial proteins and cell wall structures, reduction of nucleic acid synthesis, cytoplasmic membrane damage, energy metabolism, and interrelations with bacterial proteins and cell wall structures, have all been investigated for plant phenolics' antibacterial activity. These phenolic compounds demonstrated, in addition to their catastrophic impact on bacteria, "softer" pursuits that lead to subdued biofilm by influencing bacterial regulatory mechanisms such as QS or other global regulator systems, with no effect on bacterial growth. Ellagic acid is one such example, which can prevent the growth of *S. aureus* biofilms to a degree that is associated with increased antibiotic resistance. Inhibiting biofilm development on surfaces coated with plant products might be important in future strategies for preventing medical device biofilm-associated plants polyphenols (Slobodníková et al., 2016).

6.2 Honey

Honey's low pH, hydrogen peroxide technology from bee-derived enzyme glucose oxidase, and high sugar content all contribute to its antibacterial capabilities. With the rise in multidrug resistance among biofilm-associated organisms, understanding the mechanisms through which honey inhibits bacterial growth inside biofilm formation has become a topic of intentional consideration (Neto et al., 2015). Fortunately, there appears to be mounting data suggesting that honey shows an interest in either averting the establishment of a biofilm by interfering with adherence to host cells or by obstructing QS and breaking a well-formed biofilm. Antibiofilm action was detected in vitro for multiple honey varieties. Testing of honey's potential as an antibiofilm, wound restoration, and antibacterial alternative is mostly constricted to in vitro settings (Kamaruzzaman et al., 2018).

Previous literature has shown the capacity of honey to remove existing *P. aeruginosa* biofilms. Biofilm biomass was significantly decreased after treatment with either 16% or 32% honey. Although the two *P. aeruginosa* strains responded differently to subinhibitory concentrations (1%–4%) of honey or sugar solution, the biofilm biomass was massively improved when these conditions were used. This difference may be attributable to the fact that the two strains of *P. aeruginosa* have different biofilm-forming abilities, as evidenced by the significantly different biomass values at the 0% honey concentration (Lu et al., 2019).

6.3 Biosurfactants

Biosurfactants (BS) are able to prevent the development of biofilms because they modify the adhesion capability of cells by lowering their hydrophobicity, rupturing membranes, and disrupting the electron transport chain, hence reducing the energy requirements of cells (Abdollahi et al., 2020). A wide variety of microorganisms secrete biosurfactants with varied antimicrobial, antifungal, and antibiofilm properties (Mishra et al., 2020). This biosurfactant is assumed to be utilized as an antibiofilm agent as it can withstand high temperatures and is less hazardous than conventional alternatives, thereby incredibly challenging elimination of dermatophyte biofilms. Biosurfactant produced from the insect–parasitic fungus *Beauveria bassiana*, a lipopeptide. In ex vivo circumstances of *Microsporum canis*, it is crucial as an antibiofilm agent (Abdel-Aziz et al., 2020). To exert its effect, it compromises the struc tural integrity of cell membranes and alters their permeability. Due to its inexpensive manufacturing from strong corn liquor, *B. bassiana's* biosurfactant overcomes the drawback of high production costs. It has the potential to be an effective biosurfactant for treating resistant dermatophytosis. Surfactin is a cyclic lipopeptide that, when combined with its metal complex, is highly effective against infections caused by *Candida albicans* biofilms (Panda et al., 2020).

6.3.1 **In silico** ***tools***

Antimicrobial peptides (AMPs) are directed toward short-length peptide antibiotics (between 15 and 30 amino acids) that are gene-encoded, amphipathic, cationic, and targeted at the cell membrane. A key part of the development of complex multicellular creatures is played by AMPs. AMPs are produced by fungi and bacteria as a defensive mechanism and to obtain a competitive edge over other microorganisms, sometimes of the same species (Rossi et al., 2008). These peptides help in designing antibiofilm strategies, but there are challenges regarding production, designing, and understanding the properties. To fully comprehend AMP action and hence realize their potential as antimicrobial medicines, bioinformatic techniques are a priceless tool. To keep track of the ongoing discovery of new AMPs with varied antimicrobial potential, information systems supporting the deposition, collection, and evaluation of AMP related data, as well as the screening and in silico modeling of novel AMPs to speed up the process of antimicrobial drug discovery and design, are being developed (Jorge et al., 2012). Information storage, search, data screening, and data standardization about AMPs with details about their activity would give us a new outlook on study of peptide potential, which enables and promotes sequence specificity and develops biofilm models with this tool to understand biofilm's structure and activity.

7. Conclusion

Biofilms are a complex structure of DNA, polysaccharides, and proteins in an extracellular polymeric matrix, which allows nutrients, gases, and antimicrobial agents to permeate them, but the structures themselves change over time in response to environmental and microbial pressures from the environment; this marks biofilm formation as an origin in the progression of infectious disease. Infectious bacteria tend to colonize and form biofilms on surfaces by linking together and creating an extracellular polysaccharide matrix that encloses the earliest colonizers and biofilm formers. They consume oxygen, which increases the redox potential, permitting anaerobes to coexist, demonstrating the presence of diverse species in the biofilm. To inhibit biofilms, QS and biofilm inhibitors are used. QS prevents the AHLs molecule from binding to its receptor, resulting in no biofilm formation. The process of inhibiting biofilms is facilitated by phytochemical mechanisms mediated by polyphenols, sulfur-containing phytochemicals, tannins, and steroids, thereby playing a major role in the deformation of biofilms.

References

Abdel-Aziz, M. M., Al-Omar, M. S., Mohammed, H. A., & Emam, T. M. (2020). *In vitro* and *ex vivo* antibiofilm activity of a lipopeptide biosurfactant produced by the entomopathogenic *Beauveria bassiana* strain against *Microsporum canis*. *Microorganisms, 8*(2), 232.

Abdollahi, S., Tofighi, Z., Babaee, T., Shamsi, M., Rahimzadeh, G., Rezvanifar, H., Saeidi, E., Mohajeri Amiri, M., Saffari Ashtiani, Y., & Samadi, N. (2020). Evaluation of anti-oxidant and anti-biofilm activities of biogenic surfactants derived from Bacillus amyloliquefaciens and *Pseudomonas aeruginosa*. *Iranian Journal of Pharmaceutical Research : IJPR, 19*(2), 115–126.

Arnason, J. T., & Ta, C. A. (2015). Mini review of phytochemicals and plant taxa with activity as microbial biofilm and quorum sensing inhibitors. *Molecules (Basel, Switzerland), 21*(1), E29.

Artini, M., Papa, R., Barbato, G., Scoarughi, G., Cellini, A., Morazzoni, P., Bombardelli, E., & Selan, L. (2012). Bacterial biofilm formation inhibitory activity revealed for plant derived natural compounds. *Bioorganic & Medicinal Chemistry, 20*, 920–926.

Brackman, G., Hillaert, U., van Calenbergh, S., Nelis, H. J., & Coenye, T. (2009). Use of quorum sensing inhibitors to interfere with biofilm formation and development in burkholderia multivorans and burkholderia cenocepacia. *Res. Microbiol., 160*, 144–151.

Dincer, S., Uslu, F. M., & Delik, A. (2020). Antibiotic resistance in biofilm. In S. Dincer, M. S. Özdenefe, & A. Arkut (Eds.), *Bacterial biofilms*. IntechOpen.

Ding, X., Yin, B., Qian, L., Zeng, Z., Yang, Z., Li, H., Lu, Y., & Zhou, S. (2011). Screening for novel quorum-sensing inhibitors to interfere with the formation of *Pseudomonas aeruginosa* biofilm. *Journal of Medical Microbiology, 60*, 1827–1834.

Donlan, R. M. (2002). Biofilms: Microbial life on surfaces. *Emerging Infectious Diseases, 8*(9), 881.

Enwa, F. (2014). Mechanisms of antimicrobial actions of phytochemicals against enteric pathogens – a review. *Journal of Pharmaceutical, Chemical and Biological Sciences, 2*, 77–85.

Funari, R., & Shen, A. Q. (2022). Detection and characterization of bacterial biofilms and biofilm-based sensors. *ACS Sensors, 7*(2), 347–357.

Ganin, H., Rayo, J., Amara, N., Levy, N., Krief, P., & Meijler, M. M. (2012). Sulforaphane and erucin, natural isothiocyanates from broccoli, inhibit bacterial quorum sensing. *MedChemComm, 4*, 175–179.

Gawande, P. V., Leung, K. P., & Madhyastha, S. (2014). Antibiofilm and antimicrobial efficacy of DispersinB®-KSL-W peptide-based wound gel against chronic wound infection associated bacteria. *Current Microbiology, 6q8*(5), 635–641.

Ghosh, A., Jayaraman, N., & Chatterji, D. (2020). Small-molecule inhibition of bacterial biofilm. *ACS Omega, 5*(7), 3108–3115.

Huber, B., Eberl, L., Feucht, W., & Polster, J. (2003). Influence of polyphenols on bacterial biofilm formation and quorum-sensing. *Zeitschrift für Naturforschung C, 58*, 879–884.

Jamal, M., Ahmad, W., Andleeb, S., Jalil, F., Imran, M., Nawaz, M. A., Hussain, T., Ali, M., Rafiq, M., & Kamil, M. A. (2018). Bacterial biofilm and associated infections. *Journal of the Chinese Medical Association : JCMA, 81*(1), 7–11.

Jorge, P., Lourenço, A., & Pereira, M. O. (2012). New trends in peptide-based anti-biofilm strategies: A review of recent achievements and bioinformatic approaches. *Biofouling, 28*(10), 1033–1061.

Kamaruzzaman, N. F., Tan, L. P., Mat Yazid, K. A., Saeed, S. I., Hamdan, R. H., Choong, S. S., Wong, W. K., Chivu, A., & Gibson, A. J. (2018). Targeting the bacterial protective armour; challenges and novel strategies in the treatment of microbial biofilm. *Materials (Basel, Switzerland), 11*(9), 1705.

Kappachery, S., Paul, D., Yoon, J., & Kweon, J. H. (2010). Vanillin, a potential agent to prevent biofouling of reverse osmosis membrane. *Biofouling, 26*, 667–672.

Khan, J., Tarar, S. M., Gul, I., Nawaz, U., & Arshad, M. (2021). Challenges of antibiotic resistance biofilms and potential combating strategies: A review. *3 Biotech, 11*(4), 169.

Khare, T., Anand, U., Dey, A., Assaraf, Y. G., Chen, Z. S., Liu, Z., & Kumar, V. (2021). Exploring phytochemicals for combating antibiotic resistance in microbial pathogens. *Frontiers in Pharmacology, 12*, 720726.

Kranjec, C., Morales Angeles, D., Torrissen Mårli, M., Fernández, L., García, P., Kjos, M., & Diep, D. B. (2021). Staphylococcal biofilms: Challenges and novel therapeutic perspectives. *Antibiotics (Basel, Switzerland), 10*(2), 131.

Kumar, A., Magesh, H., Alam, A., Priyam, Sekar, U., Sumantran, V. N., & Vaidyanathan, R. (2013). Identification of natural compounds which inhibit biofilm formation in clinical isolates of Klebsiella pneumoniae. *Indian Journal of Experimental Biology, 51*, 764–772.

Le, Lee, J.-H., Park, J.-H., Cho, H. S., Joo, S. W., Cho, M. H., Lee, J., et al. (2013). Anti-biofilm activities of quercetin and tannic acid against Staphylococcus aureus. *Biofouling, 29*(5), 491–499.

Lee, J., Kim, Y., Cho, H. S., Ryu, S. Y., Cho, M. H., & Lee, J. (2014). Coumarins reduce biofilm formation and the virulence of *Escherichia coli* O157: H7. *Phytomedicine, 21*, 1037–1042.

Lee, J. H., & Lee, J. (2010). Indole as an intercellular signal in microbial communities. *FEMS Microbiology Reviews, 34*(4), 426–444.

Lin, L., Wang, J., Yu, J., Li, Y., & Liu, G. (2013). Effects of allicin on the formation of *Pseudomonas aeruginosa* biofilm and the production of quorum-sensing controlled virulence factors. *Polish Journal of Microbiology, 62*, 243–251.

Li, G., Yan, C., Xu, Y., Feng, Y., Wu, Q., Lv, X., Yang, B., Wang, X., & Xia, X. (2014). Punicalagin inhibits Salmonella virulence factors and has anti-quorum-sensing potential. *Appl. Environ. Microbiol., 80*, 6204–6211.

Liu, Liu, K., Bittner, A. N., Wang, J. D., et al. (2015). Diversity in (p) ppGpp metabolism and effectors. *Current Opinion in Microbiology, 24*, 72–79.

Lu, Lu, J., Cokcetin, N. N., Burke, C. M., Turnbull, L., Liu, M., Carter, D. A., Whitchurch, C. B., Harry, E. J., et al. (2019). Honey can inhibit and eliminate biofilms produced by Pseudomonas aeruginosa. *Scientific Reports, 9*(1), 1–13.

Mansuri, A., Lokhande, K., Kore, S., Gaikwad, S., Nawani, N., Swamy, K. V., Junnarkar, M., & Pawar, S. (2022). Antioxidant, anti-quorum sensing, biofilm inhibitory activities and chemical composition of Patchouli essential oil: In vitro and in silico approach. *Journal of Biomolecular Structure and Dynamics, 40*(1), 154–165.

Mathur, H., Field, D., Rea, M. C., Cotter, P. D., Hill, C., & Ross, R. P. (2018). Fighting biofilms with lantibiotics and other groups of bacteriocins. *NPJ Biofilms and Microbiomes, 4*, 9.

Matsunaga, T., Nakahara, A., Minnatul, K. M., Noiri, Y., Ebisu, S., Kato, A., & Azakami, H. (2010). The inhibitory effects of catechins on biofilm formation by the periodontopathogenic bacterium, Eikenella Corrodens. *Bioscience, Biotechnology, and Biochemistry, 74*, 2445–2450.

Mishra, R., Panda, A. K., De Mandal, S., Shakeel, M., Bisht, S. S., & Khan, J. (2020). Natural anti- biofilm agents: Strategies to control biofilm-forming pathogens. *Frontiers in Microbiology, 11*, 566325.

Monte, J., Abreu, A. C., Borges, A., Simões, L. C., & Simões, M. (2014). Antimicrobial activity of selected phytochemicals against *Escherichia coli* and *Staphylococcus aureus* and their biofilms. *Pathogens (Basel, Switzerland), 3*(2), 473–498.

Naresh, K., Bharati, B. K., Avaji, P. G., Chatterji, D., & Jayaraman, N. (2011). Synthesis, biological studies of linear and branched arabinofuranoside-containing glycolipids and their interaction with surfactant protein A. *Glycobiology, 21*(9), 1237–1254.

Neto, Í., Faustino, C., & Rijo, P. (2015). *Antimicrobial abietane diterpenoids against resistant bacteria and biofilms* (pp. 15–26).

Paluch, E., Rewak-Soroczyńska, J., Jedrusik, I., Mazurkiewicz, E., & Jermakow, K. (2020). Prevention of biofilm formation by quorum quenching. *Applied Microbiology and Biotechnology, 104*(5), 1871–1881.

Panda, S. K., Das, R., Lavigne, R., & Luyten, W. (2020). Indian medicinal plant extracts to control multidrug-resistant *S. aureus*, including in biofilms. *South African Journal of Botany, 128*, 283–291.

Percival, S., Malic, S., Cruz, H., & Williams, D. (2011). Introduction to biofilms. In *Biofilms and veterinary medicine* (Vol 6, pp. 41–68).

Pérez-Giraldo, C., Cruz-Villalón, G., Sánchez-Silos, R., Martínez-Rubio, R., Blanco, M., & Gómez-García, A. (2003). In vitro activity of allicin against Staphylococcus epidermidis and influence of subinhibitory concentrations on biofilm formation. *Journal of Applied Microbiology, 95*, 709–711.

Plyuta, V., Zaitseva, J., Lobakova, E., Zagoskina, N., Kuznetsov, A., & Khmel, I. (2013). Effect of plant phenolic compounds on biofilm formation by *Pseudomonas aeruginosa*. *APMIS, 121*, 1073–1081.

Qiu, J., Wang, D., Xiang, H., Feng, H., Jiang, Y., Xia, L., Dong, J., Lu, J., Yu, L., & Deng, X. (2010). Subinhibitory concentrations of thymol reduce enterotoxins A and B and α-hemolysin production in *Staphylococcus aureus* isolates. *PLoS One, 5*, e9736.

Rossi, L. M., Rangasamy, P., Zhang, J., Qiu, X. Q., & Wu, G. Y. (2008). Research advances in the development of peptide antibiotics. *Journal of Pharmaceutical Sciences, 97*(3), 1060–1070.

Sambanthamoorthy, K., Sloup, R. E., Parashar, V., Smith, J. M., Kim, E. E., Semmelhack, M. F., Neiditch, M. B., & Waters, C. M. (2012). Identification of small molecules that antagonize diguanylate cyclase enzymes to inhibit biofilm formation. *Antimicrobial Agents and Chemotherapy, 56*(10), 5202–5211.

Savinov, S. N., & Eggler, A. L. (2013). Chemical and biological mechanisms of phytochemical activation of Nrf2 and importance in disease prevention. *Recent Advances in Phytochemistry, 43*, 121–155.

Sharma, G., Raturi, K., Dang, S., Gupta, S., & Gabrani, R. (2014). Combinatorial antimicrobial effect of curcumin with selected phytochemicals on Staphylococcus epidermidis. *Journal of Asian Natural Products Research, 16*, 535–541.

Shoji, M. M., & Chen, A. F. (2020). Biofilms in periprosthetic joint infections: A review of diagnostic modalities, current treatments, and future directions. *Journal of Knee Surgery, 33*, 119–131.

Singh, R., Ray, P., Das, A., & Sharma, M. (2010). Enhanced production of exopolysaccharide matrix and biofilm by a menadione-auxotrophic *Staphylococcus aureus* small-colony variant. *Journal of Medical Microbiology, 59*, 521–527.

Singh, S., Singh, S. K., Chowdhury, I., & Singh, R. (2017). Understanding the mechanism of bacterial biofilms resistance to antimicrobial agents. *The Open Microbiology Journal, 11*, 53–62.

Slobodníková, L., Fialová, S., Rendeková, K., Kováč, J., & Mučaji, P. (2016). Antibiofilm activity of plant polyphenols. *Molecules (Basel, Switzerland), 21*(12), 1717.

Soumya, E. A., Saad, I. K., Hassan, L., Ghizlane, Z., Hind, M., & Adnane, R. (2011). Carvacrol and thymol components inhibiting *Pseudomonas aeruginosa* adherence and biofilm formation. *African Journal of Microbiology Research, 5*, 3229–3232.

Supuran, C. T. (2020). Coumarin carbonic anhydrase inhibitors from natural sources. *Journal of Enzyme Inhibition and Medicinal Chemistry, 35*(1), 1462–1470.

Syal, K., Flentie, K., Bhardwaj, N., Maiti, K., Jayaraman, N., Stallings, C. L., & Chatterji, D. (2017). Synthetic (p)ppGpp analogue is an inhibitor of stringent response in mycobacteria. *Antimicrobial Agents and Chemotherapy, 61*(6), 004433-17.

Tang, K., & Zhang, X. H. (2014). Quorum quenching agents: Resources for antivirulence therapy. *Marine Drugs, 12*(6), 3245–3282.

Thakur, M., Singh, K., & Khedkar, R. (2020). 11 - Phytochemicals: Extraction process, safety assessment, toxicological evaluations, and regulatory issues. In B. Prakash (Ed.), *Functional and preservative properties of phytochemicals* (pp. 341–361).

Upadhyay, A., Upadhyaya, I., Kollanoor-Johny, A., & Venkitanarayanan, K. (2013). Antibiofilm effect of plant derived antimicrobials on Listeria monocytogenes. *Food Microbiol, 36*, 79–89.

Vediyappan, G., Dumontet, V., Pelissier, F., & D'Enfert, C. (2013). Gymnemic acids inhibit hyphal growth and virulence in Candida albicans. *PLoS One, 8*, e74189.

Wan, Zhao, Q., Wang, C., Liu, Y., Wang, S., et al. (2007). Bacterial adhesion on the metal-polymer composite coatings. *International Journal of Adhesion and Adhesives, 27*(2), 85–91.

Wang, X., Yao, X., Zhu, Z., Tang, T., Dai, K., Sadovskaya, I., Flahaut, S., & Jabbouri, S. (2009). Effect of berberine on Staphylococcus epidermidis biofilm formation. *International Journal of Antimicrobial Agents, 34*, 60–66.

Wolcott, R. D., & Ehrlich, G. D. (2008). Biofilms and chronic infections. *JAMA, 299*(22), 2682–2684.

Yaron, S., & Römling, U. (2014). Biofilm formation by enteric pathogens and its role in plant colonization and persistence. *Microbial Biotechnology, 7*(6), 496–516.

Zhou, L., Zheng, H., Tang, Y., Yu, W., & Gong, Q. (2013). Eugenol inhibits quorum sensing at sub-inhibitory concentrations. *Biotechnology Letters, 35*, 631–637.

CHAPTER

11

Medical devices—associated biofilm infections and challenges in treatment

Sreejita Ghosh[1], *Dibyajit Lahiri*[2], *Moupriya Nag*[2] and *Rina Rani Ray*[1]

[1]Department of Biotechnology, Maulana Abul Kalam Azad University of Technology, Haringhata, West Bengal, India; [2]Department of Biotechnology, University of Engineering and Management, Kolkata, West Bengal, India

1. Introduction

With increased developments in healthcare sectors, the application of medical devices has rapidly increased. New medicines rely mostly on medical device implantation inside the bodies of humans to treat a wide variety of diseases. The increased application of medical devices and implants to treat diseases has led to the device-associated infections, almost to a total of 60%–70% of nosocomial infections (Bryers, 2008). A majority of the infections associated with medical devices are because of biofilms of microbes. Biofilms can be described as a microbial community attaching to abiotic or biotic surfaces and encapsulated within a self-produced extracellular polymeric substances (EPS) matrix (Costerton et al., 1995). Microbes residing in biofilms demonstrate a distinct property differing from their counterpart planktonic cells. Microbes residing inside biofilm are resistive toward the host immunity and against antibiotics, thereby showing modified expression of genes and hence facilitating microbial adherence and slime or EPS production followed by maturation (Bjarnsholt, 2013). Biofilms were primarily reported by Anthony van Leeuwenhoek in the year 1684, but the word "biofilm" was coined by Costerton et al. in 1978 (Costerton et al., 1995). In the year of 1993, the American Society for Microbiology accepted the importance of microbial biofilms, and in the year 1999, biofilms were described by Costerton et al. as "a structured cellular community of bacteria encased within a self-produced extra polymeric matrix, attached to a surface"(Costerton et al., 1995). The growth of microbes as

Microbial Biofilms
https://doi.org/10.1016/B978-0-323-95715-1.00021-2

biofilms onto indwelling medical devices is a big problem in medical technology (Donlan, 2002). The capacity of microbes to form biofilms on implanted medical devices is the major cause of the collapse of several medical devices, leading to the development of infections along with a number of medical complicacies (Percival et al., 2015). According to the US Center for Disease Control and Prevention, around 1.7 million persons are affected, and almost 100,000 die yearly because of biofilm-associated infections. Newly formulated medicines utilize an escalating quantity of indwelling medical devices. In the United States only, over 150 million vascular catheters and more than 5 million implants are utilized annually (Bryers, 2008). The affected intravascular catheters applied for caregiving to patients lead to the development of central line–associated bloodstream infections (CLABSIs), and in the United States only, over 250,000 instances of primary bloodstream infections are observed yearly (Haddadin & Regunath, 2018). Biofilms formed by microbes on medical devices pose severe threats to the health of the patients and, in turn, afflict the device's functioning (VanEpps & Younger, 2016). Nevertheless, the biofilms serve as bacterial reservoirs, and the bacteria living in a planktonic state can get transferred to other unaffected sites, thereby causing infections. Microbes can adhere to and consequently develop biofilms over the surfaces of indwelling implants or medical devices within the body such as artificial joints, catheters, pacemakers, heart valves, contact lenses, breast implants, and ventilation tubes. The infectious diseases caused due to biofilms lead to the elimination of the affected implants, and infection treatment is followed by affected implant substitution with an uninfected one. Hence, there is a remarkable escalation in morbidity, mortality, increased treatment expenses, and time of recovery. Numerous pathogenic bacteria such as *Streptococcus* spp., *Pseudomonas aeruginosa*, *Escherichia coli*, *Staphylococcus* spp., *Klebsiella pneumoniae*, *Enterococcus faecalis*, *Acinetobacter* spp., and *Proteus mirabilis* and fungi including *Candida* spp. have been found in the biofilms of the inserted medical devices (Pour et al., 2011). The inoculations of pathogenic microbes from the bodies of the affected individuals may be done with the help of healthcare professionals. Over the recent years, different way-outs have been fabricated to fight against biofilm of the microbes, i.e., by inhibiting the formation and maturation of biofilms and eliminating the preformed biofilms. The formation of biofilms can be inhibited by a selection of innovative materials to manufacture microbe-resistant implants or medical devices. On the other hand, the chemical constitutions and surface topology of the medical devices play a significant role in the antimicrobial action of a medical device (Neoh et al., 2017). For preventing the formation of biofilms and attachment of microbes, different surface treatments of medical devices using varied chemical and physical technologies have been applied. This chapter deals with the biofilm formed on different medical devices and the difficulties in their detection and treatment.

2. Medical devices and their uses

A medical device can be considered as an apparatus, instrument, tool, appliance, or equipment utilized for diagnosis, prevention, mitigation, treatment, generation, and/or rehabilitation of information on a medical complication or disease. The Medical Devices Bureau of Health Canada has given recognition to four types of medical devices on the basis of the extent of regulation required to ensure efficacy and safety of the devices, which are as follows:

(A) Class I: Possess low risks to individuals and do not need a license needs very low regulation norms (such as dental materials, surgical instruments, etc.)
(B) Class II: Need the declaration of device efficacy and safety from the manufacturer (includes USG machines, contact lenses, medical catheters, etc.)
(C) Class III: Possess increased risk to the individuals (includes orthopedic implants such as hip implants, bone cement, surgical meshes, machines for hemodialysis, etc.)
(D) Class IV: Possess the greatest risk to the individuals and require many detailed scrutinies and premarket approval of regulations (include pacemakers, cardiovascular implants, ventricular assist devices, etc.)

2.1 Biofilm formed on medical devices

The primary step in the formation of biofilm is triggered by the complicated communications between the microbes and the surfaces. The formation of biofilm comprises numerous stages starting with adherence and progressing to dispersion. During preliminary adherence, surface properties such as charge, hydrophobicity, time of exposure, and topography affect microbial adhesion to the surfaces of medical devices (Rochford et al., 2012). Microbial adhesion on medical devices has been observed to take place via the proteins present on the cell surfaces including autolysin protein in *Staphylococcus epidermidis*, biofilm-related proteins (fimbriae-like polymers), and encapsulated polysaccharides and also the adherence of *Staphylococci* lacking coagulase and *S. epidermidis* (Helimann et al., 1997). Proteins derived from hosts such as fibrinogen, fibronectin, and vitronectin, which are being secreted to facilitate the healing process, get absorbed on the medical device surfaces, thereby leading to the formation of a conditioned biofilm that further allows colonization of microbes via communications between the proteins of the hot and the microbes (Rochford et al., 2012). During the process of the growth of biofilm, microbes proliferate, therefore enhancing the cell–cell adherence on the surfaces being colonized. All such organized communities are henceforth encased by a self-produced EPS (Koo et al., 2017). On maturation of biofilms, they form an organized multicellular microbial community conferring protection from exterior threats such as the host immune system and treatment with antibiotics. Microbes residing within biofilms secrete autoregulators and possess a modified expression of genes, thereby stimulating the release of virulence factors and hence continuing to survive (Mangwani et al., 2012). When the cells get detached, planktonic cells are released from the surfaces, thereby leading to distal metastatic infections and/or more local formation of biofilms. Dispersed microbes revert back to an active stage in comparison with the planktonic cells, making them more vulnerable to antibiotic therapies (McDougald et al., 2011). Additionally, dispersed cells of the biofilms no longer possess their shielding actions provided by the biofilm cellular community and its organized structure. Cyclic di-GMP (C di-GMP) secondary messenger found in *P. aeruginosa*, *E. coli*, and *Salmonella enterica* provides an instance of the molecule causing the dispersal of the biofilm cells (Valentini & Filloux, 2016).

2.1.1 Biofilms formed on implants in orthodontics

With advanced healthcare techniques, every year tens to millions of indwelling medical devices are applied to treat different orthopedic problems. In contrast, in spite of advancements in biomaterial, the occurrence of implant-associated infection has escalated (Anderson et al., 2004). Infections of the bone tissues such as septic arthritis, osteomyelitis, and prosthetic joint infections (PJIs) are biofilm-associated complicacies in cases of orthopedic surgeries. Orthopedic device implantation, for example, screws, pins, external fixatives, plates, and prosthetic joints has enhanced the life quality among patients, who are suffering from bone-associated problems. Yet, the contamination with microbes on these implants is a serious concern, taking place among 1%–13% of cases, leading to the development of post-operative failure of prostheses, persistent pain, immobility, reinfections, and exorbitant medical expenses and time of recovery. 80% of implant-associated infections are caused by *Staphylococcus aureus* and *S. epidermidis* (Kalita & Verma, 2010). Osteomyelitis and septic arthritis are the two major bone-related infections caused due to *Staphylococcus*. Such infections include inflammatory bone and joint destruction, and the treatment is prevented because of small colony variants (SCVs) and staphylococcal biofilms. Occurrence of PJIs is less frequent; however, the treatment is not easy, leading to a remarkable elevation in mortality, morbidity, and time of recovery. *S. epidermidis* and *S. aureus* are mostly related to the early and late infections of hip and knee arthroplasty and are the two main bacteria causing around 65% of PJIs. *S. epidermidis* is coagulase-negative *Staphylococcus* causing hospital-associated and nosocomial infections and has been often obtained from infected medical implants (Kalita & Verma, 2010). Even though among a lot of patients, *S. epidermidis* is natural body flora, and it can cause infections among patients with compromised immunity (for example, patients receiving immunosuppressive therapies, patients suffering from AIDS, and premature newborns) and drug abusers. *S. epidermidis* is a biofilm former, and the pathogenesis of implant-related *S. epidermidis* is because of its capacity of colonizing over the implant surfaces and the formation of a thick multilayered biofilm. *S. epidermidis* biofilms are characterized by the occurrence of polysaccharide intercellular adhesin (PIA), which can shield the bacteria from adverse environmental situations, host immunity, and phagocytosis (Arciola et al., 2012). The primary pathway for infection within orthopedic implants is the patient's self-normal skin body flora or naris. The major routes of infections for septic arthritis, osteomyelitis, and PJI can be hematogenous inoculation, perioperative inoculation, or contiguous. The hematogenous entrance of bacteria is due to bacteremia, and contiguous infections transmitted through neighboring tissue, whereas direct inoculation takes place due to bone infiltration after surgery, injury, or carried by external implants.

2.1.2 *Biofilm-associated infections in catheters*

Catheter-associated urinary tract infections (CAUTIs) are one of the most widespread hospital-associated infections and contribute immensely to mortality and morbidity (Mandakhalikar et al., 2018). CAUTI is caused by *Enterococcus* spp., uropathogenic *E. coli* (UPEC), *P. mirabilis*, *K. pneumoniae*, or *P. aeruginosa*. Bacteria often get attached to the catheter surfaces and form biofilm extraluminally and intraluminally. After the biofilm gets established, it acts as the origin for the persistence of uropathogens, and the bacteria can get transmitted to the bladder. Moreover, the urease that produces *P. mirabilis* also remarkably leads to the formation of biofilm on catheters; *P. mirabilis* coats catheters via salt precipitation of magnesium and calcium phosphate, thereby increasing bacterial adhesion and biofilm (Stickler, 2008).

2.1.3 *Biofilm-associated infections on cardiovascular implants*

The application of cardiac implantable electronic devices (CIEDs) such as implantable cardioverter defibrillators (ICDs) and pacemakers remarkably reduced the morbidity and mortality statistics, thereby improving human lives. On the other hand, bacterial infection on such devices is a big problem in clinical settings. The intensity and frequency of bacterial infection alter accordingly to the site and the clinical attributes of a patient. With advancing CIEDs for the treatment of cardio diseases, the occurrence of bacterial infection has also risen. Therapies with CIED are severe concerns because of the infections, which result in remarkable morbidity along with elevated healthcare management expenses and mortality. Within human bodies, a slender pocket is needed for fitting the devices. Such pockets are the major CIED infection sites, and those preliminary infections may lead to endocarditis and secondary blood infections (Sandoe et al., 2015). Microorganisms could escape from the skin of the patients into the pockets created for fitting the medical implants, get colonized, and form biofilms. CIED infections are widely caused due to bacteria, which are parts of the normal skin flora of humans. *S. epidermidis* and *S. aureus* lead to 65%–75% of pocket infections and 89% of endocarditis. Other microbes associated with CIED infections involve *Corynebacterium* spp., *Streptococcus* spp., gram-negative *Bacillus* spp., *Cutibacterium* (previously called *Propionibacterium*) *acnes*, and *Candida* spp. Out of all the bacteria, gram-positive bacteria including *Staphylococcus* are mostly detected (Baddour et al., 2010). Due to bacterial infections and the formation of biofilms, indwelling devices need to be eliminated for treating these infections. The CIED infection management is a great challenge; diagnosis is hard, and there is a severe health risk if left untreated. There are immense risks associated with the elimination and disinfection of these devices and reimplantation, resulting in unfavorable reactions against antibiotics, leading to the growth of bacteria, which are resistant to antibiotics and probably will require long-term vascular way in.

2.1.4 *Biofilm formation on ocular implants*

Numerous ocular devices including posterior contact lenses, intraocular lenses, scleral buckles, lacrimal intubation devices, orbital implants, and conjunctival plugs are mostly utilized for overcoming the disorders and enhance the quality of life for millions of individuals. On contrary, such devices and implants are susceptible to contamination with microbes and the formation of biofilms subsequently. Cataracts and untreated refractive errors are the two most significant visual disorders all over the world requiring the application of contact lenses or intraocular lens replacement (Pascolini & Mariotti, 2012). On the other hand, the existence of the abiotic surface of the human body increases the growth of the microbes and the subsequent formation of biofilms, restricting their uses. Furthermore, the contained infections, once developed, lead to secondary permanent sequelae such as ulceration or bacterial endophthalmitis, resulting in a patient permanently losing their eyesight.

2.2 Biofilm formed on various medical devices by different bacteria

Gram-negative as well as gram-positive bacteria possess the capacities to form biofilms on medical devices; however, the most prevalent ones include *S. aureus*, *E. faecalis*, *Streptococcus viridans*, *S. epidermidis*, *K. pneumoniae*, *E. coli*, *P. aeruginosa*, and *P. mirabilis* (Chen et al., 2013).

Out of all these, *S. epidermidis* and *S. aureus* are predicted to cause around 40%–50% of infections in prosthetic heart valves, 50%–70% of biofilm-associated infections of catheters, and 87% of the infections in the bloodstream (Chen et al., 2013). The species of *Staphylococcus* consists of a varied group of gram-positive bacteria, which primarily inhabit the mucous membrane and skin of humans and other mammals. *S. epidermidis* and *S. aureus* are the major causes of nosocomial, bloodstream, and surgical site infections (Zheng et al., 2018). Two-thirds of the medical device-associated infections are due to the species of *Staphylococcus* wherein most of the infections are related to coagulase-negative *Staphylococcus* and *S. aureus* (Ribeiro et al., 2012). *P. aeruginosa* is another predominant gram-negative bacteria, which can get adapted to various stress conditions and antimicrobial agents quickly and is increasingly used in the form of an in vitro model to study the formation of biofilm.

2.2.1 *Medical device–associated infections caused by* Staphylococci

Medical devices, which are very susceptible to infections, are central venous catheters (CVCs), contact lenses, intrauterine devices (IUDs), endotracheal tubes, pacemakers, mechanical heart valves, prosthetic joints, peritoneal dialysis catheters, tympanostomy tubes, prosthetic joints, voice prostheses, and urinary catheters (Donlan, 2002). The most significant staphylococcal device–related infection (DRI) with respect to attributed morbidity, frequency, and staphylococcal involvement in comparison with the other infectious microbes is discussed in the subsequent sections.

Mechanical heart valve infections are one of the most predominant due to their increased rates of morbidity (Darouiche, 2001). *S. aureus* and *S. epidermidis* form biofilm on mechanical heart valves and in the neighboring cardiovascular tissues causing severe infections such as prosthetic valve endocarditis (Murray, 2005). The infecting microbes get involved primarily during the time of the surgical processes and infection multiplies within a year of the insertion of the valve (Rupp, 2014).

CVCs are applied to deliver nutritional solutions, blood components, and medicines and also to allow the process of dialysis. Following urinary catheters, CVCs are the mostly used implanted medical devices (Darouiche, 2001). *S. aureus* and *S. epidermidis* are the most prevalent causative organisms of CVC infections (Rupp, 2014). CVC infections are the significant causes of infections in the blood, particularly within neonates, wherein *Staphylococcus* lacking coagulase such as *S. epidermidis* is the major microbe (Cheung & Otto, 2010). Without a related infection within the bloodstream, infection of CVC with *S. epidermidis* can continue to persist without any symptoms of inflammation, whereas the prominent characteristics of CVC infections include erythema, purulence, and tenderness, which is present in case of infections with *S. aureus* (Eggimann & Pittet, 2002).

Urinary catheters are latex or silicone tubular devices used for collecting urine at the time of surgery, measuring the output of urine, urinary incontinence adjustment, and preventing retention of urine. Biofilms can quickly form on the outer or inner surfaces of the urinary catheters once inserted, and it is not easy to inhibit the colonization of bacteria via normal hygiene protocols. The microbes that are primarily obtained from such devices include *E. faecalis*, *S. epidermidis*, and *E. coli*, while later on, bacteria such as *P. mirabilis* are also found (Stickler, 2008). The prolonged the use of urinary catheters, the greater the risk of catheter-related infection of the urinary tract. In fact, it has been found that the infection risk for patients with urinary catheterization increases by around 10% with each passing day (Stickler, 2008).

Ventilator-associated pneumonia (VAP) takes place in patients using mechanical machines for ventilation in hospitals. It leads to lethal illness and death and is the second most significant nosocomial infection in pediatric intensive care units (ICUs) (Foglai et al., 2007). The endotracheal tubes present one of the leading pathways for the colonization of the bacteria in cases of VAP. Endotracheal tubes directly connect the external surrounding and the lungs, making them susceptible to external bacterial infections. In fact, biofilms can form rapidly even within a single day on the endotracheal tubes (Bauer et al., 2002). In around 20% of cases, *S. aureus* is the most common pathogen preceded by *P. aeruginosa* in VAP (Chastre & Fagon, 2002).

Prosthesis infection is another kind of DRI. PJI can be caused due to total arthroplasty of the joints, heterogenous seeding, and contamination at the time of surgeries. *S. aureus* is the leading cause of PJI with coagulase-negative *Staphylococcus* attaining almost equal rates of infections (Zimmerli et al., 2004). Biofilms get formed within the synovial fluids inside the joints, frequently in absence of surface attachment, in the form of macroscopically visible and large floating biofilms. The particular physiological condition in synovial fluid allows the development of such exceptionally large biofilms, and they are severely tolerant against any antibacterial therapies (Dastgheyb et al., 2015).

2.2.2 *Medical device–associated biofilm formed by* Pseudomonas aeruginosa

In the surroundings of a hospital, *P. aeruginosa* gets colonized at moist places such as oxygen respirators, medical ventilators, sinks, humidifiers, toilets, taps, and dialysis machines (Bassetti et al., 2018). Risk factors related to infections caused by *P. aeruginosa* are diabetes, chronic obstructive pulmonary disease (COPD), immunosuppression (preceded by bone marrow or organ transplantation) and cystic fibrosis (CF), terminal liver and kidney failures, and multiorgan failure (Bassetti et al., 2018). Infections due to *P. aeruginosa* are generally treated in ICUs such as surgical, hematological, and burn units. Clinical types of infections caused due to *P. aeruginosa* are nosocomial such as urinary tract infection (UTI), VAP, bloodstream infection (BSI) such as CLABSI, soft tissue, and skin infections, wound infections from a burn, decubitus ulcers, infections at the surgical sites, central nervous system (CNS) infections, ocular infection, joint and bone infections, and otitis interna (Moore & Flaws, 2011). The European Centre For Disease Prevention and Control (ECDC) published results indicating that the nosocomial infections caused by *P. aeruginosa* in European and Polish ICUs account for 15.9% and 20% VAP, 10.5% and 14.7% UTI, and 27.8% and 11.1% CLABSI, respectively. Global patient mortality suffering from infections of *P. aeruginosa* is around 20% and increases in cases of bacteremia (50%) and VAP (30%). Polish results from EARS-NET (2017) increased the sensitivity of the strains of *P. aeruginosa* to aminoglycosides (74.5%), carbapenems (76.8%), ceftazidime (75.4%), and multidrug resistance (MDR), which was further confirmed in 22.8% of the cases under study. In that study, the occurrence of the MDR strains of *P. aeruginosa* in Europe was 30.8%. In accordance with a new multicenter registry of medical device–associated biofilm infection records, *P. aeruginosa* plays an important role in the form of an etiological factor of VAP in Europe, indicating the significance of the problem.

2.2.3 *Medical device–associated biofilm formed by* Escherichia coli

Urinary catheter infections are one of the major medical device–associated biofilm infections having adverse effects. The capability of a microbe to develop biofilms on a surface is dependent on the conditioning of the surface. On exposure to urine, different components get absorbed on the surface of the urinary catheters, forming a conditioned film, which serves as the main interface for the communication of the microbes to occur. It is followed by the material from which the catheter is made influencing the conditioned film composition, which in turn affects the type of microbe that will grow. UTIs caused by *E. coli* have varied levels of pathogenicity, and they can lead to the development of benign asymptomatic bacteriuria (ABU), which causes few or no problems in the host, but the uropathogenic strains of *E. coli* frequently lead to severe complications and symptoms. It was found that while the strains causing ABU form better biofilms on glass and polystyrene, uropathogenic strains possess a clear advantage of competition during the growth of biofilm on surfaces of catheters (Ferrieres et al., 2007). The results of this study demonstrated that few silicone–latex and silicone catheters truly select and facilitate the formation of biofilms by the most virulent strains of *E. coli* causing UTIs and providing hardly a favorable condition for the patients being catheterized.

2.2.4 *Medical device–associated biofilm formed by* Mycobacterium chimera

Mycobacterium chimera is a slow-growing species of nontuberculous mycobacteria (NTM) widely present in the surroundings such as tap water (Glickman et al., 2020). *M. chimera* was first recognized as a portion of *Mycobacterium avium* complex (MAC) by Tortoli et al. (2004). Previously, it was reported as an opportunistic respiratory pathogenic bacterium causing pulmonary disorders among immunocompromised patients (Tortoli et al., 2004). Over recent years, *M. chimera* has been known to be causing invasive infectious diseases among individuals, who undergo cardiothoracic surgeries all over the world (Sax et al., 2015). It was demonstrated that heater–cooler devices (HCDs) utilized for thermoregulation of body temperature of patients at the time of surgery get colonized by *M. chimera* and that the bioaerosols consisting of *M. chimera* produced from HCD at the time of surgery might lead to airborne contamination of the patients with such emerging pathogenic bacteria (Chand et al., 2017).

HCDs comprise water tanks, which transfer temperature-regulated water through closed circuits to a warming blanket, cardioplegia solution, and/or oxygenator of the patient, respectively. Even though the water of the circuits of HCD is not made to be in direct connection with the blood of the patient or the patient, few studies indicated that the gaps present at the top of the tank of water and the hole series surrounding the flow of water and return pipes in the water circuits can serve as escape pathways of aerosols consisting of *M. chimera* from infected HCDs (Chand et al., 2017). The temperature of operation of HCDs ranges between 2 and 41°C and the standby units are maintained at room temperature, a situation that can allow microbial colonization in the linked tubes and the water tank if disinfection and maintenance are not done properly (Sommerstein et al., 2018). Fans available within these devices can also promote the aerosolized bacterial movement inside the surgical sterile field. A simulated operation theater smoke test indicated that the air coming out of an HCD reaches the patient if the device is operated with the fan facing the surgical site and within a few distances of the operation table (Sommerstein et al., 2018).

Biofilms formed by NTMs can overcome a varied range of pH, temperature, and limiting nutrient situations, thereby posing a disadvantage to proper disinfection. *M. chimera* biofilms are resistant to disinfectants. Additionally, other water-based medical devices such as improperly maintained waterlines for dental units and thermoregulatory devices for extracorporeal circuit membrane oxygenation (ECMO) also get colonized with *M. chimera* (Chand et al., 2017). Contamination of medical devices with NTM can take place from various sources. Hence, the risk of infections with NTM is a threat all over the medical field, particularly for HCDs.

2.2.5 Medical device biofilm formed by other bacteria

Ventilators, catheters, and medical implants get contaminated with biofilms, thereby causing diseases in plants, animals, and humans. The critical care units of any hospital carry out different interventional techniques by the use of medical devices to manage critical cases. Many bacteria can contaminate these medical devices by forming biofilms. A study was performed by Rajdev and Mulla (2012), on 100 positive bacterial cultures from medical devices, which were inserted within patients being hospitalized. The isolates of the bacteria were processed in the form of each microtiter plate. All of the isolates were subjected to a test of antibiotic vulnerability by using VITEK 2 compact automated system. It was found that out of 100 total tested isolates of bacteria, 88 of them formed biofilms. An incubation period of 16–28 h was optimal for the development of biofilm. 85% of the isolates were MDRs, and different MDR mechanisms such as carbapenemase, ESBL, and MRSA were detected in the isolates. Availability of nutrients as glucose enhances the formation of bacterial biofilms. Glucose availability and time period are also significant conditions to assess the progress of the biofilms. This is a matter of serious concern for patients with compromised immunity undergoing invasive procedures and with inserted medical devices.

2.2.6 Medical device–associated biofilm formed by viable but nonculturable bacteria

Catheter-related UTIs pose serious threats to the healthcare resource as well as to the patients. There has been much work on the development of antimicrobial catheter design. Even though these designs are successful under laboratory conditions, none of them found clinical importance. By using advanced microscopy and other microbiological techniques, an in-depth laboratory study by comparison of hydrogel latex, silicone, and hydrogel latex catheters coated with silver alloy was conducted by Wilks et al. (2021). The development of biofilms by *P. aeruginosa, E. coli,* and *P. mirabilis* on three commercially available catheters was monitored over time. The samples were then analyzed with episcopic differential interference contrast (EDIC) microscope, culture study, and staining methods for quantification of VBNC bacteria. Both quantitative and qualitative biofilm assessments found that the biofilms formed rapidly on all the three types of catheters. EDIC microscopic images showed the rough topology of the surfaces of all the materials. Differences among the counts of the culture and the quantification of the dead cells and total cells revealed the existence of the populations of the VNBC bacteria, and these bacteria retained their viability but were metabolically dormant. The application of non–culture-based strategies revealed the growth of widespread populations of VNBC bacteria. These populations of the VNBC bacteria were mostly found on the hydrogel latex catheters coated with silver alloy showing a bacteriostatic action at its best.

3. Major threatening consequences of biofilm formation on medical devices

Contamination with bacteria on prosthetics and implanted medical devices leading to infections can be threatening to life causing failure of the device, persistent infections, and high morbidity and mortality rates (Sohns et al., 2017). Treatment of implant-related infections involves implant replacement by the use of risky and expensive surgeries and/or antibiotic administration at high doses, and both of these therapies are not much effective because the strains are resistant to antibiotics and there remains a high probability of reinfection of the replaced implant. The early result regarding the healthcare-related rates of infections was published by CDC during the 1970s and constantly updated since then for including recent standardized methods and definitions (Rosenthal et al., 2016). During the initial years of 2000, hospital-acquired infections (appearing within 2–30 days of hospitalization) lead to 2 million infections and 90,000 deaths in the United Statees alone (Guggenbichler et al., 2011). Out of all implant-related infections, 50%–70% are hospital-acquired infections (VanEpps & Younger, 2016). In the year 2007, hospital-acquired infections were found to be the most widely spread adverse effects in the healthcare sectors of Canada (Leatherman & Sutherland, 2010). It is estimated that about 220,000 instances of hospital-acquired infections resulted in more than 8000 deaths annually in Canada (Leatherman & Sutherland, 2010). The most frequent postoperative complicacies are surgical site infections (SSIs), which occur at the site of the body being exposed, where the surgery has been done and these SSIs account for one-third to one-fourth of all the hospital-acquired infections reported in 6%–23% studies (Hovis et al., 2018). In the United States, it is concluded that over 500,000 SSIs take place annually at a rate of 2.8% per 100 operations (Whitehouse et al., 2002).

Data obtained from the National Cardiovascular Data Registry (NCDR) ICD registry revealed that 47% of the patients having an ICD implant have undergone surgeries repeatedly because of upgradation of the device, end of the battery life, and systemic infectious diseases yearly (Boersma et al., 2016). Even though the infections occurring inside the prosthetic joints do not cause many problems, it is suggested that the rate of mortality because of removal of affected prostheses is still about 2.7%–18% in the United States (Berbari et al., 1998). In recent years, Dutch multicenter surveillance conducted a study, where the total rate of infection in hip prostheses was observed to be 3%, and in the case of knee prostheses, it was about 4.1%. Treatment with antibiotics and removal of affected prostheses recurs a cost of about 50,000 USD. Thus, there exists a rapid requirement for antibiotic alternatives to treat such infections for regulating mortality and morbidity rates and increasing the occurrence of chronic and acute infections all over the world.

3.1 Role of biofilm in catheter-associated urinary tract infections

The initial stage in the formation of biofilm on a urinary catheter is because of the deposition of a conditioned film of urinary constituents of the host, such as electrolytes, proteins, and other organic substances (Denstedt et al., 1998). Such conditioned film can convert the urinary catheter surface, thereby neutralizing the antiadhesive characteristics (Gristina, 1987). Free-swarming bacteria adhere to the surfaces via electrostatic and hydrophobic interactions and by using the flagella (Denstedt et al., 1998). Subsequently, after attachment occurs cell division, an association of extra planktonic bacterial cells, and extracellular matrix formation. Cell–cell signaling guides the organization of the loosely packed 3D structures having fluid channels in between for allowing nutrient and waste exchange (Davies et al., 1998). Individual microbe detachment from the biofilm community completes the cycle of the process and can even send urine consisting of pathogens. The cause that biofilms are so common on the urinary catheters is that they confer an advantage of survival to the microbes, and for this reason only, biofilms formed on urinary catheters are not easily cleared.

3.2 Role of biofilm in stent-associated morbidity

The endoscopic introduction of plastic stents presents an efficient system of biliary decompression resulting in the symptomatology regression and determination of remarkable quality improvement of patients' lives, who suffer due to obstructive hepatitis related to cancerous hepatobiliary growths (tumors) or benign growths (Ballinger et al., 1994). On contrary, the major disadvantages of this palliative strategy are primarily presented by occlusion of stents, frequently followed by lethal cholangitis requiring a recurring exchange of stents and interventions. Occlusion of stents is caused by the biliary sludge deposition, which comprises calcium palmitate and bilirubinate, crystals of cholesterol, amounts of fungi and/or bacteria with cholesterol, proteins, by-products of microbes, glycoproteins, and dietary fibers (Donelli et al., 2007). Calcium salt deposition because of the biochemical actions of enzymes of the bacteria inside the biofilms, which grow on the stent surface, and intestinal content reflux back into the stents are the two major reasons for stent occlusions (Sung et al., 1993).

On the other hand, few researchers indicated that the adhesion of microbes and the formation of biofilms on the lumen and stent surface play a major role in initiating the process of clogging and subsequently blocking the stent (Guaglianone et al., 2008).

Microbes take entry into the biliary system either by ascending via the Sphincter of Oddi inside biliary–duodenum reflux or by descending through the portal venous circulation (Sung et al., 1993). Bacteria attach to the surface of the stents, and the growth of their sessile cells and production of EPS result in the development of a thick biofilm providing microbes with effective protection from phagocytic cells as well as from antibacterial substances. Enzymatic activities of the lecithinase (phospholipase C) and beta-glucuronidase of the colonizing microbes result in calcium palmitate and bilirubinate precipitation, hence leading to the accumulation of sludge inside the biliary system and subsequently to the occlusion of the stent (Leung et al., 2001).

4. Molecular mechanism behind biofilm formation on medical devices

Staphylococcal attachment to indwelling medical devices is a complicated process consisting of diverse nonspecific and specific factors. Hence, adherence can be directed on the native surface of indwelling biomaterials and on the surfaces, which get altered due to adsorption of the proteins of extracellular matrix, plasma proteins, and coagulation products such as thrombi and platelets, which are derived from the host (Vaudaux et al., 1994).

Hydrophobicity of the cell surface of *Staphylococcus* and the respective biomaterial surfaces are remarkably significant for adherence to the native polymers (Mack et al., 2001). While the extent of adherence is generally the highest on the native surfaces, particular proteins of the matrix including fibronectin, fibrinogen, thrombospondin, and vitronectin in association with the active platelets may facilitate the adherence of *S. epidermidis* and other *Staphylococcus* lacking coagulase remarkably in comparison with a surface, which is blocked with plasma proteins or albumin (Mack et al., 2001).

Particular molecules on the surfaces associated with the indwelling biomaterials are quite fascinating since they may provide a way for blocking the staphylococcal attachment completely and hence inhibit severe device-associated biofilm infections from the beginning itself.

Adhesin or autolysin proteins of *S. epidermidis* have been recognized, which, apart from their functioning in the metabolism of the cell wall, are also involved in *S. epidermidis* attachment to the native polystyrene. The most widely found autolysin of *S. epidermidis* is AtlE, which was found within a transposon mutant that has a defect for the attachment on polystyrene but can still attach and develop into a biofilm on glass (Helimann et al., 2003). AtlE probably has a dual function in *S. epidermidis* attachment because it can specifically bind with vitronectin (Helimann et al., 2003). In a CVC infection model of rats, the isogenic mutant, which has a defect in AtlE production, was impaired during its virulence (Rupp et al., 2001). Additionally, a surface-related autolysin called Aae of molecular weight 35 kDa in *S. epidermidis* has been explained that can bind with vitronectin and also possess an affinity for fibrinogen and fibronectin (Helimann et al., 2003). The function of this other autolysin in adherence to polystyrene, as described for AtlE, has not yet been reported.

A protein called Ssp1 with a molecular weight of 220 kDa present in the cell wall arranged in a structure like fimbriae and hence probably not related to adhesin/autolysin proteins facilitates adherence by a strain of *S. epidermidis*; however, further characterization of these surface proteins is not reported till date (Veenstra et al., 1996). In addition to this, a capsular polysaccharide adhesin (PS/A), which brings about preliminary attachment with unaltered surfaces of Silastic catheters, was characterized by the RP62A strain of *S. epidermidis* (Tojo et al., 1988). Apparently, PS/A does not facilitate adherence to other kinds of polymers, since no deviations in polyethylene attachment were reported in mutants lacking PS/a and the wild type (Higashi et al., 1998). Later on, studies showed that PS/A has a structural resemblance but is not identical to PIA in *S. epidermidis* (Sadovskaya et al., 2005).

Numerous *S. epidermidis* proteins have been linked functionally for adherence to surfaces altered with proteins of the extracellular matrix such as receptors for fibrinogen, fibronectin, and collagen. A huge protein was found in *S. epidermidis*, and that protein can recognize specifically fibronectin. This protein is of weight greater than 1 MDa and is coded by an open reading frame (ORF) with a length of 30.5 kb. This ORF is known as Embp and is found in *S. epidermidis* strains ATCC 12228 and RP62A (Gill et al., 2005). Embp is found in most of the clinical isolates of *S. epidermidis*; however, the antibodies against Embp are not much opsonic for enhancement of phagocytosis of numerous isolates of *S. epidermidis* (Rennermalm et al., 2004). Attachment on fibronectin-altered surfaces by *S. epidermidis* gets remarkably enhanced via teichoic acids (purified), a result that may associate with the previous observation of lipoteichoic acid involvement in adherence of *S. epidermidis* to clots of fibrin–platelet (Chugh et al., 1990).

A protein that can bind with fibrinogen also known as *sdrG* or *fbe* was identified in *S. epidermidis*, and it can correlate preferably with the characteristics of a few strains of *S. epidermidis* for binding with polymer surfaces, coated with fibrinogen (Pei & Flock, 2001). *Fbe* possesses homology with *sdrCDE* family, the receptors of cell surfaces in *S. aureus*. *Fbe/sdrG* are increasingly predominant among the clinical isolates of *S. epidermidis*, but their expression is immensely variable since main strains of *S. epidermidis* do not bind with fibrinogen avidly. The fibrinogen B beta chain binds with the fibrinogen binding protein in *S. epidermidis*, leading to the prevention of clotting induced by thrombin, via interfering with the secretion of fibrinopeptide B (Davis et al., 2001). Crystallographic analyses demonstrated a dynamic dock, lock, and latch system of binding presenting a common method of binding ligands with structurally similar proteins anchored to the cell wall of gram-positive bacteria (Ponnuraj et al., 2003). *Fbe* can qualify as a potential candidate for the vaccine since antibodies specific against *Fbe* can prevent *S. epidermidis* attachment on catheters coated with fibrinogen. Particular rat and rabbit anti-*Fbe* antibodies and human donors screened IgG formulations to inject increased doses of titers against ClfA of *S. aureus* and SdrG of *S. epidermidis*, thereby reducing infections in animal models (Rennermalm et al., 2004).

GehD, which is a lipase secreted by *S. epidermidis*, can bind specifically with collagen, thereby facilitating *S. epidermidis* attachment to collagen, which is being immobilized. Binding is dependent on inhibition with recombinant GehD and also via particular antibodies of anti-GehD. Just like the adhesin/autolysin proteins, GehD has dual roles and not just demonstrates enzymatic actions but also serves as an adhesin that is structurally not related to Cna collagen adhesion in *S. aureus*.

No particular protein factor is known to be the cause of attachment of the strains of *S. epidermidis* with immobilized thrombospondin, as in *S. aureus* (Yanagisawa et al., 2001).

5. Problems in the diagnosis of medical device–associated biofilm infections

Analytical chemistry is a significant technique to understand the biofilms formed on medical devices. Analytical procedures are required for understanding how biofilms develop, their biochemical constitution, and how much biofilm is being formed. For in vivo and in vitro conditions, it is required to identify bacteria-forming biofilms and their EPS and detect the time taken to form biofilms, how well antibiofilm strategies and prophylaxis function for preventing their maturation, and the efficacies of antimicrobial agents and coatings of the devices.

5.1 Challenges in clinical analysis of in vivo or in situ biofilm analyses

The difficulty level in clinical in vivo or in situ analysis of biofilm is dependent on the objective for analysis. In some instances, like the colonization of an easily replaceable medical device, it is adequate to eliminate the device followed by direct sampling of the biofilm present in situ. Under such circumstances, analytical procedures can be employed for detecting abundant EPS components such as proteins and polysaccharides produced from the cellular appendages, signaling molecules, or even the DNA of bacteria. In the case of indwelling devices, gaining access to the bio interface in vivo is much more difficult, and researchers now depend on body tissue/fluid sampling and diagnostic imaging or infection and inflammatory symptoms to determine if the device is required to be explanted.

More difficult than only biofilm detection is the detailed biochemical analyses of clinical samples requiring identification, surveying, and quantitation of the major microbe subpopulation within the biofilm. Such a step requires to be adopted for all important medical device—associated biofilms and probably will require the synergistic application of multiple analytical procedures.

5.2 Challenges in analysis of in vitro biofilms

In comparison with the planktonic form of the bacterial cell, biofilms grow at first as an easy target to quantify since their positions on one colonized surface are identical to a preconcentration stage. This supposed advantage on the other hand vanishes when the heterogeneity and complexity of the structure of biofilm are taken into consideration. Biofilm morphology represents a limitation to analysis since it falls anywhere between the volume of bulk material and surface-only tools. Biofilms can widely vary in thickness but are mostly explained as within thickness ranging between 10 and 100 μm. The majority of the surface analytical tools cannot provide information regarding the entire composition of these structures possessing thicknesses of up to several hundred μm. In contrast to this, the efficiency of volume analytical procedures is reduced by the low quantity of material present inside the film.

The complicated and quickly evolving characteristic of a colony with time is also a limitation for comparison and quantification of the samples. Inside biofilms, coexisting subpopulations of microbes are together held by an EPS matrix, which is a heterogenous mixture. On the maturation of the film, composition modifies with time and shows distinctions during the stages of development. Multiple species might be present, and the communication might depend on the environmental conditions and nutrient availability. A common challenge is that strains that form clinical biofilms become planktonic when repeatedly cultured by techniques, which screen only planktonic cells. In cases of in vitro studies, such strains might no longer form biofilms with the same characteristics as their corresponding parts.

Systematic in vitro investigations need reproducible biofilm formation on the surfaces of medical devices. Currently, there exist four American Society for Testing and Materials (ASTM) standardized methods to grow and measure biofilms: a flow reactor, a drip flow reactor, an altered microplate, and a CDC reactor procedure. Every method is applied to various models of in vitro devices. The CDC reactor has the capacity to hold eight big coupons of samples and is extensively utilized for testing the materials of medical devices (Donlan et al., 2002). The method of the microplate is the ancient method for incorporating miniaturized multiplexed investigations and depends on the orbital incubator for providing shear flow (Ceri et al., 1999). The last stage of this procedure needs culturing and plating for obtaining quantitative counts of bacteria, which is an offline procedure limiting actual attainable output. While these procedures are good for reproducibly growing biofilms, in vitro protocol for such formats requires to be developed, which can in a better way predict in vivo formation of biofilm on medical devices.

6. Novel approaches for combating biofilms on medical devices

6.1 Physical methods

Elimination of biofilms from medical devices inside the bodies using continuous debridement and irrigation is the most promising option to treat and remove biofilms when the available antibiotic therapies prove to be ineffective for eradicating medical device–associated infections due to biofilms (Agarwal et al., 2019). Due to high mortality and technical problems, infected implant removal such as pacemakers and prosthetic joints may not serve as the best alternative. Table 11.1 describes the various types of physical methods along with their modes of action.

6.2 Modification of surfaces

Application of antiadhesion or antibacterial substances with the medical device surface via surface engineering can efficiently prevent microbial growth and attachment, thereby inhibiting biofilm formation, and this has become an important technique to combat medical device–associated biofilms. Antibacterial coating techniques prevent biofilm formation and development on medical devices through their bactericidal and bacteriostatic actions. Conventional antibacterial coatings such as gentamicin, chlorhexidine, rifampicin, minocycline, amikacin, silver sulfadiazine, and vancomycin have been used increasingly in the clinics, and they effectively prevented catheter-associated and other implant-associated infections. The synergistic activity of multiple antibiotics such as rifampicin and minocycline has yielded

TABLE 11.1 Various physical methods for biofilm removal, their targeted medical devices and modes of action.

Physical technique	Target medical devices	Mode of action	References
Photodynamic therapy (PDT)	Root canals, biofilm-associated infections in VAP and prosthetic joints	Antibacterial PDT produces immensely cytotoxic reactive oxygen species (ROS) or excitable single oxygen for oxidizing biomolecules within or on cell membrane consisting of nucleic acids, proteins, and lipids, leading to the damage of cellular DNA, destruction of organelles and cell membrane, and also death of the cell.	Hu et al. (2018)
Low intensity ultrasound cavitation	Catheter and prosthesis-associated biofilm infections	Destroy the structure of the matrix formed by biofilms of bacteria, via cavitation to allow drug entry within the biofilm or by direct cell destruction by increasing cellular metabolic activity for bactericidal effects of the antibiotics.	Hu et al. (2018)
Water jets	Biofilms on dental implants	Wash away the loosely adhered plaques by eliminating the cells of bacteria and disrupting the structure of the matrix.	Kato et al. (2012)

TABLE 11.2 Various surface modifiers and their potential modes of action.

Surface modifiers	Targeted medical devices	Mode of action	References
Isoeugenol coating	Biofilms on polyethylene and stainless steel surfaces	Reduces bacterial attachment and inhibits the formation of biofilm	Nielsen et al. (2017)
N-acetyl cysteine	Bone-associated implants	Reduces bacterial attachment and inhibits the formation of biofilm	Costa et al. (2017)
Titanium–copper alloy	Dental implants	Kills bacteria by a cell wall and cell membrane destruction and strongly prevent the formation of biofilms and reduces the chances of reinfection	Liu et al. (2018)
Ultrahigh molecular weight polyethylene	Total knee and hip prothesis	Strong hydrophobicity prevents bacterial attachment and formation of biofilms	Lockhart et al. (2018)

good results in single drug resistivity (Li et al., 1998). However, overuse of antimicrobial coatings may lead to resistance to antibiotics among bacteria. Table 11.2 describes the different surface alterations along with their modes of action.

6.3 Antimicrobial peptides

Antimicrobial peptides (AMPs) are kinds of small molecular peptides and are also called host defense peptides, and they are abundantly available in nature as a part of the innate immune system in almost all the living entities. AMPs can also be isolated from fungi, bacteria, insects, plants, amphibians, birds, fishes, human bodies, and mammals, and these AMPs are easily produced by chemical biosynthesis. AMPs possess rapid and broad-spectrum antibacterial action and do not easily lead to drug resistance. Particularly, their bactericidal effects on specific drug-resistant pathogenic organisms have attracted a lot of attention. They can be either used in combination with other antibacterial substances or as AMPs alone. Table 11.3 describes the various AMPs isolated so far and their potential modes of action.

6.4 Nanobiotechnology

Nanobiotechnology has rapidly evolved and helps in the improvement of AMPs and can also act as good surface modifiers. Nanoparticles (NPs) on coming in close contact with cell walls of bacteria can damage the shape of the bacterial membrane, disrupt the metabolic pathways, impair the microbial enzymes, lead to protein dysfunction, and alter the expression levels of genes. NPs can show their antibiofilm or antibacterial effects only at a very low dosage. Table 11.4 lists the different types of NPs or nanomaterials (NMs) along with their modes of action.

TABLE 11.3 Different AMPs and their target microbes along with their modes of action.

AMPs	Targeted microbes	Mode of action	References
B22	*Pseudomonas aeruginosa* and *Vibrio cholerae*	Kills biofilms by penetrating deep through the sessile cells by production of ROS	Rowe- Magnus et al. (2019)
LL-37	*P. aeruginosa, Escherichia coli, Staphylococcus epidermidis, Candida albicans, Staphylococcus aureus,* and *Acinetobacter baumannii*	Reduces the expression of adhesion proteins and reduces the formation of biofilms	Geng et al. (2018)
HBD2	*P. aeruginosa*	Inhibits the biofilm formation pathway	Parducho et al. (2020)
NaI-P-113	*Porphyromonas gingivalis*	Inhibits the biosynthesis of ATP-binding proteins and ATP-binding cassette transporters	Wang et al. (2017)
Polymyxin B	*Bacillus polymyxa* and *Vibrio cholerae*	Decreases twitching, swimming, and swarming motility	Giocomucci et al. (2019)
Hepcidin 20	*S. epidermidis*	Damages the intracellular polysaccharides of the biofilm cells	Brancatisano et al. (2014)
Piscidin-3	*P. aeruginosa*	Destroys the extracellular DNA of the biofilm EPS matrix	Grassi et al. (2019)

TABLE 11.4 Different NPs and their target organisms along with their potential modes of action.

NP/NM	Targeted microbes	Mode of action	References
Silver NP (AgNPs)	*Candida albicans, Escherichia coli, Pseudomonas aeruginosa,* methicillin-resistant *Staphylococcus aureus,* and *Klebsiella pneumoniae*	Destroys biofilms by biomass reduction in preformed mature biofilms	Estevez et al. (2020)
Gold NPs (AuNPs)	*P. aeruginosa, S. aureus,* and *Burkholderia multivorans*	Reduces virulence factor production along with a decrease in swarming, swimming, and twitching motilities	Khan et al. (2019)
Zinc oxide NP (ZnONPs)	*Enterococcus faecalis*	Disrupts biofilm structure	Nair et al. (2018)
Chitosan NPs	*E. faecalis* and *S. aureus*	Destructs EPS matrix	Nair et al. (2018)
Graphene nanoplatelets	*Streptococcus mutans*	Prevents bacterial attachment, damages cell wall, and inhibits the formation of biofilms	Bregnocchi et al. (2017)

TABLE 11.5 Anti-EPS agents, their target organisms, and modes of action.

Anti-EPS agents	Target microbes	Mode of action	References
Dispersin B	*Actinobacillus actinomycetemcomitans*	Hydrolyzes *N*-acetylglucosamine (NAG) and disperses the biofilm cells	Chen & Lee (2018)
Deoxyribonuclease (DNase)	*Pseudomonas aeruginosa, Staphylococcus. aureus, Acinetobacter baumannii, Escherichia coli, Klebsiella pneumoniae,* and *Haemophilus influenzae*	Degrades eDNA in EPS matrix	Sharma & Singh (2018)
Branched polyethyleneimine (BPE)	Methicillin-resistant *Staphylococcus epidermidis*	Ruptures EPS matrix	Lam et al. (2019)
Ethylene diamine tetra acetic acid (EDTA)	*Candida* spp. and *Staphylococcus* spp.	Damages the structural arrangement of EPS	Raad et al. (2003)
Farnesol	*Candida dubliniensis* and *Fusarium keratoplasticum*	Disrupts mycelial structure	Jabra-Rizk et al. (2006)
Nicotine	*S. aureus*	Damages eDNA	Jabra-Rizk et al. (2006)

6.5 EPS destructors

EPS plays a significant role in growth, maturation, and durability of biofilms. It is not just a shielding barrier to exterior agents, but also acts as an enzyme and nutritional source and helps in linking cells. EPS destruction weakens the structure of biofilm, and Table 11.5 lists those agents, which can destroy the EPS of biofilms.

6.6 Real-time monitoring of the formation of biofilms on coated medical devices to reduce and intercept the infections caused by bacteria

Real-time monitoring of the attachment of bacteria with medical devices presents options for early detection of the formation of biofilms, thereby instigating proper intervention prior to the development of the infections due to biofilms. In a study done by Kurmoo et al. (2020), they made use of long-period grating (LPG) optical fiber sensors, introduced within the lumens of endotracheal tubes (ETT) for real-time monitoring of the colonization of *P. aeruginosa* and formation of biofilms on different surfaces. The shift in wavelength detected by the attenuation bands of LPG was monitored for a period of 24 h in comparison with the biomass of the biofilms, which can be quantified by the use of imaging techniques through confocal fluorescence microscopy (FCM). The formation of biofilms was then compared with optical fibers and ETTs that were uncoated and even on an acrylate polymer, which is resistant to biofilms after subjecting them to artificial sputum or minimum media for growth (RPMI-1640). The LPG sensor could detect biomass as low as 81 $\mu g/cm^2$ by comparison with the quantification of the CFM image. An empirical exponential function was observed to connect the inverse of the biomass of the biofilm with that of optical reduction in the shift of the wavelength, thereby facilitating biofouling quantification from the spectral response. Quantification through the sensor facilitates the interception of infections and rapid removal of the device for reducing, for instance, the risk of development of VAP.

6.7 Use of hyperspectral imaging to monitor the biofilm formed on various surfaces of medical devices

Formation of biofilms on the surfaces of medical devices even on inserted medical devices is a serious problem of contamination. Microbes develop biofilms consisting of EPS, through which they get attached irreversibly on a particular surface, thereby providing structural integrity. Microbes that can form biofilms show various types of behavior in comparison with their planktonic counterparts, threatening public health (Donlan et al., 2002). Often, these biofilms get resistant against host immunity or medicinal therapies because of the synergistic action of the characteristics of the bacteria and protective matrix functions. Hence, the detection and investigation of the biofilms adhered on the surfaces of medical devices are of utter significance.

Numerous modalities of imaging have been used for investigating the growth of biofilms in the biomedical field of research, including optical coherence tomography (OCT), laser scanning microscopy, and low coherence interferometry (Nguyen et al., 2010). Hyperspectral imaging (HSI) has been already applied in several fields of research such as monitoring biofilms on agricultural products and pieces of equipment (Jun et al., 2010). The growth of biofilm on surfaces of stainless steel has also been reported by Jun et al. (2010).

Stainless steel is one of the most abundantly used metallic materials to construct medical devices because of its resistance to heat and corrosion. Depending upon the composite materials except for iron, which consists of stainless steel nickel and chromium, stainless steel has distinct materialistic characteristics. International Organization for Standard (ISO) proposes the necessity of the composition of material for stainless steel for dental and surgical instruments in ISO 7153 and for implants in ISO 5832. In a study done by Kim et al. (2012), surgical stainless steel of type 304, which is austenitic steel comprising 8%–10% nickel and 18%–20% chromium, was used. Mirror-like polished finishing can be done on stainless steel, but mirror-like finishing is generally done by coating with chromium on the surfaces and not purely polishing the stainless steel. Brushed-finished stainless steel is generally more widely available.

Titanium is a metallic material immensely utilized in implants because of its resistance to corrosion, strength, and biocompatibility. The necessity for titanium, which is utilized in medical implants, is proposed in ISO 5832. Even though titanium has very limited applications only in microsurgical tools, titanium can also be used to construct dental and normal surgical tools. Alloys of titanium are also used in surgical tools and implants. Alloys of titanium are composed of foreign substances such as iron, aluminum, and niobium. Nonreflecting satin finishing is mostly done on titanium alloys and titanium.

Some bacteria such as *E. coli* develop biofilms leading to lethal infections. Medical device–associated infections in the duodenoscopes and bronchoscopes lead to the largest number of endoscopic infections reported so far, where the infection is mainly due to the development of microbial biofilms. Kim et al. (2012) used the HSI technique for detecting contamination due to biofilms on the surfaces of various usual materials used in medical devices. These materials include titanium, stainless steel, and stainless steel–titanium alloys. HSI method can be used to monitor the attachment of biofilms on different materialistic surfaces. By the use of HSI, the formation of biofilms could be differentiated from the background of the surface of stainless steel. However, the formation of biofilms was not observed on surfaces of an alloy of titanium and titanium. Various materials exhibited various rates of formation of biofilm mainly because of their differences in the morphology of the surfaces.

6.8 Biofilm formation assessment on device-related clinical isolates of bacteria in a hospital of tertiary level

Formation of biofilm is a developmental phenomenon having intercellular signals, which control growth. In a study carried out by Mulla & Revdiwala (2011), the formation of biofilm was quantitatively assessed in device-based clinical isolates of bacteria corresponding to different glucose concentrations with tryptic soya broth (TSB) with respect to different time intervals for incubation. This study was performed on 100 positive bacterial cultures from medical devices that were once inserted in patients at the hospitals. The isolates of the bacteria were then processed by the method of a microtiter plate with only TSB and different glucose concentrations and were observed at various time intervals. The results indicated that most of the cultures from the catheters were found to be positive. Out of the total 100 isolates of bacteria, 88 could form biofilms. An incubation period of 16–20 h was reported to be optimal for the development of biofilms. It can be concluded from this study that nutritional abundance in the form of glucose can enhance the formation of biofilms by bacteria. The formation of biofilms depends on bacterial attachment to different surfaces. Glucose abundance and time are the two major factors to be considered to assess the progression of biofilms.

6.9 Prediction of the inhibitory molecules of biofilms through the use of machine learning

Inhibitory small molecules against biofilms are promising therapeutic alternatives against the bacteria, which are biofilm formers. However, the experimental detection of these molecules is a time-taking task, and hence, the computational strategies evolved as promising alternative techniques. A study designed a tool known as "Molib" for predicting the inhibitory action of small molecules (Srivastava et al., 2020). The researchers curated a trained data set of the inhibitory molecules acting against the biofilms, and their chemical and structural attributes were utilized for characteristic selection and subsequently the optimization of algorithms and designing of machine learning–associated models of classification. On cross-validation for five times, fingerprint, random forest–based descriptor, and hybrid models of classification displayed accuracies of 0.88, 0.93, and 0.90, respectively. The efficacies of all the models were checked on two distinct data sets for validation including noninhibitory and inhibitory molecules against biofilms with an accuracy of greater than or equal to 0.90%. The Web server Molib can serve as a very reliable and useful tool to predict the inhibitory actions of the small molecules against the biofilms.

7. Conclusion

Since the treatment of different diseases depends on the insertion of indwelling medical devices, the regulation of the biofilms on the medical implants is a major challenge. Biofilms formed by microbes on the implants and the medical devices escalate the medical complexities leading to implant decontamination or surgery, thereby incurring heavy treatment expenses and prolonged time of recovery. Bacterial biofilms develop resistance against conventional antibiotics, and the sub-MICs of such antibiotics result in the increased mode of biofilm growth. Hence, there is an urgent need to discover novel antibacterial substances, which will be effective against bacterial biofilms on the inserted medical devices and can be used with ensured safety. Surface materials utilized for the manufacturing of medical devices provide a favorable abiotic surface for the growth and proliferation of biofilms. Hence, to regulate the bacterial biofilms, the implants can be made from microbe-resistant material, or a coating on the surface with antibacterial agents can also be applied. Biofilms formed by microbes are very complex microbial communities; till date, no particular methods of detection are available for diagnosing biofilms. Microbes dwelling on medical implants are detected through traditional techniques of culture like sonication following the counting of bacterial colonies. On the other hand, slow-growing or nonculturable bacteria cannot be detected through culture-associated methods. Another alternative technique is the application of molecular methods including real-time PCR or PCR. The disadvantage of PCR-related methods is that they can detect live or dead DNA of microbes giving false-positive results. Nevertheless, implant-associated biofilms must not be detected using the PCR-based techniques, and hence there exists a requirement to develop novel methods for diagnosis and biofilm detection in medical devices or samples from clinics. The application of fluorescence-associated techniques in combination with next-generation sequencing is attracting much attention to diagnose and detect biofilms, but their use in clinical conditions is not yet much established.

References

Agarwal, A., Kelkar, A., Agarwal, A. G., Jayaswal, D., Schultz, C., Jayaswal, A., Goel, V. K., Agarwal, A. K., & Gidvani, S. (2019). Implant retention or removal for management of surgical site infection after spinal surgery. *Global Spine Journal, 10*, 640–646.

Anderson, J. M., Cook, G., Costerton, J. W., Hanson, S. R., Hensten-Pettersen, A., Jacobsen, N., Johnson, R. J., Mitchell, R. N., Pasmore, M., Schoen, F. J., Shirtliff, M., & Stoodley, P. (2004). Host reactions to biomaterials and their evaluation. In B. D. Ratner, A. S. Hoffman, F. J. Schoen, & J. E. Lemons (Eds.), *Biomaterials science. An introduction to materials in medicine* (2nd ed., pp. 293–345). London: Elsevier Academic Press.

Arciola, C. R., Campoccia, D., Speziale, P., Montanaro, L., & Costerton, J. W. (2012). Biofilm formation in Staphylococcus implant infections. A review of molecular mechanisms and implications for biofilm-resistant materials. *Biomaterials, 33*, 5967–5982.

Baddour, L. M., Epstein, A. E., Erickson, C. C., et al. (2010). Update on cardiovascular implantable electronic device infections and their management. *Circulation, 121*, 458–477.

Ballinger, A. B., McHugh, M., Catnach, S. M., Alstead, E. M., & Clark, M. L. (1994). Symptom relief and quality of life after stenting for malignant bile duct obstruction. *Gut, 35*, 467–470.

Bassetti, M., Vena, A., Croxatto, A., Righi, E., & Guery, B. (2018). How to manage *Pseudomonas aeruginosa* infections. *Drugs Context, 7*, 212527. Available from: https://doi.org/10.7573/dic.212527.

Bauer, T. T., Torres, A., Ferrer, R., Heyer, C. M., Schultze-Werninghaus, G., & Rasche, K. (2002). Biofilm formation in endotracheal tubes. Association between pneumonia and the persistence of pathogens. *Monaldi Archives for Chest Disease, 57*, 84–87.

Berbari, E. F., Hanssen, A. D., Duffy, M. C., Steckelberg, J. M., Ilstrup, D. M., Harmsen, W. S., & Osmon, D. R. (1998). Risk factors for prosthetic joint infection: Case-control study. *Clin. Infect. Dis., 27*(5), 1247–1254.

Wilks, S. A., et al. (2021). *Biofilm development on urinary catheters promotes the appearance of viable but nonculturable bacteria.*

Bjarnsholt, T. (2013). The role of bacterial biofilms in chronic infections. *APMIS, 121*, 1–58.

Boersma, L., Burke, M. C., Neuzil, P., Lambiase, P., Friehling, T., Theuns, D. A., Garcia, F., Carter, N., Stivland, T., Weiss, R., & Effortless, I. D. E. S. (2016). Investigators Infection and mortality after implantation of a subcutaneous ICD after transvenous ICD extraction. *Heart Rhythm, 13*(1), 157–164.

Brancatisano, F. L., Maisetta, G., Di Luca, M., Esin, S., Bottai, D., Bizzarri, R., Campa, M., & Batoni, G. (2014). Inhibitory effect of the human liver-derived antimicrobial peptide hepcidin 20 on biofilms of polysaccharide intercellular adhesin (PIA)-positive and PIA-negative strains of Staphylococcus epidermidis. *Biofouling, 30*, 435–446.

Bregnocchi, A., Zanni, E., Uccelletti, D., Marra, F., Cavallini, D., De Angelis, F., De Bellis, G., Bossù, M., Ierardo, G., Polimeni, A., et al. (2017). Graphene-based dental adhesive with anti-biofilm activity. *Journal of Nanobiotechnology, 15*, 1–13.

Bryers, J. D. (2008). Medical biofilms. *Biotechnology and Bioengineering, 100*, 1–18.

Ceri, H., Olson, M. E., Stremick, C., Read, R. R., Morck, D., & Buret, A. (1999). *Journal of Clinical Microbiology, 37*, 1771–1776.

Chand, M., Lamagni, T., Kranzer, K., Hedge, J., Moore, G., Parks, S., et al. (2017). Insidious risk of severe *Mycobacterium* chimaera infection in cardiac surgery patients. *Clinical Infectious Diseases, 64*, 335–342.

Rajdev, B. M., & Mulla, S. (2012). *Characterization of bacterial etiologic agents of biofilm formation in medical devices in critical care setup.*

Chastre, J., & Fagon, J. Y. (2002). Ventilator-associated pneumonia. *American Journal of Respiratory and Critical Care Medicine, 165*, 867–903.

Chen, K.-J., & Lee, C.-K. (2018). Twofold enhanced dispersin B activity by N-terminal fusion to silver-binding peptide for biofilm eradication. *International Journal of Biological Macromolecules, 118*, 419–426.

Chen, M., Yu, Q., & Sun, H. (2013). Novel strategies for the prevention and treatment of biofilm related infections. *International Journal of Molecular Sciences, 14*(9), 18488–18501.

Cheung, G. Y., & Otto, M. (2010). Understanding the significance of *Staphylococcus epidermidis* bacteremia in babies and children. *Current Opinion in Infectious Diseases, 23*, 208–216.

Chugh, T. D., Burns, G. J., Shuhaiber, H. J., & Bahr, G. M. (1990). Adherence of Staphylococcus epidermidis to fibrin-platelet clots in vitro mediated by lipoteichoic acid. *Infection and Immunity, 58*, 315–319.

Costa, F., Sousa, D. M., Parreira, P., Lamghari, M., Gomes, P., & Martins, M. C. L. (2017). N-acetylcysteine-functionalized coating avoids bacterial adhesion and biofilm formation. *Scientific Reports, 7*, 1–13.

Costerton, J. W., Lewandowski, Z., Caldwell, D. E., Korber, D. R., & Lappin-Scott, H. M. (1995). Microbial biofilms. *Annual Review of Microbiology, 49*, 711–745.

Darouiche, R. O. (2001). Device-associated infections: A macroproblem that starts with microadherence. *Clinical Infectious Diseases, 33*, 1567–1572.

Dastgheyb, S. S., Villaruz, A. E., Le, K. Y., Tan, V. Y., Duong, A. C., Chatterjee, S. S., et al. (2015c). Role of phenol-soluble modulins in formation of *Staphylococcus aureus* biofilms in synovial fluid. *Infection and Immunity, 83*, 2966–2975.

Davies, D., Parsek, M., Pearson, J., Iglewski, B., Costerton, J., & Greenberg, E. (1998). The involvement of cell-to-cell signals in the development of a bacterial biofilm. *Science, 280*, 295–298.

Davis, S. L., Gurusiddappa, S., McCrea, K. W., Perkins, S., & Höök, M. (2001). SdrG, a fibrinogen-binding bacterial adhesin of the microbial surface components recognizing adhesive matrix molecules subfamily from Staphylococcus epidermidis, targets the thrombin cleavage site in the Bß chain. *Journal of Biological Chemistry, 276*, 27799–27805.

Denstedt, J., Wollin, T., & Reid, G. (1998). Biomaterials used in urology: Current issues of biocompatibility, infection, and encrustation. *Journal of Endourology, 12*, 493–500.

Donelli, G., Guaglianone, E., Di, R. R., Fiocca, F., & Basoli, A. (2007). Plastic biliary stent occlusion: Factors involved and possible preventive approaches. *Clinical Medicine & Research, 5*, 53–60.

Donlan, R. M. (2002). Biofilms: Microbial life on surfaces. *Emerging Infectious Diseases, 8*, 881–890.

Eggimann, P., & Pittet, D. (2002). Overview of catheter-related infections with special emphasis on prevention based on educational programs. *Clinical Microbiology and Infection, 8*, 295–309.

Estevez, M. B., Raffaelli, S., Mitchell, S. G., Faccio, R., & Alborés, S. (2020). Biofilm eradication using biogenic silver nanoparticles. *Molecules, 25*, 2023.

Ferrières, L., Hancock, V., & Klemm, P. (October 2007). Specific selection for virulent urinary tract infectious *Escherichia coli* strains during catheter-associated biofilm formation. *FEMS Immunology & Medical Microbiology, 51*(1), 212–219. https://doi.org/10.1111/j.1574-695X.2007.00296.x

Foglia, E., Meier, M. D., & Elward, A. (2007). Ventilator-associated pneumonia in neonatal and pediatric intensive care unit patients. *Clinical Microbiology Reviews, 20*, 409–425.

Geng, H., Yuan, Y., Adayi, A., Zhang, X., Song, X., Gong, L., Zhang, X., & Gao, P. (2018). Engineered chimeric peptides with antimicrobial and titanium-binding functions to inhibit biofilm formation on Ti implants. *Materials Science and Engineering C, 82*, 141–154.

Giacomucci, S., Cros, C. D.-N., Perron, X., Mathieu-Denoncourt, A., & Duperthuy, M. (2019). Flagella-dependent inhibition of biofilm formation by sub-inhibitory concentration of polymyxin B in *Vibrio cholerae*. *PLoS One, 14*, e0221431.

Gill, S. R., Fouts, D. E., Archer, G. L., Mongodin, E. F., Deboy, R. T., Ravel, J., Paulsen, I. T., Kolonay, J. F., Brinkac, L., Beanan, M., Dodson, R. J., Daugherty, S. C., Madupu, R., Angiuoli, S. V., Durkin, A. S., Haft, D. H., Vamathevan, J., Khouri, H., Utterback, T., … Fraser, C. M. (2005). Insights on evolution of virulence and resistance from the complete genome analysis of an early methicillin-resistant *Staphylococcus aureus* strain and a biofilm-producing methicillin-resistant Staphylococcus epidermidis strain. *Journal of Bacteriology, 187*, 2426–2438.

Glickman, C. M., Virdi, R., Hasan, N. A., Epperson, L. E., Brown, L., Dawrs, S. N., et al. (2020). Assessment of soil features on the growth of environmental nontuberculous *Mycobacterial* isolates from hawai'i. *Applied and Environmental Microbiology, 86*, 20. https://doi.org/10.1128/AEM.00121-20

Grassi, L., Batoni, G., Ostyn, L., Rigole, P., Van den Bossche, S., Rinaldi, A. C., Maisetta, G., Esin, S., Coenye, T., & Crabbé, A. (2019). The antimicrobial peptide lin-SB056-1 and its dendrimeric derivative prevent *Pseudomonas aeruginosa* biofilm formation in physiologically relevant models of chronic infections. *Frontiers in Microbiology, 10*, 198.

Gristina, A. (1987). Biomaterial-centered infection: Microbial adhesion versus tissue integration. *Science, 237*, 1588–1595.

Guaglianone, E., Cardines, R., Mastrantonio, P., Di Rosa, R., Penni, A., Puggioni, G., Basoli, A., Fiocca, F., & Donelli, G. (2008). Role of multispecies microbial biofilms in the occlusion of biliary stents. *Microbial Ecology in Health and Disease, 20*, 207–209.

Guggenbichler, P. J., Assadian, O., Boeswald, M., & Kramer, A. (2011). Incidence and clinical implication of nosocomial infections associated with implantable biomaterials – catheters, ventilator-associated pneumonia, urinary tract infections. *GMS Journal for Medical Education, 6*(1).

Haddadin, Y., & Regunath, H. (2018). Central line associated blood stream infections (CLABSI). In *StatPearls [Internet]*. Treasure Island, FL: StatPearls Publishing. NBK430891.

Heilmann, C., Hussain, M., Peters, G., & Gotz, F. (1997). Evidence for autolysin-mediated primary attachment of *Staphylococcus epidermidis* to a polystyrene surface. *Molecular Microbiology, 24*(5), 1013–1024.

Heilmann, C., Thumm, G., Chhatwal, G. S., Hartleib, J., Uekotter, A., & Peters, G. (2003). Identification and characterization of a novel autolysin (Aae) with adhesive properties from Staphylococcus epidermidis. *Microbiology, 149*, 2769–2778.

Higashi, J. M., Wang, I. W., Shlaes, D. M., Anderson, J. M., & Marchant, R. E. (1998). Adhesion of Staphylococcus epidermidis and transposon mutant strains to hydrophobic polyethylene. *Journal of Biomedical Materials Research, 39*, 341–350.

Hovis, J. P., Montalvo, R., Marinos, D., Joshi, M., Shirtliff, M. E., O'Toole, R. V., & Manson, T. T. (2018). Intraoperative vancomycin powder reduces *Staphylococcus aureus* surgical site infections and biofilm formation on fixation implants in a rabbit model. *Journal of Orthopaedic Trauma, 32*(5), 263–268.

Hu, X., Huang, Y.-Y., Wang, Y., Wang, X., & Hamblin, M. R. (2018). Antimicrobial photodynamic therapy to control clinically relevant biofilm infections. *Frontiers in Microbiology, 9*, 1299.

Jabra-Rizk, M. A., Shirtliff, M., James, C., & Meiller, T. (2006). Effect of farnesol on Candida dubliniensisbiofilm formation and fluconazole resistance. *FEMS Yeast Research, 6*, 1063–1073.

Jun, W., Kim, M. S., Cho, B. K., Millner, P. D., Chao, K., & Chan, D. E. (2010). Microbial biofilm detection on food contact surfaces by macro-scale fluorescence imaging. *Journal of Food Engineering, 99*, 314–322.

Kalita, S. J., & Verma, S. (2010). Nanocrystalline hydroxyapatite bioceramic using microwave radiation: Synthesis and characterization. *Materials Science and Engineering C, 30*, 295–303.

Kato, K., Tamura, K., & Nakagaki, H. (2012). Quantitative evaluation of the oral biofilm-removing capacity of a dental water jet using an electron-probe microanalyzer. *Archives of Oral Biology, 57*, 30–35.

Khan, F., Manivasagan, P., Lee, J.-W., Pham, D. T. N., Oh, J., & Kim, Y.-M. (2019). Fucoidan-stabilized gold nanoparticle-mediated biofilm inhibition, attenuation of virulence and motility properties in *Pseudomonas aeruginosa* PAO1. *Marine Drugs, 17*, 208.

Kim, D., Kim, M. S., & Hwang, J. (2012). Monitoring of biofilm formation on different material surfaces of medical devices using hyperspectral imaging method. *Design and Quality for Biomedical Technologies, 8215*, 821507. https://doi.org/10.1117/12.909980

Koo, H., Allan, R. N., Howlin, R. P., Stoodley, P., & Hall-Stoodley, L. (2017). Targeting microbial biofilms: Current and prospective therapeutic strategies. *Nature Reviews Microbiology, 15*, 740–755.

Kurmoo, Y., Hook, A. L., Harvey, D., Dubern, J., Williams, P., Morgan, S. P., Korposh, S., & Alexandar, M. R. (2020). Real time monitoring of biofilm formation on coated medical devices for the reduction and interception of bacterial infections. *Biomaterials Science, 5*, 1464–1477. https://doi.org/10.1039/C9BM00875F

Lam, A. K., Wouters, C. L., Moen, E. L., Pusavat, J., & Rice, C. V. (2019). Antibiofilm synergy of β-lactams and branched polyethylenimine against methicillin-resistant Staphylococcus epidermidis. *Biomacromolecules, 20*, 3778–3785.

Leatherman, S., & Sutherland, K. (2010). *Quality of healthcare in Canada*. https://www.cfhi-fcass.ca/SearchResults-News/10-02-10/42054d49-16fb-4764-be05-1d03e6ff3bbb.aspx.

Leung, J. W., Liu, Y. L., Leung, P. S., Chan, R. C., Inciardi, J. F., & Cheng, A. F. (2001). Expression of bacterial beta-glucuronidase in human bile: An in vitro study. *Gastrointest Endosc, 54*, 346–350.

Li, H., Fairfax, M. R., Dubocq, F., Darouiche, R. O., Rajpurkar, A., Thompson, M., Tefilli, M. V., & Dhabuwala, C. B. (1998). Antibacterial activity of antibiotic coated silicone grafts. *The Journal of Urology, 160*, 1910–1913.

Liu, R., Tang, Y., Zeng, L., Zhao, Y., Ma, Z., Sun, Z., Xiang, L., Ren, L., & Yang, K. (2018). In vitro and in vivo studies of anti-bacterial copper-bearing titanium alloy for dental application. *Dental Materials, 34*, 1112–1126.

Lockhart, J. N., Spoonmore, T. J., McCurdy, M. W., Rogers, B. R., Guelcher, S. A., & Harth, E. (2018). Poly(glycidol) coating on ultrahigh molecular weight polyethylene for reduced biofilm growth. *ACS Applied Materials & Interfaces, 10*, 4050–4056.

Mack, D., Bartscht, K., Fischer, C., Rohde, H., de Grahl, C., Dobinsky, S., Horstkotte, M. A., Kiel, K., & Knobloch, J. K. M. (2001). Genetic and biochemical analysis of Staphylococcus epidermidis biofilm accumulation. *Methods in Enzymology, 336*, 215–239.

Mandakhalikar, K. D., Rahmat, J. N., Chiong, E., Neoh, K. G., Shen, L., & Tambyah, P. A. (2018). Extraction and quantification of biofilm bacteria: Method optimized for urinary catheters. *Scientific Reports, 8*, 8069.

Mangwani, N., Dash, H. R., Chauhan, A., & Das, S. (2012). Bacterial quorum sensing: Functional features and potential applications in biotechnology. *Journal of Molecular Microbiology and Biotechnology, 22*(4), 215–227.

McDougald, D., Rice, S. A., Barraud, N., Steinberg, P. D., & Kjelleberg, S. (2011). Should we stay or should we go: Mechanisms and ecological consequences for biofilm dispersal. *Nature Reviews Microbiology, 10*(1), 39–50.

Moore, N. M., & Flaws, M. L. (2011). Epidemiology and pathogenesis of *Pseudomonas aeruginosa* infection. *Clinical Laboratory Science, 24*(1), 43–46. https://doi.org/10.29074/ascls24.1.43. Available from:.

Mulla, S. A., & Revdiwala, S. (2011). Assessment of biofilm formation in device-associated clinical bacterial isolates in a tertiary level hospital. *Indian Journal of Pathology & Microbiology*, 54(3), 561–564. https://doi.org/10.4103/0377-4929.85093

Murray, R. J. (2005). *Staphylococcus aureus* infective endocarditis: Diagnosis and management guidelines. *Internal Medicine Journal, 35*, S25–S44.

Nair, N., James, B., Devadathan, A., Johny, M. K., Mathew, J., & Jacob, J. (2018). Comparative evaluation of antibiofilm efficacy of chitosan nanoparticle- and zinc oxide nanoparticle-incorporated calcium hydroxide-based sealer: An in vitro study. *Contemporary Clinical Dentistry, 9*, 434–439.

Neoh, K. G., Li, M., Kang, E.-T., Chiong, E., & Tambyah, P. A. (2017). Surface modification strategies for combating catheter-related complications: Recent advances and challenges. *Journal of Materials Chemistry B, 5*, 2045–2067.

Nguyen, C. T., Tu, H., Chaney, E. J., Stewart, C. N., & Boppart, S. A. (2010). Non-invasive optical interferometry for the assessment of biofilm growth in the middle ear. *Biomedical Optics Express, 1*, 1104–1116.

Nielsen, C., Subbiahdoss, G., Zeng, G., Salmi, Z., Kjems, J., Mygind, T., Snabe, T., & Meyer, R. (2017). Antibacterial isoeugenol coating on stainless steel and polyethylene surfaces prevents biofilm growth. *Journal of Applied Microbiology, 124*, 179–187.

Parducho, K. R., Beadell, B., Ybarra, T. K., Bush, M., Escalera, E., Trejos, A. T., Chieng, A., Mendez, M., Anderson, C., Park, H., et al. (2020). The antimicrobial peptide human beta-defensin 2 inhibits biofilm production of *Pseudomonas aeruginosa* without compromising metabolic activity. *Frontiers in Immunology, 11*, 805.

Pascolini, D., & Mariotti, S. P. (2012). Global estimates of visual impairment: 2010. *British Journal of Ophthalmology, 96*, 614–618.

Pei, L., & Flock, J. I. (2001). Lack of fbe, the gene for a fibrinogen-binding protein from Staphylococcus epidermidis, reduces its adherence to fibrinogen coated surfaces. *Microbial Pathogenesis, 31*, 185–193.

Percival, S. L., Suleman, L., Vuotto, C., & Donelli, G. (2015). Healthcare-associated infections, medical devices and biofilms: Risk, tolerance and control. *Journal of Medical Microbiology, 64*, 323–334.

Ponnuraj, K., Bowden, M. G., Davis, S., Gurusiddappa, S., Moore, D., Choe, D., Xu, Y., Höök, M., & Narayana, S. V. (2003). A "dock, lock, and latch" structural model for a staphylococcal adhesin binding to fibrinogen. *Cell, 115*, 217–228.

Pour, N. K., Dusane, D. H., Dhakephalkar, P. K., Zamin, F. R., Zinjarde, S. S., & Chopade, B. A. (2011). Biofilm formation by Acinetobacter baumannii strains isolated from urinary tract infection and urinary catheters. *FEMS Immunology and Medical Microbiology, 62*, 328–338.

Raad, I., Chatzinikolaou, I., Chaiban, G., Hanna, H., Hachem, R., Dvorak, T., Cook, G., & Costerton, W. (2003). In vitro and ex vivo activities of minocycline and EDTA against microorganisms embedded in biofilm on catheter surfaces. *Antimicrobial Agents and Chemotherapy, 47*, 3580–3585.

Rennermalm, A., Nilsson, M., & Flock, J. I. (2004). The fibrinogen binding protein of Staphylococcus epidermidis is a target for opsonic antibodies. *Infection and Immunity, 72*, 3081–3083.

Ribeiro, M., Monteiro, F. J., & Ferraz, M. P. (2012). Infection of orthopedic implants with emphasis on bacterial adhesion process and techniques used in studying bacterial-material interaction. *Biomatter, 2*(4), 176–194.

Rochford, E. T., Richards, R. G., & Moriarty, T. F. (2012). Influence of material on the development of device-associated infections. *Clinical Microbiology and Infection, 18*(12), 1162–1167.

Rosenthal, V. D., Al-Abdely, H. M., El-Kholy, A. A., AlKhawaja, S. A. A., Leblebicioglu, H., Mehta, Y., Rai, V., Hung, N. V., Kanj, S. S., Salama, M. F., Salgado-Yepez, E., Elahi, N., Morfin Otero, R., Apisarnthanarak, A., De Carvalho, B. M., Ider, B. E., Fisher, D., Buenaflor, M., Petrov, M. M., , … Alvarez-Moreno, C., & Kushner-Davalos, L. (2016). International nosocomial infection control consortium report, data summary of 50 countries for 2010–2015: Device-associated module. *American Journal of Infection Control, 44*(12), 1495–1504.

Rowe-Magnus, D. A., Kao, A. Y., Prieto, A. C., Pu, M., & Kao, C. (2019). Cathelicidin peptides restrict bacterial growth via membrane perturbation and induction of reactive oxygen species. *mBio, 10*, e02021–19.

Rupp, M. E. (2014). Clinical characteristics of infections in humans due to *Staphylococcus epidermidis*. *Methods in Molecular Biology, 1106*, 1–16.

Rupp, M. E., Fey, P. D., Heilmann, C., & Götz, F. (2001). Characterization of the importance of Staphylococcus epidermidis autolysin and polysaccharide intercellular adhesin in the pathogenesis of intravascular catheter-associated infection in a rat model. *The Journal of Infectious Diseases, 183*, 1038–1042.

Sadovskaya, I., Vinogradov, E., Flahaut, S., Kogan, G., & Jabbouri, S. (2005). Extracellular carbohydrate-containing polymers of a model biofilm-producing strain, Staphylococcus epidermidis RP62A. *Infection and Immunity, 73*, 3007–3017.

Sandoe, J. A. T., Barlow, G., Chambers, J. B., et al. (2015). Guidelines for the diagnosis, prevention and management of implantable cardiac electronic device infection. Report of a joint working party project on behalf of the British Society for Antimicrobial Chemotherapy (BSAC, host organization), British Heart Rhythm Society (BHRS), British Cardiovascular Society (BCS), British Heart Valve Society (BHVS) and British Society for Echocardiography (BSE). *Journal of Antimicrobial Chemotherapy, 70*, 325–359.

Sax, H., Bloemberg, G., Hasse, B., Sommerstein, R., Kohler, P., Achermann, Y., et al. (2015). Prolonged outbreak of Mycobacterium chimaera infection after open-chest heart surgery. *Clinical Infectious Diseases, 61*, 67–75. https://doi.org/10.1093/cid/civ198

Sharma, K., & Singh, A. P. (2018). Antibiofilm effect of DNase against single and mixed species biofilm. *Foods, 7*, 42.

Sohns, J. M., Bavendiek, U., Ross, T. L., & Bengel, F. M. (2017). Targeting cardiovascular implant infection: Multimodality and molecular imaging. *Circ. Cardiovasc. Imaging., 10*(12), e005376.

Sommerstein, R., Hasse, B., Marschall, J., Sax, H., Genoni, M., Schlegel, M., et al. (2018). Global health estimate of invasive *Mycobacterium* chimaera infections associated with heater-cooler devices in cardiac surgery. *Emerging Infectious Diseases, 24*, 576–578. https://doi.org/10.3201/eid2403.171554

Srivastava, G. N., Malwe, A. S., Sharma, A. K., Shastri, V., Hibare, K., & Sharma, V. K. (2020). Molib: A machine learning based classification tool for the prediction of biofilm inhibitory molecules. *Genomics, 112*(4), 2823–2832. https://doi.org/10.1016/j.ygeno.2020.03.020

Stickler, D. J. (2008). Bacterial biofilms in patients with indwelling urinary catheters. *Nature Clinical Practice Urology, 5*, 598–608.

Sung, J. Y., Leung, J. W., Shaffer, E. A., Lam, K., & Costerton, J. W. (1993). Bacterial biofilm, brown pigment stone and blockage of biliary stents. *Journal of Gastroenterology and Hepatology, 8*, 28–34.

Tojo, M., Yamashita, N., Goldmann, D. A., & Pier, G. B. (1988). Isolation and characterization of a capsular polysaccharide adhesin from Staphylococcus epidermidis. *The Journal of Infectious Diseases, 157*, 713–722. Erratum in: 1988. *The Journal of Infectious Diseases, 158*, 268.

Tortoli, E., Rindi, L., Garcia, M. J., Chiaradonna, P., Dei, R., Garzelli, C., et al. (2004). Proposal to elevate the genetic variant MAC-A, included in the *Mycobacterium* avium complex, to species rank as *Mycobacterium* chimaera sp. nov. *International Journal of Systematic and Evolutionary Microbiology, 54*, 1277–1285. https://doi.org/10.1099/ijs.0.02777-0

Valentini, M., & Filloux, A. (2016). Biofilms and Cyclic di-GMP (c-di-GMP) Signaling: Lessons from *Pseudomonas aeruginosa* and other bacteria. *The Journal of Biological Chemistry, 291*(24), 12547–12555.

VanEpps, J. S., & Younger, J. G. (2016). Implantable device-related infection. *Shock, 46*, 597–608.

Vaudaux, P. E., Lew, D. P., & Waldvogel, F. A. (1994). Host factors predeposing to and influencing therapy of foreign body infections. In A. L. Bisno, & F. A. Waldvogel (Eds.), *Infections associated with indwelling medical devices* (pp. 1–29). Washington DC: American Society for Microbiology.

Veenstra, G. J., Cremers, F. F., van Dijk, H., & Fleer, A. (1996). Ultrastructural organization and regulation of a biomaterial adhesin of Staphylococcus epidermidis. *Journal of Bacteriology, 178*, 537–541.

Wang, H.-Y., Lin, L., Tan, L.-S., Yu, H.-Y., Cheng, J.-W., & Pan, Y.-P. (2017). Molecular pathways underlying inhibitory effect of antimicrobial peptide Nal-P-113 on bacteria biofilms formation of Porphyromonas gingivalis W83 by DNA microarray. *BMC Microbiology, 17*, 1–7.

Whitehouse, J. D., Friedman, N. D., Kirkland, K. B., Richardson, W. J., & Sexton, D. J. (2002). The impact of surgical-site infections following orthopedic surgery at a community hospital and a university hospital: Adverse quality of life, excess length of stay, and extra cost. *Infection Control & Hospital Epidemiology, 23*(4), 183–189.

Yanagisawa, N., Li, D. Q., & Ljungh, A. (2001). The N-terminal of thrombospondin-1 is essential for coagulase-negative staphylococcal binding. *Journal of Medical Microbiology, 50*, 712–719.

Zheng, Y., He, L., Asiamah, T. K., & Otto, M. (2018). Colonization of medical devices by staphylococci. *Journal of Applied & Environmental Microbiology, 20*(9), 3141–3153.

Zimmerli, W., Trampuz, A., & Ochsner, P. E. (2004). Prosthetic-joint infections. *The New England Journal of Medicine, 351*, 1645–1654.

CHAPTER

12

NMR-based metabolomics study of microbial biofilm

Bandita Dutta[1], *Dibyajit Lahiri*[2], *Moupriya Nag*[2], *Ashmita Samanta*[2] *and Rina Rani Ray*[1]

[1]Department of Biotechnology, Maulana Abul Kalam Azad University of Technology, Haringhata, West Bengal, India; [2]Department of Biotechnology, University of Engineering and Management, Kolkata, West Bengal, India

1. Introduction

Metabolomics make an appearance to the research domain for comprehensive assessment of all the metabolites along with low-molecular-weight molecules present in the biological samples. These molecules and metabolites need to be analyzed diligently to scrutinize the metabolic transition in the bacteria during biofilm formation.

In the past few years, the direct exploration of the metabolic intervening products for biological prototypes enhances the perception of metabolic processes. Mass spectroscopy (MS) and nuclear magnetic resonance (NMR) spectroscopy are the most conventional technologies. Elucidation of the molecular structures using NMR (i.e., NMR-based metabolomics) is one of the prime aspects in analyzing complex amalgamation of biological specimens.

In the small-molecule biochemistry sector, NMR is found as a potent technique. NMR including quantification of metabolites (qNMR) determines the metabolic nemesis of the drug molecule and nutrients using stable isotope tracers along with elucidation of novel metabolic pathways, several metabolic fluxes, and interactions between metabolites and proteins. These elucidations help recognize the regulation and pharmacological effects of the metabolic pathways.

2. Bacterial biofilm

Biofilm is a structured bacterial aggregation implanted within the extracellular polymeric matrix (EPS). This matrix is able to provide irreversible attachment of the bacterial cells to the

Microbial Biofilms
https://doi.org/10.1016/B978-0-323-95715-1.00002-9

abiotic and biotic surfaces with the help of pili and also helps in the quorum sensing (QS) mechanisms between the bacterial cells. It was observed that the EPS matrix is usually 0.2–1.0 μm thick, and size of the biofilm is usually 10–30 nm (Sleytr, 1997). Around 5%–35% of the bacterial biofilm comprising microorganisms and extracellular matrix saturates the remaining volume. This EPS matrix comprehended carbohydrates (1%–2%), proteins(>2%), nucleic acids (<1%), and water (97%) along with some prime nutrients, which get entrapped inside the complex network (Sutherland, 2001).

3. Steps involved in biofilm formation

Genetic analysis of biofilm formation revealed that several steps are involved in biofilm formation. QS mechanisms through cell signaling molecules are one of the main aspects of biofilm formation. A set of genes are involved in biofilm formation by regulating transition of planktonic cells to sessile cells associated with biofilms (Donlan, 2002; Federle & Bassler, 2003). Initial attachment to the surface: Bacterial cells, mainly the sessile cells, adhere to the surface with the help of flagella and pili. Some physical forces including van der Waal's forces and electrostatic interactions help the cells to initial adherence to the surface. The solid–liquid interface along with surface hydrophobicity intensifies the bacterial attachment by reducing repulsion force between bacterial cells and surface.

- Formation of microcolony: Initiation of the cell multiplication and cell division through the chemical signaling molecules enables the irreversible attachment of the bacterial cells to the solid surface and leads to the formation of microcolony. Several microcommunities inside the biofilm layer coordinate among each other through the exchange of the substrate, distribution of the metabolic products, and excretion of metabolic end products. During the anaerobic digestion complex, organic matter gets converted into CH_4 and CO_2 with the consequences of production of acid and alcohol from organic compounds via fermentative bacteria followed by consumption of these substrates by acetogenic bacteria and conversion of acetate, CO_2, and hydrogen into methane through methanogens proving them symbiotic benefits. Two or more metabolically independent bacteria are able to survive in a biofilm microenvironment via developing syntrophic association and utilize substrates explicitly for energy impetus (Davey & O'toole, 2000).
- Maturation of the biofilm: This stage of biofilm formation involves the transmission of the autoinducer signals between the bacterial cells. Autoinducer molecules help the bacterial cell to accomplish the cell density through the QS network. As the maturation step is being processed, some explicit gene products, which are involved in EPS production, get expressed. With the development of the three-dimensional network of the biofilm, some interstitial voids are assembled inside the matrix, which get filled up with water. The water played an important role in the circulatory system by distributing nutrients and removing waste products (Parsek & Singh, 2003).
- Dispersion of biofilm: Bacterial cells present in the biofilm layer multiplied rapidly in this stage and then dispersed by converting sessile cells in motile form. Several saccharolytic enzymes are produced during this detachment process to help the bacterial cells to release the surface and colonize into a new domain. It was observed that gram-

negative bacterium *Escherichia coli* produces *N*-acetyl-heparosan lyase, *Pseudomonas aeruginosa* and *Pseudomonas fluorescens* produce alginate lyase, and gram-positive *Streptococcus equi* produces *hyaluronidase* conducive to lysis of the EPS matrix and ensuring detachment (Sutherland, 1999).

4. Molecular analysis of biofilm formation

There are few pivotal statutes of biofilm formation, which include surface adhesion of metabolically active bacteria through the utilization of the sufficient nutrient for cell reproduction and production of exopolysaccharides. The sessile bacterial cells form the biofilm upon the hydrophobic surface, usually plastic surfaces or metal devices. Several nosocomial diseases are engendered via gram-positive bacteria *Enterococcus* spp. upon urinary tract, *Staphylococcus aureus* in arteries, and *S. epidermidis* in heart and via gram-negative bacterium *Escherichia coli* upon urinary tract, *P. aeruginosa* in the lung, and *Vibrio cholerae* and fungi *Candida* spp. in GI tract (Krašovec & Jerman, 2003). Identification of the critical biochemical pathways associated with biofilm formation is the preliminary step of biofilm prevention. The transition from planktonic to biofilm complex emanates the phenotypic changes with the differential expression and regulation of specific genes (Fig. 12.1).

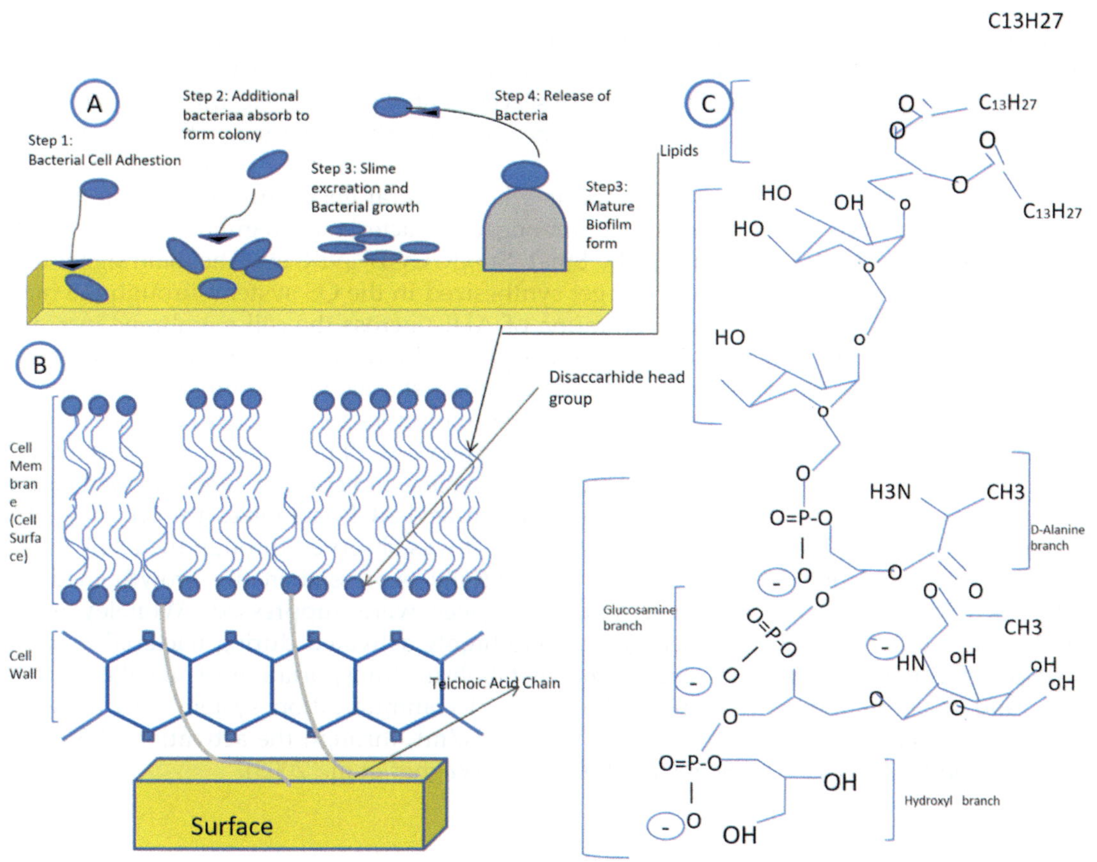

FIGURE 12.1 Molecular analysis of biofilm formation.

5. Genomic analysis of the bacterial biofilms

Genomic analysis of the bacterial biofilm formation was initiated in the early 1990s when the biofilm defective mutants were first detected. Analysis reported that several genes are associated in biofilm formation. At a recent time, DNA microarray technology has emerged as the significant one to identify upregulation and downregulation of genes in bacterial biofilm. Several pathways including stress-induced pathways, new regulatory pathways along with upregulation of genes entailed in carbohydrate production, and genes instigating stationary phase (Beloin & Ghigo, 2005).

★ Gram-negative bacteria

The gene expression in gram-negative bacteria, namely *E.coli* bacterial biofilm, was analyzed through DNA microarrays through the comparison of the gene expression of stationary-phased planktonic cells (Schembri et al., 2003). The fact was divulged that in the biofilm cells, around 600 genes were changed, only 9% of the whole genome was activated, and 4.5% of the whole genome was suppressed. A different gene expression pattern appeared while comparing transcriptional profiles of biofilm cells along with exponentially grown cells. Precisely 230 genes were distinctly expressed, and 4.8% genes were upregulated and 0.5% genes were downregulated in biofilm. In *E. coli,* at most 79 genes represent 1.84% of the whole genome, and these genes get evolved during biofilm formation. Along with the increased gene expressions in biofilms, these genes are also responsible in cell adhesion, autoaggregation, and structural proteins involving *OmpC, OmpF, OmpT, lpxC* (encoded protein is analogous to lipid A biosynthesis), and *slp* (after carbon starvation encodes outer-membrane lipoprotein) encryption. It was observed that *slp* and *OmpC* are involved in the initial stage of biofilm formation, i.e., adherence into the abiotic surface (Prigent-Combaret et al., 1999).

QS mechanism is one of the prime aspects where several genes are regulated in biofilm formation. QS circuits of the numerous gram-negative bacteria are regulated by the factors or proteins such as LasI/LasR, RhlI/RhlR, SmaI/SmaR, CviI/CviR, etc. The main signal molecules, acyl-homoserine lactones (AHL), get synthesized in the QS system through the regulation of LuxI-like proteins. The concentration of AHLs across the cell membrane increases with the increase of cell density. A cognate LuxR protein binds to AHL molecules after recognizing it followed by activating transcription of the target gene via binding the specific promoter. The LuxR regulates the production of AHL molecules by the LuxI enzymes. LuxI is integrated with the specific acyl-acyl carrier protein (acyl-ACP) inside fatty acid biosynthetic machinery via acyl-side chain of it to the homocysteine moiety of S-adenosylmetheonine (SAM). Thus methylthioadenosine is released when acyl-ACP lactonizes to form acyl-HSL (Fuqua & Greenberg, 2002). In case of *P. aeruginosa,* it was observed through the DNA microarray analysis that only 1% genes were differentially expressed in the growth medium, whereas 0.5% genes were activated and 0.5% genes were suppressed (Whiteley et al., 2001). These differentially expressed genes were highly activated during bacterial motility, surface attachment, protein translation and metabolism, transportation of metabolic products, and their regulatory functions. The bacterial communication system gets activated when the bacterial cell density rises up to 1010 cells/mL through the activation of 353 and 616 genes in the QS cascade (Schuster et al., 2003; Wagner et al., 2003).

The RpoS protein with the σS subunit of RNA polymerase plays the leading role in biofilm formation (Adams & McLean, 1999). Under the dominance of the *RpoS* protein, only 46% genes were differentially expressed and incapability of establishing sessile populations manifested via the deletion of *RpoS*. In *P. aeruginosa*, the repression of *RpoS*, i.e., *rpoS*-deficient mutants enable the healthy biofilm formation (Whiteley et al., 2001).

★ Gram-positive bacteria

Around 519 distinctly expressed genes were identified, which are responsible for biofilm formation. Over 55% of the differentially expressed genes have secular control on the biofilm-forming genes along with glycolysis, tricarboxylic acid cycle, membrane bioenergetics, and phage concomitant functions. In addition to that, few genes are involved in cells motility and chemotaxis.

Gram-positive bacteria have a different QS network from gram-negative bacteria. Gram-positive bacteria conventionally use peptide signal molecules, which are posttranscriptionally synthesized. The two-component signal transduction system has histidine kinase, which acts as a sensor element along with interacting the peptide signals (Dunny & Berntsson, 2016). The autoinducer peptides are pioneered as pro-AIPs and being secreted through the specialized transporter. Along with the posttranslational modification, the pro-AIPs became linear or circular molecules consisting 5–17 amino acids. The membrane-circumscribed sensor kinase system detects the extracellular AIPs. The sensor kinase contains histidine, which binds to AIP enabling autophosphorylation of sensor kinase. QS-target genes get controlled by the phosphorylation of the response regulator when the phosphoryl group gets passed from histidine to conserved aspartate within the system. The expression of the whole operon system encoding QS circuits that contains pro-AIP histidine kinase receptor, transporter, and response regulator gets activated through phosphorylation (Magnuson et al., 1994).

6. Analysis of bacterial biofilm using NMR-based metabolomics

Metabolomics is a dynamic mechanism to study the system biology including genomics, transcriptomics, and proteomics and quantitative correlation between the metabolites present in living system and their potent reactions. Elucidation of metabolites including amino acids, carbohydrates, and lipids and analysis of intermediate signaling molecules and final products of cellular processes are recognized as metabolomics. Metabolome is considered as the total collection of the whole metabolites present within the biological cell niche, cell tissues, organ, and organisms and analyzed as cellular extract or biofluids (Beckonert et al., 2007). 1 kDa is the stereotypical threshold to separate metabolites from macromolecules (Rubakhin et al., 2011).

NMR is one of the principal analytical methods to ascertain biomarkers for diseases associated with cardiac, liver, respiratory, cancer, and CNS disorders. NMR metabolomics has the efficiency to differentiate between healthy and unhealthy (diseased) states or drug-mediated and unmedicated states. The change of metabolome profile indicates the effectiveness of the drug molecules and thus helps in lead identification. The identification of kinase inhibitor through NMR metabolomics manifests the changes in ratio of lactate and pyruvate present in human leukemia cells (CCRF-CEM) and determine the eEF-2,

NF-κB, MK2, PKA, PKC, and PKG kinase inhibitor in human ovarian cancer cells (SKOV-3) (Tiziani et al., 2011). The in vivo study of the mode of action of chemical leads inferred that two or more drug molecules having equivalent impact on the metabolome share a specific target (Halouska et al., 2012).

To inhibit biofilm-mediated infections, the explicit study about the metabolic information of pathogenic, antibiotic resistance, and diseased form of the biofilms is required. The nondestructive, noninvasive, and no sample consumption along with utilization of the grimy and mixed samples in the form of static and dynamic state make the NMR more advantageous to study physical characteristics of biofilm. Solid-state NMR enables the molecular level study of the insoluble polymers with nominal amount of sample preparation in a gracious manner.

The one-dimensional and multidimensional heteronuclear spectroscopy along with classic 1H NMR spectroscopy divulge the molecular anomaly and its conformation in addition with chemical and functional groups present in it. The cell flow in water dynamics to study biofilm growth kinetics is analyzed via NMR flow through the diffusion of biofilm in porous medium. The one-dimensional and two-dimensional NMR, diffusion NMR, and magnetic resonance imaging (MRI) apprehended the occurrence of bacterial growth and their distribution within the biofilm layer. NMR-based metabolomics techniques permit the direct transient scan of the metabolite concentrations, flux rate, and their metabolic pathways.

7. Metabolite assays by NMR

In the past two decades, the NMR made an appearance as the prime analytical technique used in metabolomics where gas chromatography clubbed with mass spectrometry (GC-MS) and liquid chromatography conjoined to single-stage mass spectrometry (LC-MS).

7.1 1H NMR spectrometry

It was found that one-dimensional 1H NMR spectrum is generally used in metabolomics study due to its self-regulating, definitive properties, and rapidness. The collection time for 1D 1H NMR spectrum is very small, like a few minutes. Robotic sample exchanger used to load and remove the NMR samples continuously in NMR equipment that used to run for days or weeks. Around 50 to 100 metabolites can be identified and quantified from the biofluid or tissue extract using single 1D 1H NMR (Lindon et al., 2000). In the public database of 1D 1H NMR, more than 100 metabolites have been assembled and stored. Automatic and semiautomatic tools are used to process and analyze the spectra being collected within a few seconds or minutes. The identification and quantification of the metabolites using 1H NMR metabolomics, to study a broad population, have been preferred due to its expedition and linearity. 1D 1H NMR spectrum usually works under mitigated conditions and transfers the signal without any polarization. Thus, this technique is considered as the most useful to quantify metabolites and also to depict the proton distribution in molecular nuclei at different concentration degrees.

7.1.1 13C NMR spectroscopy metabolomics

1H NMR spectroscopy is usually characterized as narrow line width with narrow around 10 ppm chemical shift dispersion. Compared with this, the 13C NMR spectroscopy showed narrow line widths with 200 ppm broad chemical shift dispersion along with significant resolution. The low natural abundance of 13C with low abundance of the 13C nucleus (common nuclei observed by NMR such as 1H, 19F, or even 31P) has the significant hindrance on the 13C NMR spectroscopic application to metabolomics. Several NMR perspectives are able to subdue the natural hindrance of 13C. Distortionless enhancement by polarization transfer (DEPT) is able to enhance the 13C signal intensity (Doddrell et al., 1982). The DEPT NMR technique distinguishes the spectra between CH, CH2, and CH# signals. The CH2 peak provides the reverse intensity to the CH and CH3 peaks, while 13C NMR spectrum was examined in DEPT-135 experiment. The polarization transfer from 1H/13C NOESY nuclei can be increased up to the factor of 1.98 through the 1H signal enhancement. It was hypothesized that spin manipulation and 13C enrichment are able to enhance 13C NMR signal. Thus, categorization of microbial metabolomics and mammalian cell lines studies is feasible through 13C-enriched glucose (Buescher et al., 2015; Wushensky et al., 2018).

Plant metabolomics study is usually done through 13C-enriched CO_2 (Pang et al., 2018). In addition, the 13C acetic anhydride into the serum or urine samples enables the 13C tag to the certain metabolites, mainly the amino acids via the amine acetylation by 13C. The chemoselective labeling method is selectively used for identification and synchronized comparison of amino acids and other metabolites. The cryo-probe technology enhances the 13C signal by 2–4 fold by reducing electronic noise close to zero. This cryoprobe usually permits the accretion of the 1D 13C NMR spectra for 15 min. The greater spectral dispersion of 13C spectra is able to analyze biofluids critically. The hyperpolarization of 13C NMR amplifies the signals to track or trace the isotopes and determine the presence of carbon in several biosynthetic pathways. 13C is found to be one of the main facets in NMR and MS-based fluxomics, i.e., a branch of metabolomics in determining intracellular metabolic fluxes in living cells (Heux et al., 2017).

7.1.2 15N NMR spectroscopy for metabolomics

15N NMR has alikeness to the 13C NMR for having broad chemical shift dispersion around 100 ppm with narrow line widths. The result of direct detection of 15N has low vulnerability with low 0.37% abundance, and the 7.62 MHz/T gyromagnetic ratio is able to make 15N nucleus 262,000 times less sensitive compared with 1H nucleus. In accordance with that, enhancement of the isotopes followed by the combination with 1H-mediated enhancement makes 15N NMR feasible for indirect detection. This accredits the structural annotation of the DNA, RNA, and proteins (Dominguez et al., 2011; Pervushin et al., 1998). The 15N isotope based indirect detection enables the expansion of the pool of metabolites, i.e., quantifiable through the isotope tagging (Ye et al., 2009). The tagging method is similar to the 13C where the 15N ethanolamine tag reacts with carboxyl-containing metabolites and gets detected through 1H-15N 2D NMR. Selective labeling of specific metabolites is able for spectral amplification and peak dispersion with high sensitivity by suppressing signal transmission from nontagged metabolites. The 2D 1H-15N HSQC experiment can detect and quantify more than 100 metabolites. The newly evolved "smart" isotope tag, i.e., 15N-cholamine is able to exploit NMR and MS amalgamated strength for identification and quantification of incognito compounds. The 15N-cholamine showed dual characteristics of detecting 15N isotope-tagged metabolites using NMR with high susceptivity and chemical shift dispersion (Ye et al., 2009).

7.1.3 31P NMR spectroscopy for metabolomics

31P is found to be 100% profuse due to its wide spectral dispersion and 6.6×10^2 more relative sensitivity than 1H. 31P NMR spectroscopy has limited application, as metabolites predominantly do not contain phosphorus atoms. This technique is mainly used to analyze nucleoside metabolites such as ATP, GTP, and NADP and phospholipids, which are involved in energy catabolism. Alike 15N, this method also accompanies isotope tagging to detect hydrophobic compounds. 31P reagent, 2-chloro-4,4,5,5-tetramethyldioxaphospholane (CTMDP) tag is usually implemented to tag lipid metabolites including aldehyde, hydroxyl, and carboxyl groups. 1D 31P NMR detects and quantifies the tagged metabolites with high resolution.

7.2 2D NMR spectroscopy

Two-dimensional NMR spectroscopy is one of the effectively ratified techniques to identify molecular and structural recognition and kinetic analysis. The resonance problem due to the overlapping is feasible to be solved through the dissemination of the peaks into a second dimension depending on atomic physical properties such as covalent attachment, coupling constant, and relaxation time in 2D NMR metabolomics. Compared with the 1D NMR, the 2D NMR accedes more recognition of the metabolites. Several analogous 2D 1H-1H-NMR experiments including total correlation spectroscopy (TOCSY), correlation spectroscopy(-COSY), nuclear Overhauser effect (NOESY), diffusion-ordered spectroscopy (DOSY), two-dimensional J-resolved NMR spectroscopy (J-Res), heteronuclear 1H,13C single quantum coherence (1H-13C-HSQC), and heteronuclear multiple bond correlation (HMBC) are used to detect the metabolome profile using 2D NMR technique.

7.2.1 Correlation spectroscopy

Among all 2D NMR experiments, correlation spectroscopy (COSY) is considered as the simplest one. This technique usually provides information about the homonuclear correlations between coupled nuclei (1H-1H) and their structural interpretation (Kono, 2013). 90 radio frequency (RF) pulse is present in the COSY pulse sequence, which is able to evaluate time t1 at 90 s pulse followed by consecutive time period measurement (Keifer, 1999). Both the t1 and t2 dimensions help to relent the 2D spectrum through Fourier transformation followed by generation of cross-peaks in the 2D spectrum where paired nuclei get connected by through-bond (3JHH) couplings. The COSY is considered as the most useful in metabolomics research due to its elementary, rapid, and explicate nature (Sekiyama et al., 2011). The occurrence of cross-peak in COSY experiment indicates the presence of through-bond coupling between coupled nuclei and helps identify novel metabolites from complex biological amalgam.

7.2.2 *Total correlation spectroscopy*

Homonuclear Hartmann–Hahn (HOHAHA) experiment is proclaimed as TOCSY (total correlation spectroscopy), which is only an extension of the COSY experiment where within the spin system the chemical nuclei get harmonized. Only the short-ranged coupled protons along with a series of scalar couplings present in four or more covalent bond away provide the TOCSY spectrum. The TOCSY spectrum showed a correlation between A, B, and C where A and C are connected with proton B. Compared with 2D TOCSY, the 1D TOCSY is very rapid and induces simple analysis of the 1D NMR spectrum. The spectrum is usually generated from the nuclei of the single spin system, which characterizes the 1D TOCSY where NMR spectrum quantifies the overlapped metabolites (Keun et al., 2002). The 1D TOCSY made the first interpretation while analyzing honey samples through the strong carbohydrate signals that dominate NMR spectra (Sandusky & Raftery, 2005).

7.2.3 *2D J-resolved spectroscopy*

This is the primordial 2D NMR experiment that exhibits both the J-couplings and chemical shifts within the spectrum (Nagayama et al., 1977). Compared with the 1D NMR experiment, the spectral distribution gets rationalized with the increase of the peak dispersion in the 2D J-Res experiment (Aue et al., 1976). 2D J-Res NMR spectroscopy is the most accepted one in metabolomics study due to its relative speed. It is able to resolve metabolic resonance and identify the metabolites present in human biofluids such as blood plasma, urine, and cerebrospinal fluid (Foxall et al., 1993). Although plasma J-Res NMR spectra appeared as the elementary, it contains additional information than 1D Hahn spin-echo spectra. 2D J-Res NMR spectroscopy has the fundamental methods with optimized spectral acquisition parameters and optimal processing of data.

7.2.4 *Heteronuclear single quantum correlation spectroscopy*

Both the bond correlation spectroscopy, i.e., COSY and TOCSY-like spectroscopy, can be used in measuring not only homonuclear correlations but also the heteronuclear correlation. In case of heteronuclear correlation, signals coming from modest sensitive nuclei get enhanced through the transfer of the nuclear spin polarization from the more sensitive nucleus (NMR active different nuclei) via J-coupling. The insensitive nuclei enhanced by polarization transfer (INEPT) is a phenomenon where the nuclear spin polarization gets transferred from large Boltzmann population differences, i.e., 1H to low Boltzmann population differences, i.e., 13C and 15N nuclei. After the magnetization process, the extra sensitive nucleus gets transferred back from 13C or 15N to 1H for undeviating inspection. The INEPT-based HSQC is used to map the chemical shift from 1H nucleus to 13C nucleus through direct and indirect measurements. The 1H,13C-HSQC spectrum is able to map the chemical alterations of proton and carbon atoms, which provide bounded only one cross-peak for each H-C directly incorporated pair. Similarly 1H,15N-HSQC spectrum is able to map the chemical alterations of proton and nitrogen atoms through the yield of one cross-peak for each H-N directly incorporated pair. HSQC is mainly used to resolve and assign overlapped signals of protons and specifically metabolite signals emerging from complex biofluid mixtures (Lee et al., 2012; Yi et al., 2000).

7.2.4.1 Heteronuclear multiple bond correlation spectroscopy

Another 2D heteronuclear experimental method is HMBC, which harmonizes the chemical transition of the two different nuclei (i.e., 13C and 1H) alike HSQC or HMQC experiments. Apart from HSQC or HMQC, the connection between separated nuclei gets divulged through HMBC. A low pass filter with a small J-coupling is used to optimize the detection through the elimination of single-bond interconnection. Single CH bond connection gets eliminated through HMBC, and HSQC and HMQC experiments are not able to detect signals from quaternary and carbonyl carbons. The amalgamation of HMBC with HSQC or HMQC experiments is able to identify molecules and elucidate chemical structure through complete signal assignments (Bernini et al., 2009). 900-MHz NMR in HMBC is able to detect separate aromatic peaks of hippurate, phenylacetylglycine, and histidine present in the whole spectrum of urine sample (Fig. 12.2).

8. In vitro metabolite profiling of microbial biofilm

Extracellular polymeric substance (EPS), a viscous coating that keeps the sessile communities of bacterial cells stuck to one another and to the biotic or abiotic surface, is what is known as a biofilm. The growth of biofilm on different surfaces is influenced by a variety of external stimuli as well as the preponderance of the nutrients inside the matrix (Donlan & Costerton, 2002). It is the repository of a variety of nutrients, such as carbohydrates, proteins, and nucleic acids, which not only provide nutrients to the growing sessile communities but also aid in ensuring the stability of the residing cells, mediating their attachment to biotic or abiotic surfaces and serving as a scaffold for the attachment of antibiotics, enzymes, and cells (Stewart & William Costerton, 2001).

The concept of the numerous changes in metabolism during the mechanism of biofilm development is provided by metabolomics (Wong et al., 2018). Metabolomics is a method for identifying the metabolites that are created and consumed by biological samples when chemical or physical transformation is involved. Metabolomic research offers the idea of different metabolic processes that exist in bacterial and fungal systems. The primary goal of metabolomics is to ascertain the system's metabolic makeup (Fiehn, 2002). Metabolomics is a concept that aids in understanding how metabolic pathways evolve over time and how common small molecule metabolite intermediates are. Targeted metabolomics and their chemical impacts on the range of different metabolic processes can be used to understand the biological responses of single cells to complex systems (Nicholson & Lindon, n.d.). Metabolomics is a method for determining the metabolites and the method for discovering metabolic alterations. Metabolomics is not a novel idea, but analytical techniques such as liquid chromatography–mass spectrometry (LC-MS) and gas chromatography–associated mass spectrometry have advanced since its inception (Nag & Lahiri, 2021) (Table 12.1).

Liquid chromatography–mass spectrometry: UV or fluorescence detection can be employed with liquid chromatography (LC) techniques; however, combining LC with Mass spectroscopy (MS) offers a more sensitive method. When compared with conventional LC, ultraperformance liquid chromatography (UPLC), which uses sub-2-m particle columns, provides higher sensitivity and resolution. When coupled with tandem MS, this method further improves specificity and offers a better signal-to-noise ratio than single-stage MS

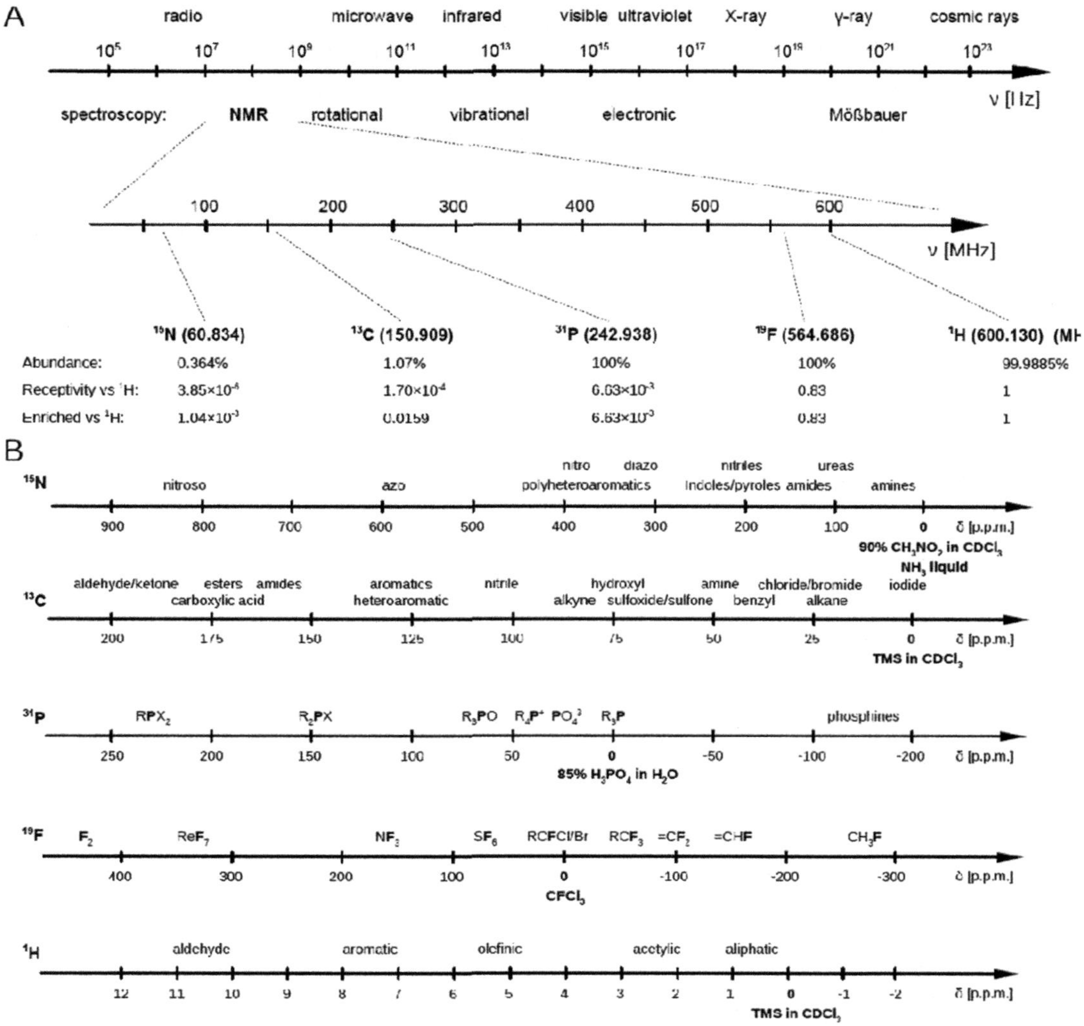

FIGURE 12.2 Depict the electromagnetic nature of the NMR spectroscopy of several nuclei where (A) represents frequency scale range and their corresponding spectroscopy and (B) represents ppm ranges with respect to the 15N, 13C,31P, 19F, and 1H nuclei present in distinct environments. *Figure adapted from Emwas, A. -H., Roy, R., McKay, R. T., Tenori, L., Saccenti, E., Gowda, G. A. N., Raftery, D., Alahmari, F., Jaremko, L., Jaremko, M., & Wishart, D. S. (2019). NMR spectroscopy for metabolomics research.* Metabolites, *9(7), 123. https://doi.org/10.3390/metabo9070123*

(Want et al., 2010). With immobilized chromatographic packing materials, high-performance liquid chromatography (HPLC), a type of column chromatography, helps pump analyte that has been dissolved in the solvent under high pressure (stationary phase). Depending on the type of analyte that transports through the mechanism of separation, the solvent's characteristics and the makeup of the stationary phase are chosen. The analytes that interact with the stationary phase with the greatest strength have the highest retention times as the sample for identification passes through the chromatographic column (Nag & Lahiri, 2021).

TABLE 12.1 Comparison between Nuclear Magnetic Resonance (NMR) Spectroscopy and Mass Spectroscopy (MS) in accordance with their advantages and applications.

	NMR	Mass spectroscopy
Reproducibility	NMR spectroscopy shows high reproducibility, which is the most beneficial side of using it.	MS spectroscopic data are comparatively less reproducible than NMR.
Sensitivity	To increase the sensitivity, we can follow multiple scans (time), cryo-cooled and microprobes, magnetic field of higher strength, and hyperpolarization methods, as the NMR is less sensitive.	MS spectroscopy is famous for its high-sensitive data, even nanomolar concentration of any metabolites is also detectable.
Selectivity	In case of nonselective analysis, NMR is the best choice.	On the other hand, MS spectroscopy is very much selective.
Sample measurement	NMR spectroscopy can able to measure relatively fast by using 1D 1H-NMR spectroscopy.	MS spectroscopy's rate of measurement is comparatively slow. To improve the rate, we can use different ionization methods.
Sample preparation	Very low amount of sample is required to be prepared, and locking solvent is used to lock the sample after transferring into the NMR tube.	Sample derivatization is required for gas chromatography–mass spectroscopy.
Sample recovery	The nondestructive nature of the NMR helps the sample recovery and storage for long time after several analysis.	The destructive nature makes the sample not to be recovered; hence a small amount of sample is used.
Quantitative analysis	Due to directly proportional relationship in between signal intensity and metabolite concentration, NMR provides the quantitative analysis data.	MS spectroscopic data are not only dependent on metabolite concentration, but also depends on ionization efficiency.
Tissue samples	NMR spectroscopy can measure the metabolite in the tissue sample by using high-resolution magic-angle sample spinning	MS spectroscopy is not so much popular to detect metabolites in the tissue sample.
Number of detectable metabolites	Not more than 200 metabolites are detected and identified simultaneously in NMR, by using different resolution of spectral.	MS spectroscopy can able to detect 1000 of metabolites at the same time but identify several hundred.
Targeted analysis	It is specially used for untargeted analysis, but targeted analysis can also be done.	MS is used for targeted analysis always.
In vivo studies	In vivo investigation can also be possible in NMR spectroscopy by using magnetic resonance spectroscopy.	Basically, MS is not so much accurate to in vivo studies, although desorption electrospray ionization can be a method to analyze tissue samples.

Gas chromatography–associated mass spectrometry: The separation of volatile chemicals in the gaseous state is accomplished by the use of gas chromatography. The polar metabolites, such as compounds containing phosphate, are mostly identified in the LC-MS study for the assessment of the bacterial metabolome because significant portions of them cannot be studied with the GC-MS. Methoxylation and *N*-methyl-*N*-trimethyl-silylfluoracetamide silylation are two mechanisms used in combination in the widely applicable derivatization technique to produce a variety of key metabolites. Using mass spectrometry, which operates on the idea of electron impact ionization, the metabolites are found and their fragments are characterized (Nag & Lahiri, 2021). GC-MS makes it easier to identify and accurately quantify the several hundred metabolites present in a single plant extract. The key benefits of this technique include the fact that robust protocols for machine setup and maintenance, chromatogram evaluation and interpretation, and metabolite profiling have been employed with it for a long time. GC-MS has a somewhat comprehensive coverage of constituent classes, including organic and amino acids, sugars, sugar alcohols, phosphorylation intermediates, and lipophilic substances, even though no one analytical technique can capture the entire metabolome. During technique validation, recovery tests on all quantifiable classes of chemicals have been performed. Recombination tests, in which extracts of two plant species are assessed both individually and after mixing, can be used to establish recovery rates for unidentified chemicals (Lisec et al., 2006).

9. Difference between NMR metabolomics of planktonic and biofilm modes of growth

Communities of bacteria that may cling to surfaces live in cells that release sticky, polymeric substances. These cells are able to tolerate antibiotics and other antimicrobial treatments thanks to their sessile lifestyle, known as a biofilm (Gjersing et al., 2007). The factors influencing a biofilm's resistance and a planktonic cell's vulnerability are still being investigated. Using NMR spectroscopy, it is possible to obtain comprehensive profiles of every metabolite species present in a raw sample. Because there are so many different chemical species (i.e., NMR peaks) present, pattern recognition methods such as principal component analysis (PCA) are frequently used to extract information from the NMR spectra. In the realm of drug development, where biological fluids from both people and animals are routinely tested to create ways for disease detection, the techniques for such analysis are well developed.

Similar to the proteome and genome, the metabolome is an inventory of all the metabolites present in a biological system. Genetic regulation causes changes in protein expression, which should at least partly be reflected in the complement of cellular metabolites. The ultimate goal of all applications using metabolome data is to comprehend cause-and-effect interactions in biological systems. Metabolite concentrations have been used as a functional genomics method to reveal phenotypes in yeast mutants and plants that, based on growth rates and fluxes, otherwise looked to be the same (Raamsdonk et al., 2001).

10. In *Pseudomonas*

P. aeruginosa is a type of bacterium that frequently lives in groups called biofilms that are adhered to surfaces and enclose themselves in a polysaccharide matrix. Due to *P. aeruginosa*'s antibiotic tolerance, biofilms have been linked to nosocomial infections and cystic fibrosis, which are both serious condition (Stewart & William Costerton, 2001).

P. aeruginosa biofilms are capable of colonizing medical implant devices including heart valves and catheters, where treatment of the infection frequently necessitates removal of the infected device. They have been linked to cystic fibrosis. This form of proliferation contrasts sharply with planktonic, free-floating cells, which are more amenable to drug eradication (Gjersing et al., 2007).

To compare fresh LB growth medium with spent medium from planktonic and biofilm-grown cultures, 1H NMR analysis of the resultant spent medium was performed. The goal was to see if various cellular growth conditions could be identified. To distinguish metabolic variations between planktonic and biofilm methods of growth as demonstrated by the composition of extracellular medium containing, Gjersing (2007) studied planktonic cells that have been grown in (continual feed) chemostats and contrasted to biofilm cells grown under comparable continuous culture conditions. The study found that under these circumstances, some metabolic components were more concentrated in the planktonic culture while others were more concentrated in the samples from biofilms. To detect the differences between the two growth patterns, PCA was successful (Gjersing et al., 2007).

11. In *Staphylococcus*

On implanted biomaterials, *Staphylococcus aureus* has the capacity to colonize and create biofilms. These biofilm structures can exhibit susceptibilities to antimicrobials that are 10–1000 times lower than corresponding populations of free-floating planktonic cells, making them inherently resistant to antimicrobial challenges and challenging to remove from the infected host (Smith et al., 2008). Based on research on *S. aureus* biofilm formation in 1000 clinical isolates, static biofilms were formed utilizing BHI as the best medium. Qualitative information from SEM imaging demonstrated that the *S. aureus* clinically derived strain employed in this study develops a biofilm.

The strength of the thick peptidoglycan cell wall causes additional problems when gram-positive bacteria and their internal metabolic intermediates were studied. The information presented here offers a novel approach for extracting metabolites from biofilms grown in 96-well format using direct bead beating and downstream analysis to compare planktonic and biofilm cells. 96-well plates are used in several previously reported methods for biofilm research, enabling bacterial culture (Stipetic et al., 2016). Studies conducted by Laurence H. Stipetic in 2016 described the low-temperature, quick mechanical technique used in conjunction with an extraction solvent to break down the peptidoglycan cell wall and extract the cellular metabolome from gram-positive cells that are planktonic or that are part of a biofilm. The study demonstrated the approach is a highly reproducible platform for researching cellular metabolism. Its usefulness for generic bacterial metabolomics is further demonstrated by comparison with other techniques for lysing planktonic bacteria. According to data, there are considerable differences between planktonic cells and biofilms in a number of metabolic processes, particularly arginine production. In contrast to a planktonic, shaking culture, the biofilm's adhesion to the well's bottom during static culture will necessarily result in reduced access to oxygen. This difference in oxygen availability may be the cause of some or all of the alterations noticed. The widespread view is that, in contrast to planktonic cells, biofilm-forming cells' metabolism is dramatically altered (Gjersing et al., 2007; Resch et al., 2006; Zhang & Powers, 2012).

12. In *Salmonella*

With an estimated 93.8 million cases of salmonellosis occurring annually throughout the world, *Salmonella*-associated infections continue to be a serious public health concern. *Salmonella* spp.'s capacity to create biofilms on surfaces has significant effects on both health and disease, especially given that biofilms may serve as ongoing sources of contamination. Numerous antimicrobial susceptibility tests have shown that biofilms are notoriously challenging to treat, even with antibiotics that are effective against the same strain produced in planktonic form (Leriche & Carpentier, 1995).

The chromatograms of planktonic and biofilm supernatants were studied by H. Wong, which showed some metabolites were discovered only in biofilm supernatant; the majority of the changes were caused by metabolites that were present in planktonic supernatant but not in biofilm supernatant. The data set was modeled using PCA, which allowed us to identify which metabolites were most responsible for the observed variance (Wong, n.d.). *Salmonella* biofilm development is a significant public health concern because it can result in recurrent sources of infection. The goal of the current investigations is to evaluate the metabolic profiles of planktonic and biofilm *Salmonella typhimurium* to ascertain whether or not they were actually "different."

Due to variations in biofilms produced under various conditions, biofilm metabolomics will need to be conducted on a single, well-defined system. Our work employed a static device, whereas many research organizations advocate the use of flow-cell systems to develop biofilms. Since not all naturally occurring biofilms are flow-cell biofilms, the study of static biofilms is crucial. Salmonella and other foodborne pathogens are more likely to form biofilms on stationary surfaces, such as those used in food preparation, equipment surfaces, and fresh fruit (Annous et al., 2005). Studies have shown that the MBEC system can successfully produce *Salmonella* biofilms. It is essential that all cells live in the same environment to compare the metabolic profiles of biofilm and planktonic cells. Studies contrasting flow cell biofilms to planktonic cultures must account for the accumulation of metabolic by-products that occurs in planktonic development but not in flow cell biofilms since the medium is constantly being changed. The fundamental goal of the MBEC experiment was to mimic the conditions found in planktonic cultures, where extracellular metabolites might accumulate rather than be "washed out." (Wong et al., 2010).

13. In *Acinetobacter baumannii*

Acinetobacter species have been detected in a wide range of habitats, including soils, water, and human clinical samples, attesting to the genus' remarkable adaptability to varied conditions. Due to the clinical significance of its multidrug-resistant (MDR) strains, *Acinetobacter baumannii* has received the majority of attention in studies on *Acinetobacter* species. An aerobic gram-negative coccobacillus called *A. baumannii* causes nosocomial human infections, especially in people with compromised immune systems. Septicemia, meningitis, endocarditis, pneumonia, wound infections, and urinary tract infections can all develop from these illnesses (Poirel et al., 2011). Compared with their planktonic counterparts, bacterial biofilm cells are extremely resilient to a variety of stressors, including antibiotics. According to this, the metabolic activity of biofilm cells is different from that of their planktonic counterparts (Høiby et al., 2010).

According to the metabolic profiles of succinate, pyruvate, and acetate, which are involved in energy metabolism, the mature biofilm produced less energy than the planktonic development stage. A biofilm has consistent slower growth rate when compared with the same bacteria's planktonic growth rate. Additionally, the biofilm enables the bacterium to endure harsh circumstances (Pysz et al., 2004). According to proteomics data on *A. baumannii* biofilm cells, the Leloir pathway is used for the synthesis of EPS matrix (Cabral et al., 2011). Yeom's research revealed that planktonic and biofilm cells may be gathered under conditions of low temperature. Numerous quenching techniques have reportedly been utilized to quickly stop metabolism (Yeom et al., 2013). Due to metabolite leakage, quick sampling and trustworthy quenching methods are particularly crucial in bacterial cells. Therefore, the drawn-out sampling processes may have compromised the study's findings. Biofilm cell analysis frequently utilized the PBS washing technique (Mashego et al., 2007).

14. In oral plaque-forming bacteria

Dietary carbohydrates can be broken down by microorganisms in the oral biofilm to produce organic acids, which lower pH and start the demineralization of tooth-hard tissues. Due to its well-known acidogenic/aciduric characteristics and its capacity to bind to enamel, *Streptococcus mutans* have long been regarded as the main pathogen causing dental caries. The focus of illness explanations has shifted in recent decades from explanations centered on particular bacteria to ones that are ecology-centered (Marsh, 1994). Numerous metabolites in oral samples, especially saliva, have been identified using H-NMR spectroscopy. Since organic acids, alcohols, and other metabolites can all be detected and quantified using NMR, tiny sample sizes are not a problem. However, metabolomic analyses of dental plaque samples would be more informative in relation to caries development because analyses of saliva samples will show metabolites produced by all oral communities, including the tongue, palate, or epithelial mucosa, among others. Thus, standardized ex vivo testing of tooth plaque's acidogenicity and acid tolerance has just recently been developed (Havsed et al., 2021).

Only 40% of our CAR samples included *S. mutans*, which is thought to be the primary cause of dental caries and represented 0.16% of the total. As a result, despite the fact that their average proportion was larger than that of the CF group, it is not a universal indicator of the condition (Mira, 2018). This is in line with the findings of Johansson and colleagues, who came to the conclusion that communities with prevention programs, like the one we studied, have a reduced prevalence of mutans streptococci as primary caries pathogens (Johansson et al., 2016). *Scardovia wiggsiae, Prevotella denticola,* and *Abiotrophia defectiva,* which have all been discovered at higher levels in caries-active children and have strong acidogenicity, are other bacteria with a higher relation to caries experience. Metabolomic results demonstrate that *S. wiggsiae* plays a crucial role in the formation of organic acids and identify this species as a key contributor to the risk of caries in adolescents, consistent with its previously reported involvement in childhood caries. However, it is not present in every caries-affected person. This indicates that the cause of caries is not only polymicrobial but also extremely individualistic (Havsed et al., 2021).

15. Evaluation of fast 2D NMR for metabolomics

One of the primary analytical tools in metabolomics has evolved from 1D 1 H NMR to include multivariate analysis in a wide range of scientific fields, including toxicology, plant biology, and clinical investigations, to mention a few (Nicholson & Lindon, n.d.). This technology has one major drawback despite its strong sensitivity and relatively simple acquisition of 1H 1D spectra: the significant degree of spectral overlap caused by the complex molecular makeup of the targeted samples combined with the narrow spectral range of 1H resonances. When signals from nondiscriminating molecules are buried beneath those from a biomarker, overlap might make the subsequent statistical analysis difficult to interpret (Bingol & Brüschweiler, 2014). Since spectral dimensionality can effectively reduce spectrum overlap, there has been an increase in interest in using 2D NMR for quantitative research over the past 10 years. The use of 2D NMR for metabolomics has garnered interest over the past several years since it makes it simpler to identify biomarkers than it is to do so with 1D spectra (Guennec et al., 2014). It has been observed that in 2D spectra, there are more sensitivity toward the spectrometer instabilities, which are shown by the t1 noise. Furthermore, it causes periodic constraints in case of high-throughput analysis of rapid clinical responses. 2D NMR is moreover not suitable for the study of unstable samples like those that are undergoing chemical or biochemical reactions.

Due to their outstanding resolution and sensitivity-improved execution of the experiment, C-H HSQC (heteronuclear single quantum coherence) spectra at C natural abundance are crucial for 2D NMR-based metabolomics (HSQCsi). Using 2D H-H TOCSY (total correlation spectroscopy) or 2D C-H HSQC-TOCSY investigations, further spin-system detail can be acquired, which is crucial for accurate metabolite identification in complicated metabolomics mixtures. 2D HSQC, TOCSY, and other methods couple COLMARm, a publicly accessible Web server for the (semi)automated detection of metabolites in a variety of hydrophilic or hydrophilic metabolomics mixes, accepts HSQC-TOCSY spectral data as input (Hansen et al., 2021). A precise and thorough list of metabolites for metabolomics research can be extracted from a single sample by collecting the entire triad of these 2D experiments.

16. 1D, 2D, and solid-state NMR techniques for metabolomics

The application of NMR-based spectroscopy can be categorized into 1D, 2D, and solid-state NMR techniques. 1D and 2D 1H NMR solution state is used for comprehensive metabolic analysis of the bacterial cell extracts (Weljie et al., 2006), whereas to analyze complete cells, solid-state NMR is used (Moestue et al., n.d.). In 1D NMR-based metabolomics analysis adding to the 1H nuclei, 1D 13C NMR is used for metabolic profiling of carbohydrates, and 1D 31P NMR is used for tissue metabolism. The several sharp lines in the 1D 1H NMR spectrum represent the low-molecular-weight metabolites existing in cell lysates. Thus, the complete 1D 1H NMR spectrum is considered as the "fingerprint" in characterizing bacterial cell state. Carr–Purcell–Meiboom–Gill (CPMG) spin–echo sequence is generally used to remove the ambiguity of the overlapped NMR resonances and interpret every broad spectrum of proteins and other biomolecules (Meiboom & Gill, 1958). The in vivo 1D 1H spectra of solid-state to solution-state NMR get originated through the utilization of high-resolution magic angle spinning (HR-MAS). This implies the direct analysis of small, untreated bacterial cells along with the speed of 4 and 12 kHz at the "magic" angle of 54.7 degrees relative to the external magnetic field (Beckonert et al., 2007). Compared with the solid samples, the chemical shift anisotropy, dipolar coupling, and magnetic susceptibility became average through spinning to reduce NMR line widths.

The 2D NMR spectroscopy is able to improve the precision of metabolic distribution via the elevation of spectral resolution and adjunction of chemical shift into a second frequency dimension. The j-coupling available in 2D NMR helps to identify network resonance of the metabolites distinctly. 2D correlation spectroscopy (COSY) and total correlation spectroscopy (TOCSY) determine the spin–spin coupling between the chemically bound carbon, hydrogen, and nitrogen pair (Xi et al., 2006). The J-coupling removes the multiple peaks in 1D projection of 2D J-resolved NMR. The peak divergence improves the precision of metabolic distinction and quantification by reducing peak number and overlap.

These NMR techniques are the most discrete for analyzing temporal and structural metabolic profiles. The imaging, transportation, and metabolite measurement methods are the most common in NMR analysis corresponding to the live and congenital surveil of biofilm. The integration of NMR with confocal laser scanning microscope (CLSM) improves the exploration of macroscopic structure and metabolic pathway along with rate data, metabolite concentrations, and water diffusion profiles within the biofilm. The NMR spectroscopy methods have the ability to detect 140–190 nL volumes of distinct metabolites within biofilms. The correlation of NMR with CLSM employs the green fluorescent protein reporter for imaging purposes and perusal of dynamic metabolic procedures within the biofilm community.

17. Conclusions

Being a novel technology in the field of system biology, NMR-based metabolomics has limited utilization in exploration of bacterial biofilms. Thus, this technology has restricted within the scientific community. Though the NMR-based metabolomics is a simpler technique, it has several complications for routine applications due to the lack of apprehension about the metabolome. A metabolomics software package is required for the automated and standardized processing of metabolic data, metabolite identification, model validation, and chemometric analysis. The coherent and precised extraction of cell metabolites is required for the investigation and optimization, which seems to be beneficial for appraisal for metabolic concentration in MS and NMR experiments.

References

Adams, J. L., & McLean, R. J. C. (1999). Impact of *rpoS* deletion on *Escherichia coli* biofilms. *Applied and Environmental Microbiology*, *65*(9), 4285–4287. https://doi.org/10.1128/AEM.65.9.4285-4287.1999

Annous, B. A., Solomon, E. B., Cooke, P. H., & Burke, A. (2005). Biofilm formation by Salmonella spp. on cantaloupe melons. *Journal of Food Safety*, *25*(4), 276–287. https://doi.org/10.1111/j.1745-4565.2005.00024.x

Aue, W. P., Karhan, J., & Ernst, R. R. (1976). Homonuclear broad band decoupling and two-dimensional J-resolved NMR spectroscopy. *The Journal of Chemical Physics*, *64*(10), 4226–4227. https://doi.org/10.1063/1.431994

Beckonert, O., Keun, H. C., Ebbels, T. M. D., Bundy, J., Holmes, E., Lindon, J. C., & Nicholson, J. K. (2007). Metabolic profiling, metabolomic and metabonomic procedures for NMR spectroscopy of urine, plasma, serum and tissue extracts. *Nature Protocols, 2*(11), 2692–2703. https://doi.org/10.1038/nprot.2007.376

Beloin, C., & Ghigo, J.-M. (2005). Finding gene-expression patterns in bacterial biofilms. *Trends in Microbiology, 13*(1), 16–19. https://doi.org/10.1016/j.tim.2004.11.008

Bernini, P., Bertini, I., Luchinat, C., Nepi, S., Saccenti, E., Schäfer, H., Schütz, B., Spraul, M., & Tenori, L. (2009). Individual human phenotypes in metabolic space and time. *Journal of Proteome Research, 8*(9), 4264–4271. https://doi.org/10.1021/pr900344m

Bingol, K., & Brüschweiler, R. (2014). Multidimensional approaches to NMR-based metabolomics. *Analytical Chemistry, 86*(1), 47–57. https://doi.org/10.1021/ac403520j

Buescher, J. M., Antoniewicz, M. R., Boros, L. G., Burgess, S. C., Brunengraber, H., Clish, C. B., DeBerardinis, R. J., Feron, O., Frezza, C., Ghesquiere, B., Gottlieb, E., Hiller, K., Jones, R. G., Kamphorst, J. J., Kibbey, R. G., Kimmelman, A. C., Locasale, J. W., Lunt, S. Y., Maddocks, O. D., … Fendt, S.-M. (2015). A roadmap for interpreting 13 C metabolite labeling patterns from cells. *Current Opinion in Biotechnology, 34*, 189–201. https://doi.org/10.1016/j.copbio.2015.02.003

Cabral, M. P., Soares, N. C., Aranda, J., Parreira, J. R., Rumbo, C., Poza, M., Valle, J., Calamia, V., Lasa, Í., & Bou, G. (2011). Proteomic and functional analyses reveal a unique lifestyle for *Acinetobacter baumannii* biofilms and a key role for histidine metabolism. *Journal of Proteome Research, 10*(8), 3399–3417. https://doi.org/10.1021/pr101299j

Davey, M. E., & O'toole, G. A. (2000). Microbial biofilms: From ecology to molecular genetics. *Microbiology and Molecular Biology Reviews, 64*(4), 847–867. https://doi.org/10.1128/MMBR.64.4.847-867.2000

Doddrell, D. M., Pegg, D. T., & Bendall, M. R. (1982). Distortionless enhancement of NMR signals by polarization transfer. *Journal of Magnetic Resonance (1969), 48*(2), 323–327. https://doi.org/10.1016/0022-2364(82)90286-4

Dominguez, C., Schubert, M., Duss, O., Ravindranathan, S., & Allain, F. H.-T. (2011). Structure determination and dynamics of protein–RNA complexes by NMR spectroscopy. *Progress in Nuclear Magnetic Resonance Spectroscopy, 58*(1–2), 1–61. https://doi.org/10.1016/j.pnmrs.2010.10.001

Donlan, R. M. (2002). Biofilms: Microbial life on surfaces. *Emerging Infectious Diseases, 8*(9), 881–890. https://doi.org/10.3201/eid0809.020063

Donlan, R. M., & Costerton, J. W. (2002). Biofilms: Survival mechanisms of clinically relevant microorganisms. *Clinical Microbiology Reviews, 15*(2), 167–193. https://doi.org/10.1128/CMR.15.2.167-193.2002

Dunny, G. M., & Berntsson, R. P.-A. (2016). Enterococcal sex pheromones: Evolutionary pathways to complex, two-signal systems. *Journal of Bacteriology, 198*(11), 1556–1562. https://doi.org/10.1128/JB.00128-16

Emwas, A.-H., Roy, R., McKay, R. T., Tenori, L., Saccenti, E., Gowda, G. A. N., Raftery, D., Alahmari, F., Jaremko, L., Jaremko, M., & Wishart, D. S. (2019). NMR spectroscopy for metabolomics research. *Metabolites, 9*(7), 123. https://doi.org/10.3390/metabo9070123

Federle, M. J., & Bassler, B. L. (2003). Interspecies communication in bacteria. *Journal of Clinical Investigation, 112*(9), 1291–1299. https://doi.org/10.1172/JCI20195

Fiehn, O. (2002). Metabolomics—the link between genotypes and phenotypes. In C. Town (Ed.), *Functional genomics* (pp. 155–171). Springer Netherlands. https://doi.org/10.1007/978-94-010-0448-0_11

Foxall, P. J. D., Parkinson, J. A., Sadler, I. H., Lindon, J. C., & Nicholson, J. K. (1993). Analysis of biological fluids using 600 MHz proton NMR spectroscopy: Application of homonuclear two-dimensional J-resolved spectroscopy to urine and blood plasma for spectral simplification and assignment. *Journal of Pharmaceutical and Biomedical Analysis, 11*(1), 21–31. https://doi.org/10.1016/0731-7085(93)80145-Q

Fuqua, C., & Greenberg, E. P. (2002). Listening in on bacteria: Acyl-homoserine lactone signalling. *Nature Reviews Molecular Cell Biology, 3*(9), 685–695. https://doi.org/10.1038/nrm907

Gjersing, E. L., Herberg, J. L., Horn, J., Schaldach, C. M., & Maxwell, R. S. (2007). NMR metabolomics of planktonic and biofilm modes of growth in *Pseudomonas aeruginosa*. *Analytical Chemistry, 79*(21), 8037–8045. https://doi.org/10.1021/ac070800t

Guennec, A. L., Giraudeau, P., & Caldarelli, S. (2014). Evaluation of fast 2D NMR for metabolomics. *Analytical Chemistry, 86*(12), 5946–5954. https://doi.org/10.1021/ac500966e

Halouska, S., Fenton, R. J., Barletta, R. G., & Powers, R. (2012). Predicting the *in vivo* mechanism of action for drug leads using NMR metabolomics. *ACS Chemical Biology, 7*(1), 166–171. https://doi.org/10.1021/cb200348m

Hansen, A. L., Kupče, E., Li, D.-W., Bruschweiler-Li, L., Wang, C., & Brüschweiler, R. (2021). 2D NMR-based metabolomics with HSQC/TOCSY NOAH supersequences. *Analytical Chemistry, 93*(15), 6112–6119. https://doi.org/10.1021/acs.analchem.0c05205

Havsed, K., Stensson, M., Jansson, H., Carda-Diéguez, M., Pedersen, A., Neilands, J., Svensäter, G., & Mira, A. (2021). Bacterial composition and metabolomics of dental plaque from adolescents. *Frontiers in Cellular and Infection Microbiology, 11*, 716493. https://doi.org/10.3389/fcimb.2021.716493

Heux, S., Bergès, C., Millard, P., Portais, J.-C., & Létisse, F. (2017). Recent advances in high-throughput 13C-fluxomics. *Current Opinion in Biotechnology, 43*, 104–109. https://doi.org/10.1016/j.copbio.2016.10.010

Høiby, N., Bjarnsholt, T., Givskov, M., Molin, S., & Ciofu, O. (2010). Antibiotic resistance of bacterial biofilms. *International Journal of Antimicrobial Agents, 35*(4), 322–332. https://doi.org/10.1016/j.ijantimicag.2009.12.011

Johansson, I., Witkowska, E., Kaveh, B., Lif Holgerson, P., & Tanner, A. C. R. (2016). The microbiome in populations with a low and high prevalence of caries. *Journal of Dental Research, 95*(1), 80–86. https://doi.org/10.1177/0022034515609554

Keifer, P. A. (1999). 90° pulse width calibrations: How to read a pulse width array. *Concepts in Magnetic Resonance, 11*(3), 165–180. https://doi.org/10.1002/(SICI)1099-0534(1999)11:3<165::AID-CMR4>3.0.CO;2-D

Keun, H. C., Beckonert, O., Griffin, J. L., Richter, C., Moskau, D., Lindon, J. C., & Nicholson, J. K. (2002). Cryogenic probe ^{13}C NMR spectroscopy of urine for metabonomic studies. *Analytical Chemistry, 74*(17), 4588–4593. https://doi.org/10.1021/ac025691r

Kono, H. (2013). 1H and 13C chemical shift assignment of the monomers that comprise carboxymethyl cellulose. *Carbohydrate Polymers, 97*(2), 384–390. https://doi.org/10.1016/j.carbpol.2013.05.031

Krašovec, R., & Jerman, I. (2003). Bacterial multicellularity as a possible source of antibiotic resistance. *Medical Hypotheses, 60*(4), 484–488. https://doi.org/10.1016/S0306-9877(02)00394-8

Lee, S.-H., Cha, E.-J., Lim, J.-E., Kwon, S.-H., Kim, D.-H., Cho, H., & Han, K.-H. (2012). Structural characterization of an intrinsically unfolded mini-HBX protein from hepatitis B virus. *Molecules and Cells, 34*(2), 165–169. https://doi.org/10.1007/s10059-012-0060-z

Leriche, V., & Carpentier, B. (1995). Viable but nonculturable *Salmonella typhimurium* in single- and binary-species biofilms in response to chlorine treatment. *Journal of Food Protection, 58*(11), 1186–1191. https://doi.org/10.4315/0362-028X-58.11.1186

Lindon, J. C., Nicholson, J. K., Holmes, E., & Everett, J. R. (2000). Metabonomics: Metabolic processes studied by NMR spectroscopy of biofluids. *Concepts in Magnetic Resonance, 12*(5), 289–320. https://doi.org/10.1002/1099-0534(2000)12:5<289::AID-CMR3>3.0.CO;2-W

Lisec, J., Schauer, N., Kopka, J., Willmitzer, L., & Fernie, A. R. (2006). Gas chromatography mass spectrometry–based metabolite profiling in plants. *Nature Protocols, 1*(1), 387–396. https://doi.org/10.1038/nprot.2006.59

Magnuson, R., Solomon, J., & Grossman, A. D. (1994). Biochemical and genetic characterization of a competence pheromone from B. subtilis. *Cell, 77*(2), 207–216. https://doi.org/10.1016/0092-8674(94)90313-1

Marsh, P. D. (1994). Microbial ecology of dental plaque and its significance in health and disease. *Advances in Dental Research, 8*(2), 263–271. https://doi.org/10.1177/08959374940080022001

Mashego, M. R., Rumbold, K., De Mey, M., Vandamme, E., Soetaert, W., & Heijnen, J. J. (2007). Microbial metabolomics: Past, present and future methodologies. *Biotechnology Letters, 29*(1), 1–16. https://doi.org/10.1007/s10529-006-9218-0

Meiboom, S., & Gill, D. (1958). Modified spin-echo method for measuring nuclear relaxation times. *Review of Scientific Instruments, 29*(8), 688–691. https://doi.org/10.1063/1.1716296

Mira, A. (2018). Oral microbiome studies: Potential diagnostic and therapeutic implications. *Advances in Dental Research, 29*(1), 71–77. https://doi.org/10.1177/0022034517737024

Moestue, S., Sitter, B., Bathen, T. F., Tessem, M.-B., & Gribbestad, I. S. (n.d.). *HR MAS MR spectroscopy in metabolic characterization of cancer* (Vol. 25).

Nagayama, K., Wiithrich, K., & Bachmann, P. (1977). Biochemical and biophysical research communications. *Biochemical and Biophysical Research Communications, 78*(1), 7.

Nag, M., & Lahiri, D. (Eds.). (2021). *Analytical methodologies for biofilm research*. Springer US. https://doi.org/10.1007/978-1-0716-1378-8

Nicholson, J. K., & Lindon, J. C. (n.d.). Organisms often respond in complex and unpredictable ways to stimuli that cause disease or injury. By measuring and mathematically modelling changes in the levels of products of metabolism found in biological fluids and tissues, metabonomics offers fresh insight into the effects of diet, drugs and disease (Vol. 3).

Pang, Q., Zhang, T., Wang, Y., Kong, W., Guan, Q., Yan, X., & Chen, S. (2018). Metabolomics of early stage plant cell–microbe interaction using stable isotope labeling. *Frontiers in Plant Science, 9*, 760. https://doi.org/10.3389/fpls.2018.00760

Parsek, M. R., & Singh, P. K. (2003). Bacterial biofilms: An emerging link to disease pathogenesis. *Annual Review of Microbiology, 57*(1), 677–701. https://doi.org/10.1146/annurev.micro.57.030502.090720

Pervushin, K., Ono, A., Fernández, C., Szyperski, T., Kainosho, M., & Wüthrich, K. (1998). NMR scalar couplings across Watson–Crick base pair hydrogen bonds in DNA observed by transverse relaxation-optimized spectroscopy. *Proceedings of the National Academy of Sciences, 95*(24), 14147–14151. https://doi.org/10.1073/pnas.95.24.14147

Poirel, L., Bonnin, R. A., & Nordmann, P. (2011). Genetic basis of antibiotic resistance in pathogenic Acinetobacter species. *IUBMB Life, 63*(12), 1061–1067. https://doi.org/10.1002/iub.532

Prigent-Combaret, C., Vidal, O., Dorel, C., & Lejeune, P. (1999). Abiotic surface sensing and biofilm-dependent regulation of gene expression in *Escherichia coli*. *Journal of Bacteriology, 181*, 10.

Pysz, M. A., Conners, S. B., Montero, C. I., Shockley, K. R., Johnson, M. R., Ward, D. E., & Kelly, R. M. (2004). Transcriptional analysis of Biofilm Formation processes in the anaerobic, hyperthermophilic bacterium *Thermotoga maritima*. *Applied and Environmental Microbiology, 70*(10), 6098–6112. https://doi.org/10.1128/AEM.70.10.6098-6112.2004

Raamsdonk, L. M., Teusink, B., Broadhurst, D., Zhang, N., Hayes, A., Walsh, M. C., Berden, J. A., Brindle, K. M., Kell, D. B., Rowland, J. J., Westerhoff, H. V., van Dam, K., & Oliver, S. G. (2001). A functional genomics strategy that uses metabolome data to reveal the phenotype of silent mutations. *Nature Biotechnology, 19*(1), 45–50. https://doi.org/10.1038/83496

Resch, A., Leicht, S., Saric, M., Pásztor, L., Jakob, A., Götz, F., & Nordheim, A. (2006). Comparative proteome analysis of Staphylococcus aureus biofilm and planktonic cells and correlation with transcriptome profiling. *Proteomics, 6*(6), 1867–1877. https://doi.org/10.1002/pmic.200500531

Rubakhin, S. S., Romanova, E. V., Nemes, P., & Sweedler, J. V. (2011). Profiling metabolites and peptides in single cells. *Nature Methods, 8*(S4), S20–S29. https://doi.org/10.1038/nmeth.1549

Sandusky, P., & Raftery, D. (2005). Use of semiselective TOCSY and the pearson correlation for the metabonomic analysis of biofluid mixtures: Application to urine. *Analytical Chemistry, 77*(23), 7717–7723. https://doi.org/10.1021/ac0510890

Schembri, M. A., Kjaergaard, K., & Klemm, P. (2003). Global gene expression in *Escherichia coli* biofilms: Gene expression in *E. coli* biofilms. *Molecular Microbiology, 48*(1), 253–267. https://doi.org/10.1046/j.1365-2958.2003.03432.x

Schuster, M., Lostroh, C. P., Ogi, T., & Greenberg, E. P. (2003). Identification, timing, and signal specificity of *Pseudomonas aeruginosa* quorum-controlled genes: A transcriptome analysis. *Journal of Bacteriology, 185*(7), 2066–2079. https://doi.org/10.1128/JB.185.7.2066-2079.2003

Sekiyama, Y., Chikayama, E., & Kikuchi, J. (2011). Evaluation of a semipolar solvent system as a step toward heteronuclear multidimensional NMR-based metabolomics for ^{13}C-labeled bacteria, plants, and animals. *Analytical Chemistry, 83*(3), 719–726. https://doi.org/10.1021/ac102097u

Sleytr, U. B. (1997). I. Basic and applied S-layer research: An overview. *FEMS Microbiology Reviews, 20*(1–2), 5–12. https://doi.org/10.1111/j.1574-6976.1997.tb00301.x

Smith, K., Perez, A., Ramage, G., Lappin, D., Gemmell, C. G., & Lang, S. (2008). Biofilm formation by Scottish clinical isolates of *Staphylococcus aureus*. *Journal of Medical Microbiology, 57*(8), 1018–1023. https://doi.org/10.1099/jmm.0.2008/000968-0

Stewart, P. S., & William Costerton, J. (2001). Antibiotic resistance of bacteria in biofilms. *The Lancet, 358*(9276), 135–138. https://doi.org/10.1016/S0140-6736(01)05321-1

Stipetic, L. H., Dalby, M. J., Davies, R. L., Morton, F. R., Ramage, G., & Burgess, K. E. V. (2016). A novel metabolomic approach used for the comparison of *Staphylococcus aureus* planktonic cells and biofilm samples. *Metabolomics, 12*(4), 75. https://doi.org/10.1007/s11306-016-1002-0

Sutherland, I. W. (1999). Polysaccharases for microbial exopolysaccharides. *Carbohydrate Polymers, 38*(4), 319–328. https://doi.org/10.1016/S0144-8617(98)00114-3

Sutherland, I. (2001). The biofilm matrix – an immobilized but dynamic microbial environment. *Trends in Microbiology, 9*(5), 222–227. https://doi.org/10.1016/S0966-842X(01)02012-1

Tiziani, S., Kang, Y., Choi, J. S., Roberts, W., & Paternostro, G. (2011). Metabolomic high-content nuclear magnetic resonance-based drug screening of a kinase inhibitor library. *Nature Communications, 2*(1), 545. https://doi.org/10.1038/ncomms1562

Wagner, V. E., Bushnell, D., Passador, L., Brooks, A. I., & Iglewski, B. H. (2003). Microarray analysis of *Pseudomonas aeruginosa* quorum-sensing regulons: Effects of growth phase and environment. *Journal of Bacteriology, 185*(7), 2080–2095. https://doi.org/10.1128/JB.185.7.2080-2095.2003

Want, E. J., Coen, M., Masson, P., Keun, H. C., Pearce, J. T. M., Reily, M. D., Robertson, D. G., Rohde, C. M., Holmes, E., Lindon, J. C., Plumb, R. S., & Nicholson, J. K. (2010). Ultra performance liquid chromatography-mass spectrometry profiling of bile acid metabolites in biofluids: Application to experimental toxicology studies. *Analytical Chemistry, 82*(12), 5282–5289. https://doi.org/10.1021/ac1007078

Weljie, A. M., Newton, J., Mercier, P., Carlson, E., & Slupsky, C. M. (2006). Targeted profiling: Quantitative analysis of ^{1}H NMR metabolomics data. *Analytical Chemistry, 78*(13), 4430–4442. https://doi.org/10.1021/ac060209g

Whiteley, M., Bangera, M. G., Bumgarner, R. E., Parsek, M. R., Teitzel, G. M., Lory, S., & Greenberg, E. P. (2001). Gene expression in *Pseudomonas aeruginosa* bio®lms. *Binding Protein, 413*, 5.

Wong, H. S. (n.d.). *Comparative evaluation of planktonic and biofilm modes of growth in Salmonella typhimurium*, 180.

Wong, E. H. J., Ng, C. G., Goh, K. L., Vadivelu, J., Ho, B., & Loke, M. F. (2018). Metabolomic analysis of low and high biofilm-forming *Helicobacter pylori* strains. *Scientific Reports, 8*(1), 1409. https://doi.org/10.1038/s41598-018-19697-0

Wong, H. S., Townsend, K. M., Fenwick, S. G., Maker, G., Trengove, R. D., & O'Handley, R. M. (2010). Comparative susceptibility of *Salmonella* Typhimurium biofilms of different ages to disinfectants. *Biofouling, 26*(7), 859–864. https://doi.org/10.1080/08927014.2010.527959

Wushensky, J. A., Youngster, T., Mendonca, C. M., & Aristilde, L. (2018). Flux connections between gluconate pathway, glycolysis, and pentose–phosphate pathway during carbohydrate metabolism in Bacillus megaterium QM B1551. *Frontiers in Microbiology, 9*, 2789. https://doi.org/10.3389/fmicb.2018.02789

Xi, Y., de Ropp, J. S., Viant, M. R., Woodruff, D. L., & Yu, P. (2006). Automated screening for metabolites in complex mixtures using 2D COSY NMR spectroscopy. *Metabolomics, 2*(4), 221–233. https://doi.org/10.1007/s11306-006-0036-0

Ye, T., Mo, H., Shanaiah, N., Gowda, G. A. N., Zhang, S., & Raftery, D. (2009). Chemoselective ^{15}N tag for sensitive and high-resolution nuclear magnetic resonance profiling of the carboxyl-containing metabolome. *Analytical Chemistry, 81*(12), 4882–4888. https://doi.org/10.1021/ac900539y

Yeom, J., Shin, J.-H., Yang, J.-Y., Kim, J., & Hwang, G.-S. (2013). 1H NMR-based metabolite profiling of planktonic and biofilm cells in acinetobacter baumannii 1656-2. *PLoS One, 8*(3), e57730. https://doi.org/10.1371/journal.pone.0057730

Yi, Q., Scalley-Kim, M. L., Alm, E. J., & Baker, D. (2000). NMR characterization of residual structure in the denatured state of protein L. *Journal of Molecular Biology, 299*(5), 1341–1351. https://doi.org/10.1006/jmbi.2000.3816

Zhang, B., & Powers, R. (2012). Analysis of bacterial biofilms using NMR-based metabolomics. *Future Medicinal Chemistry, 4*(10), 1273–1306. https://doi.org/10.4155/fmc.12.59

CHAPTER

13

Isotope labeling LC-MS for metabolomics of biofilm study and tracer-based biofilm metabolomics analysis

Yuvaraj Sivamani[1], *Karan Narasimha Murthy*[2], *Harshit Sajal*[2] *and Sumitha Elayaperumal*[2]

[1]**Department of Pharmaceutical Chemistry, Cauvery College of Pharmacy, Mysore, Karnataka, India;** [2]**Department of Biotechnology and Bioinformatics, JSS Academy of Higher Education and Research, Mysore, Karnataka, India**

1. Introduction

A biofilm is a group of microbial cells that are enclosed in a polysaccharide-based matrix and are firmly attached to a surface that cannot be removed easily by simple rinsing. Biofilm on tooth surfaces observed by Van Leeuwenhoek was first reported as microbial biofilms (Donlan, 2002). Heukelekian and Heller discovered that incorporating a surface for microorganisms to cling to cause dramatically increases in bacterial growth and activity (Heukelekian & Heller, 1940). Additionally, it was found that there were significantly more organisms on surfaces than in the surrounding medium. Biofilms can develop on a wide range of material surfaces, which include living tissues, medical equipment, pipelines of industrial or potable water systems, and aquatic ecosystems in nature (Coughlan et al., 2016; Dobretsov et al., 2006).

The majority of microbial cells are found in nature as biofilms to withstand unfavorable conditions such as inadequate nutrients, high temperatures, acid and alkali, UV radiation, disinfectants, and antibiotics (Yin et al., 2019). They can be useful or have a harmful impact, particularly when formed in industrial settings or on medical devices (Khatoon et al., 2018). A detailed picture and analysis of biofilms requires an electron microscope, which offers

Microbial Biofilms
https://doi.org/10.1016/B978-0-323-95715-1.00022-4

high-resolution photomicroscopy at far greater magnifications than the light microscope and other technologies such as NMR and liquid chromatography–mass spectrometry (LC-MS) (Khatoon et al., 2018).

Biofilm metabolomics can be revealed through the analysis of the metabolites, thereby which biofilm activity can be elucidated. Annotating and quantifying hundreds of metabolites is now possible due to the developments in high-throughput bioanalytical platforms, advances in computers for data analysis and interpretation, and access to metabolite databases of the specific organism (Chokkathukalam et al., 2014). The study of metabolomics should offer precise flux measurement, absolute quantification, and metabolite identification of biofilm.

Microbial metabolites are produced in vivo by inward nutrients through a network of coupled enzymatic processes. The metabolic study allows the measurement of metabolite concentrations and helps to understand the pathway activity of the biofilm, which can be quantified in terms of material flow per unit time called metabolic flux (Sauer, 2006). Metabolic flux describes in vivo enzyme activity inside biofilm cells based on carbon and energy metabolism (Wan et al., 2018).

Fluxes cannot be measured directly; however, by the use of isotope tracers, they can be inferred. The organization of metabolic networks and pathways has long been studied using isotope labeling to analyze metabolic fluxes. Additionally, metabolic networks derived from isotope-labeling methods can serve as a framework for combining and understanding other postgenomic data sets, including transcriptomics and proteomics profiles (Birkemeyer et al., 2005). The development of mass spectrometry (MS) has permitted the use of stable isotopes for these purposes. For several decades, stable isotopes have been used in MS by pharmacologists as an internal standard to quantify drugs and metabolites. The latest development in metabolomics with ultrahigh-resolution MS has established stable isotope tracer-based metabolomics strategies to quantify and identify different pathways involved in the biofilm (Birkemeyer et al., 2005). Stable isotope labeling is used to provide crucial details regarding metabolic flux that could not be analyzed by traditional label-free metabolomics research.

2. Stable isotopes in metabolomics

The use of isotopic tracers to study cellular metabolism dates back to the 1930s. Numerous well-known metabolic pathways were extensively categorized using ^{3}H and ^{14}C radioisotopes (Crown & Antoniewicz, 2013). Till the 1980s, metabolic research was dominated by the use of radioactive tracers, but as MS, and computational methodologies, have advanced, stable isotopes became a better alternative.

Stable isotopes have the same number of protons and differ in the number of the neutron. They differ in mass but share the same physicochemical characteristics. Stable isotopes, which include heavy isotopes ^{13}C, ^{2}H, ^{18}O, ^{15}N, and ^{34}S of common elements, are used to detect metabolites that elute with higher mass LC-MS peaks.

The majority of investigations on metabolism and fluxes today use stable-isotope methods. Stable-isotope methods rapidly displaced radioisotope methods because they were simpler to carry out, permitted numerous measurements of isotopic enrichment from a single experiment, were safer to carry out, and offered more useful information for flux estimates.

3. Strategies used in stable isotope labeling to assist metabolomics using LC-MS

Metabolites can be of two types: targeted and nontargeted (Roberts et al., 2012). LC-MS is one of the most effective methods for the determination of a large number of metabolites in biofilm over different concentration ranges.

Targeted techniques involve identifying the abundances of metabolites from a predefined group of well-known compounds. This method can give absolute quantification of metabolites as they are compared with reference standards. Nontargeted techniques, in contrast, find the mass of the unknown metabolites, which are not known during the time of sample measurement. Therefore, this method offers the benefit of probing the entire metabolites and obtaining relative abundances of thousands of known and unknown metabolites. Fig. 13.1 depicts the general flow of processes involved in metabolomics using LC-MS.

Different strategies are used for stable isotope labeling to assist metabolomics studies (Bueschl et al., 2013). Generally, ^{13}C-, ^{15}N-, and ^{34}S-enriched substances are not separated from natural isotopologues by chromatography; thus, they elute at the same retention time with identical peaks.

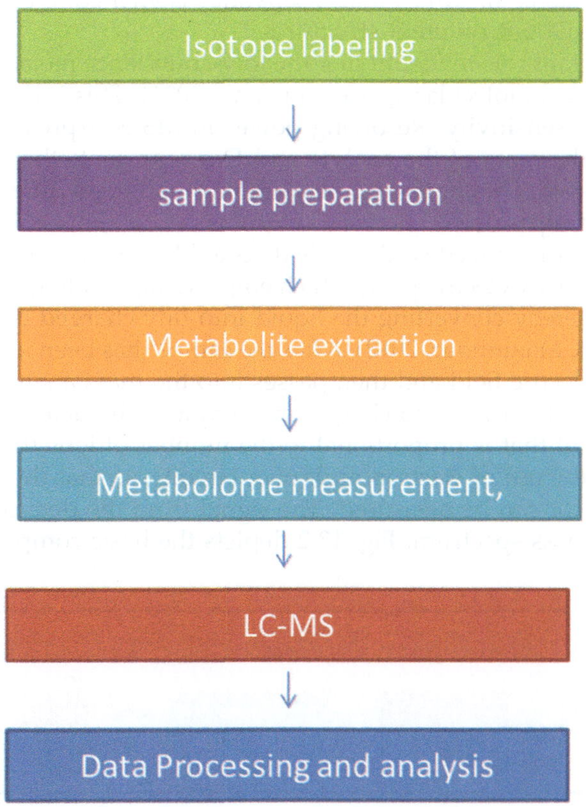

FIGURE 13.1 Process involved in metabolomics using LC-MS.

Method A: In nontargeted metabolites, metabolomes in the organism are cultivated parallel using different isotopologue-enriched nutrition sources (e.g., ^{13}C glucose as a carbon source). The extracts are subsequently mixed and measured with LC-MS. The resulting spectra can interpret the true biological signals.

Method B: This method uses labeled sample extract from a global stock of the same organism for interexperiment comparison to quantify absolute compounds using authentic, labeled standards.

Method D: This study uses fully labeled, natural tracer substances to distinguish between biological signals and other metabolic products.

Method E: Metabolites are identified by using nonlabeled and labeled derivatization agents (dansylation chemistry), which belong to the same chemical class.

4. Liquid chromatography—mass spectrometry

Separation of analytes by column chromatography before MS analyses improves metabolome coverage and increases the quantitative accuracy of MS (Zhou et al., 2012).

Chromatographic separation reduces ion suppression and improves the detection of low-abundance species (Xiao et al., 2012). It also separates isomer compounds, which are having same molecular formula but different structures.

MS plays an important role in identifying the low-abundance metabolites without interference from closely related molecular species (Li et al., 2021). This is due to its resolving capability and exceptional sensitivity. Resolving power in MS is expressed as the ratio m/Dm, where m represents the mass of the analyte and Dm represents the lowest mass difference that can be distinguished. Due to its high resolution, small mass differences between metabolites can be detected separately.

Metabolite extracts are ionized so that it is detected by MS. It is done by different ionization technique generally by electrospray, which employs high voltage to liquid as it flows out the tip of a needle, thereby converting the liquid into tiny-charged droplets that lead to gas phase ions (Banerjee & Mazumdar, 2012). After the sample has been ionized, the beam of ions is accelerated by an electric field and then passes into the mass analyzer, where the ions are separated according to their mass-to-charge (m/z) ratios. The detector consists of a counter that produces a current that is proportional to the number of ions that strike it. The current can be measured by the use of electron multiplier circuits, so that the current caused by one ion striking the detector can be measured. The signal from the detector is fed to a recorder, which produces the mass spectrum. Fig. 13.2 depicts the basic components of LC-MS.

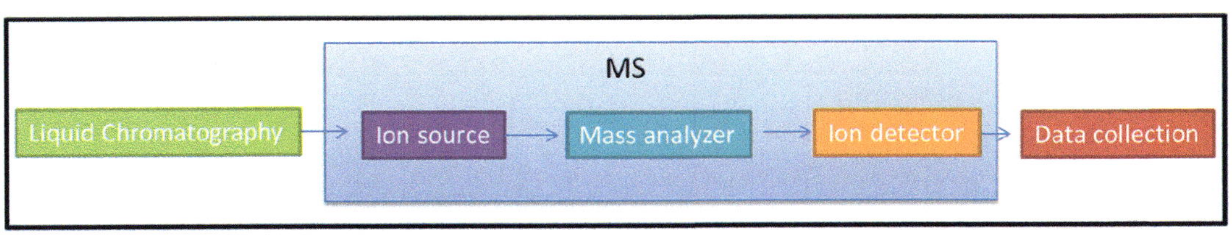

FIGURE 13.2 Basic components of LC-MS.

5. LC-MS in metabolomics of biofilm

Deborah MB et al. studied the components of the *Haemophilus influenzae* in biofilm matrix and planktonic population using stable isotope labeling of amino acids in cell culture combined with mass spectrometry (Post et al., 2014). Following proteolytic digestion, the amino acids in respective studies are typically labeled with heavy stable isotopes and that was incorporated into the cell culture of *Haemophilus influenzae*. Any peptides containing the labeled amino acid will travel differently in mass spectrometry based on the specified mass. Following mixing and LC/MS analysis, this mass shift enables a direct comparison of the protein levels in both populations. Based on this study, it has been found that the nontypeable *Haemophilus influenzae* biofilm could manage oxidative stress and acquire crucial cofactors for biofilm persistence while being in a semidormant state with reduced energy metabolism and protein synthesis.

Atshan et al. (2015) studied the extracellular proteins expressed by various clonal types of *Staphylococcus aureus* during planktonic growth and biofilm development at different time points of 12, 24, and 48 h by combining 2D-gel electrophoresis and LC-MS/MS analysis. The study revealed extracellular proteins secreted at 48 h incubation have significant changes in protein expression as compared with the exponential growth at 12 h incubation.

Munuswamy et al. (2021) used LC-MS techniques to study the biological and metabolic processes essential for *Candida albicans* biofilm growth. The ability of *C. albicans* to form biofilms is essential for its pathogenesis and drug resistance. At first, the planktonic culture is established by inoculating the synthetically defined medium incubated, and biofilm culture is established by suspending *C. albicans* in SD synthetically defined medium and extracted by mechanical methods such as cell scrapping and pipetting. Then the proteins are extracted from both types of cells by using *PROTEOSPIN* total protein isolation kit, and proteins are concentrated by using various methods such as filtration, trypsin digestion, and centrifugation. The dried samples are then suspended in HPLC grade water, and concentration is estimated using a nanodrop spectrophotometer at 280 nm wavelength. The protein concentration of samples is then adjusted to a final concentration of 100 μg/mL. The samples are then loaded into a 1260 infinity HPLC-chip/MS system instrument. HPLC grade water and acetonitrile were used as mobile phases. The nanoflow gradient is adjusted to various percentages, and an accurate-mass q-TOF LC-MS system operated in positive ion mode was used for mass detection. The sequence homology search tool (SPIDER) of the PEAKS software was used to detect the proteins in both planktonic and biofilm proteomes. This study revealed that there was increased expression of proteins involved in two biological processes, i.e., cellular homeostasis and vesicle-mediated transport, and several metabolic pathways such as folate metabolism, fatty acid metabolism, and biosynthesis of serine, lysine, and glycine during biofilm growth of *C. albicans* show differential protein expression and metabolism during biofilm growth.

6. Tracer-based biofilm metabolomics analysis

Wan et al. (2018) investigated the metabolism of *Pseudomonas aeruginosa* during biofilm development using microscopy imaging, gene expression analysis, and isotope labeling using ^{13}C. Cells at different locations inside the biofilms may exhibit different metabolic processes and rates. To study these variations in metabolic processes, various technologies were used, which include isotope labeling.

In the current study to understand the flux distribution in biofilm cells, two different isotopic tracers were used. ^{13}C fingerprinting of protein-forming amino acids was used to track carbon fluxes for biomass synthesis and substrate utilization. Then dynamic labeling uses ^{13}C glucose in pulses to measure the speed of the ^{13}C isotope entering into central metabolic pathways in biofilms and planktonic cells. For tracing flux distribution, both the planktonic and biofilm cells in a tubular biofilm reactor were grown in a medium containing ^{13}C glucose. Substrate concentrations including glucose, acetate, and lactate were measured using HPLC, and proteinogenic amino acids were analyzed using GC-MS measurement. Glucose uptake in the biofilm and planktonic was measured by finding the rate at which ^{13}C is incorporated into two key metabolites: glucose-6-P and glutamate. To label planktonic cells, the unlabeled planktonic cells were grown in the shake flask containing nonlabeled glucose, and then ^{13}C labeled glucose is added. For dynamic labeling of biofilm cells, biofilms are prepared in glass slides and then soaked in a medium containing ^{13}C glucose for various durations. Then metabolism of cells is quenched by treating with ethanol and methanol/chloroform for planktonic and biofilm cells, respectively. The metabolites are then extracted and then analyzed using LC-MS technology.

The results of dynamic labeling showed that the labeling rates of G6P (an important metabolite formed at the first step of glucose uptake) were faster in planktonic cells compared with biofilm cells. But the final labeling percentages reached more than 85% in biofilm cells indicating biofilm cells were metabolically active despite the slow rate of glucose uptake. But in contrast to dynamic labeling, there were few variations in metabolic fluxes between biofilm and planktonic cells.

Pisithkul et al. (2019) used isotopic tracer techniques to study metabolic remodeling during the development of biofilms of bacteria *Bacillus subtilis. B. subtilis* is a gram-positive bacterium often used as a model system for investigating biofilm formation and development. *B. subtilis* can form biofilms called pellicles at air–liquid interfaces when grown on solid surfaces and in standing liquid cultures in vitro. This process of pellicle formation is used to determine and characterize metabolic alterations that occur during biofilm formation and development. They combined metabolic techniques such as LC-MS with transcriptome techniques such as RNA sequencing and proteomics techniques such as LC-tandem MS (LC-MS/MS). During the analysis, they used dynamic labeling with tracer isotope ^{13}C glycerol to investigate changes in the Krebs cycle during biofilm formation and found carbon flux into the Krebs cycle via both enzyme citrate synthase and anaplerotic reactions. ^{13}C tracer experiments also revealed a high rate of incorporation of ^{13}C into purines and pyrimidines showing a high rate of de novo synthesis of nucleotides during biofilm formation. Further ^{13}C tracer isotope experiments revealed that during biofilm formation, there were an increase in the synthesis of components of the extracellular matrix, an increase in amino acid synthesis except for glycine and asparagine, increase in the synthesis of bacillibactin and its precursor 2,3-dihydroxybenzoate for better iron acquisition and metabolism, and increase in the synthesis of antibiotics such as bacilysin and subtilosin A. It also revealed the shift from fatty acid biosynthesis to fatty acid oxidation during biofilm formation.

Philips et al. (2012) used a ^{13}C isotope-labeled arginine to analyze the shift from aerobic to anaerobic respiration in pathogens *Neisseria gonorrhoeae. N. gonorrhoeae* can form biofilms in vitro and in vivo. At first two populations of *N. gonorrhoeae* strain, 1291 auxotrophics for arginine were grown in continuous flow chambers: one with ^{13}C-arginine for planktonic and the other with unlabeled arginine for biofilm cells. The proteins were then extracted separately and mixed and then separated by SDS-PAGE. The extracts are then analyzed using LC-MS techniques. A total of 757 proteins were identified out of which 152 unique proteins were expressed in biofilms. In comparison with planktonic cells, biofilm cells had 73 upregulated and 54 downregulated proteins. Upregulated proteins are those that were involved in energy metabolism, and downregulated proteins were those that were involved in protein synthesis. During this study, it was found that nitrite reductase and cytochrome c peroxidase, which are key enzymes involved in anaerobic metabolism, were highly upregulated in biofilm cells indicating a shift from aerobic to anaerobic respiration.

7. Conclusion

LC-MS method can be used for the quantitative metabolome profiling of biofilm with relatively high metabolome coverage. The specific changes in the metabolome of biofilm at different environmental conditions can be tested for functional studies of cellular metabolisms and networks. Analysis of the biofilm by LC-MS can be provided insight into the growth of biofilm and its influence on factors such as drug resistance and device-related infections. An improved approach to patient treatment can be obtained from a better understanding of biofilm dynamics, thereby resulting in the development of new, efficient control mechanisms for biofilm management.

References

Atshan, S. S., Shamsudin, M. N., Sekawi, Z., Lung, L. T. T., Barantalab, F., Liew, Y. K., Ali Alreshidi, M., Abduljaleel, S. A., & Hamat, R. A. (2015). Comparative proteomic analysis of extracellular proteins expressed by various clonal types of Staphylococcus aureus and during planktonic growth and biofilm development. *Frontiers in Microbiology, 6*. https://doi.org/10.3389/FMICB.2015.00524

Banerjee, S., & Mazumdar, S. (2012). Electrospray ionization mass spectrometry: A technique to access the information beyond the molecular weight of the analyte. *International Journal of Analytical Chemistry, 2012*, 1–40. https://doi.org/10.1155/2012/282574

Birkemeyer, C., Alexander, L., Wagner, C., Alexander, E., & Kopka, J. (2005). Metabolome analysis: The potential of in vivo labeling with stable isotopes for metabolite profiling. *Trends in Biotechnology, 23*(1), 28–33. https://doi.org/10.1016/J.TIBTECH.2004.12.001

Bueschl, C., Krska, R., Kluger, B., & Schuhmacher, R. (2013). Isotopic labeling-assisted metabolomics using LC–MS. *Analytical and Bioanalytical Chemistry, 405*(1), 27. https://doi.org/10.1007/S00216-012-6375-Y

Chokkathukalam, A., Kim, D. H., Barrett, M. P., Breitling, R., & Creek, D. J. (2014). Stable isotope-labeling studies in metabolomics: New insights into structure and dynamics of metabolic networks. *Bioanalysis, 6*(4), 511–524. https://doi.org/10.4155/BIO.13.348

Coughlan, L. M., Cotter, P. D., Hill, C., & Alvarez-Ordóñez, A. (2016). New weapons to fight old enemies: Novel strategies for the (bio)control of bacterial biofilms in the food industry. *Frontiers in Microbiology, 7*, 1641. https://doi.org/10.3389/FMICB.2016.01641/BIBTEX

Crown, S. B., & Antoniewicz, M. R. (2013). Parallel labeling experiments and metabolic flux analysis: Past, present and future methodologies. *Metabolic Engineering, 16*(1), 21. https://doi.org/10.1016/J.YMBEN.2012.11.010

Dobretsov, S., Dahms, H. U., & Qian, P. Y. (2006). Inhibition of biofouling by marine microorganisms and their metabolites. *Biofouling, 22*(1–2), 43–54. https://doi.org/10.1080/08927010500504784

Donlan, R. M. (2002). Biofilms: Microbial life on surfaces. *Emerging Infectious Diseases, 8*(9), 881. https://doi.org/10.3201/EID0809.020063

Heukelekian, H., & Heller, A. (1940). Relation between food concentration and surface for bacterial growth. *Journal of Bacteriology, 40*(4), 547–558. https://doi.org/10.1128/JB.40.4.547-558.1940

Khatoon, Z., McTiernan, C. D., Suuronen, E. J., Mah, T. F., & Alarcon, E. I. (2018). Bacterial biofilm formation on implantable devices and approaches to its treatment and prevention. *Heliyon, 4*(12). https://doi.org/10.1016/J.HELIYON.2018.E01067

Li, C., Chu, S., Tan, S., Yin, X., Jiang, Y., Dai, X., Gong, X., Fang, X., & Tian, D. (2021). Towards higher sensitivity of mass spectrometry: A perspective from the mass analyzers. *Frontiers in Chemistry, 9*, 1148. https://doi.org/10.3389/FCHEM.2021.813359/BIBTEX

Munusamy, K., Loke, M. F., Vadivelu, J., & Tay, S. T. (2021). LC-MS analysis reveals biological and metabolic processes essential for Candida albicans biofilm growth. *Microbial Pathogenesis, 152*. https://doi.org/10.1016/J.MICPATH.2020.104614

Phillips, N. J., Steichen, C. T., Schilling, B., Post, D. M. B., Niles, R. K., Bair, T. B., Falsetta, M. L., Apicella, M. A., & Gibson, B. W. (2012). Proteomic analysis of Neisseria gonorrhoeae biofilms shows shift to anaerobic respiration and changes in nutrient transport and outermembrane proteins. *PLoS One, 7*(6). https://doi.org/10.1371/JOURNAL.PONE.0038303

Pisithkul, T., Schroeder, J. W., Trujillo, E. A., Yeesin, P., Stevenson, D. M., Tai, C., Coon, J. J., Wang, J. D., & Amador-Noguez, D. (2019). Metabolic remodeling during biofilm development of Bacillus subtilis. *MBio, 10*(3). https://doi.org/10.1128/MBIO.00623-19

Post, D. M. B., Held, J. M., Ketterer, M. R., Phillips, N. J., Sahu, A., Apicella, M. A., & Gibson, B. W. (2014). Comparative analyses of proteins from Haemophilus influenzae biofilm and planktonic populations using metabolic labeling and mass spectrometry. *BMC Microbiology, 14*(1). https://doi.org/10.1186/S12866-014-0329-9

Roberts, L. D., Souza, A. L., Gerszten, R. E., & Clish, C. B. (2012). Targeted metabolomics. *Current Protocols in Molecular Biology Chapter*, (SUPPL.98). Unit30.2 https://doi.org/10.1002/0471142727.MB3002S98.

Sauer, U. (2006). Metabolic networks in motion: 13C-Based flux analysis. *Molecular Systems Biology, 2*(1), 62. https://doi.org/10.1038/MSB4100109

Wan, N., Wang, H., Ng, C. K., Mukherjee, M., Ren, D., Cao, B., & Tang, Y. J. (2018). Bacterial metabolism during biofilm growth investigated By13C tracing. *Frontiers in Microbiology, 9*. https://doi.org/10.3389/FMICB.2018.02657/FULL

Xiao, J. F., Zhou, B., & Ressom, H. W. (2012). Metabolite identification and quantitation in LC-MS/MS-Based metabolomics. *Trends in Analytical Chemistry: TRAC, 32*, 1. https://doi.org/10.1016/J.TRAC.2011.08.009

Yin, W., Wang, Y., Liu, L., & Jin, H. (2019). Biofilms: The microbial 'protective clothing' in extreme environments. *International Journal of Molecular Sciences, 20*(14). https://doi.org/10.3390/IJMS20143423

Zhou, B., Xiao, J. F., Tuli, L., & Ressom, H. W. (2012). LC-MS-Based metabolomics. *Molecular BioSystems, 8*(2), 470. https://doi.org/10.1039/C1MB05350G

CHAPTER

14

Lipidomics profiling of microbial biofilm

Gowtham Kumaraguru[1,2], *Saruniyadevi Moorthy*[1,3], *Ashok Kumar Krishna Kumar*[4] *and Jayanthi Malaiyandi*[1]

[1]Department of Biotechnology, Vels Institute of Science Technology and Advanced Studies, Chennai, Tamil Nadu, India; [2]Plant Biotechnology Lab, Department of Biotechnology, Bharathiar University, Coimbatore, Tamil Nadu, India; [3]MTA Infotech, Varanasi, Uttar Pradesh, India; [4]Department of Lifescience, University of Greenwich, London, United Kingdom

1. Introduction

For more than 150 years, microbiologists have used direct methods to examine bacteria grown on the surface to analyze natural habitats. Claude Zobell investigated wild marine populations using direct microscopy over a decade later; in 1934, he discovered that these bacteria are attracted to the surfaces to which they regularly adhere, creating microcolonies, a process known as "biofilm." Microorganism adhesion develops a slimy substance known as a biofilm or extracellular polysaccharides (EPS). The biofilm is made of microcolonies consisting of 10%–25% cells and 75%–90% of EPS and also includes proteins, polysaccharides, glycolipids, glycoproteins, and DNA (Fig. 14.1). EPS immobilizes the cell within the biofilm, allowing intense interactions such as cell–cell interaction and horizontal gene transfer and forming synergistic microconsortia for extended periods (Flemming & Wingender, 2010). The biofilm thickens and grows over time even though microbes reproduce and additional polymers are secreted. The habitual location and condition where biofilm formed affect the microbes of the biofilm, which leads to the cell detachment and releases into the running water system (O'Toole et al., 2000).

2. Why should we concerned about biofilms?

The impact of biofilms on health devices is also significant. Hence, it is crucial to focus on health device contamination, leading to microbiological and chronic disorders in the body. For many human diseases, biofilm is the primary source of infection; cells residing within

https://doi.org/10.1016/B978-0-323-95715-1.00007-8

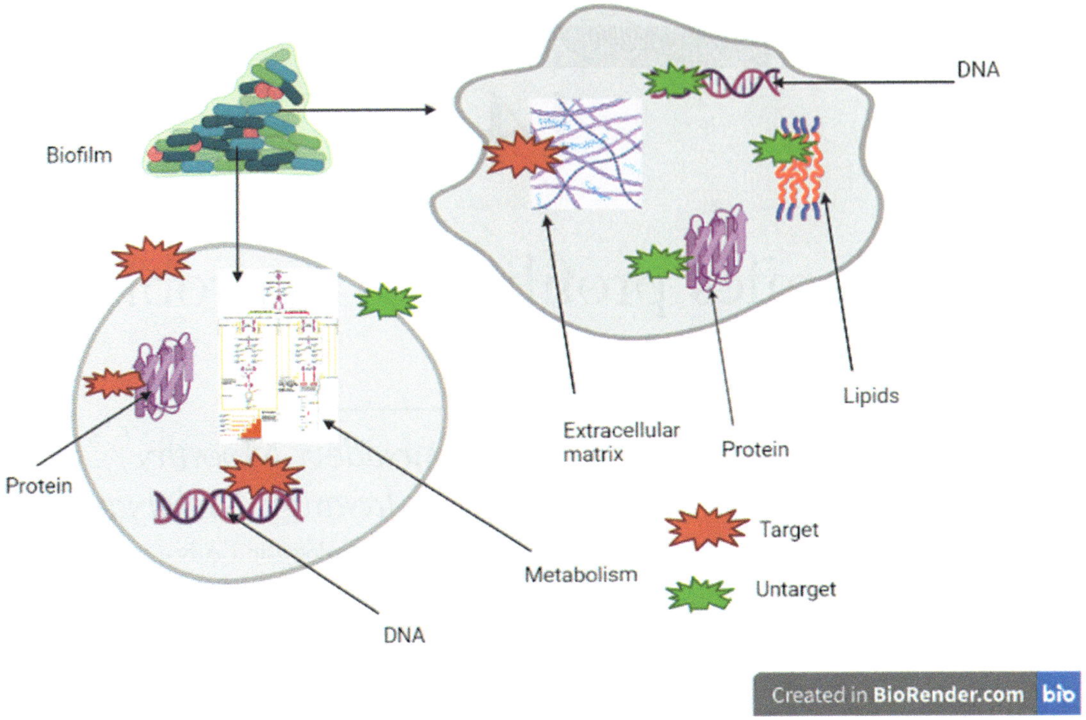

FIGURE 14.1 Composition of biofilm.

the biofilm are more resistant to antibiotics and disinfectants, making conditions extremely difficult to cure, thus causing significant sickness, leading to surgical failure and treatments (O'Toole et al., 2000; Nielsen, 2015; Del Pozo, 2018).

Biofilm formation, prevention, and treatment studies improve medical and dental practice due to biofilms' significant human health implications. Understanding the consequence of how biofilm contributes to disease pathogenesis is essential for designing efficient biofilm-related infection treatments. Despite the sessile state of microbial life and its implication for disease pathology, researchers do not entirely comprehend sessile bacteria's physiology (Monds & O'Toole, 2009). Therefore, understanding the causes of biofilm formation and its traits is vital. Biofilms are resistant to antimicrobial drugs due to three possible mechanisms (Stewart & Costerton, 2001).

➢ **First hypothesis**: The polymer matrix of the biofilm controls inhibitors from reaching biological sites.
➢ **Second hypothesis**: Cells present in the deeper layer of biofilm proliferate slowly as they have lesser access to nutrients and antibiotics.
➢ **Third hypothesis**: It is assumed that a "biofilm phenotype" occurs.

The differentiation in the biofilm development depends upon the habitat and the protein regulating gene. In the biofilm, 75%–90% is an extracellular polysaccharide in which 15% content is lipids. Lipids in the biofilm also influenced the developing new strategies against infection caused by a microorganism. Nevertheless, understanding the role of lipids in fungal biofilm formation would help us to understand the disease and also help in therapeutic importance.

3. What is lipidomics?

Lipids are the fundamental components of a biofilm and play vital roles in biological systems:

1. Lipid bilayer structure isolates cells from external environment.
2. Lipids of the biofilm provide a suitable hydrophobic medium for the functional implementation of membrane protein and their interaction.
3. Lipid molecules of the biofilm produce second messenger by enzymatic reaction.

Research on biofilm lipids helps understand the fungal pathogenicity and its virulence mechanism. The presence of sphingolipid–ergosterol determines the virulence properties of the biofilm, and it also affects the physical properties of the cell membrane. A vital component of EPS is lipids, which act as a barrier against biofilm formation by fungi. There is nearly 1.8% lipid content in the biofilm formed by yeast, fungi, and sulfate-reducing bacteria, as well as smaller amounts in mycobacteria. Surfactin, viscosin, and emulsan are lipids with surface-active EPS that can disperse and utilize hydrophobic substances. Lipidomics is the study to characterize the phospholipids profile and identify the associated molecular classes produced differently in biofilms and planktonic cells (Lattif et al., 2011).

4. Importance of lipidomics in biofilms

Lipids are a vital component of the extracellular polysaccharide (EPS), which protects fungal biofilms (Alim et al., 2018). Microbiological studies significantly influenced cell adhesion to surfaces and biofilm formation (Rowlett et al., 2017). The biofilm's resistance mechanism is relatively different from the traditional antibiotic resistance mechanism. The physiological characteristics of sessile cells in biofilm cells were recently studied due to their importance in industry, the environment, and human health. Genomic research was conducted earlier by screening mutants with biofilm defects (Folsom et al., 2010). Transcriptomics or proteomics method is used in biofilm research to find up- or downregulated genes or proteins of sessile cells of the microorganism (Junter & Jouenne, 2004; Zhang & Powers, 2012). A recent approach to the metabolic state in biofilm is metabolomic studies (Denich et al., 2003). Typically, it is reported that the inner membrane (IM) is a critical component for microbe survival when exposed to external stimuli of chemical or physical condition (O'Toole et al., 1999). Interestingly, few investigations on lipidomics of biofilm species have been published to date. The prevalence of the biofilm toward environmental conditions depends on the lipid content distribution, as it changes during the growth, which reflects its adaptation.

5. Lipid nomenclature

Lipids play a vital role in energy storage in addition to their membrane property and also in signaling pathways. Fatty acids are classified according to IUPAC's standard nomenclature and categorized based on three naming system:

➢ **Delta nomenclature**

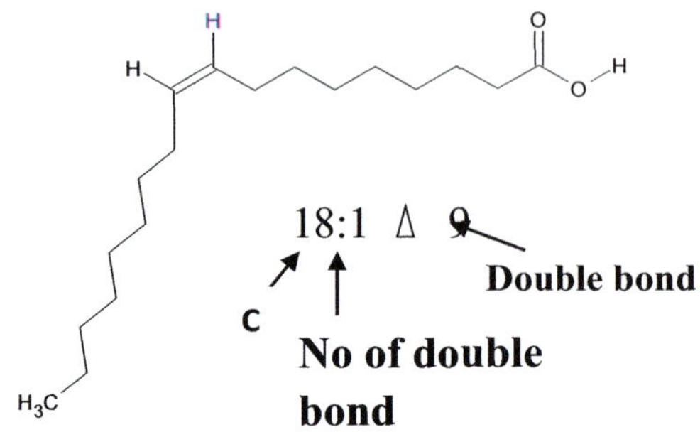

➢ **Omega nomenclature**

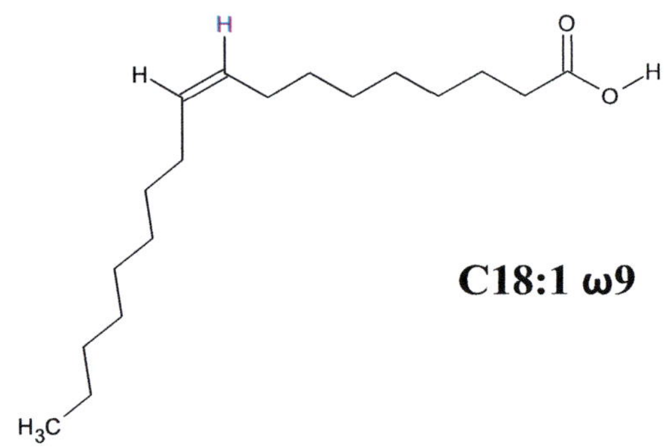

➢ **Common names**

Nomenclature	Systemic name	Common name
Saturated fatty acids		
C4:0	Butanoic	Butiric
C6:0	Hexadecanoic	Caproic
C8:0	Octanoic	Caprilic
C10:0	Decanoic	Capric
C12:0	Dodecanoic	Lauric

—cont'd

Nomenclature	Systemic name	Common name
C14:0	Tetradecanoic	Miristic
C16:0	Hexadecanoic	Palmitic
C18:0	Octadecanoic	Estearic
C20:0	Eicosanoic	Arachidic
C22:0	Docosanoic	Behenic
C24:0	Tetracosanoic	Lignoceric
Unsaturated fatty acid		
C16:1	9-Hexadecenoic	Palmitoleic
C18:1	9-Octadecenoic	Oleic
C18:1	11-Octodecenoic	Vaccenic
C18:2	9,12-Octadecadienoic	Linoleic
C18:3	9,12,15-Octadecatrienoic	Linolenic
C20:4	5,8,11,14-Eicosatetraenoic	Arachidonic

6. Extraction method

It is an ecologically friendly method of lipid extraction that eliminates the energy-intensive drying process. EPS plays a significant role in the biofilm's structural integrity; hence the extraction of EPS is carried by physical or chemical, or hydrolytic enzyme treatment, destabilizing the biofilm. Douglas et al. (2000) initiated the extraction methodology for biofilm matrix; since then, various methods have been analyzed for the isolation and analysis of intracellular components of the biofilm matrix. The matrix composition is ideal for biochemical and functional investigation (Fig. 14.2). Many biofilm-forming microalgae species have a thick cell wall, which prevents access to internal lipids; hence, it requires cell lysis technology to extract lipid. Cell lysis is usually conducted before lipid extraction; it is possible to combine cell lysis and extraction in one step (Carvalho et al., 2020). Limiting cell lysis is an essential aspect of a good EPS extraction process. Biofilm matrix cells are quantified before and after extraction to assess the cell lysis extraction procedure. Primarily, different intracellular enzymes are used to determine cell lysing capability.

Glucose-6-phosphate dehydrogenase activity is found to be a promising biomarker of intracellular component, when cell of biofilm is lysed, although it does not apply to some

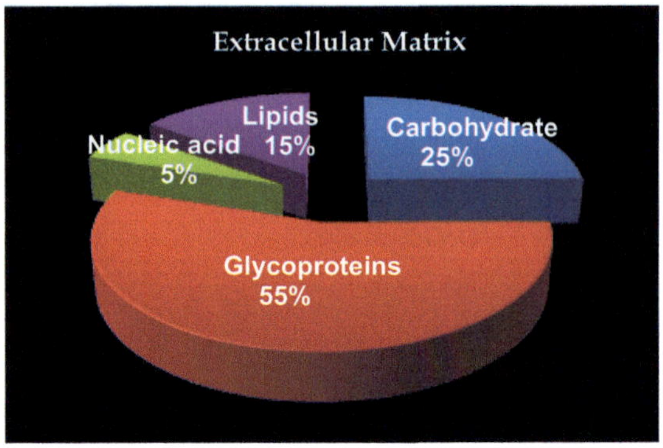

FIGURE 14.2 Extracellular matrix composition of biofilm.

EPS extraction procedures (Frolund et al., 1996; Ras et al., 2008). To interact with neutral lipid molecules, conventional lipid extraction method uses organic solvents (Grima et al., 2013). Lipid molecules are separated from the cell matrix and dissolved in organic solvents when neutral lipid interactions dominate; organic solvents are used based on their potential neutral lipid selectivity. The penetration of extraction solvents through the cell membrane allows them to interact preferentially with internal lipids. Integrating unit activities reduces the number of processing stages required for lipid extraction. Extraction solvents help prevent the degradation of target products and reduce contaminants in crude extracts by removing target lipid molecules selectively. To isolate microalgal lipids, several cell lysis approaches have been investigated using organic solvents, SC-CO_2, and ILs as extract ants, and various extraction methods were also studied, such as ultrasound, microwave, bead-beating, surfactant, and PEF-assisted extractions. When compared with other approaches, microwave-assisted extraction has shown promising results. Qv et al. (2014) found that optimized microwave-assisted extraction produced a 57.02% greater lipid yield than ultrasonic-assisted extraction from *Dunaliella tertiolecta* (45.94%). Wahidin et al. (2014) used a 1:2 methanol/hexane mixture as an extraction solvent in *Nannochloropsis* sp. using the microwave irradiation method. The lipid yield was about 38.31% compared with water bath–aided extraction, which yielded 23.01%.

7. Separation technique

The biofilm composition was explored by applying different experimental approaches such as vibration spectroscopy and infrared spectroscopy, which provides information on the chemical composition and quantities of polysaccharides, proteins, and other EPS content in the biofilm matrix. FT-IR spectroscopy is a simple technique for detecting and identifying organic compounds and studying microbial adhesion and biofilm formation (Nivens et al., 1993; Suci et al., 1998). About eight different types of lipids were discovered in the biofilm matrix using gas chromatography, with glycerolipids accounting for 99.5% of the total matrix lipids and sphingolipids accounting for only 0.5% (Andes et al., 2004). In small amounts, mass spectroscopy confirmed the presence of ergosterol and prostaglandin E2 (Zarnowski et al., 2014).

8. Mass spectrometry–based lipidomics analysis

Mass spectrometry (MS)–based analytical techniques are the most comprehensive quantification tool for lipid profiling. Lipidomics is the study of profiling and quantifying biogenic lipid molecules; it helps in the pathway interpretation, which also influences physiological significance (Watson, 2006; Wenk, 2005). Lipidomics has been boosted by advanced technique such as MS, NMR spectroscopy, and chromatography; MS has excellent qualitative and quantitative capabilities and thus has been the most widely used lipidomics method. Targeted and nontargeted lipidomics are two approaches that aim to identify and quantify all the lipids in the system. Nontargeted lipidomic methods have been a promising technique for analyzing all the lipids in the system, while targeted lipidomics profiling helps in solving specific and complex biological problems. Advanced approaches of targeted and nontargeted developed recently by MS innovation and data analysis method. The three predominant techniques of lipidomics are MS coupled with TLC, GC, and LC; direct infusion shotgun MS; and mass spectrometry imaging (MSI). Among the analytical methods, LC-MS is particular for lipidomic research, which provides excellent separation efficiency, high sensitivity, and strong specificity (Wang et al., 2017). GC-MS is the best technique for analyzing lipids, such as free fatty acids (FFAs) and steroids. Still, it involves a derivatization method to convert steroids and FFAs into volatile esters. The imaging mass spectrometry technique has examined lipid metabolites in tissues as potential disease biomarkers. Matrix-assisted laser desorption/ionization is primarily used in imaging MS to ionize lipid metabolites within samples (Wu et al., 2014). The application of MS coupled with chromatography technique would be the most promising technique for measuring small molecules of the vaginal biofilm and other clinical biofilm studies.

9. Application in dental plaque, *Candida albicans*, vaginal biofilm

Dental plaque is a biofilm with a specific structural organisation made up of different bacterial colonies and extracellular matrix. Dental biofilm, often known as plaque, has been studied extensively in dental health and disease. The dental plaque contains TAGs, cholesterol, CEs, DAGs, and phospholipid. The lipidomic profiling approach in dental plaque would tabulate biomarkers, which helps track the biofilm phenotype. The biomarkers are the indicators of the diseased state of microorganism colonies, which converts microbial homeostasis into dysbiosis, i.e., change in biofilm composition (Drotleff et al., 2020). Compared with chronic periodontitis with type II diabetes, lipid profile analysis revealed significant effects on TC, VLDLc, and TGL among chronic periodontitis patients with type II diabetes. However, HDLc and LDLc were not significantly altered (Thomas et al., 2017).

Phospholipid and sphingolipid levels were higher in biofilms. Lipid profile changes may influence the antifungal resistance in biofilm of planktonic Candida cells. Cell shape changes can lead to surface adherence problems, reducing *Candida albicans'* ability to produce biofilms. Further studies on phospholipid composition or ECM are required, which influence the structure and function of cell wall/membrane and affect cell shape. *C. albicans* could provide additional insight into this problem by identifying and systematically deleting P4-type lipid translocases and their regulators. Membrane lipid composition investigation on biofilm of mutat strain aids in analyzing changes in lipid, which subsequently affect cell shape and biofilm formation (Alim et al., 2018). Integrating environmental signals and phospholipid detection regulates biofilm formation, which helps develop an antifungal drug. There is a lack in finding and targeting lipid-dependent cellular pathway, which is essential for biofilm formation and essential biofilm-specific antifungal, which would have clinical significance.

Bacterial vaginosis (BV) is characterized by diverse communities of biofilm-producing anaerobes that are produced by modifications in metabolomic composition, such as lipids, polyamines, carbohydrates, amino acids, and organic acids. Arachidonate is less prevalent, and 12-hydroxyeicosatetraenoic acid (12-HETE) is more abundant in BV, indicating a change in lipid composition. Similarly, a precursor to carnitine, deoxycarnitine, was elevated in BV, while carnitine was lower. Likewise, acyl-carnitines such as acetylcarnitine, propionyl carnitine, and butyrylcarnitine decreased in BV. The levels of 4-hydroxybutyrate and 13- hydroxyoctadecadienoic acid (13-HODE) were higher in BV. Consequently, the levels of an enzyme involved in glycerol metabolisms, such as glycerol and glycerol-3-phosphate, were lower (Srinivasan, 2015).

References

Alim, D., Sircaik, S., & Panwar, S. L. (2018). The significance of lipids to biofilm formation in *Candida albicans*: An emerging perspective. *Journal of Fungi, 4*(4), 140.

Andes, D., Nett, J., Oschel, P., Albrecht, R., Marchillo, K., & Pitula, A. (2004). Development and characterization of an in vivo central venous catheter *Candida albicans* biofilm model. *Infection and Immunity, 72*(10), 6023–6031.

Baillie, G. S., & Douglas, L. J. (2000). Matrix polymers of Candida biofilms and their possible role in biofilm resistance to antifungal agents. *Journal of Antimicrobial Chemotherapy, 46*(3), 397–403.

de Carvalho, J. C., Magalhaes, A. I., Jr., de Melo Pereira, G. V., Medeiros, A. B. P., Sydney, E. B., Rodrigues, C., & Soccol, C. R. (2020). Microalgal biomass pre-treatment for integrated processing into biofuels, food, and feed. *Bioresource Technology, 300*, 122719.

Del Pozo, J. L. (2018). Biofilm-related disease. *Expert Review of Anti-infective Therapy, 16*(1), 51–65.

Denich, T. J., Beaudette, L. A., Lee, H., & Trevors, J. T. (2003). Effect of selected environmental and physico-chemical factors on bacterial cytoplasmic membranes. *Journal of Microbiological Methods, 52*(2), 149–182.

Drotleff, B., Roth, S. R., Henkel, K., Calderon, C., Schlotterneek, J., Neukamm, M. A., & Lammerhofer, M. (2020). Lipidomic profiling of non-mineralized dental plaque and biofilm by untargeted UHPLC-QTOF-MS/MS and SWATH acquistion. *Analytical and Bioanalytical Chemistry, 412*(10), 2303–2314. https://doi.org/10.1007/s00216-019-02364-2

Flemming, H., & Wingender, J. (2010). The biofilm matrix. *Nature Reviews Microbiology, 8*, 623–633.

Folsom, J. P., Richards, L., Pitts, B., Roe, F., Ehrlich, G. D., Parker, A., & Stewart, P. S. (2010). Physiology of *Pseudomonas aeruginosa* in biofilms as revealed by transcriptome analysis. *BMC Microbiology, 10*(1), 1–17.

Frølund, B., Palmgren, R., Keiding, K., & Nielsen, P. H. (1996). Extraction of extracellular polymers from activated sludge using a cation exchange resin. *Water Research, 30*(8), 1749–1758.

Grima, E. M., González, M. J. I., & Giménez, A. G. (2013). Solvent extraction for microalgae lipids. In *Algae for biofuels and energy* (pp. 187–205). Dordrecht: Springer.

Junter, G. A., & Jouenne, T. (2004). Immobilized viable microbial cells: From the process to the proteomeor the cart before the horse. *Biotechnology Advances, 22*(8), 633–658.

Lattif, A. A., Mukherjee, P. K., Chandra, J., Roth, M. R., Welti, R., Rouabhia, M., & Ghannoum, M. A. (2011). Lipidomics of *Candida albicans* biofilms reveals phase-dependent production of phospholipid molecular classes and role for lipid rafts in biofilm formation. *Microbiology, 157*(Pt 11), 3232.

Monds, R. D., & O'Toole, G. A. (2009). The developmental model of microbial biofilms: Ten years of a paradigm up for review. *Trends in Microbiology, 17*(2), 73–87.

Nivens, D. E., Chambers, J. Q., Anderson, T. R., Tunlid, A., Smit, J., & White, D. C. (1993). Monitoring microbial adhesion and biofilm formation by attenuated total reflection/Fourier transform infrared spectroscopy. *Journal of Microbiological Methods, 17*(3), 199–213.

O'Toole, G., Kaplan, H. B., & Kolter, R. (2000). Biofilm formation as microbial development. *Annual Review of Microbiology, 54*, 49.

O'Toole, G. A., Pratt, L. A., Watnick, P. I., Newman, D. K., Weaver, V. B., & Kolter, R. (1999). Genetic approaches to study of biofilms. *Methods in Enzymology, 310*, 91–109.

Qv, X. Y., Zhou, Q. F., & Jiang, J. G. (2014). Ultrasound-enhanced and microwave-assisted extraction of lipid from *Dunaliella tertiolecta* and fatty acid profile analysis. *Journal of Separation Science, 37*(20), 2991–2999.

Ras, M., Girbal-Neuhauser, E., Paul, E., Spérandio, M., & Lefebvre, D. (2008). Protein extraction from activated sludge: An analytical approach. *Water Research, 42*(8–9), 1867–1878.

Rowlett, V. W., Mallampalli, V. K., Karlstaedt, A., Dowhan, W., Taegtmeyer, H., Margolin, W., & Vitrac, H. (2017). Impact of membrane phospholipid alterations in *Escherichia coli* on cellular function and bacterial stress adaptation. *Journal of Bacteriology, 199*(13), 16–008499.

Srinivasan, S., Morgan, M. T., Fiedler, T. L., Djukovic, D., Hoffman, N. G., Raftery, D., & Fredricks, D. N. (2015). Metabolic signatures of bacterial vaginosis. *American Society for Microbiology, 6*(2), e00204–e00215.

Stewart, P. S., & Costerton, J. W. (2001). Antibiotic resistance of bacteria in biofilms. *The Lancet, 358*(9276), 135–138.

Suci, P. A., Vrany, J. D., & Mittelman, M. W. (1998). Investigation of interactions between antimicrobial agents and bacterial biofilms using attenuated total reflection Fourier transform infrared spectroscopy. *Biomaterials, 19*(4–5), 327–339.

Thomas, B., Prasad, R. B., Shetty, S., & Vishakh, R. (2017). Comparative evaluation of the lipid profile in the serum of patients with type II diabetes mellitus and healthy individuals with periodontitis. *Contemporary Clinical Dentistry, 8*(1), 96.

Tolker-Nielsen, T. (2015). Biofilm development. *Microbial Spectrum, 3*. MB-0001-2014.

Wahidin, S., Idris, A., & Shaleh, S. R. M. (2014). Rapid biodiesel production using wet microalgae via microwave irradiation. *Energy Conversion and Management, 84*, 227–233.

Wang, M., Wang, C., & Han, X. (2017). Selection of internal standards for accurate quantification of complex lipid species in biological extracts by electrospray ionization mass spectrometry What, how and why? *Mass Spectrometry Reviews, 36*(6), 693–714.

Watson, A. D. (2006). Thematic review series: Systems biology approaches to metabolic and cardiovascular disorders. Lipidomics: A global approach to lipid analysis in biological systems. *Journal of Lipid Research, 47*(10), 2101–2111.

Wenk, M. R. (2005). The emerging field of lipidomics. *Nature Reviews Drug Discovery, 4*(7), 594–610.

Wu, Z., Shon, J. C., & Liu, K. H. (2014). Mass spectrometry-based lipidomics and its application to biomedical research. *Journal of Lifestyle Medicine, 4*(1), 17.

Zarnowski, R., Westler, W. M., Lacmbouh, G. A., Marita, J. M., Bothe, J. R., Bernhardt, J., & Andes, D. R. (2014). Novel entries in a fungal biofilm matrix encyclopedia. *American Society for Microbiology, 5*(4), 14–013333.

Zhang, B., & Powers, R. (2012). Analysis of bacterial biofilms using NMR-based metabolomics. *Future Medicinal Chemistry, 4*(10), 1273–1306.

CHAPTER

15

Functional metabolomics approaches in determining the inhibition of biofilm

Bandita Dutta[1], *Dibyajit Lahiri*[2], *Moupriya Nag*[2] *and Rina Rani Ray*[1]

[1]Department of Biotechnology, Maulana Abul Kalam Azad University of Technology, Haringhata, West Bengal, India; [2]Department of Biotechnology, University of Engineering and Management, Kolkata, West Bengal, India

1. Introduction

In many habitats, microbial populations frequently take the form of organized and structured aggregates called bacterial biofilms at different interfaces such as solid–air, solid–liquid, liquid–air, and liquid–liquid. Their discovery of the mucosal edges of many organs, medical equipment, and wounds sparked studies into these bacterial survival tactics. The existence of biofilms can be linked to over 80% of human illnesses. They can be made up of a single microbial cell or a coculture (10%–15% of the total volume), embedded in a highly hydrated, self-produced exopolymeric matrix with microbial biopolymers (polysaccharides, proteins and glycoproteins, nucleic acids, and lipids) serving as essential building blocks. Depending on the genetic makeup of the microorganisms involved, polysaccharides, typically synthesized as virulence factors and structural components of the bacterial cell wall, might be released into the media (extracellular polymeric substance, EPS) (Carradori et al., 2020). The control of biofilm infections is particularly challenging since biofilms frequently show significant intrinsic resistance to antimicrobial medications. If the pathogenic microbe is resistant to numerous treatments, this difficulty may increase (Shen et al., 2020).

Biofilm formation is a native propensity of bacteria; for example, biofilm is an etymology of fouling in the food industry sectors of fresh harvesting, poultry, dairy farm, and processing of red meat. Food contamination is frequently caused by biofilm, which is frequently the problem. Numerous pieces of data suggest that the biofilm network of life precedes to an

Microbial Biofilms
https://doi.org/10.1016/B978-0-323-95715-1.00004-2

increase in antimicrobial agent resistance. Biofilms are more resistant to antibacterial agents than planktonic cells are. It is crucial to find new methods to prevent biofilm (Lou et al., 2015). Studies have concluded that biofilms are self-fecund with structurally assembled communities that can form under different stressful habitats by diverse microorganisms and adhere to surfaces or liquid—air interfaces. Biofilms cause outbreaks of difficult issues in a variety of industries; for example, they can reduce agricultural output, compromise food safety, contaminate industrial processes, and even so result in antibiotic resistance in the biomedical sciences. The mechanism of biofilm formation is still not fully known, despite the scientific communities considerable effort over the. It has taken the past few decades to comprehend the biochemistry of biofilm formation (Rui & Xilin, n.d.). However, the methods that are now in place to address these issues and eliminate biofilms in various domains are insufficient. A new window opened by metabolism enables us to examine the distinction mechanism to avoid and alter biofilm formation in the materialistic world and to surpass the recognition of biofilm formation.

Metabolomics is a powerful tool for analyzing the dynamic alternations that all tiny molecular metabolites in biological apparatus undergo in response to external impulses. Examining the changes in the metabolome following the administration of medications or plant integrant can yield intricate data for the entire research system. Statistical analysis assists to identify their internal relationships, identify the important biomarkers, and determine the actual raw materials used to make plant components. As a result, metabolomics can be used to thoroughly as well as precisely select active substances, eliminating the difficult processes of activity evaluation and separation and purification (Lou et al., 2015).

2. Formation of biofilm

Due to changes in availability of nutrient and varying cell populations, biofilms are seen as aggregation of bacterial cells enclosed inside various microenvironments. The bacteria that make up the biofilm core are physiologically different from planktonic bacteria because they are in a static or latent developmental phase. As a result, antibiotics that target biological pathways linked to bacterial growth are less effective against biofilms (Zhang & Powers, 2012). Biofilms have an asymmetrical spatial organization that is influenced by a wide range of ecological, biological, chemical, and physical phenomena. Numerous computer simulations have been used to thoroughly study how these elements affect the production of biofilms. Bacterial cells change their morphologies in response to the presence of a surface layer during the intricate adhesion process. As single-species and multispecies microcolonies arise in the early phases of biofilm genesis, sessile bacteria perceive themselves in a substantial juxtaposition accompanied by the cells of the identical species and with those of other species (Costerton et al., n.d.). Each biofilm bacterium's microenvironment is affected by these cellular contiguity and the brisk exopolysaccharide matrix synthesis in the growing biofilm. A structurally complex mature biofilm progressively emerges as distinct biofilm bacteria adapt to their unique microenvironmental conditions with different growth paradigm.

Some general principles revolve around the formation of biofilm, which metabolically allows the active bacteria to adhere to the surface. There is an adequate amount of nutrient supplement required for cellular replication and producing extracellular polymeric substances (EPS) for the surface to be encrusted with organic substances to instigate biofilm formation. *Escherichia coli* in the urinary system and *Enterococcus* spp. in the urinary tract, *Pseudomonas aeruginosa* in the lung, *Staphylococcus aureus* in arteries and *Staphylococcus epidermidis* in the heart, *Vibrio cholera* in the gastrointestinal (GI) tract, and fungus-like *Candida* spp. in the GI tract are a few examples of prevalent bacterial biofilm infections (Zhang & Powers, 2012). A crucial initial step in being apt to stop these bacterial biofilm-mediated infections distinguishes several metabolic pathways essential for development of the biofilm. The underlying mechanisms that cause the transition of planktonic cells to a sessile cell—associated biofilm are still unknown, despite the common grasp of the fundamental structure and growth of bacterial biofilms. Accordingly, the transition from planktonic to sessile biofilm is a difficult and tightly controlled approach that brings about phenotypic alterations (Zhang & Powers, 2012).

Due to extensive phenotypic alterations, controlled by a sigma factor related to the particular facet that controls sporulation, starvation, and survival along with rough—smooth phase fluctuations, biofilm cells drastically divaricate from planktonic cells of identical species. In the abundance of proteins resolved by modern gel chromatography, preliminary evidence propounds that the cell envelope subsets of biofilm and planktonic cells of the same species divaricate profoundly. It thus implies that the gene cassette, i.e., controlled by the sigma factor is triggered by adherence including several facet that affects cell wall permeability. The fundamental structural and functional component of the microbial biofilm is the bacterial microcolony (Costerton, 1995). Microcolonies can consist of cells from one species or multiple species, but they can be distinguished by their EPS matrix, which keeps the cells in stable juxtaposition and controls how effectively they interact with the fluid phase.

3. What is metabolomics?

The postgenome era's emerging field of metabolomics is a part of systems biology, which also includes genomics, transcriptomics, and proteomics. The field of study known as metabolomics studies the quantitative relationships between the metabolite composition of coherent living systems and their dynamic reactions to commute in both visceral (such as physiology and development) and extrinsic (such as environmental) elements (including environmental factors and xenobiotics). The total set of all metabolites found in a biological cell compartment, cell, tissue, organ, or organism and analyzed as a cellular extract or biofluid is known as the metabolome. The average molecular weight threshold that distinguishes macromolecules from metabolites is 1 kDa (Zhang & Powers, 2012).

Urine, plasma, and serum are the basic biological samples used in metabolomics, which focuses on an array of primarily low-molecular-weight metabolites (MW $<$ 1 kDa) that act as the substrates as well as metabolic pathways analogous end product. The collection followed by pretreatment of samples, the gathering and processing of the collection of data, and the interpretation of metabolic variance all contribute to the integrity of metabonomic processes (Lao et al., 2009). In general, nuclear magnetic resonance (NMR)—based methods, chromatography-based methods, and mass spectroscopy (MS)—based methods are the key technologies metabolomics relies upon. Based on a series of analyses of various typical spectra combined with chemical pattern recognition techniques, metabolomics can be exploited to identify pathophysiological states, gene functions, drug efficacy, and toxicity in the organisms.

A thorough investigation of metabolites has revealed their enormous dynamics and complexity, which aids in the improvement and development of various types of analytical techniques for versatile analysis. At present, numerous such analytical strategies are being used that are broadly categorized in chromatography, MS, and NMR based for significant detection techniques. Several analytical methods are employed in TCM contemporary research, including LC for quality assurance of TCM, highly efficient TCM fingerprint along with qualitative and quantitative estimation of active modules in TCM, NMR for biological dactylogram and elaboration of complex TCM science hypothesis, GC for the inception of experimental animal models, CE for studying pharmacokinetic of TCM, and MS (generally combined with other techniques, such as GC and LC) for analysis of TCM (Lao et al., 2009).

4. Metabolic remodulation triggers biofilm formation

Metabolomics is considered as omics technology, which is system biology directed. This is mainly used to design the global profile of all the small molecules associated with metabolites. The change of intensity of each molecules is able to capture different biological events inside the cells (Kendall & Sperandio, 2014).

Several metabolisms are critically involved in biofilm formation, and remodulation of this metabolomic program revealed the difference between biofilm formation of planktonic cells and sessile cells (Hobley et al., 2014). Biofilm formation of planktonic cells is significantly different from sessile cells, which was evidenced by PLS-DA analysis of metabolomic data using LC-MS/MS system. Analysis revealed the metabolic divergence and deviation of extracellular polymeric substance (EPS) secretion during biofilm formation in case of planktonic cells and sessile cells (Sharma et al., 2019).

Metabolic remodulation is one of the prime aspects in biofilm formation. This remodulation intricates the association of multilayer molecules in formation of biofilms. With the combination of open-source and local databases, some significant metabolites and analogous metabolic pathways have revealed the biofilm formation by sessile group of bacterial cells. Around 38 distinctive metabolites, which include carbohydrates, amino acids and organic acids, uridine, and polyamine, are present upon the biofilm crust. Determination of glycerol-derived metabolites, through the significant reduction of it, especially 2-phosphoglyceric acid, glycerol 3-phosphate, and D-glyceraldehyde 3-phosphate, was found as phenomenal in upregulation of biofilm formation. Apart from glycerol metabolites, the aggregations of carbohydrates get increased with the significant decrease of D-maltose and D-glucose. Most of the metabolism including glycolipid, amino acids, and carbohydrate get reprogrammed in biofilm formation with the upregulation of the differential metabolites. According to the study, carbohydrates and proteins appeared to be the most predominant compounds of biofilm formation. Combined D-amino acids are responsible for *Bacillus subtilis*–mediated biofilm dispersion followed by YqxM protein–arbitrated disintegration of the biofilm. The attachment and disattachment of the amyloid fibers to the host cells get synchronized by the YqxM protein (Kolodkin-Gal et al., 2010). The level of L-amino acids compared with the D-amino acids gets changed distinctly, which stipulate the involvement of the L-amino acids in UTI8911-induced biofilm formation and D-amino acids in providing nutrition for bacterial growth (Fernández & Zúñiga, 2006).

Metabolic reprogramming of the amino acids substantiates the of biofilm formation. Autoinducer peptide AI-2, which is the quorum sensing (QS) molecule for the gram-positive bacteria, is being processed by lsr protein. High accumulation of glycerol-3-phosphate (G-3-P) leads to the inhibition of the lsr protein followed by arrest of biofilm formation (Willias et al., 2014). Consumption of glycerol for biosynthesis of G 3-P regulates the AI-2 expression for biofilm formation and keeps away from dispersion of biofilm. The adhesiveness of the biofilm and EPS formation get increased through the assimilation of carbohydrates into FimH (mannose) adhesion (Schwartz et al., 2013). The microbes get the required nutrient and energy through the carbohydrate and glycolipid metabolism by the principle precursor of glyceraldehyde-3-phosphate, i.e., G-3-P (Fig. 15.1). Bacterial cells basically utilize amino acids, organic acids, lipids, sugars, and uridines through the amino acid, carbohydrate, and glycolipid metabolism for the EPS production (Larsen & Engelsen, 2015). This enables the formation of biofilm network, which provides an anaerobic microenvironment to the bacterial cells, which are embedded in the epidermis of the biofilm for their secure survival.

5. Inhibition of biofilm

The formation of biofilm is a multifactorial infection process that can also influence the physicochemical properties of the substrate dopant. Especially, a dental material's wettability affects various processes, including exposure with intraoral fluids (mainly saliva), the adsorption of salivary proteins, and the adherence of bacterial biofilms with frictional forces caused by rubbing against oral tissues or food particles. They get in touch with the orthodontic device (Okada et al., 2008). Pathogens that possess the QS control ability to form biofilm are a great solution for the treatment of several infections. With the aid of tiny secreted signaling molecules called autoinducers, QS is defined as the capacity of bacterial cells to detect and respond quickly to changes in cell population density (AI). The creation of biosurfactants, exopolysaccharide (EPS) synthesis, surging, and swarming motility, among other QS-based variables, have all been linked to the development of biofilms (Packiavathy et al., 2014). Biosurfactants have tensioactive qualities that can improve swarming movement and consequently affect the growth of bacterial pathogens' biofilms. EPS serves as a protective barrier by limiting the entrance of chemotherapeutic chemicals and is crucial for the formation of the biofilm architecture. Because the bacteria that populate the biofilm are immune to phagocytes, antibodies, and antibiotics, the human defense system and antimicrobial chemotherapy are unable to eradicate the pathogens that live there.

Long utilized to treat microbial illnesses, chemicals originating from plants are now attracting recognition as an origin of anti-QS agents for the treatment of biofilms. Due to their nontoxicity, dietary phytochemicals with a prolonged history of medical usage in humans are

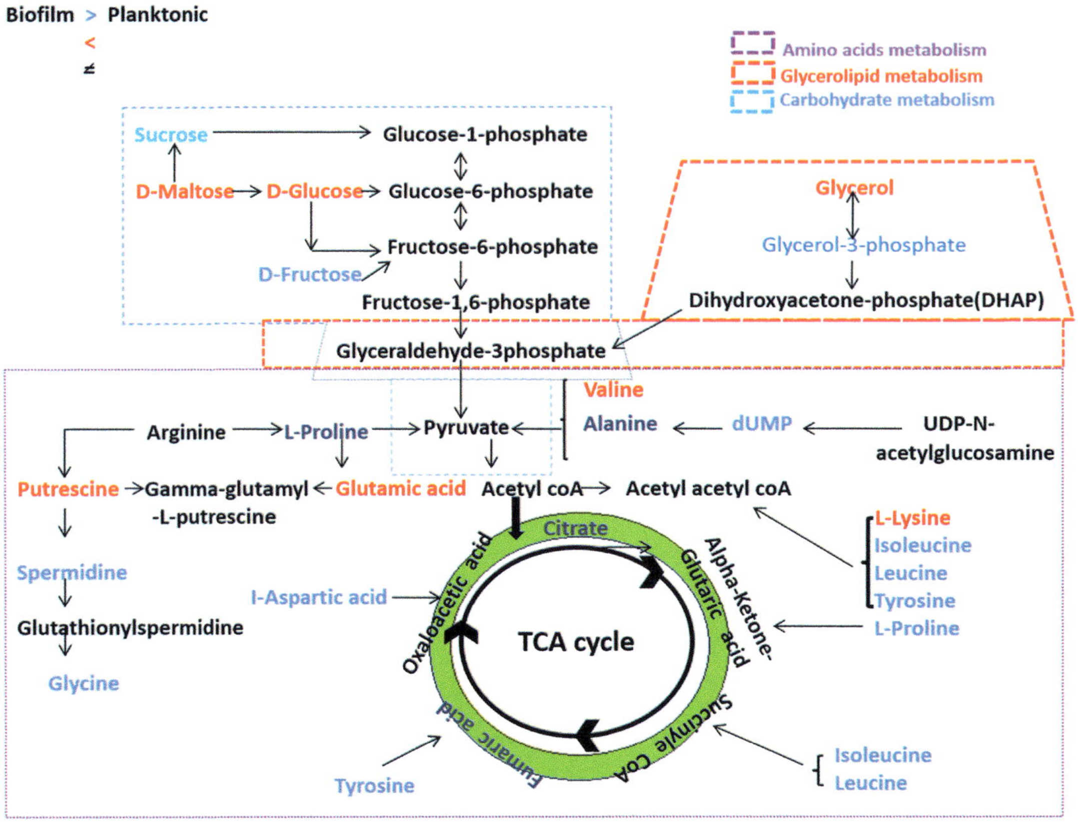

FIGURE 15.1 The substantially affected metabolic pathways due to the metabolic remodulation. Amino acid metabolism, glycerolipid metabolism, and TCA cycle are major metabolic functions of biofilm formation that gets affected significantly.

currently being researched with the goal of reducing biofilms. In this study, curcumin, a key component of turmeric rhizomes, was chosen as one such dietary phytochemical. Although data on turmeric's anti-QS qualities are limited, it has long been utilized as an antiinflammatory, antimicrobial, and antifungal agent (Packiavathy et al., 2014).

We need a prospective antibiofilm medicine that either promotes the dispersal of existing biofilms or prevents the development of new biofilms in vivo. Many possible antibiofilm agents with distinctive structures that were derived from natural products have been produced and have demonstrated efficacy in dispersing biofilms or inhibiting the formation of biofilms by bacteria in vitro. The majority of the recently created antibiofilm compounds do not directly affect bacterial life in comparison with conventional antibiotics, so it is anticipated that resistance to these molecules will not happen easily. It is envisaged that some of these lead compounds may be developed into antibiofilm medications in the upcoming years. Passivating the surfaces of medical implants with chemicals that inhibit biofilm formation is a supplementary method for preventing biofilm formation (Rabin et al., 2015).

6. Prevention of biofilm formation through targeting quorum sensing mechanism

QS is an ability of the bacterial cells to detect cell population solidity by small molecules—mediated gene regulation. Prevention of QS is found to be effective in biofilm inhibition. It was proclaimed that AHL-lactonase extracted from endophytic *Enterobacter* (cell free supernatant) is able to inhibit *Aeromonas hydrophila* biofilm formation by degrading N-AHL (Shastry et al., 2019). *Lactobacillus crustorum* ZHG 2−1 is reported as pioneered quorum quenching (QQ) bacteria due to its ability of degrading *N*-butyryl-DL-homoserine lactone (C4-HSL) and *N*-3-oxododecanoyl-DL-homoserine lactone (3-oxo-C12-HSL) of *P. aeruginosa* (Cui et al., 2020). Various QQ molecules were isolated from different sources, which have been reported recently. Ethyl acetate extract of *Natrinemaversi forme* cell free supernatant has antibiofilm effect against *P. aeruginosa* (Basaran et al., 2020). The QS network is mainly disrupted through inhibition followed by degradation of signaling molecules and imitation of signaling molecules to inhibit the binding into the receptor molecule using antibiofilm agents (Kalia, 2013). Quorum quencher molecules are mainly species specific, which enables elimination of mixed species biofilm through their combined actions. A sulfur-rich molecule present in garlic, i.e., ajoene is able to decrease the articulation of the small regulatory RNAs (sRNAs) in both the gram-positive and gram-negative bacteria. This molecule usually drops the RNAIII gene expression in case of *S. aureus* and RsmY and RsmZ gene expressions followed by repression of Pel and Psl, i.e., biofilm matrix polysaccharides and T6SS type VI secretion system in case of *P. aeruginosa* (Jakobsen et al., 2017; Scoffone et al., 2019). T6SS is an important system in *P. aeruginosa*, which controls formation of biofilm, expression of several virulence factors, pyocyanin production, and degree of pathogenicity (Li et al., 2020). A phytocompound, ajoene, is able to reduce regulatory RNA and RNAIII expression along with inhibition of RNAIII-dependent virulence factors including alpha hemolysinin, lipase, and protease on *S. aureus*. *Polygonum cuspidatum* (Asian knotweed) and *Rheum palmatum* (Chinese rhubarb) are the source of anthraquinone derivative is emodin (1, 2, 8-trihydroxy-6-methyl anthraquinone). It is able to downregulate genes of gram-positive bacteria, i.e., agrA, icaA, and sarA genes in case of *S. aureus* and *luxS* gene in case of *Streptococcus suis* (Yan et al., 2017; Yang et al., 2015).

Human cathelicidin LI-37 is an antibiofilm peptide that inhibits *P. aeruginosa* biofilm formation though arresting cell signaling system and downregulating QS cascade (Di Somma et al., 2020). Usually, AMPs interact with the bacterial membranes to activate QS genes. Membrane vesicles help the QS autoinducers to get entered through plasma membrane and process the activation of QS-related virulence gene activation. Interspecies signal transduction is mediated by autoinducers. One of the example of this is AIP from *Lactobacilli*, i.e., small autoinducing peptide that suppresses microbial viability and suppresses bacterotoxin production though restraining agr QS system (Vasilchenko & Rogozhin, 2019).

It has been widely used in biological sample analysis and TCM modern research because it is affordable, simple to use, extremely sensitive, not constrained by sample volatility and stability, and has favorable separating power. These analyses, however, almost always target component analyses of the entire samples combined with chemometrics rather than fingerprint analyses. We must create general methodologies based on HPLC in an effort to perform high-throughput fingerprint analysis to boost analytical effectiveness (Yang et al., 2002). For TCM fingerprint analysis and other areas of contemporary TCM research, improved HPLCs have recently emerged one after the other as technological advancements. Examples include capillary HPLC (CHPLC) for qualitative and quantitative measurement of functional components in TCM, hydrophilic interaction chromatography for thorough quality inspection of TCM, ultrahigh-performance liquid chromatography (UPLC) for recognition of active compounds in TCM, and monolithic HPLC (Jin et al., 2009).

NMR has long been regarded as one of the primary analytical methods for studying metabolism. The NMR-based approach has advantages over MS and HPLC. It is possible to conduct in vivo and in situ studies because of the following advantages: (1) noninvasive and nondestructive to samples; (2) quantitative and simultaneous detection methods; (3) high throughput (flow injection technology can measure up to 400 samples per 24 h); (4) profuse in effectual molecular information, such as metabolite structure, its concentration, molecular aggressiveness, interactions, pH, and compartmentation when diffusion rephrase techniques are used; (5) requirement of small to no sample preparation; (6) satisfactory resolution; and (6) reproduciblity (Lao et al., 2009).

In particular, for current TCM research, the NMR-based approach is ideal for the investigation of complicated components in metabolites. To forecast the antiplasmodial activity of TCM, analysis of the extracts of *Artemisia annua* uses 1H NMR spectroscopy and chemometrics, such as partial least squares discriminant analysis (PLS-DA) (Bailey et al., 2004). Without being aware of this activity beforehand, our method allowed the classification of TCM samples from various origin according to their antiplasmodial activity. Recent efforts to produce drugs using an NMR-based approach have been quite successful.

The main benefit of GC is its great vulnerability of detection for virtually all volatile or nonvolatile chemical compounds, which can be easily derivatized. Considering volatile chemicals in TCM and biological samples such as urine, blood plasma, blood serum, etc. has been one of the major uses of GC (Lao et al., 2009). For instance, TCM scientists used a high amount of hydrocortisone to create an animal model with "kidney deficiency syndrome," a precursor to obesity and diabetes. They used GC-MS and multivariate statistical methods as part of a metabonomic approach to determine the urine metabolic patterns of the treated rats (Chen et al., 2005). The findings suggest that chemometric analysis might be used to spot distinct and consistent biochemical alterations brought on by hydrocortisone administration in well-monitored settings. This work demonstrates the potential utility of the metabonomic approach as a dominant tool for initial diagnosis to examine the biochemical alterations of several physiopathologic disorders such as metabolic syndrome.

7. Metabolomics in drug targeting

The important functions that small compounds can play in bacterial biofilms are best shown by autoinducers, whose methods of action might be mimicked or inhibited as a potential drug development target. Similar functions may be played by some additional inter- or intracellular small-molecular-weight compounds in the development, growth, and persistence of bacterial biofilms. Biofilms are specifically a geographical dispersion of diverse cell types that exist in a variety of metabolic states to maximize survival. Therefore, a thorough characterization of the many metabolic states within an intricate biological community is necessary to comprehend biofilms. Likewise, metabolomics offers a methodical explanation for this complicated system (Zhang & Powers, 2012). The idea that metabolism is an assembly of functional modules connected by shared enzymes and substrates that collectively synchronize cell's biochemical processes is growing in popularity. Many pathways share common enzymes. By treating metabolism, it is possible to apply advanced graph theoretic analysis to find the network's core organizational principles (Peregrín-Alvarez et al., 2009).

An initiative to create and approve 10 new antibiotics by the year 2020 has been put up by the Infectious Diseases Society (IDSA) (Infectious Diseases Society of America, 2010). However, it is possible that current drug discovery techniques cannot handle these difficulties. Target-based high-throughput screening (HTS) of vast chemical libraries, followed by lead optimization, is a key component of drug development efforts. Unfortunately, this strategy has been shown to have a very high failure rate and provide false leads. Even if a genuine HTS hit is discovered, it is unknown whether this chemical lead will enter the bacterial cell and exhibit in vivo activity. NMR metabolomics is becoming an important part of the drug discovery process and provides a low-cost way to help researchers get beyond the many obstacles they face (Halouska et al., 2012).

Proteomics and transcriptomics can both benefit from using metabolomics as an additional technique for finding disease biomarkers. Visualizing the biological state of an organism is made feasible by the connection between metabolomics, mRNA, and protein expression. The recent development of metabolome analysis has also made it possible for NMR to contribute to the stage of clinical validation. This stage of the drug development process is by far the most difficult and expensive, and it is also the stage where the majority of failures take place. Using biofluids, tissues, and cell extracts as a starting point, NMR can assess metabolome alterations brought on by the biological activity of a pharmacological lead (Powers, 2009). The metabolome is made up of tons to thousands of small-molecular-weight molecules, and the correlative concentration and flux of these chemicals reflect the state of the system. Human health would greatly benefit from the use of NMR to build precise and noninvasive techniques for early disease diagnosis by finding biomarkers.

8. Metabolomics based on nuclear magnetic resonance

The application of NMR-based metabolomics has been observed in the identification of cardiac arrest, liver disease, respiratory disease, cancer, along with CNS disorders. NMR metabolomics can help us compare disease and healthy states, as well as treated and untreated drugs. The metabolomic transition from an ailment to a healthy state or by simple observation of compound changes allows drug discovery or identification of chemical lead (Zhang & Powers, 2012). The changes in metabolome can be measured by NMR that results from the biological activity of drug leads. The metabolome is made up of hundreds to thousands of small-molecular-weight molecules, and the corresponding concentration and flux of these chemicals reflect the state of the system.

NMR-based metabolomics has advantage over the proficiency and lucidity of the methodology. Generally, biofluids or cell lysates can be simply added to a deuterated aqueous buffer for the maintenance of the pH and to provide a lock signal for the transfer of the NMR sample tube to a one-dimensional 1H NMR spectrum. The study of tiny molecules discovered in living cells or biofluids is the foundation of the relatively new subject of metabolomics. The identity and concentration of metabolites serve as biochemical indicators for monitoring the physiological effects of antibiotic effectiveness, selectivity, and toxicity since small molecules are downstream products of biomolecular activities. The identification of these biochemical fingerprints requires the worldwide determination of a large number of endogenous small molecules, followed by multivariate analysis for pattern recognition. With little sample handling and highly reproducible results, such complete biochemical knowledge can be easily obtained using 1H NMR spectroscopy (Halouska et al., 2012).

The rapid approach to the identification of global trends and relationships is provided by PC analysis of NMR metabolomics. A detailed study of the identity and concentration flow of metabolites makes possible the identification of illness biomarkers along with the location of impacted metabolic pathways. On account of the complexity of the metabolome and the absence of reference NMR spectra, this is a really difficult task. The metabolome varies between species, is not fully defined, and may hold an unlimited number of molecules (Kell, 2004).

9. Comparison of NMR-and MS-based metabolomics

NMR and MS offer complementary methods for the examination of metabolomic data along with mass spectroscopy (MS) and have historically been employed to find metabolomic disturbances. The comparatively high sensitivity of MS and its capacity to track concentration alteration for minute components that are generally undetectable by NMR are advantages. Because metabolites tend to have modest molecular weight distributions, MS frequently necessitates a mixed method. To separate substances with identical molecular weights (MWs), gas or liquid chromatography may be used. This may cause the relative concentration of metabolites to be removed or altered. Additionally, MS can only detect metabolites that can ionize effectively (Wilson et al., 2005).

In comparison with MS, NMR spectroscopy produces measurements with comparatively modest sensitivity, with detection limits on the order of 10 M or a few nmol at high fields when employing modern cryoprobes. Nevertheless, because NMR is highly quantitative and repeatable, NMR-based metabolic profiling can be carried out successfully. These characteristics make up for the NMR's poor sensitivity and are particularly crucial when multivariate statistical approaches are used to examine metabolomic data (Pan & Raftery, 2007). Additionally, because NMR sensitivity is independent of metabolite pKa or hydrophobicity, it is a superb choice for broad-based investigations and appropriate for samples under various conditions. Promising research involving the diagnosis of ovarian cancer and inborn metabolic abnormalities are recent clinical examples of this strategy. The J-resolved experiment and other 2-D techniques have helped to advance the resolution and sensitivity of NMR spectroscopy (Pan & Raftery, 2007; Powers, 2009).

The two main analytical methods for metabolite detection are MS and NMR. To ascertain the elemental makeup and clarify the chemical structure of molecules, MS examines the mass-to-charge ratio of charged molecules. While a precise mass can be established by MS, chromatography is frequently needed due to the metabolome's low-molecular-weight dispersion. Usual separation methods used in MS-based metabolomics include GC, HPLC, and CE (Kell, 2004).

Despite the fact that the usage of cryogenic probes has considerably boosted the susceptivity of NMR in four factors, MS is still considerably more sensitive than NMR and encloses a larger diversity of the metabolome (Zhang & Powers, 2012). NMR essentially only picks up the most prevalent metabolites at concentrations of over 1–5 M. Since 1H NMR is 64 times more sensitive than 13C NMR, metabolomics commonly uses 1H NMR. Even though there is just 1.1% of naturally occurring 13C, NMR cryoprobes even so can be used to find metabolites (Keun et al., 2002). Although MS is a destructive method, it uses a far smaller sample volume (100 L) than NMR (600 L). Accordingly, these methods are complementary to one another and bring different information to the study of a metabolome. Numerous metabolomic investigations combining both MS and NMR have proven the complementary nature of the two methods (Pan & Raftery, 2007).

In practice, a 3D score plot can be produced by combining the NMR and MS data. The additional resolution required distinguishing between many classes or groups may be provided by the supplemental data's enhanced dimensionality. The six chemicals that distinguish the biological samples are represented as subregions from the NMR and MS spectra corresponding to their respective peaks. In a 3D score plot, the NMR PC1 values are directly added to the DESI-MS PC values because the 2D scoring plots of PCs from the NMR are independent of the DESI-MS data. Higher biological sample separation in the PC space is the end consequence (Powers, 2009; Zhang & Powers, 2012).

10. Determination of biofilms using metabolomics based on nuclear magnetic resonance

The metabolome differences between planktonic and biofilm-forming sessile cells have been studied using NMR-based metabolomics. The metabolome differences between chemostat planktonic and biofilm cells of *P. aeruginosa* were investigated using 1H HR-MAS (Gjersing et al., 2007). When both the planktonic and biofilm-forming cells were cultivated with a constant feeding chemostat mode, there was no discernible variation in the extracellular metabolic framework. On the other hand, the intracellular metabolome's 2D PCA scores plot showed a strong separation between batch, chemostat, and planktonic and biofilm cells. The matching loadings figure reveals that the two metabolomes differ significantly and in a complex manner. Although the differences in metabolites were not thoroughly examined, it was noted that biofilm-mediated metabolites were typically found in lower concentrations, probably because the cells closest to the substrate had slower metabolic rates. Similar to this, 1H NMR was utilized to describe the phenotype of various *Pseudomonas fluorescens* biofilm colonies (Workentine et al., 2010).

The distinction between planktonic cells and biofilms can be demonstrated by NMR, which is a crucial step in the development of a drug discovery application. Questions are raised about whether chemicals such as lead can block the metabolic processes involved in the production of biofilms or whether medication therapy can derive metabolome that is more comparable with planktonic cells. These important problems might be resolved by comparing the metabolomes of planktonic cells treated and untreated with drugs. Additionally, the comparison of planktonic cells with biofilms has previously revealed alterations in the activity of particular cellular processes linked to the production of biofilms. To avert the biofilm formation, it may be possible to target proteins required for exopolysaccharide synthesis and the retaliation to oxidative stress. Additionally, the distinctions in metabolome between planktonic cells and biofilms can be employed as a diagnostic aid and in the development of therapeutic strategies (Zhang & Powers, 2012).

The removal of the contaminated device is frequently necessary for treating biofilm-mediated infections that have been colonized on medical implants. Researching alterations in bacterial metabolomes responding to various surfaces (such as metals and polymers) may aid in the creation of innovative substances that are resistant to the formation of biofilms (Kim et al., n.d.). Antibiotic film or immersing on medical devices is a typical method of preventing biofilm infections. However, using too many antibiotics runs the danger of hastening the emergence of resistance (Shunmugaperumal, 2010).

Metabolomics can be used for the construction of metabolic pathways along with the contributions of proteomics and genomics information. Studies have shown time-consuming analysis of *S. aureus* exponential growth in glucose starvation during the transition (Liebeke et al., 2011). More than 500 proteins are actively working and followed by the concentration of 94 metabolites. Compounds in the media before inoculation and at specific times during cell growth were quantified using 1D 1H NMR. By using GC-MS/LC-MS, intracellular metabolites were evaluated. Changes in the proteome generally followed changes in the metabolome, while the latter showed a greater dynamic range. The amino acids saw the most pronounced modifications. Glycolysis and protein synthesis were very active during the early stages of cell growth, but when glucose became depleted, gluconeogenesis and the TCA cycle became active. Once more, this is in line with the idea that changes in glucose concentrations regulate biofilm production through controlling TCA cycle activity.

As a valuable approach toward the characterizations of the state of a system, metabolomics can also incorporate additional complementary data such as proteomics, significant enhancement of reliable information. The observation of a relationship between a protein's level of expression and a metabolite's concentration further supports the significance and relevance of the protein into the biofilm-forming network. It can, therefore, offer strong support for a potential novel drug development target from the perspective of drug discovery (Zhang & Powers, 2012).

11. Effect of xylitol and fluoride on metabolome profile of oral biofilm

Ubiquitously, fluoride and xylitol appeared as prophylactic in oral disease. Fluoride stimulates the remineralization upon the tooth surface by inhibiting in vitro production of bacterial acid and in vivo production of plaque acid and demineralization (Cate, 1999; Hamilton,

1990; Marquis, 1990). In vitro studies manifested the inhibition of an enzyme enolase in the EMP pathway of streptococci, although inhibitory activity of fluoride against multiple bacterial species has not been established (Kaufmann, 1992). Moreover, a nonfermentative sugar alcohol xylitol is able to suppress the acid production from glucose by the *Streptococcus mutans*. The glycolytic enzymes get inhibited by the xylitol 5-phosphate, which is produced via phosphoenolpyruvate–fructose phosphotransferase system, leading to the restrain of the acid production (Miyasawa et al., 2003). Subsequent dephosphorylation of xylitol 5-phosphate to xylitol leads to the emergence of a "futile cycle," i.e., an energy wasting cycle that usually suppresses the growth of *S. mutans* (Pihlanto-Leppälä et al., 1990).

Seven periodontally robust volunteers with 0.14–0.38 teeth decay and without any antibiotic consumption were selected for supragingival plaque accumulation. Each of the volunteers was asked to rinse with 10% of glucose or 10% glucose–10% xylitol mixture or sodium fluoride (225 or 990 ppm F) in 10 mL volume for 60 s in due course. Those who had washed with glucose or glucose–xylitol mixture were put through to the plaque sample collection after 10 min. The volunteers who had rinsed with sodium fluoride were subjected to the subsequent rinse with 10 mL of 10% glucose followed by the collection of plaque. Analysis of these plaque samples indicated the effect of fluoride and xylitol on the in vivo glucose metabolism in supragingival plaque. Change of metabolic and lactate aggregation in dental plaque stipulates the effect of these dental washes. It was observed that addition of xylitol does not bring any considerable changes to the metabolite profile in supragingival plaque in vivo as well as into the lactate production. Xylitol 5-phosphate is usually produced via the bacterial phosphotransferase system, which is dependent on phosphoenolpyruvate (Miyasawa et al., 2003). Although formation of xylitol 5-phosphate is involved in "futile cycle," it does not affect the in vivo glucose fermentation in supragingival plaque due to the limited number of mutans streptococci cell accumulation upon the healthy tooth surface (Bowden et al., 1975). A nonfermentative sugar substitute xylitol-mediated prevention is usually carried out without producing any lactate that enables null effect on the metabolome profile. Rinsing with fluoride before glucose evolves the accumulation of 3-phosphoglycerate into the oral cavity, which results in the inhibition of a metabolic step, i.e., production of phosphoenolpyruvate catalyzed by glycolytic enzyme enolase (Takahashi & Washio, 2011). Significant reduction of lactate production has similarity toward the in vivo fluoride-based inhibition of enolase in *S. mutans* and *Streptococcus sanguinis*. Notable change was observed in EMP pathway where aldolase, glyceraldehyde3-phosphate dehydrogenase, and phosphoglycerate kinase get inhibited with the increase of glucose 6-phosphate,fructose 6-phosphate, and fructose 1,6-bisphosphate and decrease of dihydroxyacetone phosphate. However, this phenomenon is yet to be clarified against *S. mutans* and *S. sanguinis* (Maehara et al., 2005).

12. Metabolic dysregulation of *Staphylococcus aureus* biofilm by carnosol

A common polyphenol carnosol is a dietary diterpene, which is usually extracted from Lamiaceae (mint) family. Few plants such as lavender, oregano, rosemary and sage belong to the Lamiaceae group (Horiuchi et al., 2007). Carnosol and carnosic acid are usually extracted from these plants due to their antioxidant activity and potency in food preservation, detoxification, and antiinflammatory reactions (Sanchez et al., 2015). Multiple pharmacological activities such as antifungal, antifibrosis, antiinflammatory, antitumor, and neuroprotective activities along with antioxidant pursuit of this carnosol have been reported in several cell lines and animal models affecting their molecular pathways (Johnson, 2011).

Antibiofilm activities of carnosol have been delineated against *B. subtilis, S. aureus, S. epidermidis,* and *Streptococcus pyogenes* (de Oliveira, 2016). A positive equivalence was perceived between the carnosol and inhibition of *S. aureus* biofilm. Subinhibitory concentration of ($^{1}/_{8}$ X to 1 X MIC) carnosol was accountable for the inhibition of *S. aureus* ATCC 29213, MRSA and MSSA sessile bacterial cell attachment on the biotic or abiotic surfaces. EPS production is reduced by 78.2% for $^{1}/_{2}$ X MIC treatment followed by restriction of EPS production in case of MRSA 4670. The metabolic profile of *S. aureus* ATCC 29213 showed a significant alteration under carnosol stress. Around 13 metabolites were upregulated, and 25 metabolites were downregulated with respect to the control batch. Carnosol treatment induces 1.9 and 1.7 times reduction of L-ornithine and L-glutamic acid, respectively, in *S. aureus* biofilms. According to the KEGG pathway, the metabolic profiles of *S. aureus* planktonic cells and biofilms get changed due to the disruption of arginine and proline alanine, aspartate and glutamate metabolism, glyoxylate and dicarboxylate, glycine, serine and threonine metabolism, and arginine and aminoacyl-tRNA biosynthesis. The alteration of D-glutamine, D-glutamate, and glutathione metabolism in biofilms has been perceived (Chabán et al., 2019; Ferreira et al., 2016).

13. Conclusion

Metabolomics is a cytological method, which combined the metabolome analysis of the biofilm along with bacterial growth and pathogenicity. Several metabolomic data revealed the metabolic reprogramming during the transition of the planktonic population to the sessile biofilm formation. Various small molecular metabolites are expository important in biofilm formation and biofilm-associated metabolic pathways. Metabolomic analysis also helps to develop the biofilm-based treatments against novel targets and discovery of antibiotics against nosocomial biofilm diseases at metabolic level. It also helps to monitor the physiological conditions of the biofilm and determine the balance of symbiosis inside the biofilm. The effect of pH, nutrition, reduction–oxidation potential, and QQ molecules on the metabolomic profile of the pathogenic biofilm influences to the novel metabolic approaches in the symbiotic microbiota and metabolome analysis.

References

Bailey, N. J. C., Wang, Y., Sampson, J., Davis, W., Whitcombe, I., Hylands, P. J., Croft, S. L., & Holmes, E. (2004). Prediction of anti-plasmodial activity of Artemisia annua extracts: Application of NMR spectroscopy and chemometrics. *Journal of Pharmaceutical and Biomedical Analysis, 35*(1), 117–126. https://doi.org/10.1016/j.jpba.2003.12.024

Basaran, T. I., Berber, D., Gökalsın, B., Tramice, A., Tommonaro, G., Abbamondi, G. R., Erginer Hasköylü, M., Toksoy Öner, E., Iodice, C., & Sesal, N. C. (2020). Extremophilic natrinema versiforme against *Pseudomonas aeruginosa* quorum sensing and biofilm. *Frontiers in Microbiology, 11,* 79. https://doi.org/10.3389/fmicb.2020.00079

Bowden, G. H., Hardie, J. M., & Slack, G. L. (1975). Microbial variations in approximal dental plaque. *CariesRes, 9*, 253–277.

Carradori, S., Di Giacomo, N., Lobefalo, M., Luisi, G., Campestre, C., & Sisto, F. (2020). Biofilm and quorum sensing inhibitors: The road so far. *Expert Opinion on Therapeutic Patents, 30*(12), 917–930. https://doi.org/10.1080/13543776.2020.1830059

Cate, J. M. T. (1999). Current concepts on the theories of the mechanism of action of fluoride. *Acta Odontologica Scandinavica, 57*(6), 325–329. https://doi.org/10.1080/000163599428562

Chabán, M. F., Karagianni, C., Joray, M. B., Toumpa, D., Sola, C., Crespo, M. I., Palacios, S. M., Athanassopoulos, C. M., & Carpinella, M. C. (2019). Antibacterial effects of extracts obtained from plants of Argentina: Bioguided isolation of compounds from the anti-infectious medicinal plant Lepechinia meyenii. *Journal of Ethnopharmacology, 239*, 111930. https://doi.org/10.1016/j.jep.2019.111930

Chen, M., Zhao, L., & Jia, W. (2005). Metabonomic study on the biochemical profiles of a hydrocortisone-induced animal model. *Journal of Proteome Research, 4*(6), 2391–2396. https://doi.org/10.1021/pr050158o

Costerton, J. (1995). Overview of microbial biofilms. *Journal of Industrial Microbiology, 15*(3), 137–140. https://doi.org/10.1007/BF01569816

Costerton, J. W., Lewandowski, Z., Caldwell, D. E., Korber, D. R., & Lappin-Scott, H. M. (n.d.). *Microbial biofilms* (Vol. 35).

Cui, T., Bai, F., Sun, M., Lv, X., Li, X., Zhang, D., & Du, H. (2020). Lactobacillus crustorum ZHG 2-1 as novel quorum-quenching bacteria reducing virulence factors and biofilms formation of *Pseudomonas aeruginosa*. *LWT, 117*, 108696. https://doi.org/10.1016/j.lwt.2019.108696

Di Somma, A., Moretta, A., Canè, C., Cirillo, A., & Duilio, A. (2020). Antimicrobial and antibiofilm peptides. *Biomolecules, 10*(4), 652. https://doi.org/10.3390/biom10040652

Fernández, M., & Zúñiga, M. (2006). Amino acid catabolic pathways of lactic acid bacteria. *Critical Reviews in Microbiology, 32*(3), 155–183. https://doi.org/10.1080/10408410600880643

Ferreira, D. S., Tozatti, M. G., Souza, M. G. M., Martins, C. H. G., Januário, A. H., Pauletti, P. M., Silva, M. L. A., & Cunha, W. R. (2016). Antimicrobial activity of compounds isolated from Rosmarinus officinalis and semi-synthetic derivatives against multidrug-resistant bacteria. *Latin American Journal of Pharmacy, 8*.

Gjersing, E. L., Herberg, J. L., Horn, J., Schaldach, C. M., & Maxwell, R. S. (2007). NMR metabolomics of planktonic and biofilm modes of growth in *Pseudomonas aeruginosa*. *Analytical Chemistry, 79*(21), 8037–8045. https://doi.org/10.1021/ac070800t

Halouska, S., Fenton, R. J., Barletta, R. G., & Powers, R. (2012). Predicting the *in vivo* mechanism of action for drug leads using NMR metabolomics. *ACS Chemical Biology, 7*(1), 166–171. https://doi.org/10.1021/cb200348m

Hamilton, I. R. (1990). Biochemical effects of fluoride on oral bacteria. *Journal of Dental Research, 69*(2_Suppl. 1), 660–667. https://doi.org/10.1177/00220345900690S128

Hobley, L., Kim, S. H., Maezato, Y., Wyllie, S., Fairlamb, A. H., Stanley-Wall, N. R., & Michael, A. J. (2014). Norspermidine is not a self-produced trigger for biofilm disassembly. *Cell, 156*(4), 844–854. https://doi.org/10.1016/j.cell.2014.01.012

Horiuchi, K., Shiota, S., Kuroda, T., Hatano, T., Yoshida, T., & Tsuchiya, T. (2007). Potentiation of antimicrobial activity of aminoglycosides by carnosol from salvia officinalis. *Biological and Pharmaceutical Bulletin, 30*(2), 287–290. https://doi.org/10.1248/bpb.30.287

Infectious Diseases Society of America. (2010). The 10 × '20 initiative: Pursuing a global commitment to develop 10 new antibacterial drugs by 2020. *Clinical Infectious Diseases, 50*(8), 1081–1083. https://doi.org/10.1086/652237

Jakobsen, T. H., Warming, A. N., Vejborg, R. M., Moscoso, J. A., Stegger, M., Lorenzen, F., Rybtke, M., Andersen, J. B., Petersen, R., Andersen, P. S., Nielsen, T. E., Tolker-Nielsen, T., Filloux, A., Ingmer, H., & Givskov, M. (2017). A broad range quorum sensing inhibitor working through sRNA inhibition. *Scientific Reports, 7*(1), 9857. https://doi.org/10.1038/s41598-017-09886-8

Jin, Y., Liang, T., Fu, Q., Xiao, Y.-S., Feng, J.-T., Ke, Y.-X., & Liang, X.-M. (2009). Fingerprint analysis of Ligusticum chuanxiong using hydrophilic interaction chromatography and reversed-phase liquid chromatography. *Journal of Chromatography A, 1216*(11), 2136–2141. https://doi.org/10.1016/j.chroma.2008.04.010

Johnson, J. J. (2011). Carnosol: A promising anti-cancer and anti-inflammatory agent. *Cancer Letters, 305*(1), 1–7. https://doi.org/10.1016/j.canlet.2011.02.005

Kaufmann, B. P. (1992). Purification,characterization and inhibition by fluoride enolase from Streptococcus mutans DSM 320523. *CariesRes, 26*, 110–116.

Kalia, V. C. (2013). Quorum sensing inhibitors: An overview. *Biotechnology Advances, 31*(2), 224–245. https://doi.org/10.1016/j.biotechadv.2012.10.004

Kell, D. (2004). Metabolomics and systems biology: Making sense of the soup. *Current Opinion in Microbiology, 7*(3), 296–307. https://doi.org/10.1016/j.mib.2004.04.012

Kendall, M. M., & Sperandio, V. (2014). Cell-to-cell signaling in *Escherichia coli* and *Salmonella. EcoSal Plus, 6*(1). https://doi.org/10.1128/ecosalplus.ESP-0002-2013. ecosalplus.ESP-0002-2013.

Keun, H. C., Beckonert, O., Griffin, J. L., Richter, C., Moskau, D., Lindon, J. C., & Nicholson, J. K. (2002). Cryogenic probe ^{13} C NMR spectroscopy of urine for metabonomic studies. *Analytical Chemistry, 74*(17), 4588–4593. https://doi.org/10.1021/ac025691r

Kim, Y. H., An, E. S., Song, B. K., Kim, D. S., & Chelikani, R. (n.d.). *Polymerization of cardanol using soybean peroxidase and its potential application as anti-biofilm coating material* (Vol. 4).

Kolodkin-Gal, I., Romero, D., Cao, S., Clardy, J., Kolter, R., & Losick, R. (2010). d -amino acids trigger biofilm disassembly. *Science, 328*(5978), 627–629. https://doi.org/10.1126/science.1188628

Lao, Y.-M., Jiang, J.-G., & Yan, L. (2009). Application of metabonomic analytical techniques in the modernization and toxicology research of traditional Chinese medicine. *British Journal of Pharmacology, 157*(7), 1128–1141. https://doi.org/10.1111/j.1476-5381.2009.00257.x

Larsen, F. H., & Engelsen, S. B. (2015). Insight into the functionality of microbial exopolysaccharides by NMR spectroscopy and molecular modeling. *Frontiers in Microbiology, 6*. https://doi.org/10.3389/fmicb.2015.01374

Li, Y., Chen, L., Zhang, P., Bhagirath, A. Y., & Duan, K. (2020). ClpV3 of the H3-type VI secretion system (H3-T6SS) affects multiple virulence factors in *Pseudomonas aeruginosa. Frontiers in Microbiology, 11*, 1096. https://doi.org/10.3389/fmicb.2020.01096

Liebeke, M., Dörries, K., Zühlke, D., Bernhardt, J., Fuchs, S., Pané-Farré, J., Engelmann, S., Völker, U., Bode, R., Dandekar, T., Lindequist, U., Hecker, M., & Lalk, M. (2011). A metabolomics and proteomics study of the adaptation of *Staphylococcus aureus* to glucose starvation. *Molecular BioSystems, 7*(4), 1241. https://doi.org/10.1039/c0mb00315h

Lou, Z., Tang, Y., Song, X., & Wang, H. (2015). Metabolomics-based screening of biofilm-inhibitory compounds against *Pseudomonas aeruginosa* from burdock leaf. *Molecules, 20*(9), 16266–16277. https://doi.org/10.3390/molecules200916266

Maehara, H., Iwami, Y., Mayanagi, H., & Takahashi, N. (2005). Synergistic inhibition by combination of fluoride and xylitol on glycolysis by mutans streptococciand its biochemical mechanism. *Caries Research, 39*(6), 521–528. https://doi.org/10.1159/000088190

Marquis, R. E. (1990). Diminished acid tolerance of plaque bacteria caused by fluoride. *Journal of Dental Research, 69*(2_Suppl. l), 672–675. https://doi.org/10.1177/00220345900690S130

Miyasawa, H., Iwami, Y., Mayanagi, H., & Takahashi, N. (2003). Xylitol inhibition of anaerobic acid production by *Streptococcus mutans* at various pH levels: Effect of pH on xylitol inhibition. *Oral Microbiology and Immunology, 18*(4), 215–219. https://doi.org/10.1034/j.1399-302X.2003.00068.x

Okada, A., Nikaido, T., Ikeda, M., Okada, K., Yamauchi, J., Foxton, R. M., Sawada, H., Tagami, J., & Matin, K. (2008). Inhibition of biofilm formation using newly developed coating materials with self-cleaning properties. *Dental Materials Journal, 27*(4), 565–572. https://doi.org/10.4012/dmj.27.565

de Oliveira, M. R. (2016). The dietary components carnosic acid and carnosol as neuroprotective agents: A mechanistic view. *Molecular Neurobiology, 53*(9), 6155–6168. https://doi.org/10.1007/s12035-015-9519-1

Packiavathy, I. A. S. V., Priya, S., Pandian, S. K., & Ravi, A. V. (2014). Inhibition of biofilm development of uropathogens by curcumin – an anti-quorum sensing agent from Curcuma longa. *Food Chemistry, 148*, 453–460. https://doi.org/10.1016/j.foodchem.2012.08.002

Pan, Z., & Raftery, D. (2007). Comparing and combining NMR spectroscopy and mass spectrometry in metabolomics. *Analytical and Bioanalytical Chemistry, 387*(2), 525–527. https://doi.org/10.1007/s00216-006-0687-8

Peregrín-Alvarez, J. M., Sanford, C., & Parkinson, J. (2009). The conservation and evolutionary modularity of metabolism. *Genome Biology, 10*(6), R63. https://doi.org/10.1186/gb-2009-10-6-r63

Pihlanto-Leppälä, A., Söderling, E., & Mäkinen, K. K. (1990). Expulsion mechanism of xylitol 5-phosphate in Streptococcus mutans. *European Journal of Oral Sciences, 98*(2), 112–119. https://doi.org/10.1111/j.1600-0722.1990.tb00949.x

Powers, R. (2009). NMR metabolomics and drug discovery: NMR metabolomics and drug discovery. *Magnetic Resonance in Chemistry, 47*(S1), S2–S11. https://doi.org/10.1002/mrc.2461

Rabin, N., Zheng, Y., Opoku-Temeng, C., Du, Y., Bonsu, E., & Sintim, H. O. (2015). Agents that inhibit bacterial biofilm formation. *Future Medicinal Chemistry, 7*(5), 647–671. https://doi.org/10.4155/fmc.15.7

Rui, G., & Xilin, L. (n.d.). *Novel Functional metabolites that affect biofilm formation are regulated by bioavailable iron with siderophore-dependent pathway*, 71.

Sanchez, C., Horcajada, M.-N., Membrez Scalfo, F., Ameye, L., Offord, E., & Henrotin, Y. (2015). Carnosol inhibits pro-inflammatory and catabolic mediators of cartilage breakdown in human osteoarthritic chondrocytes and mediates cross-talk between subchondral bone osteoblasts and chondrocytes. *PLoS One, 10*(8), e0136118. https://doi.org/10.1371/journal.pone.0136118

Schwartz, D. J., Kalas, V., Pinkner, J. S., Chen, S. L., Spaulding, C. N., Dodson, K. W., & Hultgren, S. J. (2013). Positively selected FimH residues enhance virulence during urinary tract infection by altering FimH conformation. *Proceedings of the National Academy of Sciences, 110*(39), 15530–15537. https://doi.org/10.1073/pnas.1315203110

Scoffone, T., Chiarelli, B., & Buroni, S. (2019). Quorum sensing as antivirulence target in cystic fibrosis pathogens. *International Journal of Molecular Sciences, 20*(8), 1838. https://doi.org/10.3390/ijms20081838

Sharma, D., Misba, L., & Khan, A. U. (2019). Antibiotics versus biofilm: An emerging battleground in microbial communities. *Antimicrobial Resistance & Infection Control, 8*(1), 76. https://doi.org/10.1186/s13756-019-0533-3

Shastry, R. P., Rekha, P. D., & Rai, V. R. (2019). Biofilm inhibitory activity of metalloprotein AHL-lactonase from the cell-free lysate of endophytic Enterobacter species isolated from Coscinium fenestratum Gaertn. *Biocatalysis and Agricultural Biotechnology, 18*, 101009. https://doi.org/10.1016/j.bcab.2019.01.047

Shen, F., Ge, C., & Yuan, P. (2020). Metabolomics study reveals inhibition and metabolic dysregulation in *Staphylococcus aureus* planktonic cells and biofilms induced by carnosol. *Frontiers in Microbiology, 11*, 538572. https://doi.org/10.3389/fmicb.2020.538572

Shunmugaperumal, T. (2010). Microbial colonization of medical devices and novel preventive strategies. *Recent Patents on Drug Delivery & Formulation, 4*(2), 153–173. https://doi.org/10.2174/187221110791185006

Takahashi, N., & Washio, J. (2011). Metabolomic effects of xylitol and fluoride on plaque biofilm *in vivo*. *Journal of Dental Research, 90*(12), 1463–1468. https://doi.org/10.1177/0022034511423395

Vasilchenko, A. S., & Rogozhin, E. A. (2019). Sub-inhibitory effects of antimicrobial peptides. *Frontiers in Microbiology, 10*, 1160. https://doi.org/10.3389/fmicb.2019.01160

Willias, S. P., Chauhan, S., & Motin, V. L. (2014). Functional characterization of *Yersinia pestis* aerobic glycerol metabolism. *Microbial Pathogenesis, 76*, 33–43. https://doi.org/10.1016/j.micpath.2014.08.010

Wilson, I., Plumb, R., Granger, J., Major, H., Williams, R., & Lenz, E. (2005). HPLC-MS-based methods for the study of metabonomics. *Journal of Chromatography B, 817*(1), 67–76. https://doi.org/10.1016/j.jchromb.2004.07.045

Workentine, M. L., Harrison, J. J., Weljie, A. M., Tran, V. A., Stenroos, P. U., Tremaroli, V., Vogel, H. J., Ceri, H., & Turner, R. J. (2010). Phenotypic and metabolic profiling of colony morphology variants evolved from *Pseudomonas fluorescens* biofilms. *Environmental Microbiology*. https://doi.org/10.1111/j.1462-2920.2010.02185.x

Yan, X., Gu, S., Shi, Y., Cui, X., Wen, S., & Ge, J. (2017). The effect of emodin on *Staphylococcus aureus* strains in planktonic form and biofilm formation in vitro. *Archives of Microbiology, 199*(9), 1267–1275. https://doi.org/10.1007/s00203-017-1396-8

Yang, Y.-B., Wang, S., Wang, C., Huang, Q.-Y., Bai, J.-W., Chen, J.-Q., Chen, X.-Y., & Li, Y.-H. (2015). Emodin affects biofilm formation and expression of virulence factors in Streptococcus suis ATCC700794. *Archives of Microbiology, 197*(10), 1173–1180. https://doi.org/10.1007/s00203-015-1158-4

Yang, J., Xu, G., Kong, H., Zheng, Y., Pang, T., & Yang, Q. (2002). Artificial neural network classification based on high-performance liquid chromatography of urinary and serum nucleosides for the clinical diagnosis of cancer. *Journal of Chromatography B, 780*(1), 27–33. https://doi.org/10.1016/S1570-0232(02)00408-7

Zhang, B., & Powers, R. (2012). Analysis of bacterial biofilms using NMR-based metabolomics. *Future Medicinal Chemistry, 4*(10), 1273–1306. https://doi.org/10.4155/fmc.12.59

CHAPTER

16

Reprogramming the metabolomics of biofilms

Harshit Sajal[1], *Yuvaraj Sivamani*[2], *Nimisha Joly Chennakadan*[1], *Yadhu Krishna Menon Kakkanattu*[1] *and Sumitha Elayaperumal*[1]

[1]**Department of Biotechnology and Bioinformatics, JSS Academy of Higher Education and Research, Mysore, Karnataka, India;** [2]**Department of Pharmaceutical Chemistry, Cauvery College of Pharmacy, Mysore, Karnataka, India**

1. Introduction

A biofilm is a conglomeration of surface-attached microbial cells enclosed in a self-produced matrix of extracellular polymeric substances (EPS) with extracellular proteins, DNA, and polysaccharides, formed under stressful environments (Hung et al., 2013). Bacterial biofilms are particularly ubiquitous in the natural environment and are important in a variety of industrial, clinical, and ecological settings (Costerton et al., 1995). Nevertheless, biofilms are responsible for many clinical infections, food, and environmental contamination. Many human infections with high-frequency antibiotic resistance have been observed to be associated with biofilm formation, which allows bacteria to survive and prolong their life span, supporting the recurrence of infection. *Pseudomonas aeruginosa* biofilms are common in a variety of illnesses, including surgical site infections brought on by medical devices, persistent wounds, and respiratory diseases such as cystic fibrosis. A common, frequently recurrent, or even fatal infection, urinary tract infection (UTI), is typically brought on by uropathogenic *Escherichia coli* (UPEC) (Hall-Stoodley et al., 2004). Several other bacterial biofilms such as *Staphylococcus aureus* and *Candida albicans* are associated with catheter-related bloodstream infection and biliary tube–related infection, respectively (Del Pozo, 2018). Nonetheless, there is a scarcity of effective biofilm diagnosis, prevention, and mitigation strategies.

Microbial Biofilms
https://doi.org/10.1016/B978-0-323-95715-1.00015-7

Metabolomics is gaining importance as an effective tool for investigating metabolic processes, discovering critical biomarkers accountable for metabolic features, and revealing the metabolic process associated with metabolism. The molecular mechanisms of biofilm formation have been vastly studied using metabolomics. Metabolites reflect downstream gene and enzyme activity; metabolomics can capture a snapshot of cell activity directly related to growth mode. Complex life activity across the species is depicted by the universal language, metabolomics (Peng et al., 2015). Currently, profiling of metabolomics in a high-throughput manner is primarily centered on the discovery of biomarkers and the study of metabolic mechanisms, but one potential direction is to reprogram metabolomics to prevent and treat diseases. Using reprogramming metabolomics, a framework could be established for developing targeted tools to deal with the changes, such as boosting host immunity against pathogen infection and controlling and/or preventing infection, based on the metabolic mechanisms that respond to altered external and internal environmental factors. Identification, characterization, and quantification of small molecules or metabolites (cell/organism) could be achieved using metabolomics and high-throughput analytical technologies. Gas chromatography—mass spectrometry (GC-MS), capillary electrophoresis coupled to mass spectrometry (CE-MS), liquid chromatography—mass spectrometry (LC-MS), and a combination of GC-MS and nuclear magnetic resonance (NMR) can be used for the separation, dramatization, and characterization of small molecules/metabolites. Data derived from the metabolomics are recorded in the form of peaks unassigned of different intensities including time of retention, masses or mass fragments (mass spectrometric data), or chemical shifts (NMR data) (Peng et al., 2015). The data matrix is normalized followed by statistical analysis, for biomarkers identification. A metabolome can be modified using crucial biomarkers, developing a specific metabolic strategy to deal with changing internal and external conditions. These exogenous or engineered metabolites act as both a substrate for enzymes and a regulator of other molecules in cellular activity, including proteins. Thus, the detailed explanation of significant metabolic pathway changes between planktonic and biofilm phenotypes has the potential to reveal new targets for prevention and mitigation as well as distinct metabolic signatures for biofilm diagnosis.

1.1 Metabolomics: workflow and experimental design

This chapter will discuss the analytic workflows and the processing that are routinely utilized in high-throughput untargeted metabolomic research. In untargeted metabolic research, large numbers of metabolites from each sample are concurrently measured. This method, also referred to as the top-down approach, examines the entire metabolic profile rather than requiring prior hypotheses on a specific group of metabolites. As a result, this research generates a tremendous amount of data. The data generated are distinguished not just by its size but also due to its complexity, necessitating the use of high-performance bioinformatic techniques and tools. The essential steps in the metabolomics process flow include sample pretreatment, analytical analysis and data acquisition, spectral processing, data analysis, metabolic identification, and finally the interpretation of results to create biological models (Fig. 16.1).

Optimized samples are subjected to high-throughput analytical analysis, and spectral data are generated. The vast majority of metabolomics data consist of unassigned peaks of varying intensities at different times of retention, fragments of mass/masses (MS data), or chemical shifts (NMR data). The evaluation and interpretation of these data reveal that many MS-

FIGURE 16.1 The workflow of analysis in untargeted metabolomic studies.

based platforms share various properties. Using the software, the area of each peak in an individual peak is derived from the total ion chromatograms. Overlapped peaks are deconvolved using single ions. Based on comparisons between retention time and mass spectral data, compounds are identified. Before doing statistical analysis, the data matrix needs to be normalized. Specific needs are met by the normalization parameters. Quality control samples are derived from the entire data set (internal standards/statistical models, such as the overall peak intensities, the range of quartiles, or metabolite fragments that match). After normalization, the gathered data are handled using multivariate statistical methods or employing pattern recognition techniques to find the distinguishing features and the biomarkers from the tested samples (Gao & Xu, 2015).

1.2 Metabolomics sample preparation

There is no one-size-fits-all approach to metabolomics. Customized and designed sample preparation methods are necessary due to variations in metabolite physicochemical properties (such as turnover rates of various metabolite classes) and the physical and biological structure of the bacteria itself (such as permeability of cell wall and plasma membrane) (Segers et al., 2019). The difficulties could be brought on by the abundance of metabolites, intracellular leakage, cell disintegration, and low extraction reproducibility (León et al., 2013; Siegel et al., 2014). To accurately reflect the actual changes in cellular metabolism, a reliable, effective, simple, and repeatable approach is preferred. The steps involved in preparing a metabolite sample are cell cultivation and harvesting, cell quenching, cell disruption, and metabolite extraction.

1.2.1 Cell cultivation and harvesting

The experimental settings for cell growth and harvesting can be very diverse since biofilms can develop on a variety of substrates or habitats. Proper growth of bacteria is important for metabolomics studies.

Research is improvised by different culture conditions, i.e., minimal or rich media including, Luria–Bertani (LB) (Meyer et al., 2013), Mueller–Hinton broth (Maifiah et al., 2016), and nutritional broth (Marcinowska et al., 2011). To identify any statistically significant metabolite differences across groups, sample replication is crucial, especially for biological repetitions (Weckwerth, 2003). The number of duplicates should be chosen based on a power calculation that takes into account baseline technical reproducibility, biological variability, and expected differences, according to Roberts et al. (2012). In the majority of the microbial metabolomics studies, biological replicates ($n = 3$) were used in the sample group (Mohd Kamal et al., 2022). Apart from it, several studies also employed crystal violet staining for biofilm formation quantification, visualization, and further statistical analysis.

To ensure the detection of enough metabolites, bacterial culture should be harvested at ideal density and volume at a specific time point and condition. The timing of sample collection is extremely important since cell densities frequently shift between growth phases (Halouska et al., 2013). Bacteria cultures produce different kinds of metabolite at different concentrations during growth phases, which reflect intracellular physiological changes occurring within the cell. Bacterial samples in the majority of the investigations were harvested in the early and late exponential growth phases. For the sample replicates, the cell density of the samples must be the same as that which may be standardized using the CFU/mL or optical density (for example, OD600 $\sim$ 0.5).

1.2.2 Cell quenching

To obtain a precise "snapshot" of the metabolome, it is essential to quickly and effectively quench all biological and enzymatic activities during sample processing. Metabolites with a high turnover rate include pyruvate, phosphoenolpyruvate, oxoglutarate, fructose-6-phosphate, fumarate, and others (Canelas et al., 2008). It is also essential to prevent inducing a stress response or cell death as this would render the research useless. Cold methanol-60% (−48°C to −20°C) is used often for quenching due to its low freezing effect and less toxicity in comparison with other solvents. Quenching can be done before, during, or after sample collection. However, due to the risk of leakage in the cell and the loss of metabolites during the process of quenching and washing procedure, methanol may only be suitable for gram-positive bacteria or fungi. In *Pseudomonas fluorescens*, it was observed that quenching carried out in the presence of cold glycerol−saline (3:2) was found to have higher intracellular metabolite recovery in comparison with cold methanol (Mohd Kamal et al., 2022). In a different investigation, it was discovered that methanol/glycerol (−50°C) dramatically reduced ATP leakage in *E. coli* in comparison with methanol or water (Link et al., 2008).

The metabolomic samples can easily be contaminated due to the inappropriate removal of cell growth media and washing the cells. Before collecting the metabolome, the culture medium is typically removed using filtration or centrifugation. Although the lengthy centrifugation process raises concerns about induced stress and metabolome alterations, sample preparation is more consistently done in centrifugation. Filtering, on the other hand, is faster and easier to quench cells on a filter membrane. The removal of all frozen cells from the filter paper consistently and uniformly raises some practical issues, nevertheless. As of right now, there are many different ways to separate and wash cells, which suggests that any metabolomics study must go through an optimization process.

1.2.3 Metabolite extraction

Occasionally, metabolite extraction and cell lysis can be done concurrently. Techniques based on organic solvents frequently used include (1) disruption by mechanical means and (2) Fast-Prep system (Maharjan & Ferenci, 2003; Shaw et al., 2004). Trichloroacetic acid (TCA) is the common method employed for lysing the cells in the filter paper. Although it is highly efficient, it deteriorates the filter paper whereby it creates a considerable background for metabolomics data (Canelas et al., 2008). Characteristics of an ideal extraction buffer are the ability to extract the most metabolites, possess nonselectivity, i.e., not eliminate the molecules with physical or chemical features, nondestructivity, and do not alter the metabolites physically and chemically (Wittmann et al., 2004). Because metabolites are frequently dissolved in a CDCl3 or D2O buffer for NMR, alternative extraction solvents are not very advantageous. Methanol, chloroform, and water mixed in a ratio of 5:2:2 v/v are frequently used as extraction solvents (Weckwerth et al., 2004). For lipid-containing metabolite extractions, water and chloroform are used, as is a 1:1 v/v mixture of methanol and chloroform (Wu et al., 2008).

1.2.4 Analytical analysis and data acquisition

The study of metabolic phenotypes in life science research has become a priority because of the development of metabolomics. Metabolomics is a broad method for detecting and quantifying endogenous/exogenous small molecules/metabolites with low molecular weight (1 kDa) in biological systems. Biological samples are subjected to two sequential processes in metabolomic studies: (1) detection of a comprehensive spectrum of metabolites by high-throughput analysis; and (2) model-recognizable analysis for deciphering the resulting data.

1.2.5 Comparison between NMR and MS

Nuclear magnetic resonance (NMR) and mass spectrometry (MS) are the most commonly used technologies in metabolomics. To determine the elemental composition and investigate the structure of compounds, MS examines the mass-to-charge ratio of charged molecules. Although MS is capable of determining a precise mass, chromatography is typically necessary due to the metabolome's limited molecular weight dispersion (Kell, 2004). Common separation methods in MS-based metabolomics include GC, CE, and HPLC (Blow, 2008). However, chromatography invariably affects the metabolome and depicts change. The amount obtained by chromatographic separation will differ for each metabolite to be identified since some metabolites will be lost or undergo chemical modification. Furthermore, for the detection, MS requires prior ionization of the molecule, which in turn raises the possibility that a particular metabolite will not ionize. Moreover, it can be difficult to estimate the concentration of each metabolite using MS (Gao & Xu, 2015).

On the other hand, NMR uses measurements of nuclear chemical shifts in a magnetic field to ascertain a molecule's structure. A single peak in an NMR spectrum can yield three important pieces of information. The chemical shift and peak splitting (J-coupling), both of which are utilized to determine the chemical structure of the metabolite (Freeman, 2003), are associated with the local chemical environment of that particular nucleus (e.g., 1H, 13C, and 15N). It is significant to note that the peak intensity is correlated with the concentration of the metabolite; the greater the intensity, the higher the concentration. Typically, multiple different NMR resonances are observed with each molecule, which removes the requirement for chromatographic separation and improves the precision of metabolite identification. Although the use of cryogenic probes has greatly boosted the sensitivity of NMR by a multiple of 4 (Moskau et al., 2003), MS is noticeably more sensitive than NMR and covers a larger diversity of the metabolome. NMR essentially only picks up the most prevalent metabolites at concentrations over 1–5 μM.

1.2.6 Analytical methodologies

LC-MS, GC-MS, and CE-MS are the most commonly used analytical procedures in MS. These four analytical methodologies serve as technical cornerstones for high-throughput metabolomics discovery. Each of these methods has benefits and drawbacks of its own. NMR offers a highly reproducible and less interlaboratory variable in situ method for screening certain classes of metabolites. NMR enables the quantitative detection of numerous different categories of metabolites and the quick, nondestructive, automated analysis of crude samples. However, the disadvantage is the reduced sensitivity and dynamic range (Wolfender et al., 2015). With the standardized MS electron ionization energy of 70 eV, GC-MS unites the benefits of high resolution, separation repeatability, and stable metabolite fragmentation patterns, making it the ideal tool for the investigation of both nonvolatile and volatile compounds (or its derivatives) (Gao & Xu, 2015). For a very long time, biological fluid metabolites have been identified using GC-MS. However, the volatility or capacity to produce suitable volatile derivatives of the molecules limits the range of metabolites that may be examined using this method. The examination of nonvolatile thermally labile substances is possible, however, with LC-MS. Molecules such as polar sugars, various lipids, and nonaromatic organic acids are all detectable by LC-MS (Commisso et al., 2013). Due to its complementarity with specific chemical information that LC-MS and GC-MS do not provide, CE-MS has become a popular technology for the profiling of polar and charged molecules in biological samples and for application in metabolomics (Ramautar et al., 2015). Small molecules and metabolites are frequently separated, characterized, and dramatized using the combination of GC-MS with NMR or/and LC-MS. This enhanced the structural analysis and the elucidation of targeted metabolomes (Kumar et al., 2014; Gao & Xu, 2015).

1.3 Spectral processing

To precisely identify and quantify the features in a metabolomics sample spectra investigation, a methodological approach called "spectral processing" is generally used. Also to acquire a complete set of feature quantifications, metabolic spectra are processed sequentially or cooperatively. To ensure that every final measurement refers to the same metabolomic characteristic across all samples, spectral processing is also necessary. The information obtained by spectrum processing is often organized in a feature quantification matrix (FQM), which contains the metabolic features of all the examined samples and serves as the input for future statistical analysis. In this data structure, the samples are represented by rows, while the various metabolomic properties are represented by columns. The concentration of a certain metabolite is fundamentally linked to each metabolomic characteristic.

1.3.1 Spectral preprocessing

Several preprocessing processes are typically used to enhance the signal quality and minimize any biases that may be present in the raw data. Baseline correction is used to eliminate low-frequency artifacts and sample variances in NMR- and MS-based spectra that are brought on by experimental and technical variation (Dietrich et al., 1991; Smith et al., 2006). After that, it could be essential to apply high-frequency filters to remove the electronic noise that was previously included in the data, which was produced by the measurement equipment.

1.3.2 Feature detection

Finding and quantifying the characteristics that are present in the spectra is the goal of the feature detection step. In MS-based investigations, the most widely used algorithmic solution for feature detection is the peak-based approach. Throughout the spectrum, these techniques detect the peaks and then combine the regions to calculate the quantity of the metabolite at the core. Spectral alignment is frequently used in this method, either prior or postpeak detection (Rafiei & Sleno, 2015).

1.3.3 Peak detection

Each sample spectrum is examined separately by the most popular peak detection techniques. Two analytical steps form the foundation of these techniques. The spectra are rounded out in the initial stage. Several alternative filters are available for this goal. However, the wavelet transform–based filters outperformed the others. Using one or more detection thresholds, the various metabolite peaks are recognized in the second phase. These thresholds are applied to various metrics, including the signal-to-noise ratio, intensity, or area of each peak from the filtered spectra that are produced (Yang et al., 2009).

1.3.4 Spectral alignment

One of the key processing processes in multisample metabolomic investigations involves spectral alignment. Nonlinear shifts in the position of peaks belonging to the same metabolic characteristic may be observed when examining multiple spectra. In NMR-based research, these fluctuations in the ppm axis are typically caused due to the alterations in the sample's chemical environment, such as the ionic strength, pH, or concentration of the protein in the sample. Peak shifts across the retention time axis are frequently seen in MS-based investigations and are typically correlated with changes in the chromatographic column's stationary phase (Alonso et al., 2015).

1.4 Feature normalization

In general, a data normalization step is necessary to accurately quantify the characteristics in metabolomic analysis. By eliminating unwanted systematic biases, normalization aims to leave only physiologically meaningful differences in the data. This process is essential for evaluating complex biofluids like blood because the introduction of internal standards can be challenging and sample to sample variations in metabolite concentration can be substantial.

1.5 Deconvolution methods in targeted analysis

The overlap between peaks from various metabolites is a major drawback for quantifying metabolomic characteristics. These biases are very common in NMR and GC-MS spectra. Several methodological strategies have been created to address this technological difficulty. These strategies are based on spectral deconvolution when several peaks overlap within the same spectral region, a signal processing method that calculates the relative area associated with each peak (Hao et al., 2014) in that sample analysis. The fact that prior knowledge of the metabolites present in the mixture is highly essential is a prerequisite for the restriction of deconvolution techniques. Additionally, due to computational intractability, it is still not practical to utilize these methods in studies of untargeted metabolites.

2. Data analysis

Numerous univariate and multivariate statistical techniques can be applied to carry out the necessary research analysis once the metabolite properties have been robustly measured. Commonly referred to as "chemometric approaches," these classes of techniques typically call for some level of expertise to be effectively utilized.

2.1 Univariate and multivariate analysis methods

Individual metabolomic characteristics are examined by univariate approaches. Since they are widespread statistical analysis techniques, their main benefit is that they are simple to apply and analyze. Their primary flaw, however, is that they fail to account for the interactions that may exist between various metabolic variables. Furthermore, these techniques do not account for the impact of potential confounding variables such as sex, diet, or BMI, which increases the risk of receiving false-positive or false-negative results on analysis (Alonso et al., 2015).

Approaches to multivariate analysis, as opposed to univariate approaches, consider all of the metabolomic variables simultaneously and, as a result, can spot patterns of relationships between them. The two categories of these pattern recognition techniques are supervised and unsupervised techniques. Patterns of data similarity are identified using unsupervised analysis, where the similarity of the data patterns is taken into account regardless of the study samples' type or classification. Sample labels are used to determine which features or a combination of features are most frequently associated with a particular phenotype of interest.

2.2 Unsupervised and supervised methods

Unsupervised techniques are frequently used to summarize complicated metabolomic data. Data patterns related to biological and/or experimental variables are identified by this approach.

The widely used unsupervised technique in metabolic studies is the principal component analysis (PCA). This method relies on the linear transformation of the metabolic features into a set of principal components, which are variables that are orthogonal and linearly uncorrelated (Bro & Smilde, 2014). Applied to metabolomic data are other unsupervised techniques such as self-organizing maps (SOMs) and hierarchical clustering analysis (HCA). These techniques may be especially useful for identifying nonlinear data trends that PCA does not easily cover.

Using supervised techniques, metabolic patterns associated with the target phenotypic variable can be identified, while other sources of variance are downweighted. These techniques are also the foundation for metabolomic feature-based classifiers. In metabolomics, partial least squares (PLS) is one of the most popular supervised techniques. It can be applied as a binary classifier or a regression analysis, depending on the quantitative variable of interest (Fonville et al., 2010). The covariance between the relevant variable and the metabolomic data is maximized using PLS components, as opposed to PCA, which maximizes the explained data set variance (Alonso et al., 2015). As a result, the feature coefficients (loadings) of PLS components serve as an indicator of how much a feature aids in the separation of the various sample groups.

2.3 Metabolite identification

One of the most promising applications of metabolomics in the medical sciences is the identification of biomarkers. Since supervised analytic models may combine the evidence of many metabolites, they are typically used to identify new metabolomic biomarkers. Metabolite identification is one of the most difficult tasks in high-throughput metabolomic studies.

This phase is crucial for giving the related features in a metabolomic analysis in a biological context. Metabolite identification requires metabolite spectral databases. The effectiveness of identification algorithms depends on the quality of the data that have been saved and the number of metabolite spectra that are accessible in these records.

2.4 Metabolite identification from MS-based spectra

The typical method for metabolite identification in MS-based investigations is to use a tolerance window to query metabolomic databases for the neutral molecular mass values of the detected peaks. The calculated neutral molecular mass depends on the chemical composition of the identified peak and is derived from the peak m/z value (i.e., ionization mode and ionization adduct). Each peak m/z value can result in a variety of plausible neutral molecular weights that can represent various ionization adducts (Na+, H+, K+, ...), assuming no prior knowledge. Many false-positive identifications are generated by this multiplicity. Several techniques have been developed to decrease false positives. AStream and Camera are tools made to recognize isotopic and adduct patterns to lessen the complexity of the data in MS investigations (Alonso et al., 2011; Kuhl et al., 2012). These methods only infer one neutral mass from each defined pattern while estimating the chemical composition of each picked ion peak. Using these techniques also improves the identification of actual biological substances.

2.5 Metabolite identification from NMR-based spectra

Automatic metabolite identification is frequently carried out in NMR-based investigations by comparing the recorded NMR peaks to a collection of reference metabolite patterns. Each metabolite reference spectrum is composed of one or more peaks, which are identified by their relative intensities and ppm locations. By comparing the reference peak locations to the list of discovered peak positions, the online program MetaboHunter can identify chemicals (Tulpan et al., 2011). Due to the fact that this method only employs one peak parameter to match reference peaks, it may result in significant false-positive rates. Newer techniques built on the reliable cluster notion have replaced the MetaboHunter methodology. These techniques use peak intensities and intersample intensity correlation as criteria for matching data peaks to reference peaks in addition to the ppm position. This metabolite identification method is followed by the NMR analysis workflow employed in FOCUS, with the added benefit that it also takes into consideration the presence of missing peaks caused by spectral overlaps (Alonso et al., 2015).

3. Pathway and network analysis of metabolomic data

The amount of information produced by metabolomic research is increased by pathway and network analysis techniques. Both strategies make use of the relational characteristics found in metabolomic data. Pathway analysis examines metabolite patterns from an integrated perspective using prior biological information. In contrast, network analysis builds metabolic networks that characterize the complex relationships present in the set of detected metabolites by utilizing the high degree of correlation present in metabolomics data.

3.1 Pathway analysis

Comprehensive information on a large number of metabolic pathways can be found in biological databases such as Kyoto Encyclopedia of Genes and Genomes (KEGG), small molecule pathway database (SMPDB), EHMN, WikiPathways, and MetaCyc. The use of pathway-based techniques in metabolomics is thus made possible by the availability of these data. These techniques, known as metabolite set enrichment analysis (MSEA), are based on the gene set enrichment analysis (GSEA) approach and were created to analyze pathways in gene expression data. To evaluate metabolomic data at the pathway level, metabolomics researchers today have access to a wide range of software tools. Numerous tools for visualizing pathways are offered by programs such as Paintomics, Vanted, and Cytoscape.

3.2 Correlation-based network analysis

Correlation-based approaches, in contrast to pathway analysis, construct metabolite networks in accordance with the relationship patterns seen in the experiment data. Each metabolite is represented as a network node in the final network, but unlike pathway analysis, the links between nodes show the degree of mathematical correlation between each pair of metabolites (Krumsiek et al., 2011). High correlation values are common in metabolomics data because systemic and indirect correlations exist. Using traditional correlation coefficients results in dense networks that fail to discern between direct and indirect links. Partial correlation can be used to solve this issue (Alonso et al., 2015).

4. Methodologies for reprogramming metabolomics

While reprogramming metabolomics discovers key biomarkers and pathways, discovery metabolomics offers prospective metabolic modulators, reprogramming metabolites, for changing an existing metabolome to a targeted metabolome. Reconstruction of the metabolic strategy results from this. Using functional assays and metabolomics analyses, the reconstitution may be proven. By calculating the mass isotopomer distribution for all labeled compounds, the metabolomics analysis is utilized to validate the impact of exogenous metabolite boosting on various metabolites and metabolic pathways and discover the metabolic flux magnitudes into each metabolite pool. Functional experiments give evidence of whether a metabolic approach that has been recreated might exhibit the same symptoms and mechanisms of action (Peng et al., 2015) Therefore, by analyzing the response and reconstituted metabolomes, respectively, reprogramming metabolomics discloses the response and reconstituted mechanisms, in which methodologies on molecular biology, cell biology, biochemistry, and immunology will become basic tools to interpret these mechanisms. Fig. 16.2 presents the fundamental procedures of reprogramming metabolomics.

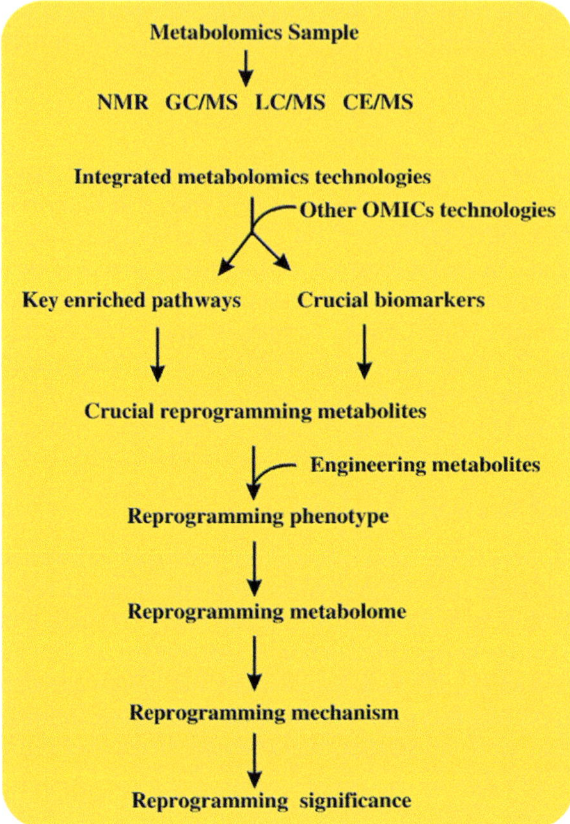

FIGURE 16.2 Outline of metabolome reprogramming.

References

Alonso, A., Julià, A., Beltran, A., Vinaixa, M., Díaz, M., Ibañez, L., Correig, X., & Marsal, S. (2011). AStream: an R package for annotating LC/MS metabolomic data. *Bioinformatics (Oxford, England), 27*(9), 1339–1340. https://doi.org/10.1093/bioinformatics/btr138

Alonso, A., Marsal, S., & Julià, A. (2015). Analytical methods in untargeted metabolomics: State of the art in 2015. *Frontiers in Bioengineering and Biotechnology, 3*, 23. https://doi.org/10.3389/fbioe.2015.00023

Blow, N. (2008). Metabolomics: Biochemistry's new look. *Nature, 455*(7213), 697–700. https://doi.org/10.1038/455697a

Bro, R., & Smilde, A. K. (2014). Principal component analysis. *Analytical Methods, 6*(9), 2812–2831. https://doi.org/10.1039/C3AY41907J

Canelas, A. B., Ras, C., ten Pierick, A., van Dam, J. C., H eijnen, J. J., & van Gulik, W. M. (2008). Leakage-free rapid quenching technique for yeast metabolomics. *Metabolomics, 4*(3), 226–239. https://doi.org/10.1007/s11306-008-0116-4

Commisso, M., Strazzer, P., Toffali, K., Stocchero, M., & Guzzo, F. (2013). Untargeted metabolomics: An emerging approach to determine the composition of herbal products. *Computational and Structural Biotechnology Journal, 4*, e201301007. https://doi.org/10.5936/csbj.201301007

Costerton, J. W., Lewandowski, Z., Caldwell, D. E., Korber, D. R., & Lappin-Scott, H. M. (1995). Microbial biofilms. *Annual Review of Microbiology, 49*, 711–745. https://doi.org/10.1146/annurev.mi.49.100195.003431

Del Pozo, J. L. (2018). Biofilm-related disease. *Expert Review of Anti-Infective Therapy, 16*(1), 51–65. https://doi.org/10.1080/14787210.2018.1417036

Dietrich, W., Rüdel, C., & Neumann, M. (1991). Fast and precise automatic baseline correction of one- and two-dimensional nmr spectra. *Journal of Magnetic Resonance, 91*, 1–11.

Fonville, J. M., Richards, S. E., Barton, R. H., Boulange, C. L., Ebbels, T. M. D., Nicholson, J. K., Holmes, E., & Dumas, M. E. (2010). The evolution of partial least squares models and related chemometric approaches in metabonomics and metabolic phenotyping. *Journal of Chemometrics, 24*(11–12), 636–649. https://doi.org/10.1002/cem.1359

Freeman, R. (2003). *Magnetic resonance in chemistry and medicine*. Oxford University Press.

Gao, P., & Xu, G. (2015). Mass-spectrometry-based microbial metabolomics: Recent developments and applications. *Analytical and Bioanalytical Chemistry, 407*(3), 669–680. https://doi.org/10.1007/s00216-014-8127-7

Hall-Stoodley, L., Costerton, J. W., & Stoodley, P. (2004). Bacterial biofilms: From the natural environment to infectious diseases. *Nature Reviews Microbiology, 2*(2), 95–108. https://doi.org/10.1038/nrmicro821

Halouska, S., Zhang, B., Gaupp, R., Lei, S., Snell, E., Fenton, R. J., Barletta, R. G., Somerville, G. A., & Powers, R. (2013). Revisiting protocols for the NMR analysis of bacterial metabolomes. *Journal of Integrated OMICS, 3*(2), 120–137. https://doi.org/10.5584/jiomics.v3i2.139

Hao, J., Liebeke, M., Astle, W., De Iorio, M., Bundy, J. G., & Ebbels, T. M. D. (2014). Bayesian deconvolution and quantification of metabolites in complex 1D NMR spectra using BATMAN. *Nature Protocols, 9*(6), 1416–1427. https://doi.org/10.1038/nprot.2014.090

Hung, C., Pinkner, J. S., Dodson, K. W., Crowley, J. R., Heuser, J., Chapman, M. R., Zhou, Y., Hadjifrangiskou, M., Hultgren, S. J., & Henderson, J. P. (2013). *Escherichia coli* biofilms have an organized and complex. *MBio, 4*(5), 1–10. https://doi.org/10.1128/mBio.00645-13

Kell, D. B. (2004). Metabolomics and systems biology: Making sense of the soup. *Current Opinion in Microbiology, 7*(3), 296–307. https://doi.org/10.1016/j.mib.2004.04.012

Krumsiek, J., Suhre, K., Illig, T., Adamski, J., & Theis, F. J. (2011). Gaussian graphical modeling reconstructs pathway reactions from high-throughput metabolomics data. *BMC Systems Biology, 5*, 21. https://doi.org/10.1186/1752-0509-5-21

Kuhl, C., Tautenhahn, R., Böttcher, C., Larson, T. R., & Neumann, S. (2012). Camera: An integrated strategy for compound spectra extraction and annotation of liquid chromatography/mass spectrometry data sets. *Analytical Chemistry, 84*(1), 283–289. https://doi.org/10.1021/ac202450g

Kumar, B., Prakash, A., Ruhela, R. K., & Medhi, B. (2014). Potential of metabolomics in preclinical and clinical drug development. *Pharmacological Reports : PR, 66*(6), 956–963. https://doi.org/10.1016/j.pharep.2014.06.010

León, Z., García-Cañaveras, J. C., Donato, M. T., & Lahoz, A. (2013). Mammalian cell metabolomics: Experimental design and sample preparation. *Electrophoresis, 34*(19), 2762–2775. https://doi.org/10.1002/elps.201200605

Link, H., Anselment, B., & Weuster-Botz, D. (2008). Leakage of adenylates during cold methanol/glycerol quenching of *Escherichia coli*. In *Metabolomics* (Vol. 4). Boston: Springer US.

Maharjan, R. P., & Ferenci, T. (2003). Global metabolite analysis: The influence of extraction methodology on metabolome profiles of *Escherichia coli*. *Analytical Biochemistry, 313*(1), 145–154. https://doi.org/10.1016/s0003-2697(02)00536-5

Maifiah, M. H. M., Cheah, S.-E., Johnson, M. D., Han, M.-L., Boyce, J. D., Thamlikitkul, V., Forrest, A., Kaye, K. S., Hertzog, P., Purcell, A. W., Song, J., Velkov, T., Creek, D. J., & Li, J. (2016). Global metabolic analyses identify key differences in metabolite levels between polymyxin-susceptible and polymyxin-resistant Acinetobacter baumannii. *Scientific Reports, 6*, 22287. https://doi.org/10.1038/srep22287

Marcinowska, R., Trygg, J., Wolf-Watz, H., Mortiz, T., & Surowiec, I. (2011). Optimization of a sample preparation method for the metabolomic analysis of clinically relevant bacteria. *Journal of Microbiological Methods, 87*(1), 24–31. https://doi.org/10.1016/j.mimet.2011.07.001

Meyer, H., Weidmann, H., & Lalk, M. (2013). Methodological approaches to help unravel the intracellular metabolome of Bacillus subtilis. *Microbial Cell Factories, 12*, 69. https://doi.org/10.1186/1475-2859-12-69

Mohd Kamal, K., Mahamad Maifiah, M. H., Abdul Rahim, N., Hashim, Y. Z. H.-Y., Abdullah Sani, M. S., & Azizan, K. A. (2022). Bacterial metabolomics: Sample preparation methods. *Biochemistry Research International, 2022*, 9186536. https://doi.org/10.1155/2022/9186536

Moskau, D., Ritcher, C., Kovaks, H., Salzmann, M., Baselgia, L., Haeberli, M., Marek, D., & Schett, O. (2003). Highest sensitivity for cutting-edge NMR applications: 600 MHz CryoProbes. *Spectra Analyse, 32*, 39–41.

Peng, B., Li, H., & Peng, X.-X. (2015). Functional metabolomics: From biomarker discovery to metabolome reprogramming. *Protein & Cell, 6*(9), 628–637. https://doi.org/10.1007/s13238-015-0185-x

Rafiei, A., & Sleno, L. (2015). Comparison of peak-picking workflows for untargeted liquid chromatography/high-resolution mass spectrometry metabolomics data analysis. *Rapid Communications in Mass Spectrometry: RCM, 29*(1), 119–127. https://doi.org/10.1002/rcm.7094

Ramautar, R., Somsen, G. W., & de Jong, G. J. (2015). CE-MS for metabolomics: Developments and applications in the period 2012–2014. *Electrophoresis, 36*(1), 212–224. https://doi.org/10.1002/elps.201400388

Roberts, L. D., Souza, A. L., Gerszten, R. E., & Clish, C. B. (2012). Targeted metabolomics. *Current Protocols in Molecular Biology, Chapter 30.* Unit 30.2.1–24 https://doi.org/10.1002/0471142727.mb3002s98.

Segers, K., Declerck, S., Mangelings, D., Heyden, Y. V., & Van Eeckhaut, A. (2019). Analytical techniques for metabolomic studies: A review. *Bioanalysis, 11*(24), 2297–2318. https://doi.org/10.4155/bio-2019-0014

Shaw, L., Golonka, E., Potempa, J., & Foster, S. J. (2004). The role and regulation of the extracellular proteases of *Staphylococcus aureus. Microbiology (Reading, England), 150*(Pt 1), 217–228. https://doi.org/10.1099/mic.0.26634-0

Siegel, D., Permentier, H., Reijngoud, D.-J., & Bischoff, R. (2014). Chemical and technical challenges in the analysis of central carbon metabolites by liquid-chromatography mass spectrometry. *Journal of Chromatography. B, Analytical Technologies in the Biomedical and Life Sciences, 966*, 21–33. https://doi.org/10.1016/j.jchromb.2013.11.022

Smith, C. A., Want, E. J., O'Maille, G., Abagyan, R., & Siuzdak, G. (2006). XCMS: Processing mass spectrometry data for metabolite profiling using nonlinear peak alignment, matching, and identification. *Analytical Chemistry, 78*(3), 779–787. https://doi.org/10.1021/ac051437y

Tulpan, D., Léger, S., Belliveau, L., Culf, A., & Cuperlović-Culf, M. (2011). MetaboHunter: An automatic approach for identification of metabolites from 1H-NMR spectra of complex mixtures. *BMC Bioinformatics, 12*, 400. https://doi.org/10.1186/1471-2105-12-400

Weckwerth, W. (2003). Metabolomics in systems biology. *Annual Review of Plant Biology, 54*, 669–689. https://doi.org/10.1146/annurev.arplant.54.031902.135014

Weckwerth, W., Wenzel, K., & Fiehn, O. (2004). Process for the integrated extraction, identification and quantification of metabolites, proteins and RNA to reveal their co-regulation in biochemical networks. *Proteomics, 4*(1), 78–83. https://doi.org/10.1002/pmic.200200500

Wittmann, C., Krömer, J. O., Kiefer, P., Binz, T., & Heinzle, E. (2004). Impact of the cold shock phenomenon on quantification of intracellular metabolites in bacteria. *Analytical Biochemistry, 327*(1), 135–139. https://doi.org/10.1016/j.ab.2004.01.002

Wolfender, J.-L., Marti, G., Thomas, A., & Bertrand, S. (2015). Current approaches and challenges for the metabolite profiling of complex natural extracts. *Journal of Chromatography. A, 1382*, 136–164. https://doi.org/10.1016/j.chroma.2014.10.091

Wu, H., Southam, A. D., Hines, A., & Viant, M. R. (2008). High-throughput tissue extraction protocol for NMR- and MS-based metabolomics. *Analytical Biochemistry, 372*(2), 204–212. https://doi.org/10.1016/j.ab.2007.10.002

Yang, C., He, Z., & Yu, W. (2009). Comparison of public peak detection algorithms for MALDI mass spectrometry data analysis. *BMC Bioinformatics, 10*, 4. https://doi.org/10.1186/1471-2105-10-4

CHAPTER

17

Dysbiosis of microbiome: a risk factor for cancer, metabolic and inflammatory diseases

Yuvaraj Sivamani[1], *Harshit Sajal*[2] *and Sumitha Elayaperumal*[2]

[1]Department of Pharmaceutical Chemistry, Cauvery College of Pharmacy, Mysore, Karnataka, India; [2]Department of Biotechnology and Bioinformatics, JSS Academy of Higher Education and Research, Mysore, Karnataka, India

1. Introduction

The road to health is paved with good intestines! ***Sherry A. Rogers***

Metabolic syndrome (MetS) gained global prominence when it was observed to be significantly linked to the emergence of metabolic disorders (e.g., diabetes, etc.), where its incidence increases rapidly in children and adolescents (Wang et al., 2020). Under certain conditions, metabolic diseases are reversible and can be prevented. Hence, there arises a need for early diagnosis and treatment to prevent the premature onset of the disease (Zhu et al., 2020). Current research focuses on microbiome-based interventions to prevent the onset of metabolic syndrome (Scheithauer et al., 2020). Metabolic syndrome is a complex disorder characterized by interconnected factors including clinical, biochemical, and physiological factors. Alteration in the metabolic factors could lead to an increased risk of diabetes mellitus and cardiovascular diseases. As per International Diabetes Federation (IDF), metabolic syndromes are characterized by features including dyslipidemia, a rise in fasting glucose, blood pressure, and obesity (Zhu et al., 2020). The composition of the gut varies depending on the geographical origin where the prevalence was found to vary from <10% to 84% (Cornier et al., 2008). Since the economic and social burden is on an increase, current research focuses to understand the complex pathogenicity of metabolic disorders (Fukuda & Ohno, 2014). The gut microbiota exerts functional properties that

Microbial Biofilms
https://doi.org/10.1016/B978-0-323-95715-1.00011-X

influence both the physiological and pathological status in humans (Fukuda & Ohno, 2014). Production of vitamins and digestion of the dietary components mediated by fermentation modifies the host nutrition and energy harvest; the impact of homeostasis in the intestinal epithelial tissues; the development of the host immune system; infection resistance; and metabolism of drugs (Fan & Pedersen, 2020; Fukuda & Ohno, 2014; Zheng et al., 2020). Studies on the gut microbiota and its function are hampered by the challenges of growing intestinal microorganisms. Recent improvements in various analyzing methods including metagenome analysis and bacterial sequencing have led to an understanding of the knowledge of the microbes, which cannot be cultured, cross-talk between the host and pathogenicity, and an understanding of the functions of the microbes (Liu et al., 2022; Zheng et al., 2020).

Bacteria are widespread in the environment including soil, water, and air. Large amounts of bacteria reside in the skin of humans and mostly in the gastrointestinal tract (GIT) of humans (Liu et al., 2022; Zheng et al., 2020). Bacteria come in a variety of shapes including coccid, rod-shaped, spirillum, and budding bacteria (van Teeseling et al., 2017). They are classified as aerobic, anaerobic, or microaerophilic based on their usage of oxygen and tolerance to it. Bacteria are classed based on their form, cell membrane features, and energy utilization, such as heterotrophic, phototrophic, or lithotrophic. Under optimal conditions, bacteria multiply every 20 min resulting in doubling the population in a short span of time and possess the ability to adapt according to the environmental conditions. For a long time, the primary focus has been on bacteria that are pathogenic to humans, e.g., *Mycobacterium tuberculosis* causing tuberculosis in humans, *Streptococcus*, *Pseudomonas*, and *Pneumococcus* lead to pneumonia, tetanus caused by bacterium *Clostridium tetani*, etc. But in case of microbiota they are not harmful and they are essential for the host. The microbiota has a significant impact on food breakdown and adsorption, pathogen defense, immune system stimulation, and gut health. Studies have shown that high variability of microbiota is observed in individuals, which is highly influenced by genetic makeup of the individual, medication, illness, diet, and personal hygiene (Cho & Blaser, 2012; O'Hara & Shanahan, 2006; Sommer & Bäckhed, 2013).

1.1 Colonization—diversity of microbes

The microbiome plays an important role in immune system development (Zheng et al., 2020). The gut microbiome, also known as the human body's "second genome," is the sum of the genetic information of the gut microbiota and performs extremely complex functions (Zhu et al., 2010). The microbiota's complexity and composition are critical for intestinal health homeostasis (Zheng et al., 2020). Trillions of microbes that inhabit the gut are mammalian symbionts. GM is often classified into two major kinds: (1) mucosal and (2) luminal (Zheng et al., 2020). Bacteria account for 99% of total GM, with fungi, yeast, and archaea accounting for the remaining 1% (Khan et al., 2014). Two major taxa the bacteria belong to are (1) *Firmicutes* (typically gram-positive) and (b) Bacteroidetes (usually gram-negative), with exceptions with a smaller proportion of *Actinobacteria*, *Proteobacteria*, *Fusobacteria*, and *Verrucomicrobia* (Ferreira et al., 2011; Khan et al., 2014). The structure of the GM is influenced by both internal and external factors including age, the genetic makeup of the individual, age, dietary habits, and usage of antibiotics. A diet high in fat, carbohydrates, or protein has an impact on an individual's healthy GM composition. *Endophytes* predominant in the high-fat diet include *Bacteroides* and *Enterobacteriaceae*, whereas in the vegetarian diet *Bacteroidetes* and *Prevotella* were strongly linked (Conlon & Bird, 2014; Simpson & Campbell, 2015). *Clostridium* cluster XIVa (*Roseburia-E. rectale* and *Clostridium coccoides-Eubacterium*) is abundant in omnivorous individuals. Single nucleotide polymorphisms (SNPs) were found to be associated with various bacterial taxa identified in, 48 irritable bowel disease (Becattini et al., 2017; Knights et al., 2014). Becattini et al. demonstrated that GM has a resistant activity (through colonization resistance) against the foodborne pathogen (*Listeria monocytogenes*) demonstrating that GM acts as a potential natural probiotic. Depletion of GM was observed in mice treated with chemotherapeutic agents or antibiotic-treated mice with the predominance of *L. monocytogenes*.

The GM works as a metabolite factory, converting host-based substrates including proteins, carbohydrates, and lipids into a variety of metabolites such as bile acids (BAs), amino acids, and short-chain fatty acids (SCFAs) butyrate, acetate, and propionate (Rajilić-Stojanović, 2013). The physiological homeostasis of GM aids in the regulation of the immune system, avoids sickness, and controls the consumption of energy (Lee & Lee, 2014).

2. Host—microbiota interaction

The gastrointestinal (GI) system serves as a link between the host and the environment. The small intestine, which is defined by a large surface area, plays a vital role in the digestion and absorption of nutrients. Because the intestine has the largest bacterial concentration, the host was compelled to create a strategy for tolerance to useful and harmless microorganisms, as well as an effective defense system against infections and bacterial overgrowth (Sommer & Bäckhed, 2013). Mucus represents the first line of defense mechanism. The mucus lines the intestine (inner dense layer), which prevents the bacteria from penetration, the thickest being in the colon, thereby isolating the epithelial layer from luminal microbiota (Juge, 2012). Mucins are rich in glycosylation and display higher order of conservation, which poses a crucial role in the selection of commensal gut bacteria. Due to the possession of the lectins, the microorganisms can adhere to the mucus, e.g., *Lactobacillus reuteri*. Human histoblood group antigens present in mucins have been proposed as receptors (Juge, 2012). Interestingly, healthy colons had a significant abundance of mucin-degrading *Akkermansia muciniphila*, but individuals with inflammatory bowel disease had lower quantities. Thus, intestinal integrity is maintained by *A. muciniphila* (Juge, 2012). Regarding this context, animal studies have shown that in the formation of mucins, the bacteria possess a great impact (Sommer & Bäckhed, 2013). The observation was confirmed in germ-free (GF) mice, which possessed fewer goblet cells that were proportional to the thin mucus layer. The mice were vulnerable to frequent infections with alterations in the immunoglobulin and cytokine levels (Shanahan, 2002).

The intestinal epithelium layer acts as a physical barrier that restricts the paracellular transport of molecules composed of paneth, goblet, enterocytes, and endoenterocrine cells. The integrity of epithelium is maintained by both cytosolic (e.g., occludin, claudins) and transmembrane proteins (Chelakkot et al., 2018). An increase in paracellular and vascular permeability affects Crohn's disease (CD), where the area around injury was dominated by *Escherichia* and a lower proportion of the phylum *Firmicutes*, including *Blautia*, *Lachnospira*, and *Faecalibacterium*, compared with adjacent mucosal sites without injury (Libertucci et al., 2018). In patients diagnosed with CD, endothelial lesions are observed, which is favored by dysbiosis. Currently, dysbiosis is also depicted in patients with ankylosing spondylitis where the invasion of bacteria has been described (Ciccia et al., 2017). Although animal studies have concluded the role of commensal microbiota in the maintenance of intestinal homeostasis, further studies need to be carried out to rule out whether dysbiosis remains the cause of several diseases.

The gut-associated lymphoid tissue present in the intestine ensures the maintenance of homeostasis, which regulates and plays a key role in the host defense mechanism. The intestinal immune system faces challenges including antigens, opportunistic microorganisms, and potential pathogens. Antigens, commensal and opportunistic microorganisms, as well as potential pathogens, are constantly bombarding the intestinal immune system. As a result, it is constantly forced to differentiate between noninvasive bacteria and pathogens. Recognition of bacteria by the epithelial cell and dendritic cell is mediated by pattern recognition receptors (PRRs) such as Toll-like receptors 4 (TLR4) or nod-like receptors. TLR4 is mostly activated by gram-negative bacteria's lipopolysaccharide, whereas TLR2 is activated by gram-positive bacteria's peptidoglycan and lipoteichoic acids. Activation of the immune cells is mediated by PPRs, leading to the recruitment of immune cells, e.g., T cells, macrophages, etc. at the vicinity of infection, resulting in the secretion of chemokines and cytokines (Magrone & Jirillo, 2013; O'Hara & Shanahan, 2006). Protection against bacterial sepsis is mediated by the increased concentration of IgA secreted by the plasma cells inferred by the mouse studies (Wilmore et al., 2018). An individual cytokine pattern is provoked by each bacterium, which includes the anti- and proinflammatory cytokines (12p70, IL-23, IL-10), thereby deciding inflammation or tolerance tuned by the microbiota and the host immune system (Manuzak et al., 2012).

Existing lines of evidence suggest the interaction between the microbiota and brain in a bidirectional manner. In contrast, gut microbes have an impact on the central nervous system via neuroendocrine, neuronal, and immune-mediated processes (Martin et al., 2018). Short-chain fatty acids, which play a vital role in intestinal barrier function, are maintained by the gut microbiota (Feng et al., 2018). Furthermore, gut bacteria influence mucosal immune function and alter afferent sensory nerves, for example, by inhibiting calcium-dependent potassium channels (Kunze et al., 2009). The brain influences gut microbiota by modulating GI motility and mucus production, as well as altering intestinal permeability and immune function. The interaction between the microbiota and the immune system is well addressed by preclinical studies (Kelly et al., 2016; Sudo et al., 2004). Alterations in the gut–brain axis are well elucidated in the depressive and psychiatric disorders in autism spectrum disorders (Foster & McVey Neufeld, 2013): also associated with neurological disorders including Parkinson's disease and fibromyalgia (Foster & McVey Neufeld, 2013). Other conditions associated with the altered gut rain axis include obesity (Moser et al., 2018) and even irritable bowel syndrome (Moser et al., 2018).

2.1 Causes of dysbiosis

The mode of delivery is the first aspect determining microbe acquisition (Pinto Coelho et al., 2021). The composition of microorganisms is also influenced by early nutrition (breast- or formula-feeding) and the sanitary environment (lifestyle and geographic location) in which a child is raised (Pinto Coelho et al., 2021). Later in life, the exposome, which includes environmental stimuli such as nutrition, stress, sanitary milieu (interaction with disinfectants or animals), medications, and geography (e.g., air pollution), helps to shape the gut ecology. Daily food has a huge influence on the gut environment (Su & Liu, 2021). A high-fat diet, in particular, appears to alter the microbiota toward a more dysbiotic pattern, which is linked to an increased risk of intestinal inflammation. Animal and human research both revealed quick adaptation to changing food patterns (Valdes et al., 2018). By feeding mice an entirely animal- or plant-based diet for a short period, David et al. revealed the plasticity and stability of the intestinal ecosystem in response to short-term perturbations (Lindsay et al., 2020). However, other studies have shown that significant changes in the variety and composition of the intestinal ecosystem are not generated by short-term changes in dietary habits or the administration of specific nutrients, but rather by long-term pressure on the ecosystem (Zheng et al., 2020).

During birth, the newborn comes in contact with microbes, which gradually increases in the due course of the first month of life. Colonization of the microbes occurs at the skin, oral, gut, and nasopharyngeal membrane, which is strikingly different from the mother's body habitat. Studies have shown that accusation of microbiota was dominated by *Lactobacillus, Sneathia* spp., and *Prevotella* in vaginally delivered babies; in contrast, babies born via cesarean section were found to have dominant microbiota in skin, e.g., *Propionibacterium* spp., etc. that are not similar to maternal skin microbiota. It is to be noted that a substantial variation in the microbial pattern exists in the babies. Modification of the gut microbiota occurs successively during the first year of life with the maintenance of individual gut microorganisms, which offers a competitive advance over early-settled bacteria. Diversity associated with colonization is highly essential for well-balanced interaction between the host and microbiota. Stabilization of the microbiota occurs at the end of 2 years (Bokulich et al., 2016) (Fig. 17.1). Existing studies suggest sequential variations exist in the microbiota in individuals, which explains that environmental conditions play a vital role in the establishment of gut microflora (Gross, 2007; Palmer et al., 2007). Application of antibiotics possesses a negative impact on the colonization of gut bacteria, e.g., *Enterobacteriaceae* and *Erysipelotrichaceae* (Bokulich et al., 2016). The concentration of microbiota exists in small numbers in the stomach and inversely in the colon, which implies a steady rise down the GIT. It is to be

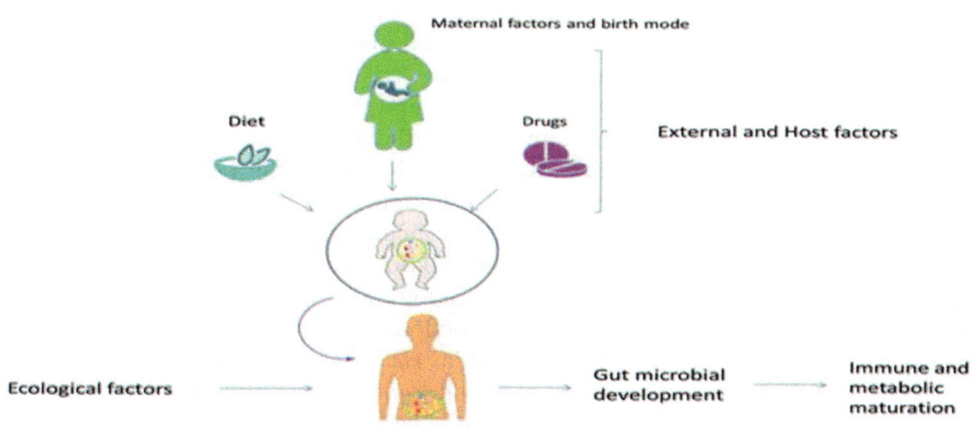

FIGURE 17.1 Factors affecting the gut microbiota development in infant.

noted that the gut microorganisms are anaerobic belonging to *phyla Proteobacteria, Bacteroidetes*, and *Firmicutes*. Other gut bacteria in the healthy gut account for 1% including *Acidobacteria, Actinobacteria*, etc. Longitudinal variations exist between the microbiota between the gut lumen and epithelium (Dieterich et al., 2019; Sommer & Bäckhed, 2013).

3. Microbiota—health and diseased condition

The ability to resist change under physiological stress conditions coupled with high diversity defines healthy microbiota (Geng et al., 2020). Under diseased conditions, lower species diversity is observed in beneficial microbes, and an increase in the pathobionts was observed. Disruption of the microbiota has been found to disrupt the metabolism in the host and immunity, leading to disease of the GIT (Belkaid & Hand, 2014). The functions of the GIT include absorption of nutrients, removal of waste, and acting as a barrier, thereby preventing invasion of pathogens, maintaining homeostasis, balancing inflammatory conditions, and preventing the translocation of antigens or food in the bloodstream. Disruption of the microbiota in the gut makes the GIT susceptible to diseased conditions (Belkaid & Hand, 2014; Rajilić-Stojanović, 2013).

In humans, three distinct enterotypes have been described, where long-term studies have depicted clusters belonging to phylum Bacteroidetes including *Bacteroides* and *Prevotella* (Wu et al., 2011). *Enterotypes* were found to be associated with the type of diet, *Prevotella* being increased in a diet rich in carbohydrates and *Bacteroides* in a diet rich in high protein and fat. It is highly interesting to note that the composition of bacteria varies within 24 h, but the endophytes were found to be stable during a 10-day diet (Wu et al., 2011). Substantial conservation in the microbiota was observed after dietary changes (gluten-free diet, low fermentable carbohydrates) according to Wu et al. Susceptible and vulnerable microbial composition was observed in patients with nonceliac gluten sensitivity in par with the healthy controls. Altered composition of gut microbiota was observed in patients with inflammatory diseases including psoriasis, systemic lupus erythematosus (SLE), and rheumatoid arthritis (RA) leading to dysbiosis (Fig. 17.2) (Forbes et al., 2016).

In recent years, the prevalence of GI disorders has increased, leading to physiological and morphological abnormalities characterized by altered microbiota, and hypersensitivity of the visceral, mucosal, and immune dysfunction (Fan & Pedersen, 2020). Serious GI illnesses that affect the inner wall of the colon include IBDs, ulcerative colitis (UC), and Crohn's disease (CD), which results in patches of inflammation in the GI tract (McDowell et al., 2022). Frequent idiopathies of the GI include diverticular disease (small pockets of the bowl) and irritable bowel syndrome (IBS) characterized by suppressor T cell dominance (McDowell et al., 2022). Factors contributing to GI disorders include predisposition to genetic disease, exposure to chemotherapeutic agents, and antiinflammatory drugs (e.g., aspirin, etc.), unhealthy lifestyle, lack of physical inactivity, smoking, and low fiber diet (De Filippis et al., 2020). Other factors contributing to IBD and functional disorders of GI affecting 50% of the population are caused by gram-negative bacteria, *Helicobacter pylori* (Franzosa et al., 2019). Symptoms of the GIT include discomfort of the abdomen, distention of the abdomen, acidity of the stomach, constipation, malnutrition, reflux disease of the gastro intestine, hemorrhage of the GIT, obstruction of the intestine, and malabsorption (Fukuda & Ohno, 2014).

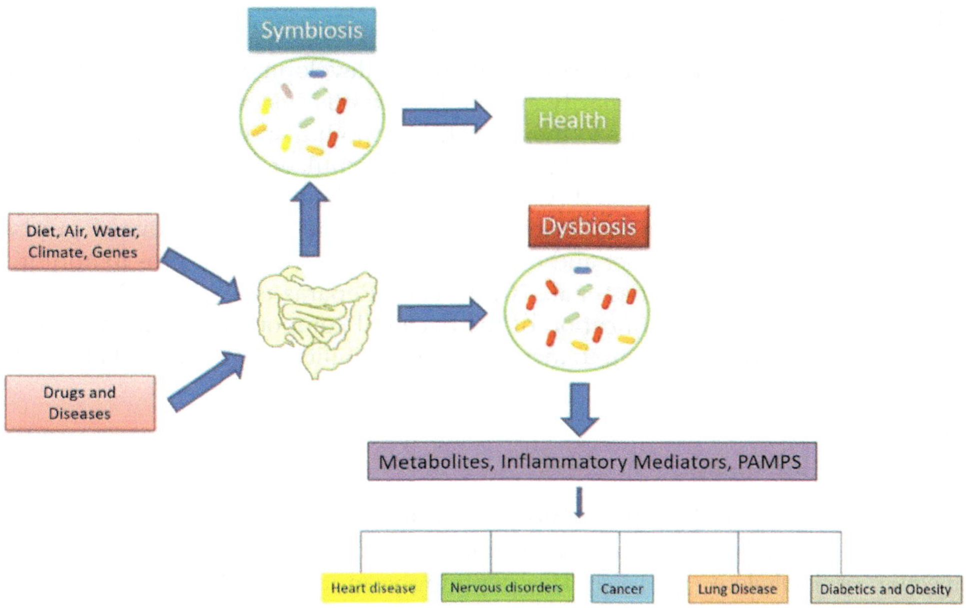

FIGURE 17.2 The human microbial dysbiosis in diseases.

3.1 Inflammatory bowel disease

Factors contributing to pathogenesis in irritable bowel syndrome include diet and microbiota. Increased hydrogen absorption is observed in these patients due to impairment in the absorption of dietary carbohydrates. Also, an increased amount of methane is produced, which was found to correlate with the disease symptoms (Chong et al., 2019). Alteration of both protein and carbohydrate metabolism was observed in these patients with a change in the diversity of the gut bacteria, with reduced presence of *Bacteroidetes* and enriched *Firmicutes* observed in a subset of IBS patients (Chong et al., 2019). Twin studies suggest the predominance of *Actinobacteria* and *Proteobacteria* in healthy siblings with a concordant decrease in *Bacteroidetes* (Lepage et al., 2011). An increase in the production of *Desulfovibrio* subspecies (sulfide-producing) and *Fusobacterium varium* was observed in UC, where both these bacteria possess the ability to invade epithelium (Lepage et al., 2011). However, it remains elusive whether dysbiosis contributes to pathogenicity (Magne et al., 2020). Several studies have shown that the immune system is not only triggered by bacteria but also impacts the composition of microbiota in the gut (Zheng et al., 2020). Derangement of the immune defense caused by pathogens affects the immunological situation. An increased proportion of *Enterobacteriaceae* was observed in patients diagnosed with Crohn's disease with minor bacterial diversity (Zheng et al., 2020). In patients with inflammatory bowel diseases, *Roseburia* and *Faecalibacterium* were found to be diminished (Forbes et al., 2016; Kostic et al., 2014; Rajilić-Stojanović et al., 2011; Sheehan et al., 2015).

3.2 Multiple sclerosis

Environmental and genetic factors contribute to the development of the condition termed multiple sclerosis (MS), which primarily affects the nervous system (Boziki et al., 2020). Under pathological conditions, it leads to the activation of both B and T lymphocytes, leading to the degradation of the myelin sheath of the cells (Wingerchuk et al., 2001). Animal models on MS (autoimmune encephalomyelitis [EAE]) have shown the role of gut bacteria contributing to MS. Mice when treated with oral antibiotics were found to have reduced gut bacteria, which drastically delayed the onset and disease severity. A shift from pro- to antiinflammatory cytokine was observed (Forbes et al., 2016; Ochoa-Repáraz et al., 2009). New therapeutic options could open on the identification of the bacterial components or disease-specific microorganisms, which is contributing to increased susceptibility in MS (Berer et al., 2011; Kirby & Ochoa-Repáraz, 2018).

3.3 Rheumatoid arthritis

Rheumatoid is a chronic inflammatory disorder caused by the gut microorganisms involved in the elicitation of an immune response. Animal studies on GF rats after intradermal injection of *Staphylococcus epidermidis*, peptidoglycan layer, or *Mycobacterium bovis* were found to develop adjuvant-induced arthritis, whereas the conventional rats were found to be protected, which concludes that commensal gut bacteria do not play a vital role in the development of adjuvant-induced arthritis. Interestingly, it was found that they suppress and modulate the immune system mediated by the predominance of suppressor T cells (Kohashi et al., 1979). Also, it was found that on colonization with segmented filamentous bacteria, expansion of inflammatory Th17 cells was observed with an imbalance response between the regulatory and inflammatory T cells. Severe autoimmune arthritis was observed mediated by bacterial polysaccharides, which upregulate TLR4, resulting in cytokine storm observed in the mice synovial tissues on extraction (Abdollahi-Roodsaz et al., 2008). A decrease in arthritis was observed in mice induced with adjuvants colonized with *Bacteroides* and *Escherichia coli* (gram-negative bacteria) via lipopolysaccharide; conversely, colonization of gram-positive bacteria, including *Lactobacillus*, *Propionibacterium acnes*, and Bifidobacterium, was found to intensify the disease via peptidoglycan (Kohashi et al., 1985). Mansson et al. in 1965, elucidated dysbiosis in patients with RA; the predominance of *Clostridium perfringens* observed in stool samples. Later, it was found that many other bacteria were also contributing factors in the development of RA (Gullberg, 1978). The premise that the role of Enterobacteriaceae in pathogenicity was supported by the presence of antibodies in the joint and fluid samples. Although the role of gut bacteria in the pathogenesis of RA is illustrated by various studies, the mechanism of gut microorganisms on RA remains unanswered (Vaahtovuo et al., 2008)

3.4 Systemic lupus erythematosus

Systemic lupus erythematosus (SLE) is an autoimmune condition where antibodies are produced against the double-stranded DNA involving the heart, kidneys, lungs, blood cells, and skin. Phenotypic heterogeneity exists between individuals diagnosed with SLE. In addition to genetic predisposition, environmental factors do play a key role, for example, bacterial dysbiosis, a condition attributed due to vitamin D deficiency, is also considered a factor to stimulate SLE. Recently, studies have shown that the ratio of Bacteroidetes to firmicutes was found to be lower in comparison with the healthy controls. *Lachnospiraceae* and *Ruminococcaceae* were found to be positively related to the healthy controls (Hevia et al., 2014). Though these results were on a concluding mark, contrasting results were reported by Zhang et al. in murine lupus (Zhang et al., 2014). During disease progression, it was observed that in female mice, there is depletion of *lactobacilli* in the gut with increased levels of *Clostridiaceae* and *Lachnospiraceae* in comparison with the healthy controls. Significant improvement in the symptoms was observed in the lupus mice when these mice were fed with retinoic acid, leading to restoration of *lactobacilli*. Hence the interaction between the immune cells and gut microorganisms was found to impart a vital role in asserting tolerance to self-antigens. However, detailed study must be carried out to justify the following: (1) a shift in gut microbiota caused due to inappropriate immune response, and (2) dysbiosis is a major factor for the causality of SLE (López et al., 2016).

3.5 Psoriasis

An immune-mediated inflammatory condition caused due to hyperproliferation of the keratinocytes. A significant difference in the microbial pattern coupled with diversity and heterogeneity was described by Chang et al. where reduced stability was observed in comparison with the healthy controls. *Staphylococcus aureus* was found to be on an increase, whereas *P. acnes* and *S. epidermidis* were found to be reduced. Studies on GF mice have demonstrated increased colonisation with pathogens like *Staphylococcus aureus* may result from the loss of community stability and reduction in immunoregulatory bacteria like *Staphylococcus epidermidis* and *Propionibacterium acnes*, which might worsen cutaneous inflammation along the Th17 axis (Chang et al., 2018). Concordant results were observed when human skin biopsies were analyzed. In comparison with the healthy controls, Proteobacteria were found to be on an increase in the trunk skin and with a significant drop in *Propionibacteria* and *Staphylococci* colonization. Studies have shown that intestinal bacteria were also found to be associated with psoriasis (Huang et al., 2019). Dysbiosis fortified with the enriched form of phylum *Bacteroidetes* with fewer phylum *Firmicutes* was observed in the gut microbiota of psoriasis patients in comparison with that of the controls. An increased abundance of *Bacteroides* was observed; however, under severe conditions, *Akkermansia muciniphila* (lower abundance) and *Faecalibacterium prausnitzii* (lower content) were found to be on a decrease. Interestingly, *E.coli* was observed on an increase, which was evidenced by an altered microbiota in the skin axis in patients with psoriasis (Tan et al., 2018).

3.6 Obesity

Intestinal bacteria were found to play a profound role in obesity, which was first demonstrated in GF mice and conventionally raised mice (CONV-R). This was supported by the notion that when the human gut bacteria were transferred from an obese twin to GF mice, abnormal weight gain was observed and vice versa from a lean twin where the metabolic phenotype associated with obesity and overweight was not observed where the mice were fed with low-fat and high-fiber diet (Ridaura et al., 2013). Firmicutes and Bacteroidetes are two phyla that predominate in the adult gut microbiota. Many researchers define "obesogenic microbiota" as a relative decline in the abundance of Bacteroidetes and a relative increase in Firmicutes. Studies have shown that long-chain carbohydrates are digested by phylum Firmicutes where the host is infested with an additional source of nutrients, leading to obesity (Sanmiguel et al., 2015). Also, in individuals with obesity, altered metabolic pathway coupled with reduced microbial diversity was observed (Sanmiguel et al., 2015). Mechanisms proposed for the underlying condition include the following: (1) insulin sensitivity increased secretion of leptin, and (2) increased hepatic triglyceride synthesis increased monosaccharide absorption and activation of carbon response element protein-1 binding to protein due to the upregulation of acetyl-CoA carboxylase and fatty acid synthase (Canfora et al., 2019). Storage of fat in the adipocytes with increased LPL activity was found to be increased due to suppression of fasting-induced adipose factor (FIAF) (Bäckhed et al., 2004). Hence, this concludes the role of microbiota in increased adipogenesis and obesity. Also, successive studies have reported a decrease in LPL activity with an increase in fatty acid oxidation. With reduced fatty acid storage and LPL activity, there is increased energy absorption due to hydrolysis, and fermentation of polysaccharides resulting in the generation of short-chain fatty acids (SCFAs) (Zwartjes et al., 2021). These SCFAs play a vital role in influencing the host appetite, energy absorption and harvest, lipogenesis, and intestinal transit time (Den Besten et al., 2013). The CONV-R mice were found to have a 40% increase in body fat on par with GF mice, independent of food uptake (Moretti et al., 2021). On colonization of the GF mice with the gut bacteria from CONV-R mice, a proportional increase (60%) was observed. Other factors that increase obesity include the high-fat diet due to a proportional increase in gram-negative species, thereby leading to increased absorption of LPS, i.e., bacterial fragments leading to endotoxemia (Moretti et al., 2021). Interestingly, *Enterobacter cloacae* B29 (obese human gut) on morbidity when injected into GF mice was found to acquire an obese phenotype coupled with increased circulating levels of endotoxin (Fei & Zhao, 2013), e.g., reduced TLR4 activation. Another way by which gut homeostasis is regulated is through the endocannabinoid system (Acharya et al., 2017).

3.7 Colorectal cancer

Altered fecal microbial profiles were observed in patients diagnosed with CRC when the western diets were adopted by African-Americans (Rebersek, 2021). Increased profiles of *Bacteroides fragilis, Enterococcus, Escherichia/Shigella, Klebsiella, Streptococcus, Peptostreptococcus, and Roseburia* were observed in lower abundance (Rebersek, 2021). Existing lines of evidence link the risk of developing CRC with vegetable, dietary fiber, and fruit consumption (Wang et al., 2012). Microbiota enterotype (Firmicutes and Proteobacteria) was found to be increased with long-term fiber consumption with the decrease in *Actinobacteria*, *Bifidobacteria*, and *Bacteroidetes* (Wang et al., 2012). In patients undergoing colectomy, fewer complications were observed possibly due to improved barrier integrity mediated by the beneficial microbes. The risk of CRC can be minimized by increased intake of dietary fiber resulting in the enrichment of SCFA production (Zeng et al., 2014). Current research suggests that colon tumorigenesis could be attributed to changes in the microbiota of the gut (Fan et al., 2021).

3.8 Nonalcoholic fatty liver disease

Nonalcoholic fatty liver disease (NAFLD) occurs globally of which nonalcoholic steatohepatitis (NASH), cirrhosis, and liver cancer account for 10%–20%. NAFLD was found to be associated with small intestine bacterial outgrowth, resulting in increased levels of the tumor, necrosis factor, trimethylamine, and acetaldehyde (Khan et al., 2021). The liver is highly vulnerable to translocation of bacterial products or endotoxins due to the portal venous system, which connects the liver and gut (Alexopoulou et al., 2017). An increase in hepatic triglyceride levels with alteration in lipid metabolism is due to obesogenic microbiota posing an indirect effect on the fatty acid storage in the liver (Dai et al., 2020). In humans, obese patients were found to be associated with hepatic steatosis and leaky gut associated with SIBO. In humans, chronic endotoxin exposure is linked to the severity of NAFLD (Kessoku et al., 2021). Changes in the abundance of *Proteobacteria* and *Erysipelotrichia* were found to be associated with deficiency of choline and fatty liver. Hence the identification of bacterial abundance of this type will help the prediction of the risk of developing fatty liver (Mu et al., 2020).

3.9 Breast cancer

Globally, breast cancer accounts for the third in the world, which remains the leading cause of mortality in women (Azamjah et al., 2019). Studies have shown that tumorogenesis could be prevented when normal interactions exist between the gut microbes and the host, whereas dysregulation could lead to an alteration of the environment of the microbes, leading to the origin of the disease and progression (Jiang et al., 2020). A study conducted by Minelli EB et al. uncovered a significant difference in the gut flora of premenopausal breast cancer patients and healthy women in 1990. Dysbiosis of the intestinal flora was observed by microscopic examination of fecal extracts and culture (Fukuda and Ohno, 2014). Furthermore, it has been proposed that gut flora may be associated with the stages of breast cancer. For example, it has been shown that in patients diagnosed with breast cancer, an increased number of *Blautia* sp. were observed. When compared to women with lower grade (I/II) or lower stage breast cancers, two *Firmicutes* taxa *(g_Clostridium, g_Veillonella)* were more prevalent in women with higher grade (III) or higher stage breast cancers. Moreover, higher grade was linked to a higher *Actinobacteria (g_Eggerthella)* abundance and reduced levels of *Firmucutes (f_Lachnospiraceae, g_Anaerostipes, f_Ruminococcaceae)* and other *Actinobacteria (f_Coriobacteriaceae)* taxa (Wu et al., 2020). It has recently been demonstrated that increased HER2 and increased alpha diversification of microbiota were observed in HER2+ breast cancer patients, with a more distinct bacterial observed (Modica et al., 2022). In patients diagnosed with an invasive form of breast, Fernandez MF et al. observed that the counts of *Methylobacterium* were found to be associated (Fernández et al., 2018).

3.10 Gut microbiome—depressive disorder

Major depressive disorder (MDD) remains the leading cause of death worldwide. Patients with depressive disorder were found to have changes in the metabolic and immune profile including CNS abnormalities (Beurel et al., 2020). Aside from nervous system abnormalities, patients with depression were found to have changes in their metabolic, immune, and endocrine systems. Microbe-specific metabolites are encoded by proteins. For example, these microbiotas play a vital role in the immune system priming, thereby assisting the breakdown of indigestible fibers, which in turn drives the development of the brain, thereby shaping the behavior. It has been shown that a bidirectional network exists between the gut and brain, known as the gut–brain axis, where the role of microbiota and its metabolites play a crucial role (Schächtle & Rosshart, 2021). Studies have shown the relation between the development of the central nervous system with the colonization of the microbiota in maintaining homeostasis. As a result, disruption of the microbiota in early life can impair the process associated with neurogenesis and synapsis formation, and limitation in both axonal and dendritic growth, which overall affects mental health (Schächtle & Rosshart, 2021). Existing lines of evidence suggest that the microbiota was found to play a role in the pathogenicity of the disease. In comparison with the fecal microbiota in MDD and healthy subjects, differences were observed based on taxonomic association studies. Variations in phyla including *Proteobacteria, Actinobacteria,* and *Bacteroidetes* and *Enterobacteriaceae, Alistipes,* etc. were observed although the association between the bacterial taxa and disease was found to be contradictory (Dash et al., 2022).

3.11 Gut microbiota and dyslipidemia

Dyslipidemia is defined as the abnormal metabolism of lipids, which relates to the concentration of the lipid in the blood (Matey-Hernandez et al., 2018). Results from in vitro studies have confirmed the imbalance in the microbiota of the gut leading to dysbiosis leading to aggravation of metabolic disorders of lipids due to acquired or congenital factors (Ma et al., 2018). Both in vitro and in vivo studies have found that the imbalance in the microbiota of the gut can lead to the generation of metabolic disorders of lipids (Wang et al., 2020). In a parallel study, C57BL/6J mice when fed with increased concentration of glucose or fructose preceded by 16S rRNA analysis were found to significantly alter the gut microbiota (Bech-Nielsen et al., 2012). An increased abundance of taxa of *Actinobacteria* and decreased concentration of *Bacteroidetes* and *proteobacteria* were observed (Bech-Nielsen et al., 2012; Li et al., 2021). This supports the cross talk between the lipids in the blood and the gut microbiome.

3.12 Gut microbiota—hypertension

The pathogenesis of hypertension involves a multitude of factors including genetic and environmental factors, inflammatory forces, hormonal factors, etc. (Yang et al., 2015). Hypertension constitutes metabolic syndrome as one of its components. Based on the results from animal models and human studies, dysbiosis was found to be linked to the gut. Risk factors for hypertension include diseases of the cerebral and cardiovascular systems. Yang et al. in the spontaneously hypertensive rat have shown that a marked decrease in microbial diversification and richness was observed (Yang et al., 2015). An increased ratio of firmicutes to bactericides was observed. Also in the rat model of chronic angiotensin II infusion model, a similar increase in microbiota was observed. This concluded the fact that dysbiosis was found to be associated with hypertension in both animals and humans (Yang et al., 2015). In both prehypertensive and hypertensive populations, there was a decrease in healthy microbiota with the predominance of *Prevotella* and *Klebsiella* observed.

3.13 Gut microbiota—hyperuricemia

Hyperuricemia is a purine metabolic disorder with decreased uric acid secretion. Hyperuricemia acts as the primary cause of the gout and manifestation of metabolic syndrome (Jin et al., 2012). A study carried out by Xu et al. in a mouse model of hyperuricemia showed that there was a decrease in the frequency of *firmicutes* at the phylum level (Wang et al., 2020). In contrast, the prevalence of the *Bacteroides* was found to be increased. At the family level in the hyperuricemic mouse model, *Prevotellaceae, Rikenellaceae, Bacteroidaceae,* and *Bacteroidales* were found to be in abundance (Wang et al., 2020). Bacterial populations common in the hyperuricemia group include *Lactobacillus, Clostridium, Ruminococcaceae,* and *Clostridium* (Wang et al., 2020). According to Guo et al., a significant difference was observed in terms of both the organism and functional structure when a comparison of microbiota was carried out between the gout patients with healthy people (Guo et al., 2016). In the gout patients, *Bacteroides caccae* and *Bacteroides xylanisolvens* were found in more abundance, while *F. prausnitzii* and *Bifidobacterium pseudocatenulatum* were found to be on a decrease (Guo et al., 2016).

3.14 Obstructive sleep apnea hypopnea syndrome

Obstructive sleep apnea hypopnea syndrome (OSAHS) is a sleep condition characterized by apneas associated with increased morbidity and mortality associated with cerebral and cardiovascular systems (Zhang & Si, 2012). Various studies have shown the association of the gut microbiota with dysbiosis linked with the underlying condition. *Firmicutes* were observed in mice when exposed to intermittent hypoxic conditions when compared with the control group (Ko et al., 2019). *Bacteroidetes* and *Proteobacteria* were found to be markedly decreased. The main risk factor for developing OSAHS was found to be *Ruminococcus*, results generated on risk stratification analysis (Ko et al., 2019).

3.15 Alzheimer's disease

Alzheimer's disease affects globally 50 million people worldwide, data derived from epidemiological studies (Zhang et al., 2021). The incidence was found to range from 5/1000 spanning 65–70 years of age to 80. Dementia was found to be associated with a reduction in microbial diversification with a relative increase in *Proteobacteria* and markedly decreased *Bifidobacteria* (Zhang et al., 2021). Both these bacteria were found to be contributing factors in the development of dementia. Also, the microbiota was found to impair the process associated with lipid metabolism in turn reducing the short-chain fatty acids. Probiotic formulation containing *Lactobacilli* and *Bifidobacteria* can be administered to patients based on the results from a double-blind controlled trial (Zhang et al., 2021). It was observed that the serum triglyceride profiles were improved, which in turn impoved cognitive performance. This confirms that the microbiota plays a crucial role in maintaining the metabolism of lipids.

3.16 Necrotizing enterocolitis

Necrotizing enterocolitis (NEC), with an incidence rate of 3% and 15%, is the most common severe GI disease with increased severity, usually present with infants of very low birth weight (Niemarkt et al., 2019). The mortality rate of this disease was found to range from 3% to 15%. The factors contributing to the disease remain multifactorial. Contributing factors to this disease include (1) enteral feeding, (2) immaturity of the gut, (3) hypoxia–ischemia, and (4) dysbiosis of the microbes, leading to cascade events associated with inflammation response of the gut leading to inflammatory conditions resembling sepsis. The inflammatory response was found to be mediated by TLR4 signaling. Infants with this disease present with distended abdomen and hematochezia (Niemarkt et al., 2019). Although some infants recovered, they were found to be associated with impairment of the nervous system and short-bowel syndrome. A markedly increased presence of *Proteobacteria* was identified with a parallel decrease in Firmicutes levels (Niemarkt et al., 2019). Although the primary causative pathogen of this condition remains elusive, there has been a clear understanding between the microbiota alterations leading to dysbiosis and NEC.

3.17 *Clostridium difficile*-associated diarrhea

Globally, *Clostridioides difficile*, anaerobic spore-forming bacterium, remains the causative agent for antibiotic-associated diarrhea (AAD) in the healthcare sector (Berkell et al., 2021). Symptomatic features of these conditions include inflammation mediated by toxins and watery stool. Risk factors including (1) age, (2) hospitalization, (3) colonization by *C. difficile* and antibiotic exposure (carbapenems, cephalosporins, etc.) in combination with inhibitors of beta-lactamase were found to be associated with the development of *C. difficile*-associated infection(CDI) (Berkell et al., 2021). In both CDI and AAD, the diversity in the gut microbiota was altered by elevated levels of *Enterococcus* with a markedly decrease in the phylum including *Bacteroidetes*, the *Lachnospiraceae* and *Ruminococcaceae* families, and *Prevotella* spp. (Berkell et al., 2021). A single-center study in determining the risk factors for CDI was due to the absence or reduction in *Clostridiales Incertae Sedis* XI (Berkell et al., 2021). Damage induced by the antibiotic was found to decrease the alpha diversity and also the antibiotic resistance genes. However, studies are yet carried out to understand the role of antibiotics in the development of the dysbiosis of CDI.

4. Healthy microbial profile—bacteriotherapy

Diet is thought to represent an adaptive intervention. Since the dietary factors are frequently changed or disputed, it remains a challenge in modifying the diet. This can be overcome by bacteriotherapy (Leeming et al., 2019). This novel therapy involves replacing the beneficial intestinal microbes. Although bacteriotherapy remains an alternative approach, its potency in humans remains elusive.

4.1 Probiotics, prebiotics, and symbiotics

Probiotics are living microorganisms (functional food) that provide therapeutic effects to the host if ingested in sufficient quantities (Kechagia et al., 2013). Prebiotics are nonliving nondigestible forms of fiber or carbohydrates that benefit the host by promoting the development of beneficial colonic bacteria. A combination of pro- and prebiotics makes up the symbiotics (Kechagia et al., 2013). Since the probiotics were found to be strain-specific, adequate doses must be administered to be effective (McFarland et al., 2018). Various types of probiotics have been tested and used in both localized and systematic diseases. Although probiotics were found to have a therapeutic impact on GI diseases, there is lack of the supportive fact that the gut microbiota mediates a positive impact. Similarly, the role of probiotics in the context of systemic diseases was found to be variable.

4.2 Antibiotics

Host microbiota gets altered on exposure to antibiotic treatment. Although it is presumed that the restoration of microbiota occurs within weeks on discontinuation of antibiotics (Ramirez et al., 2020), this belief is being called into question by new evidence. It was found that on the administration of clindamycin (7 days), significant decrease in *Bacteroides* was observed even after 2 years (Ramirez et al., 2020). Hence the antibiotic can be regarded as a double-edged sword for dysbiosis. Although these antibiotics elicit a critical role in the alleviation of harmful pathogens, they were also found to disrupt the diversity of microbes, leading to the colonization of opportunistic bacteria in the gut (Jernberg et al., 2007). This was supported by the observation that in a patient who received clindamycin in the year 1970, the presence of *C. difficile* was documented. Long-term antibiotic use in children has been linked to an increased risk of IBD in adulthood (Belkaid & Hand, 2014; Jernberg et al., 2007; Khan et al., 2014). This raises serious concerns in a clinical setting about the appropriate use or avoidance of antibiotics. A more specific antimicrobial approach must be carried out to prevent the disruption of the microbiota of the gut.

5. Fecal transplantation

The normal microbiota is established in a symptomatic host by fecal transplantation approach, i.e., by transplantation of healthy donor stool, which serves as a promising approach for dysbiosis-associated diseases (Ser et al., 2021). It has been shown that the FT approach has been safer in comparison with the standard antibiotic treatment in the resolution of recurrent *C. difficile* infections (Ser et al., 2021). Although the current process of fecal transplant lacks standardization, these methods have been found to have a success rate of 95% (Soo et al., 2020). As various studies have confirmed the role of microbiota in immunity, microbiota manipulation therapies are carried out.

6. Conclusions

Gut microbiota plays a key role in the maintenance of homeostasis of the gut, immunity, regulation of endocrine and neurological function, and absorption of nutrients (Rowland et al., 2018). Imbalance in the microbiota, i.e., gut microbes can affect human health, leading to increased susceptibility to systematic and GI disorders (Vijay & Valdes, 2021). According to clinical and experimental evidence, gut microbiota is one of the most important pathogenic factors in MetS. Existing lines of evidence suggest the role of microbiota in the elicitation of immune and metabolic responses, thereby probing for potential targets by dietary or by pharmacological approach (Vijay & Valdes, 2021). An imbalance in the microbiota of the gut elicits a low-grade inflammatory response, thereby affecting the gut barrier and resulting in resistance to insulin-mediated by the metabolites, which affect both the metabolism and release of hormones, leading to a chain of events leading to the progression of metabolic syndrome. Hence, there arises a need to restore the microbiota in a safer and feasible and cost-effective manner for the treatment of various diseases. The key therapeutic approach mainly includes using dietary supplements such as prebiotics, probiotics, and postbiotics and fecal transplantation to maintain intestinal integrity and functions (Simpson & Campbell, 2015). Various therapeutic options exist through an understanding of the manipulation of microbiota to prevent and treat metabolic syndromes, thereby paving way for a novel therapeutic approach.

References

Abdollahi-Roodsaz, S., Joosten, L. A. B., Koenders, M. I., Devesa, I., Roelofs, M. F., Radstake, T. R. D. J., Heuvelmans-Jacobs, M., Akira, S., Nicklin, M. J. H., Ribeiro-Dias, F., & Van Den Berg, W. B. (2008). Stimulation of TLR2 and TLR4 differentially skews the balance of T cells in a mouse model of arthritis. *The Journal of Clinical Investigation, 118*(1), 205–216. https://doi.org/10.1172/JCI32639

Acharya, N., Penukonda, S., Shcheglova, T., Hagymasi, A. T., Basu, S., & Srivastava, P. K. (2017). Endocannabinoid system acts as a regulator of immune homeostasis in the gut. *Proceedings of the National Academy of Sciences of the United States of America, 114*(19), 5005–5010. https://doi.org/10.1073/PNAS.1612177114/-/DCSUPPLEMENTAL

Alexopoulou, A., Agiasotelli, D., Vasilieva, L. E., & Dourakis, S. P. (2017). Bacterial translocation markers in liver cirrhosis. *Annals of Gastroenterology, 30*(5), 486. https://doi.org/10.20524/AOG.2017.0178

Azamjah, N., Soltan-Zadeh, Y., & Zayeri, F. (2019). Global trend of breast cancer mortality rate: A 25-year study. *Asian Pacific Journal of Cancer Prevention: APJCP, 20*(7), 2015. https://doi.org/10.31557/APJCP.2019.20.7.2015

Bäckhed, F., Ding, H., Wang, T., Hooper, L. V., Gou, Y. K., Nagy, A., Semenkovich, C. F., & Gordon, J. I. (2004). The gut microbiota as an environmental factor that regulates fat storage. *Proceedings of the National Academy of Sciences of the United States of America, 101*(44), 15718–15723. https://doi.org/10.1073/PNAS.0407076101/SUPPL_FILE/07076TABLE4.PDF

Becattini, S., Littmann, E. R., Carter, R. A., Kim, S. G., Morjaria, S. M., Ling, L., Gyaltshen, Y., Fontana, E., Taur, Y., Leiner, I. M., & Pamer, E. G. (2017). Commensal microbes provide first line defense against Listeria monocytogenes infection. *The Journal of Experimental Medicine, 214*(7), 1973–1989. https://doi.org/10.1084/JEM.20170495

Bech-Nielsen, G. V., Hansen, C. H. F., Hufeldt, M. R., Nielsen, D. S., Aasted, B., Vogensen, F. K., Midtvedt, T., & Hansen, A. K. (2012). Manipulation of the gut microbiota in C57BL/6 mice changes glucose tolerance without affecting weight development and gut mucosal immunity. *Research in Veterinary Science, 92*(3), 501–508. https://doi.org/10.1016/J.RVSC.2011.04.005

Belkaid, Y., & Hand, T. W. (2014). Role of the microbiota in immunity and inflammation. *Cell, 157*(1), 121–141. https://doi.org/10.1016/J.CELL.2014.03.011

Berer, K., Mues, M., Koutrolos, M., AlRasbi, Z., Boziki, M., Johner, C., Wekerle, H., & Krishnamoorthy, G. (2011). Commensal microbiota and myelin autoantigen cooperate to trigger autoimmune demyelination. *Nature, 479*(7374), 538–541. https://doi.org/10.1038/NATURE10554

Berkell, M., Mysara, M., Xavier, B. B., van Werkhoven, C. H., Monsieurs, P., Lammens, C., Ducher, A., Vehreschild, M. J. G. T., Goossens, H., de Gunzburg, J., Bonten, M. J. M., Malhotra-Kumar, S., Engbers, A., de Regt, M., Biehl, L. M., Cornely, O. A., Jazmati, N., Bouverne, M. N., Sablier-Gallis, F., … Launay, O. (2021). Microbiota-based markers predictive of development of Clostridioides difficile infection. *Nature Communications, 12*(1). https://doi.org/10.1038/S41467-021-22302-0

Beurel, E., Toups, M., & Nemeroff, C. B. (2020). The bidirectional relationship of depression and inflammation: Double trouble. *Neuron, 107*(2), 234. https://doi.org/10.1016/J.NEURON.2020.06.002

Bokulich, N. A., Chung, J., Battaglia, T., Henderson, N., Jay, M., Li, H., Lieber, A. D., Wu, F., Perez-Perez, G. I., Chen, Y., Schweizer, W., Zheng, X., Contreras, M., Dominguez-Bello, M. G., & Blaser, M. J. (2016). Antibiotics, birth mode, and diet shape microbiome maturation during early life. *Science Translational Medicine, 8*(343), 343ra82. https://doi.org/10.1126/SCITRANSLMED.AAD7121

Boziki, M. K., Kesidou, E., Theotokis, P., Mentis, A. F. A., Karafoulidou, E., Melnikov, M., Sviridova, A., Rogovski, V., Boyko, A., & Grigoriadis, N. (2020). Microbiome in multiple sclerosis: Where are we, what we know and do not know. *Brain Sciences, 10*(4). https://doi.org/10.3390/BRAINSCI10040234

Canfora, E. E., Meex, R. C. R., Venema, K., & Blaak, E. E. (2019). Gut microbial metabolites in obesity, NAFLD and T2DM. *Nature Reviews Endocrinology, 15*(5), 261–273. https://doi.org/10.1038/s41574-019-0156-z

Chang, H. W., Yan, D., Singh, R., Liu, J., Lu, X., Ucmak, D., Lee, K., Afifi, L., Fadrosh, D., Leech, J., Vasquez, K. S., Lowe, M. M., Rosenblum, M. D., Scharschmidt, T. C., Lynch, S. V., & Liao, W. (2018). Alteration of the cutaneous microbiome in psoriasis and potential role in Th17 polarization. *Microbiome, 6*(1). https://doi.org/10.1186/S40168-018-0533-1

Chelakkot, C., Ghim, J., & Ryu, S. H. (2018). Mechanisms regulating intestinal barrier integrity and its pathological implications. *Experimental & Molecular Medicine, 50*(8). https://doi.org/10.1038/S12276-018-0126-X

Cho, I., & Blaser, M. J. (2012). The human microbiome: At the interface of health and disease. *Nature Reviews. Genetics, 13*(4), 260–270. https://doi.org/10.1038/NRG3182

Chong, P. P., Chin, V. K., Looi, C. Y., Wong, W. F., Madhavan, P., & Yong, V. C. (2019). The microbiome and irritable bowel syndrome - a review on the pathophysiology, current research and future therapy. *Frontiers in Microbiology, 10*(JUN), 1136. https://doi.org/10.3389/FMICB.2019.01136/BIBTEX

Ciccia, F., Guggino, G., Rizzo, A., Alessandro, R., Luchetti, M. M., Milling, S., Saieva, L., Cypers, H., Stampone, T., Di Benedetto, P., Gabrielli, A., Fasano, A., Elewaut, D., & Triolo, G. (2017). Dysbiosis and zonulin upregulation alter gut epithelial and vascular barriers in patients with ankylosing spondylitis. *Annals of the Rheumatic Diseases, 76*(6), 1123–1132. https://doi.org/10.1136/ANNRHEUMDIS-2016-210000

Conlon, M. A., & Bird, A. R. (2014). The impact of diet and lifestyle on gut microbiota and human health. *Nutrients, 7*(1), 17–44. https://doi.org/10.3390/NU7010017

Cornier, M. A., Dabelea, D., Hernandez, T. L., Lindstrom, R. C., Steig, A. J., Stob, N. R., Van Pelt, R. E., Wang, H., & Eckel, R. H. (2008). The metabolic syndrome. *Endocrine Reviews, 29*(7), 777–822. https://doi.org/10.1210/ER.2008-0024

Dai, X., Hou, H., Zhang, W., Liu, T., Li, Y., Wang, S., Wang, B., & Cao, H. (2020). Microbial metabolites: Critical regulators in NAFLD. *Frontiers in Microbiology, 11*, 2373. https://doi.org/10.3389/FMICB.2020.567654/BIBTEX

Dash, S., Syed, Y. A., & Khan, M. R. (2022). Understanding the role of the gut microbiome in brain development and its association with neurodevelopmental psychiatric disorders. *Frontiers in Cell and Developmental Biology, 10.* https://doi.org/10.3389/FCELL.2022.880544

De Filippis, A., Ullah, H., Baldi, A., Dacrema, M., Esposito, C., Garzarella, E. U., Santarcangelo, C., Tantipongpiradet, A., & Daglia, M. (2020). Gastrointestinal disorders and metabolic syndrome: Dysbiosis as a key link and common bioactive dietary components useful for their treatment. *International Journal of Molecular Sciences, 21*(14), 4929. https://doi.org/10.3390/IJMS21144929

Den Besten, G., Van Eunen, K., Groen, A. K., Venema, K., Reijngoud, D. J., & Bakker, B. M. (2013). The role of short-chain fatty acids in the interplay between diet, gut microbiota, and host energy metabolism. *Journal of Lipid Research, 54*(9), 2325. https://doi.org/10.1194/JLR.R036012

Dieterich, W., Schuppan, D., Schink, M., Schwappacher, R., Wirtz, S., Agaimy, A., Neurath, M. F., & Zopf, Y. (2019). Influence of low FODMAP and gluten-free diets on disease activity and intestinal microbiota in patients with non-celiac gluten sensitivity. *Clinical Nutrition (Edinburgh, Scotland), 38*(2), 697–707. https://doi.org/10.1016/J.CLNU.2018.03.017

Fan, X., Jin, Y., Chen, G., Ma, X., & Zhang, L. (2021). Gut microbiota dysbiosis drives the development of colorectal cancer. *Digestion, 102*(4), 508–515. https://doi.org/10.1159/000508328

Fan, Y., & Pedersen, O. (2020). Gut microbiota in human metabolic health and disease. *Nature Reviews Microbiology, 19*(1), 55–71. https://doi.org/10.1038/s41579-020-0433-9

Fei, N., & Zhao, L. (2013). An opportunistic pathogen isolated from the gut of an obese human causes obesity in germfree mice. *ISME Journal, 7*(4), 880–884. https://doi.org/10.1038/ISMEJ.2012.153

Feng, Y., Wang, Y., Wang, P., Huang, Y., & Wang, F. (2018). Short-chain fatty acids manifest stimulative and protective effects on intestinal barrier function through the inhibition of NLRP3 inflammasome and autophagy. *Cellular*

Physiology and Biochemistry: International Journal of Experimental Cellular Physiology, Biochemistry, and Pharmacology, 49(1), 190–205. https://doi.org/10.1159/000492853

Fernández, M. F., Reina-Pérez, I., Astorga, J. M., Rodríguez-Carrillo, A., Plaza-Díaz, J., & Fontana, L. (2018). Breast cancer and its relationship with the microbiota. *International Journal of Environmental Research and Public Health, 15*(8), 1747. https://doi.org/10.3390/IJERPH15081747

Ferreira, R. B. R., Gill, N., Willing, B. P., Antunes, L. C. M., Russell, S. L., Croxen, M. A., & Finlay, B. B. (2011). The intestinal microbiota plays a role in Salmonella-induced colitis independent of pathogen colonization. *PloS One, 6*(5). https://doi.org/10.1371/JOURNAL.PONE.0020338

Forbes, J. D., Van Domselaar, G., & Bernstein, C. N. (2016). The gut microbiota in immune-mediated inflammatory diseases. *Frontiers in Microbiology, 7*. https://doi.org/10.3389/FMICB.2016.01081

Foster, J. A., & McVey Neufeld, K. A. (2013). Gut-brain axis: How the microbiome influences anxiety and depression. *Trends in Neurosciences, 36*(5), 305–312. https://doi.org/10.1016/J.TINS.2013.01.005

Franzosa, E. A., Sirota-Madi, A., Avila-Pacheco, J., Fornelos, N., Haiser, H. J., Reinker, S., Vatanen, T., Hall, A. B., Mallick, H., McIver, L. J., Sauk, J. S., Wilson, R. G., Stevens, B. W., Scott, J. M., Pierce, K., Deik, A. A., Bullock, K., Imhann, F., Porter, J. A., … Xavier, R. J. (2019). Gut microbiome structure and metabolic activity in inflammatory bowel disease. *Nature Microbiology, 4*(2), 293–305. https://doi.org/10.1038/S41564-018-0306-4

Fukuda, S., & Ohno, H. (2014). Gut microbiome and metabolic diseases. *Seminars in Immunopathology, 36*(1), 103–114. https://doi.org/10.1007/S00281-013-0399-Z

Geng, S., Yang, L., Cheng, F., Zhang, Z., Li, J., Liu, W., Li, Y., Chen, Y., Bao, Y., Chen, L., Fei, Z., Li, X., Hou, J., Lin, Y., Liu, Z., Zhang, S., Wang, H., Zhang, Q., Wang, H., … Zhang, J. (2020). Gut microbiota are associated with psychological stress-induced defections in intestinal and blood–brain barriers. *Frontiers in Microbiology, 10*, 3067. https://doi.org/10.3389/FMICB.2019.03067/FULL

Gross, L. (2007). Microbes colonize a baby's gut with distinction. *PLoS Biology, 5*(7), e191. https://doi.org/10.1371/journal.pbio.0050191

Gullberg, R. (1978). Possible role of alterations of the intestinal flora in rheumatoid arthritis. *Rheumatology and Rehabilitation, 17*(Suppl), 5–10. https://doi.org/10.1093/RHEUMATOLOGY/XVII.SUPPL.5

Guo, Z., Zhang, J., Wang, Z., Ang, K. Y., Huang, S., Hou, Q., Su, X., Qiao, J., Zheng, Y., Wang, L., Koh, E., Danliang, H., Xu, J., Lee, Y. K., & Zhang, H. (2016). Intestinal microbiota distinguish gout patients from healthy humans. *Scientific Reports, 6*. https://doi.org/10.1038/SREP20602

Hevia, A., Milani, C., López, P., Cuervo, A., Arboleya, S., Duranti, S., Turroni, F., González, S., Suárez, A., Gueimonde, M., Ventura, M., Sánchez, B., & Margolles, A. (2014). Intestinal dysbiosis associated with systemic lupus erythematosus. *MBio, 5*(5). https://doi.org/10.1128/MBIO.01548-14

Huang, L., Gao, R., Yu, N., Zhu, Y., Ding, Y., & Qin, H. (2019). Dysbiosis of gut microbiota was closely associated with psoriasis. *Science China Life Sciences, 62*(6), 807–815. https://doi.org/10.1007/s11427-018-9376-6

Jernberg, C., Löfmark, S., Edlund, C., & Jansson, J. K. (2007). Long-term ecological impacts of antibiotic administration on the human intestinal microbiota. *The ISME Journal, 1*(1), 56–66. https://doi.org/10.1038/ismej.2007.3

Jiang, Y., Chen, X., Fu, S., Jiang, Y., Chen, X., & Fu, S. (2020). Advances in the correlation between intestinal microbiota and breast cancer development. *Journal of Cancer Therapy, 11*(12), 758–771. https://doi.org/10.4236/JCT.2020.1112066

Jin, M., Yang, F., Yang, I., Yin, Y., Luo, J. J., Wang, H., & Yang, X. F. (2012). Uric acid, hyperuricemia and vascular diseases. *Frontiers in Bioscience: A Journal and Virtual Library, 17*(2), 656. https://doi.org/10.2741/3950

Juge, N. (2012). Microbial adhesins to gastrointestinal mucus. *Trends in Microbiology, 20*(1), 30–39. https://doi.org/10.1016/J.TIM.2011.10.001

Kechagia, M., Basoulis, D., Konstantopoulou, S., Dimitriadi, D., Gyftopoulou, K., Skarmoutsou, N., & Fakiri, E. M. (2013). Health benefits of probiotics: A review. *ISRN Nutrition, 2013*, 1–7. https://doi.org/10.5402/2013/481651

Kelly, J. R., Borre, Y., O' Brien, C., Patterson, E., El Aidy, S., Deane, J., Kennedy, P. J., Beers, S., Scott, K., Moloney, G., Hoban, A. E., Scott, L., Fitzgerald, P., Ross, P., Stanton, C., Clarke, G., Cryan, J. F., & Dinan, T. G. (2016). Transferring the blues: Depression-associated gut microbiota induces neurobehavioural changes in the rat. *Journal of Psychiatric Research, 82*, 109–118. https://doi.org/10.1016/J.JPSYCHIRES.2016.07.019

Kessoku, T., Kobayashi, T., Imajo, K., Tanaka, K., Yamamoto, A., Takahashi, K., Kasai, Y., Ozaki, A., Iwaki, M., Nogami, A., Honda, Y., Ogawa, Y., Kato, S., Higurashi, T., Hosono, K., Yoneda, M., Okamoto, T., Usuda, H., Wada, K., … Nakajima, A. (2021). Endotoxins and non-alcoholic fatty liver disease. *Frontiers in Endocrinology, 12*. https://doi.org/10.3389/FENDO.2021.770986

Khan, A., Ding, Z., Ishaq, M., Bacha, A. S., Khan, I., Hanif, A., Li, W., & Guo, X. (2021). Understanding the effects of gut microbiota dysbiosis on nonalcoholic fatty liver disease and the possible probiotics role: Recent updates. *International Journal of Biological Sciences, 17*(3), 818. https://doi.org/10.7150/IJBS.56214

Khan, I., Yasir, M., Azhar, E., Kumosani, T., Barbour, E., Bibi, F., & Kamal, M. (2014). Implication of gut microbiota in human health. *CNS & Neurological Disorders Drug Targets, 13*(8), 1325–1333. https://doi.org/10.2174/1871527313666141023153506

Kirby, T., & Ochoa-Repáraz, J. (2018). The gut microbiome in multiple sclerosis: A potential therapeutic avenue. *Medical Sciences (Basel, Switzerland), 6*(3), 69. https://doi.org/10.3390/MEDSCI6030069

Knights, D., Silverberg, M. S., Weersma, R. K., Gevers, D., Dijkstra, G., Huang, H., Tyler, A. D., Van Sommeren, S., Imhann, F., Stempak, J. M., Huang, H., Vangay, P., Al-Ghalith, G. A., Russell, C., Sauk, J., Knight, J., Daly, M. J., Huttenhower, C., & Xavier, R. J. (2014). Complex host genetics influence the microbiome in inflammatory bowel disease. *Genome Medicine, 6*(12). https://doi.org/10.1186/S13073-014-0107-1

Kohashi, O., Kohashi, Y., Shigematsu, N., Takahashi, T., & Ozawa, A. (1985). Reverse effect of gram-positive bacteria vs. gram-negative bacteria on adjuvant-induced arthritis in germfree rats. *Microbiology and Immunology, 29*(6), 487–497. https://doi.org/10.1111/J.1348-0421.1985.TB00851.X

Kohashi, O., Kuwata, J., Umehara, K., Uemura, F., Takahashi, T., & Ozawa, A. (1979). Susceptibility to adjuvant-induced arthritis among germfree, specific-pathogen-free, and conventional rats. *Infection and Immunity, 26*(3), 791–794. https://doi.org/10.1128/IAI.26.3.791-794.1979

Ko, C. Y., Liu, Q. Q., Su, H. Z., Zhang, H. P., Fan, J. M., Yang, J. H., Hu, A. K., Liu, Y. Q., Chou, D., & Zeng, Y. M. (2019). Gut microbiota in obstructive sleep apnea–hypopnea syndrome: Disease-related dysbiosis and metabolic comorbidities. *Clinical Science, 133*(7), 905–917. https://doi.org/10.1042/CS20180891

Kostic, A. D., Xavier, R. J., & Gevers, D. (2014). The microbiome in inflammatory bowel disease: Current status and the future ahead. *Gastroenterology, 146*(6), 1489–1499. https://doi.org/10.1053/J.GASTRO.2014.02.009

Kunze, W. A., Mao, Y. K., Wang, B., Huizinga, J. D., Ma, X., Forsythe, P., & Bienenstock, J. (2009). Lactobacillus reuteri enhances excitability of colonic AH neurons by inhibiting calcium-dependent potassium channel opening. *Journal of Cellular and Molecular Medicine, 13*(8B), 2261–2270. https://doi.org/10.1111/J.1582-4934.2009.00686.X

Lee, K. N., & Lee, O. Y. (2014). Intestinal microbiota in pathophysiology and management of irritable bowel syndrome. *World Journal of Gastroenterology, 20*(27), 8886–8897. https://doi.org/10.3748/WJG.V20.I27.8886

Leeming, E. R., Johnson, A. J., Spector, T. D., & Roy, C. I. L. (2019). Effect of diet on the gut microbiota: Rethinking intervention duration. *Nutrients, 11*(12). https://doi.org/10.3390/NU11122862

Lepage, P., Hösler, R., Spehlmann, M. E., Rehman, A., Zvirbliene, A., Begun, A., Ott, S., Kupcinskas, L., Doré, J., Raedler, A., & Schreiber, S. (2011). Twin study indicates loss of interaction between microbiota and mucosa of patients with ulcerative colitis. *Gastroenterology, 141*(1), 227–236. https://doi.org/10.1053/J.GASTRO.2011.04.011

Libertucci, J., Dutta, U., Kaur, S., Jury, J., Rossi, L., Fontes, M. E., Shajib, M. S., Khan, W. I., Surette, M. G., Verdu, E. F., & Armstrong, D. (2018). Inflammation-related differences in mucosa-associated microbiota and intestinal barrier function in colonic Crohn's disease. *American Journal of Physiology. Gastrointestinal and Liver Physiology, 315*(3), G420–G431. https://doi.org/10.1152/AJPGI.00411.2017

Lindsay, E. C., Metcalfe, N. B., & Llewellyn, M. S. (2020). The potential role of the gut microbiota in shaping host energetics and metabolic rate. *Journal of Animal Ecology, 89*(11), 2415–2426. https://doi.org/10.1111/1365-2656.13327

Li, J., Si, H., Du, H., Guo, H., Dai, H., Xu, S., & Wan, J. (2021). Comparison of gut microbiota structure and Actinobacteria abundances in healthy young adults and elderly subjects: A pilot study. *BMC Microbiology, 21*(1), 1–10. https://doi.org/10.1186/S12866-020-02068-Z/FIGURES/4

Liu, S., Moon, C. D., Zheng, N., Huws, S., Zhao, S., & Wang, J. (2022). Opportunities and challenges of using metagenomic data to bring uncultured microbes into cultivation. *Microbiome, 10*(1), 1–14. https://doi.org/10.1186/S40168-022-01272-5

López, P., Sánchez, B., Margolles, A., & Suárez, A. (2016). Intestinal dysbiosis in systemic lupus erythematosus: Cause or consequence? *Current Opinion in Rheumatology, 28*(5), 515–522. https://doi.org/10.1097/BOR.0000000000000309

Magne, F., Gotteland, M., Gauthier, L., Zazueta, A., Pesoa, S., Navarrete, P., & Balamurugan, R. (2020). The firmicutes/bacteroidetes ratio: A relevant marker of gut dysbiosis in obese patients? *Nutrients, 12*(5). https://doi.org/10.3390/NU12051474

Magrone, T., & Jirillo, E. (2013). The interplay between the gut immune system and microbiota in health and disease: Nutraceutical intervention for restoring intestinal homeostasis. *Current Pharmaceutical Design, 19*(7), 1329–1342. https://doi.org/10.2174/138161213804805793

Ma, N., Liu, X. W., Kong, X. J., Li, S. H., Jiao, Z. H., Qin, Z., Yang, Y. J., & Li, J. Y. (2018). Aspirin eugenol ester regulates cecal contents metabolomic profile and microbiota in an animal model of hyperlipidemia. *BMC Veterinary Research, 14*(1). https://doi.org/10.1186/S12917-018-1711-X

Manuzak, J., Dillon, S., & Wilson, C. (2012). Differential interleukin-10 (IL-10) and IL-23 production by human blood monocytes and dendritic cells in response to commensal enteric bacteria. *Clinical and Vaccine Immunology: CVI, 19*(8), 1207–1217. https://doi.org/10.1128/CVI.00282-12

Martin, C. R., Osadchiy, V., Kalani, A., & Mayer, E. A. (2018). The brain-gut-microbiome axis. *Cellular and Molecular Gastroenterology and Hepatology, 6*(2), 133–148. https://doi.org/10.1016/J.JCMGH.2018.04.003

Matey-Hernandez, M. L., Williams, F. M. K., Potter, T., Valdes, A. M., Spector, T. D., & Menni, C. (2018). Genetic and microbiome influence on lipid metabolism and dyslipidemia. *Physiological Genomics, 50*(2), 117–126. https://doi.org/10.1152/PHYSIOLGENOMICS.00053.2017

McDowell, C., Farooq, U., & Haseeb, M. (2022). Inflammatory bowel disease. *StatPearls*. https://www.ncbi.nlm.nih.gov/books/NBK470312/.

McFarland, L. V., Evans, C. T., & Goldstein, E. J. C. (2018). Strain-specificity and disease-specificity of probiotic efficacy: A systematic review and meta-analysis. *Frontiers in Medicine, 5*, 124. https://doi.org/10.3389/FMED.2018.00124/XML/NLM

Di Modica, M., Arlotta, V., Sfondrini, L., Tagliabue, E., & Triulzi, T. (2022). The link between the microbiota and HER2+ breast cancer: The new challenge of precision medicine. *Frontiers in Oncology, 12*. https://doi.org/10.3389/FONC.2022.947188

Moretti, C. H., Schiffer, T. A., Li, X., Weitzberg, E., Carlström, M., & Lundberg, J. O. (2021). Germ-free mice are not protected against diet-induced obesity and metabolic dysfunction. *Acta Physiologica (Oxford, England), 231*(3). https://doi.org/10.1111/APHA.13581

Moser, G., Fournier, C., & Peter, J. (2018). Intestinal microbiome-gut-brain axis and irritable bowel syndrome. *Wiener Medizinische Wochenschrift (1946), 168*(3–4), 62–66. https://doi.org/10.1007/S10354-017-0592-0

Mu, H., Zhou, Q., Yang, R., Zeng, J., Li, X., Zhang, R., Tang, W., Li, H., Wang, S., Shen, T., Huang, X., Dou, L., & Dong, J. (2020). Naringin attenuates high fat diet induced non-alcoholic fatty liver disease and gut bacterial dysbiosis in mice. *Frontiers in Microbiology, 11*, 2903. https://doi.org/10.3389/FMICB.2020.585066/BIBTEX

Niemarkt, H. J., De Meij, T. G., Van Ganzewinkel, C. J., De Boer, N. K. H., Andriessen, P., Hütten, M. C., & Kramer, B. W. (2019). Necrotizing enterocolitis, gut microbiota, and brain development: Role of the brain-gut axis. *Neonatology, 115*(4), 423–431. https://doi.org/10.1159/000497420

Ochoa-Repáraz, J., Mielcarz, D. W., Ditrio, L. E., Burroughs, A. R., Foureau, D. M., Haque-Begum, S., & Kasper, L. H. (2009). Role of gut commensal microflora in the development of experimental autoimmune encephalomyelitis. *Journal of Immunology (Baltimore, Md.: 1950), 183*(10), 6041–6050. https://doi.org/10.4049/JIMMUNOL.0900747

O'Hara, A. M., & Shanahan, F. (2006). The gut flora as a forgotten organ. *EMBO Reports, 7*(7), 688–693. https://doi.org/10.1038/SJ.EMBOR.7400731

Palmer, C., Bik, E. M., DiGiulio, D. B., Relman, D. A., & Brown, P. O. (2007). Development of the human infant intestinal microbiota. *PLoS Biology, 5*(7), 1556–1573. https://doi.org/10.1371/JOURNAL.PBIO.0050177

Pinto Coelho, G. D., Arial Ayres, L. F., Barreto, D. S., Henriques, B. D., Cardoso Prado, M. R. M., & Dos Passos, C. M. (2021). Acquisition of microbiota according to the type of birth: An integrative review. *Revista Latino-Americana de Enfermagem, 29*. https://doi.org/10.1590/1518.8345.4466.3446

Rajilić-Stojanović, M. (2013). Function of the microbiota. *Best Practice & Research. Clinical Gastroenterology, 27*(1), 5–16. https://doi.org/10.1016/J.BPG.2013.03.006

Rajilić-Stojanović, M., Biagi, E., Heilig, H. G. H. J., Kajander, K., Kekkonen, R. A., Tims, S., & De Vos, W. M. (2011). Global and deep molecular analysis of microbiota signatures in fecal samples from patients with irritable bowel syndrome. *Gastroenterology, 141*(5), 1792–1801. https://doi.org/10.1053/J.GASTRO.2011.07.043

Ramirez, J., Guarner, F., Bustos Fernandez, L., Maruy, A., Sdepanian, V. L., & Cohen, H. (2020). Antibiotics as major disruptors of gut microbiota. *Frontiers in Cellular and Infection Microbiology, 10*. https://doi.org/10.3389/FCIMB.2020.572912

Rebersek, M. (2021). Gut microbiome and its role in colorectal cancer. *BMC Cancer, 21*(1), 1–13. https://doi.org/10.1186/S12885-021-09054-2/FIGURES/5

Ridaura, V. K., Faith, J. J., Rey, F. E., Cheng, J., Duncan, A. E., Kau, A. L., Griffin, N. W., Lombard, V., Henrissat, B., Bain, J. R., Muehlbauer, M. J., Ilkayeva, O., Semenkovich, C. F., Funai, K., Hayashi, D. K., Lyle, B. J., Martini, M. C., Ursell, L. K., Clemente, J. C., … Gordon, J. I. (2013). Gut microbiota from twins discordant for obesity modulate metabolism in mice. *Science (New York, N.Y.), 341*(6150). https://doi.org/10.1126/SCIENCE.1241214

Rowland, I., Gibson, G., Heinken, A., Scott, K., Swann, J., Thiele, I., & Tuohy, K. (2018). Gut microbiota functions: Metabolism of nutrients and other food components. *European Journal of Nutrition, 57*(1), 1. https://doi.org/10.1007/S00394-017-1445-8

Sanmiguel, C., Gupta, A., & Mayer, E. A. (2015). Gut microbiome and obesity: A plausible explanation for obesity. *Current Obesity Reports, 4*(2), 250–261. https://doi.org/10.1007/S13679-015-0152-0

Schächtle, M. A., & Rosshart, S. P. (2021). The microbiota-gut-brain axis in health and disease and its implications for translational research. *Frontiers in Cellular Neuroscience, 15*, 256. https://doi.org/10.3389/FNCEL.2021.698172/BIBTEX

Scheithauer, T. P. M., Rampanelli, E., Nieuwdorp, M., Vallance, B. A., Verchere, C. B., van Raalte, D. H., & Herrema, H. (2020). Gut microbiota as a trigger for metabolic inflammation in obesity and type 2 diabetes. *Frontiers in Immunology, 11*, 2546. https://doi.org/10.3389/FIMMU.2020.571731/BIBTEX

Ser, H. L., Letchumanan, V., Goh, B. H., Wong, S. H., & Lee, L. H. (2021). The use of fecal microbiome transplant in treating human diseases: Too early for poop? *Frontiers in Microbiology, 12*, 1005. https://doi.org/10.3389/FMICB.2021.519836/XML/NLM

Shanahan, F. (2002). The host-microbe interface within the gut. *Best Practice & Research. Clinical Gastroenterology, 16*(6), 915–931. https://doi.org/10.1053/BEGA.2002.0342

Sheehan, D., Moran, C., & Shanahan, F. (2015). The microbiota in inflammatory bowel disease. *Journal of Gastroenterology, 50*(5), 495–507. https://doi.org/10.1007/S00535-015-1064-1

Simpson, H. L., & Campbell, B. J. (2015). Review article: Dietary fibre-microbiota interactions. *Alimentary Pharmacology & Therapeutics, 42*(2), 158–179. https://doi.org/10.1111/APT.13248

Sommer, F., & Bäckhed, F. (2013). The gut microbiota–masters of host development and physiology. *Nature Reviews. Microbiology, 11*(4), 227–238. https://doi.org/10.1038/NRMICRO2974

Soo, W. T., Bryant, R. V., & Costello, S. P. (2020). Faecal microbiota transplantation: Indications, evidence and safety. *Australian Prescriber, 43*(2), 36–38. https://doi.org/10.18773/AUSTPRESCR.2020.014

Sudo, N., Chida, Y., Aiba, Y., Sonoda, J., Oyama, N., Yu, X. N., Kubo, C., & Koga, Y. (2004). Postnatal microbial colonization programs the hypothalamic-pituitary-adrenal system for stress response in mice. *The Journal of Physiology, 558*(Pt 1), 263–275. https://doi.org/10.1113/JPHYSIOL.2004.063388

Su, Q., & Liu, Q. (2021). Factors affecting gut microbiome in daily diet. *Frontiers in Nutrition, 8*, 218. https://doi.org/10.3389/FNUT.2021.644138/BIBTEX

Tan, L. R., Zhao, S., Zhu, W., Wu, L., Li, J., Shen, M. X., Lei, L., Chen, X., & Peng, C. (2018). The Akkermansia muciniphila is a gut microbiota signature in psoriasis. *Experimental Dermatology, 27*(2), 144–149. https://doi.org/10.1111/EXD.13463

van Teeseling, M. C. F., de Pedro, M. A., & Cava, F. (2017). Determinants of bacterial morphology: From fundamentals to possibilities for antimicrobial targeting. *Frontiers in Microbiology, 8*, 1264. https://doi.org/10.3389/FMICB.2017.01264/BIBTEX

Vaahtovuo, J., Munukka, E., Korkeamäki, M., Luukkainen, R., & Toivanen, P. (2008). Fecal microbiota in early rheumatoid arthritis. *The Journal of Rheumatology, 35*(8), 1500–1505. https://pubmed.ncbi.nlm.nih.gov/18528968/.

Valdes, A. M., Walter, J., Segal, E., & Spector, T. D. (2018). Role of the gut microbiota in nutrition and health. *BMJ, 361*, 36–44. https://doi.org/10.1136/BMJ.K2179

Vijay, A., & Valdes, A. M. (2021). Role of the gut microbiome in chronic diseases: A narrative review. *European Journal of Clinical Nutrition, 76*(4), 489–501. https://doi.org/10.1038/s41430-021-00991-6

Wang, T., Cai, G., Qiu, Y., Fei, N., Zhang, M., Pang, X., Jia, W., Cai, S., & Zhao, L. (2012). Structural segregation of gut microbiota between colorectal cancer patients and healthy volunteers. *The ISME Journal, 6*(2), 320. https://doi.org/10.1038/ISMEJ.2011.109

Wang, P. X., Deng, X. R., Zhang, C. H., & Yuan, H. J. (2020). Gut microbiota and metabolic syndrome. *Chinese Medical Journal, 133*(7), 808–816. https://doi.org/10.1097/CM9.0000000000000696

Wilmore, J. R., Gaudette, B. T., Gomez Atria, D., Hashemi, T., Jones, D. D., Gardner, C. A., Cole, S. D., Misic, A. M., Beiting, D. P., & Allman, D. (2018). Commensal microbes induce serum IgA responses that protect against polymicrobial sepsis. *Cell Host & Microbe, 23*(3), 302–311.e3. https://doi.org/10.1016/J.CHOM.2018.01.005

Wingerchuk, D. M., Lucchinetti, C. F., & Noseworthy, J. H. (2001). Multiple Sclerosis: Current pathophysiological concepts. *Laboratory Investigation, 81*(3), 263–281. https://doi.org/10.1038/labinvest.3780235

Wu, G. D., Chen, J., Hoffmann, C., Bittinger, K., Chen, Y. Y., Keilbaugh, S. A., Bewtra, M., Knights, D., Walters, W. A., Knight, R., Sinha, R., Gilroy, E., Gupta, K., Baldassano, R., Nessel, L., Li, H., Bushman, F. D., & Lewis, J. D. (2011). Linking long-term dietary patterns with gut microbial enterotypes. *Science (New York, N.Y.), 334*(6052), 105–108. https://doi.org/10.1126/SCIENCE.1208344

Wu, A. H., Tseng, C., Vigen, C., Yu, Y., Cozen, W., Garcia, A. A., & Spicer, D. (2020). Gut microbiome associations with breast cancer risk factors and tumor characteristics: A pilot study. *Breast Cancer Research and Treatment, 182*(2), 451. https://doi.org/10.1007/S10549-020-05702-6

Yang, T., Santisteban, M. M., Rodriguez, V., Li, E., Ahmari, N., Carvajal, J. M., Zadeh, M., Gong, M., Qi, Y., Zubcevic, J., Sahay, B., Pepine, C. J., Raizada, M. K., & Mohamadzadeh, M. (2015). Gut dysbiosis is linked to hypertension. *Hypertension, 65*(6), 1331–1340. https://doi.org/10.1161/HYPERTENSIONAHA.115.05315

Zeng, H., Lazarova, D. L., & Bordonaro, M. (2014). Mechanisms linking dietary fiber, gut microbiota and colon cancer prevention. *World Journal of Gastrointestinal Oncology, 6*(2), 41. https://doi.org/10.4251/WJGO.V6.I2.41

Zhang, H., Liao, X., Sparks, J. B., & Luo, X. M. (2014). Dynamics of gut microbiota in autoimmune lupus. *Applied and Environmental Microbiology, 80*(24), 7551–7560. https://doi.org/10.1128/AEM.02676-14

Zhang, W., & Si, L. Y. (2012). Obstructive sleep apnea syndrome (OSAS) and hypertension: Pathogenic mechanisms and possible therapeutic approaches. *Upsala Journal of Medical Sciences, 117*(4), 370. https://doi.org/10.3109/03009734.2012.707253

Zhang, X. X., Tian, Y., Wang, Z. T., Ma, Y. H., Tan, L., & Yu, J. T. (2021). The epidemiology of alzheimer's disease modifiable risk factors and prevention. *Journal of Prevention of Alzheimer's Disease, 8*(3), 313–321. https://doi.org/10.14283/JPAD.2021.15/FIGURES/2

Zheng, D., Liwinski, T., & Elinav, E. (2020). Interaction between microbiota and immunity in health and disease. *Cell Research, 30*(6), 492–506. https://doi.org/10.1038/s41422-020-0332-7

Zhu, L., Spence, C., Yang, W. J., & Ma, G. X. (2020). The IDF definition is better suited for screening metabolic syndrome and estimating risks of diabetes in asian American adults: Evidence from NHANES 2011–2016. *Journal of Clinical Medicine, 9*(12), 1–13. https://doi.org/10.3390/JCM9123871

Zhu, B., Wang, X., & Li, L. (2010). Human gut microbiome: The second genome of human body. *Protein and Cell, 1*(8), 718–725. https://doi.org/10.1007/S13238-010-0093-Z

Zwartjes, M. S. Z., Gerdes, V. E. A., & Nieuwdorp, M. (2021). The role of gut microbiota and its produced metabolites in obesity, dyslipidemia, adipocyte dysfunction, and its interventions. *Metabolites, 11*(8). https://doi.org/10.3390/METABO11080531

CHAPTER

18

Strategies to reduce microbial biofilm in medical prosthesis and other devices

Juhita Dhar, Anupriya Das, Soumyadip Biswas, Talaha Nishat Ahmed, Sanjana Ghosh and Arup Kumar Mitra

Department of Microbiology, St. Xavier's College (Autonomous), Kolkata, West Bengal, India

1. Introduction

A medical device could be any instrument, tool, implant, or gadget designed and produced specifically for healthcare applications, as opposed to only pharmaceutical or dietary ones (Aronson et al., 2020). In other words, a medical device could be considered as any tool, apparatus, implement, machine, appliance, implant, reagent for in vitro usage, software, material, or other similar or related product recommended by the manufacturer to be used, for a therapeutic treatment, either separately or in combination. As compared with clinical devices, medical implants could be synthetic devices composed of a diverse range of substances such as metal alloys, hydrogels, ceramics, plastic polymers, and composites (Barchowsky, 2020). Implants have been used to pervade some types of bodily components, which are absent with prosthetic devices. Implants could be completely placed or retrieved when they are no longer needed. For example, hip implants and stents (placed into the lumen) could be removed at a later date but are intended to be lifelong.

Contrary to drugs, medical equipment was not proposed to have any immunogenic, pharmacological, or metabolic effects on the human body. Advanced technology has increased the healthcare potential and the economic benefits to spur growth. In addition to clinical technology becoming more specialized and sophisticated, a variety of medical items have left hospitals or healthcare facilities and have entered local surroundings for use by the commoners (Bitterman, 2011).

https://doi.org/10.1016/B978-0-323-95715-1.00006-6

Apart from clinical settings, nonclinical settings where medical devices have been employed include houses, institutions, airports, theaters, and many more. Some common household medical devices could cover a broad range of goods, including heating pads, gloves, dressings, condoms, syringes, contact lenses, antiseptics, urine testing kits, thermometers, and blood testing kits. An issue concerning these household devices was that they frequently were not similar versions to those used at healthcare facilities. The government authorities have categorized the medical devices according to the safety guidelines and standards such as the length of interaction, the balance between both local and systemic effects, and the level of invasiveness (Shah & Goyal, 2008).

Medical implants for prostheses could be categorized as devices or tissues; these devices have been used to replace a missing body part or deliver medicine. These medical implants have also been known as medical devices. FDA (Food and Drug Administration) regulated them and have categorized them into three classes: Class I devices (could generally control very little health hazard associated); Class II (controlled with special controls, to ensure safety selected tools are used); and Class III (general controls with special controls and premarket approval could be required) (van Eck et al., 2009). The types of implants include the following:

Tooth implant
Bone implant
Pacemaker
Naturally dissolving suture
Synthetic skin tissue ligament

A biomaterial could be any substance that was specially designed to interact and associate with biological systems. A biomaterial has also been known for supporting and enhancing biological functions. It was also used to replace damaged tissue and biomembranes. A variety of biomaterials have been known. These could be used primarily in implants and the common types of biomaterials that have been used in metal, polymers, ceramics, and composites.

Metals such as stainless steel were used extensively. Other metals such as titanium, chromium, and cobalt alloys could also be used owing to their property of having high tensile strength, ductility, good biocompatibility, low cost, resistance to corrosion, and good biocompatibility (Nandakumar et al., 2013). Metallic implants coated with calcium phosphate ceramics were found to promote bone integration of implants.

Polymers that have served as a biomaterial for implants mainly include polyamides, polyesters, polyacrylates, silicon rubber, and polyether. These were of low cost; however, these were incompetent in recent times. This was because these could generate immunogenic reactions in the biological system. Another disadvantage of using polymers was their poor mechanical strength, due to which it has been quite susceptible to fracture during function. Moreover, polymers could only be sterilized using gamma radiation, which might cause discoloration, embrittlement, and even degradation in some cases (Ong et al., 2015).

Ceramics used were, namely, alumina, zirconia (zirconium dioxide), and calcium phosphate ceramics. These were biodegradable and resistant to corrosion and had high tensile strength (Shuai et al., 2019). Resorbable materials were also used in cases of sutures. These were absorbed by the body over the course of time.

The term "biofilm" was first coined in 1975 to understand the interface association between microbial communities (Jakubovics et al., 2021). The extracellular polymeric matrix (EPS) produced by biofilm have been composed of macromolecules such as lipids, proteins, carbohydrates, polysaccharides, and nucleic acids. The major feature of this matrix was to protect the cells from external factors such as antimicrobial treatments, oxidation, and host defense mechanisms by forming a three-dimensional wall (Wi & Patel, 2018). The structure of a biofilm has been dependent on different species and on environmental factors, which include nutrient level, temperature, cell-to-cell contact, and second messengers such as cyclic antimicrobial peptide (AMP) and cyclic di-GMP (Tolker-Nielsen, 2015). Locomotion of cells, synthesis of extracellular substances, and cellular proteins could all be regulated by the second messengers.

The microbial biofilm formation primarily involved many stages, where at first the planktonic cells got adhered to the surfaces, known as cell surface adherence (Toyofuku et al., 2016). Upon attachment, bacterial cells could start developing microcolonies by cell division and assembling previously attached cells. This has been known as microcolony formation. These microcolonies upon maturation produced EPS to form a structured layer, known as maturation. Finally, these maturated cells get dispersed from the biofilm. The dispersed cells could then explore the environment and again get adhered to a new surface, which marked the start of a new life cycle (McDougald et al., 2012). The process has been represented in Fig. 18.1.

Biofilms can be formed by many microorganisms such as both gram-positive bacteria and gram-negative bacteria, cyanobacteria, and fungi (Orell et al., 2017). Biofilm-forming bacteria can cling onto multiple surfaces and form biofilm where nutrition is available and corrodes the surfaces they adhere to (Yadav and Sanyal, 2019). *Staphylococcus aureus*, *Pseudomonas fluorescens*, and *Bacillus subtilis*, among others, are capable of forming biofilms in food-processing environments (Di Ciccio et al., 2015). The development of biofilm is one of the many reasons for food-associated outbreaks all around the world (Srey et al., 2013). Several *Enterococcus* spp. such as *Enterococcus faecalis* and *Enterococcus faecium* form biofilm on dairy

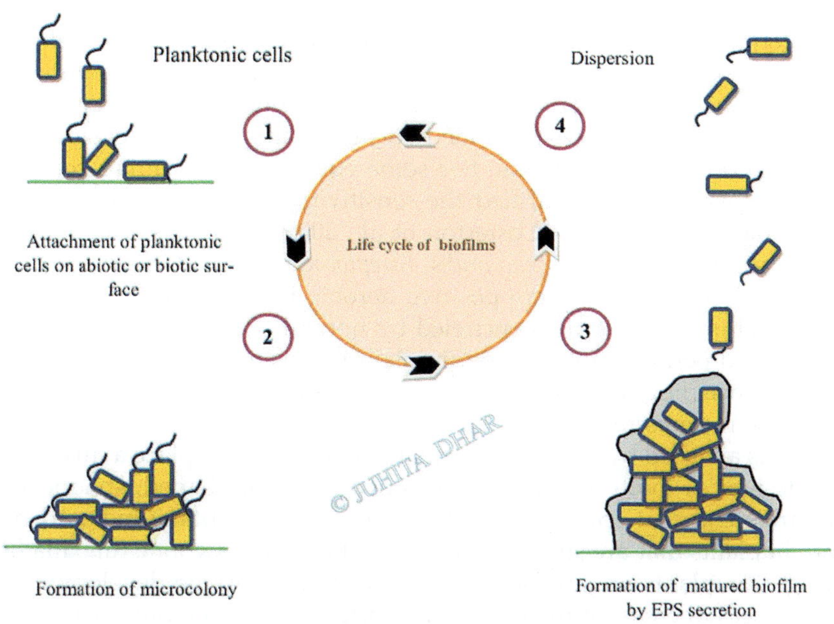

FIGURE 18.1 Formation of biofilm.

sectors such as raw milk, thermometers, milk apparatuses, etc. (de Castro et al., 2017). Apart from the aforementioned microorganisms, *Vibrio* spp. and *Shigella* spp. are also involved in seafood-related infection.

Biofilms serve an important part in the dissemination as well as in persistence of human diseases notably for diseases that are associated with neutral surfaces, such as illnesses linked to medical implants for both internal and external use; these tools include bone implants, dental implants, naturally dissolving sutures, breast implants, pacemakers, synthetic skin or tissue or ligaments, and catheters (Das et al., 2019). Medical implants are primarily affected by a limited quantity of microorganisms that are frequently introduced into the implants through wastewater, the skin of patients, and several external factors. The capability of clinical strains to produce biofilms is correlated with their resistance to harsh environmental settings and also to implantable biomaterials such as ceramics or metals. Biofilm-linked infection of medical devices poses a risk to a patient's overall health as well as adversely affects devices' efficiency. Gram-positive and gram-negative bacteria such as *E. faecalis, Escherichia coli, S. aureus,* and *Pseudomonas aeruginosa* can cause healthcare-related disorders in hospitals and homes (von Gotz et al., 2004). Illnesses linked with ventilators, urinary tract, and lower respiratory tract are among the most frequent healthcare-related diseases.

2. Tooth implant

2.1 Nature and types

A diverse group of microorganisms that are enmeshed in a polymer matrix located on the surfaces of teeth is known as dental plaque. The concept of plaque now includes biofilms on all oral surfaces. Contrary to any other region of the body, the mouth provides these microbes with the best conditions for their proliferation. These pathogenic microorganisms can grow and spread easily on the nonshedding surface of the mouth. Some common oral diseases caused by bacteria include dental decay, cavities, thrush, gingivitis, and periodontal infection. Over the past 20 years, researchers have focused significantly on the development of biofilm on a broad range of medical implants. For some equipment, including contact lenses and catheters, researchers have also clarified the sensitivity of different substances to bacterial adherence and biofilm formation (Costerton et al., 2005). *S. aureus* was the first organism to infect dental implants. *Streptococcus oralis, Streptococcus mitis, Neisseria* spp., and *Streptococcus sanguis* are some of the gram-positive aerobic bacteria that have been previously detected on dental implant surfaces encircled by normal oral habitat. Biofilm formation on the implant leads to periimplantitis or periimplant mucositis. Mucositis is the inflammation of delicate tissue encircling the tooth implant, which can be identified by changes in the color and texture of the mucous membrane.

Tooth implants are used by dentists as a treatment option to replace a missing tooth. There are multiple designs of tooth implants such as oval- or cone-shaped, spherical-shaped, and threads that are somewhat spread apart. In recent days, the titanium-based biomaterial is used as tooth implants that are surgically implanted into the jawbone (maxilla or mandible). The anchoring of implants on jawbone has been enhanced by the development of implants with coarse surfaces. Increasing the developed surface at the micrometer level increases the interlocking between the graft and bone during growth (Becker et al., 2000; Guillaume, 2016).

2.2 Surface adherence of bacteria

Skin and the mouth cavity are two examples of human body parts that are colonized by the resident microflora; each of these habitats is home to a diverse population of microorganisms with characteristics that set them apart from one another. Microbial plaque is an ideal example of a synergistic relationship between microbes and the host organism that provides a platform for complex microbial colonization (Minkiewicz-Zochniak et al., 2021). Numerous external factors, including the composition of biofilm, the prosthetic elements, chemical features, and the texture of the implant, all contribute to biofilm production on tooth implants.

Surface nature: The nature of the surface considerably affects osseointegration (a strong connection between implant and bone). An increase in surface coarseness facilitates bacterial adhesion, as the contact area between that surface and bacterial cell rises. Han et al. in (2016) reported that coarse surfaces demonstrated stronger bone incorporation, but smooth and barely coarse surfaces did not, while in some experiments, relatively rough surfaces exhibited stronger bone responses than rough surfaces. Furthermore, a thin film of saliva that coats every surface inside the mouth may also change the nanotopography, which may have a significant impact on the surface nature (Siqueira et al., 2012).

Surface charge: It has long been known to influence the structure and growth of biofilms and plays a significant role in defining the binding nature between bacteria and surfaces. The physical or chemical characteristics of a biomaterial can also influence the arrangement and density of the thin salivary film (Teughels et al., 2006; Xing et al., 2015). The antimicrobial properties of cations such as polyethyleneimine or quaternary ammonium can kill the adhering cells when present on surfaces.

Surface energy: Free energy of the surface affects the attachment of bacteria to the surfaces of the implant. Hydrophobic surfaces develop fewer biofilm than hydrophilic surfaces in oral settings (Song et al., 2015; Rzhepishevska et al., 2013). Strains of *Streptococcus salivarius*, *Streptococcus mutans*, and *S. sanguis* have inferior adhesion properties on hydrophobic surfaces (Han et al., 2016). The existence of a salivary layer on microbial cells and tooth surfaces may also have a significant impact on how hydrophobic nature influences the association of microbes and oral implants.

2.3 Harmfulness

Dental plaque is one of the best-known biofilms in the human body. The plaque can be categorized into supragingival and subgingival plaque. Gram-positive organisms dominate in supragingival plaque, while gram-negative organisms prevail in subgingival plaque (Kornman et al., 1991). Treatment of periodontal disease caused by *Aggregatibacter actinomycetemcomitans*, *Porphyromonas gingivalis*, among others, and gingivitis infection caused by nonspecific bacterial flora may get complicated in subgingival plaque due to the development of bacterial biofilm.

The structure of the dental implants supports biofilm development. The success and endurance rate of dental implants is likewise 90% after a decade of medical service. A wide range of bacteria reside in saliva that gradually settle and multiply on the implant materials at first by nonspecific physicochemical processes followed by distinctive physicochemical association with salivary proteins (Lasserre et al., 2018). Lang and co-workers (2011) reported a link between ceased dental health in implant recipients and the onset of periimplant mucositis, hence proving that biofilms on implantable devices are the primary etiological agent for periimplant mucositis and that the host reaction to biofilms does not fluctuate significantly between implantable devices and teeth. Hahnel et al. (2017) reported that periimplant infections occur not only due to microbial colonization and biofilm formation on the implant surfaces but also due to alteration in environmental conditions such as implant cracking or additional plaster, which can stimulate the proliferation of microorganisms by expressing their pathogenic factors (Mombelli & Décaillet, 2011).

Furthermore, Darveau in 2010 revealed that biofilm-forming bacteria have assisted us in comprehending by what means microorganisms might evade host–immune responses as well as specific traditional antiseptic treatments. This limited inflammation is complicated because it involves innate and acquired immunity, in addition to the complement pathway.

3. Bone implant

3.1 Nature and types

The materials that have generally been used in orthopedic implants are stainless steel, alloys (cobalt chromium, titanium), and polyethylene (Nandakumar, 2013).

Titanium has been a common material used in orthopedic implants. Most total hip femoral stem and total shoulder arthroplasty stem components were made of a titanium alloy. It predictably oxidized when implanted, and had a modulus of elasticity (the ability of a material to deform under stress) which is comparable to cortical bone (Ong, 2015).

The polymers, alloys, and polymers so far made weak attraction with the bone because they did not directly attach to it (LeGeros, 1992).

Bulk material such as cobalt chrome has been biologically inactive. It contained nickel that might lead to metal hypersensitivity reactions and had a very high modulus of elasticity. Stainless steel has provided a high modulus of elasticity. This material did not demonstrate corrosion in the oxygen-rich environment of the body, but it could cause allergic reaction. Polyethylene's most common application has been in the arthroplasty, and it could generate a macrophage-mediated response, which could lead to bone resorption (Ong, 2015).

3.2 Surface adherence of bacteria

The implant-based infection has contributed to the mortality, morbidity, and extended hospitalization (Davide Campoccia et al., 2006).

The main microorganisms that adhered to orthopedic implants were *S. aureus*, methicillin-resistant *S. aureus* (Marta Ribeiro et al., 2012), and *Staphylococcus epidermidis*. *S. aureus* and *S. epidermidis* were 70%, while others were 22%. *S. aureus* was mainly associated with rheumatoid arthritis. Furthermore, it has been the main causative agent of hip arthroplasty. In addition, it has been found that approximately 12% of perinatal hip infections were caused by *E. faecalis*, *Corynebacterium*, and *Streptococcus pyogenes*. Anaerobic bacteria, namely, *Peptococcus saccharides* and *Peptococcus acnes* were also causative agents (Nandakumar et al., 2013), while in the prosthetic joint, they are caused by coagulase-negative *Staphylococcus* and *Propionibacterium acnes*. The most commonly found etiological agents of osteomyelitis were *E. coli, S. epidermidis, S. aureus, and P. aeruginosa* (Marta Ribeiro et al., 2012).

3.3 Mechanism

Bacterial cells were moved to the surface of the implants by means of physical forces such as gravitational force, Brownian motion, van der Waals attraction, hydrophobic interaction, and surface electrostatic charge force (Nandakumar et al., 2013). Microcolony forms, when planktonic bacteria adhere to the implant surface. It began to secrete EPS (exopolysaccharide, DNA, protein)-cementing substance that has enabled bacterial intercellular aggregation (Montanaro et al., 2011), thus forming a monolayer, also called a slime layer that has made it difficult for the host—immune system to protect the implant surface and susceptible to antibiotics. Adhesin molecules are secreted by the bacteria for the attachment with the receptor of the host surface. Pili acted as an adhesin. This implied a firmer adhesion (Nandakumar et al., 2013). For example, individual *Staphylococcus* secretes positively charged homopolymers (McConoughey et al., 2014). The adhesion process has been influenced by different factors including environmental factors (presence of the host proteins and antibiotics), the substrate on the implants, and the properties of bacteria. Once attached to the surface, the bacteria have undergone a transition from planktonic movement to type IV pilus motility. In the early stage, the biofilm has been unstable. In the maturation stage, the bacterial cells communicated with each other through quorum sensing (QS) (Nandakumar et al., 2013).

Hydrophobic surfaces of the implant favor biofilm formation. Adherence of the bacteria to the implant surface has increased with the increase in surface energy and surface roughness. Some host serum proteins such as fibronectin, fibrinogen, etc. could influence the adhesion process of bacteria on the implant surface by binding to the implants or the bacterial cell. For example, the rate of adhesion of the bacteria *S. aureus* got promoted in the presence of fibronectin of the host (Marta Ribeiro et al., 2012).

3.4 Harmfulness

Detachment of the prosthesis from the bone, fracture of the bone, and infection at the site of implants have been some of the major causes of failure of implants (Kurtz Steven et al., 2007).

The major problem with the orthopedic implant failure was infection. Treatment of orthopedic implant infections was a major challenge nowadays. In severe cases, infections might result in mortality. These orthopedic implant infections were caused by several organisms. *Staphylococcus* was the predominant of all pathogens that cause orthopedic infections in implant. Among them, the most frequently isolated organism was *S. epidermidis*, which belong to the group of coagulase-negative *staphylococci* (CoNS) (Nandakumar et al., 2013).

Both septic arthritis and osteomyelitis diseases were involved in the inflammation as well as destruction of joints and bones. Implant infections were very hard to treat, primarily because of the ability of the organisms to form stable biofilms. Other infections might include soft tissue infections such as diabetic foot infections (Nandakumar et al., 2013).

The percentage of the occurrence of late infection of total joint prostheses was 0.6%. In a study, it has been seen that the most common pathogens responsible were both *S. aureus* and *S. epidermidis* (54%), even though dental region was the origin of infection. The most common origins were skin, soft tissue, and dental (Maderazo et al., 1988).

Implant-associated infections could be classified based on the beginning of the infection as early (1–2 weeks), delayed (2–10 weeks), and late (more than <10 weeks). It could also be classified based on the infection route, which could be perioperative, hematogenous, and contiguous (Trampuz et al., 2006).

4. Pacemaker

4.1 Nature

Biomaterials required for development of implantable pacemakers are alloplastic, which are nonbiological in origin, mostly related to skeletomuscular systems. It includes metals, ceramic glasses, and polymers. The main difference in their physical character is brought about by their internal chemical bonds that hold them together. Metals have metallic bonds typically, which hold the material characterized by the free electrons in the lattice and positive atomic core/cations at the center, which makes them highly conductive. Besides orthopedic implantable devices, metals are employed almost exclusively in pacemaker housing and stimulation electrodes (Schaldach, 1992). There are several parts to the pacemaker: outer casing or main body is generally made of titanium, whereas the battery, electrode, and circuit board are made using lithium, titanium, lithium–carbon, lithium–iodine, and mercury (Varshini et al., 2021).

4.2 Surface adherence of bacteria

Study made from pacemaker aseptically removed from patient suffering from three sequential episodes of *S. aureus* bacteremia; its outer surface was swabbed using a cotton swab, used to inoculate onto blood agar media; and a small part of the pacemaker lead was removed to study the intracardiac portion of pacemaker, and scanning electron microscopy (SEM) was used to study the sample. *S. aureus* was isolated from several sites on the surface of pacemaker lead; SEM study revealed differential colonization of pacemaker lead: metal tip, inner surface, and inner wires were covered with heavy biofilms, whereas the outer silastic surface had no biofilm adherent to it. The observation is evident to show the reason for infection persisting on microbial devices despite antibiotic chemotherapy (Marrie et al., 1982).

4.3 Types

Pr. acnes and *Propionibacterium* species are gram-positive bacilli that form branches. These are a part of normal human microflora of skin, conjunctiva, and external ear. It has a low virulence nature; it is usually identified as an etiological agent in many infections, including CNS shunt infection, brain abscesses, endophthalmitis. etc.; there are cases where infection endocarditis due to *Pr. acnes* with pacemaker is seen (Chua et al., 1998; Zedtwitz-Liebenstein et al., 2002). *Pr. acnes* is usually isolated from the surface of an explanted cardiac pacemaker device exhibiting no clinical signs of infection. Culture test using stamp culture method, pressing the surface of pacemaker on surface of agar plates—frequent colony growth of *Pr. acnes* on surface of pacemaker, was found from 7 out of 31 devices that were used in isolation (Okuda et al., 2018).

4.4 Mechanism

The bacterial biofilms of *S. aureus*, *S. epidermidis*, extracellular DNA(e-DNA), polysaccharide intracellular adhesion (PIA), and proteins are main components of ECM (extracellular matrix). PIA has been important for staphylococcal film development, while some can form biofilms independent of PIA. *P. aeruginosa* produces alginate exopolysaccharides such as Pel and Psl involved in establishment of biofilms; alginate plays an important role in initial biofilm development and in maturation of biofilms, which leads to establishment of the infection. CdrA, a secreted protein and constitutinng structural component of *P. aeruginosa* biofilm, was seen to directly bind Psl that forms a component of the biofilm structure. Extracellular DNA (eDNA) required for intracellular connection and stability of the biofilm allows cells to come close and ultimately form the biofilm structure (Okuda et al., 2018).

4.5 Harmfulness

Since there are extravascular and intravascular segments related to a pacemaker implantation, infections can affect the device, the intravascular region of lead in form of device-associated endocarditis, and the infraclavicular site of device implantation in form of localized pocket infection. The localized infection accounts for a large proportion of the disease; generally exhibited clinical symptoms are of redness, warmth, and swelling (Döring et al. 2018).

5. Naturally dissolving suture

Absorbable sutures were made from materials that the body could naturally absorb over time. The enzymes (Nandakumar et al., 2013) and the body fluids were essential for the degradation. They are ideal for healing internal wounds and surgical sites. They generally lost their tensile strength over periods ranging from a week to months.

5.1 Nature

The most commonly used materials in absorbable suture were animal tissue (collagen), bovine intestine (purified serosa), catgut that had good tensile strength, and sheep submucosa. They' are called natural absorbable suture. Although the materials used in synthetic absorbable suture are polydioxanone (PDS) that repairs various kinds of soft-tissue wounds and abdominal closures, and poliglecaprone (MONOCRYL), which are used in case of vascular anastomosis procedures that connect blood vessels; polyglactin (Vicryl) too is used in vascular anastomosis procedures (Mahesh et al., 2019).

5.2 Surface adherence of bacteria

Surgical suture had been an excellent surface for the adherence of pathogenic bacteria, followed by colonization and formation of biofilm. *P. aeruginosa* (Daniela Vieira. et al., 2022) *Streptococcus* spp., methicillin-resistant *S. aureus*, and *S. aureus* were the main biofilm-producing bacteria found on the surface of naturally dissolving suture (Reinbold et al., 2017).

5.3 Types

Naturally dissolving sutures were of two types—natural (catgut) and synthetic (PDS). The naturally dissolving sutures were degraded by proteolysis in the human body, whereas the synthetic dissolving suture was degraded by hydrolysis and it was less inflammatory compared with that of the naturally dissolving suture. They were subdivided into braided suture and microfilament suture (Mahesh et al., 2019).

5.4 Mechanism

Surgical site infections have overall morbidity and mortality. Sutures being foreign material have actively contributed to the development of surgical site infections (Ercan et al., 2018), which helps in holding the open wound end together and helps in wound closing.

At the time of operation, the superficial skin harbored bacterial colony. The bacteria from the suture material penetrate from the superficial skin surface into the deeper tissues during wound closure. The suture material served itself as a nidus for bacterial adherence and proliferation. There was generation of hypoxic environment, which has contributed to the delayed healing, which further leads to wound decontamination and biofilm (Mahesh et al., 2019).

5.5 Harmfulness

Sutures provided an excellent surface to which bacteria adhere, proliferate, and induce surgical site infection. The suture material being a foreign material or substance could cause a reaction in the form of inflammation, scarring in the superficial tissues (Lascelles & Claringbold et al., 1960). Also, by using materials such as sheep and beef intestine, the risk of spreading zoonoses got increased. The signs of infection include redness, oozing, pain, fever, swelling etc. There were complications—necrotizing fasciitis and sepsis, which were mainly caused by *Streptococcus* bacteria. However, this was rare.

6. Synthetic skin/tissue/ligament

6.1 Nature

Synthetic skin can also be referred to as a substance that can reinstate the functions of the skin either permanently or temporarily. These are usually used when the body endures a severe skin injury such as burns, skin disease, or disorders. Besides their use in the medical field, synthetic skin is also being made use of in large industries such as pharmaceuticals and cosmetics (Mohd Noor & Mahmud, 2014).

Synthetic tissues can be explained as an alternative to natural tissues and can also possess added properties. They are composed of a 3D-patterned assembly of sections that can communicate with each other and with the surroundings. Thus, they are strictly artificial and contain no live cells (Bayley et al., 2019).

Synthetic tissues can act as a substitute for skin, which are privileged of being comparably plentiful in supply and cheap. They can be produced on demand and can be modified for a particular requirement with an enhanced control over scaffold composition (Halim et al., 2010).

These substances should possess balance, must be biodegradable, and must have a suitable environment for the renewal of tissues. Their three-dimensional structure must be retained for at least 3 weeks to promote the ingrowth of blood vessels, fibroblast, and coverage by epithelial cells. After this period, biodegradation of the synthetic tissue should take place without having a major nonnative reaction in the body, which would escalate the inflammatory response, which would lead to intense scarring. Thus, the synthetic material should contain some immune compatible substance to reduce the chances of an immunoreactive response (Halim et al., 2010).

6.2 Surface adherence of bacteria

Modern medicine always faces a threat of infection in biomedical implants due to microbes. The main cause of nonsuccess in biomedical implants is due to an infection caused by formation of a biofilm. Colonization by microbes can be observed in all types of biomedical devices, and they are responsible for about 60%–70% of hospital-acquired infections particularly in chronically ill patients. Biofilms formed by bacteria can occupy the surfaces of tissues and the medical devices implanted (Veerachamy et al., 2014).

The matrix of the extracellular polysaccharide of a biofilm is composed of a multilayered cell cluster, which helps in the adhesion of the microorganisms, which protects them from the immune system of the host and antimicrobial therapy. The surface of the many synthetic materials, such as polyester vascular grafts and many more, which are used by microbiologists and surgeons, binds these molecules and promotes biofilm formation (Cirioni et al., 2010).

Reconstruction of the aorta and the lack of autologous vein material in the peripheral bypass operations are mainly where vascular grafts are used. Vascular graft infections (VGIs), which occur in 2%–4% of cases despite the highest hygienic precautionary measures that are taken during surgery, have high mortality and morbidity rates. The aortic graft infections are usually caused by gram-positive bacteria such as *S. aureus* and *S. epidermidis* (Herten et al., 2017).

The EPS of the biofilm on vascular graft surfaces consist of glycolipids, proteins, and bacterial DNA. This EPS protects the bacteria by restricting the entry of macromolecules such as antibiotics and the immune system's inflammatory cells into the matrix of the biofilm as well as by presenting a barricade of diffusion for molecules having characteristic antimicrobial properties. Inactivation of metabolism and formation of persister cells occur within the biofilm. Since antibiotics act only against the bacteria that are active in metabolism, it may show no effect for the resting bacteria (Herten et al., 2017).

The biofilm formation done by *Staphylococcus* species is accompanied with a composite pathway, which includes attachment, maturation, and dispersion, in which the adhesion surface and the kind of nutrients available, which are factors of the environment, also plays a role (Simonetti et al., 2021). Roy et al. (2020) in a study had shown that the biofilm, which was established by *S. aureus* in a porcine skin wound model, had surged the time of healing of the wound and increased the cost of care. Hence the biofilm compromises the wound tissue and promotes the recurrence of the local infection.

Coagulase-negative staphylococci (CNS) such as *S. epidermidis* and *Staphylococcus lugdunensis* in addition to *S. aureus* are also responsible for maintenance of the chronic wound structure (Simonetti et al., 2021).

7. Prevention and control

According to an estimation by the National Institute of Health, pathogens within the biofilms were responsible for approximately 80% of the diseases such as nosocomial infections, cystic fibrosis, periodontitis, ocular diseases, and meningitis, among others (Khatoon et al., 2018). Biofilm from the dental plaque was difficult to remove, although the overall microbial content and the various virulent isolates within the plaque could be controlled, which helped in preserving the normal dental flora. One of the main issues with the microbial biofilms was their antimicrobial resistance (AR) (Stewart & William Costerton, 2001). Biofilm communities were much more antibiotic resistant as compared with planktonic cells. An increase in drug concentration by 1000 times is desired to eradicate the pathogens linked with biofilms (Balaure & Grumezescu, 2020). Numerous studies have demonstrated that factors such as the pH, nutrition content, temperature, growth rate, and prior subjection of bacteria to antimicrobial doses could influence their sensitivity to antibiotics and biocontrol agents. The process of biofilm formation and control has been represented in Fig. 18.2.

Strategies involved in prevention and control of biofilm development:

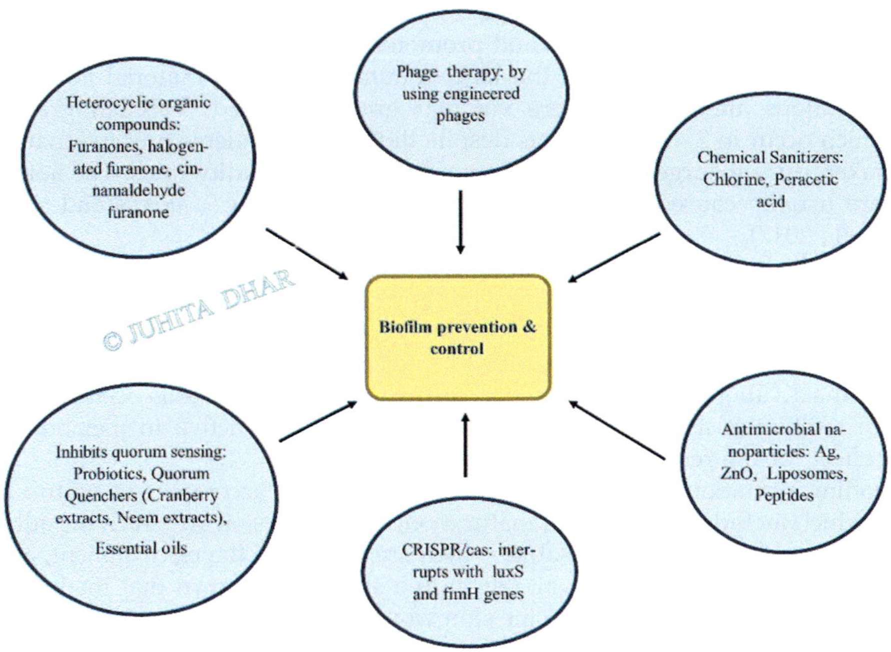

FIGURE 18.2 Biofilm prevention and control.

Nanoparticles as antibiofilm: Nanoparticles (NPs) could function as antimicrobial drugs or as carriers of drugs to enhance the efficacy of antibiotic drugs. They have prohibited electrostatic interaction with EPS matrix constituents such as polysaccharides or DNA and deactivated enzymes such as β-lactamases to protect antibiotic substances from enzymes (De Jong et al., 2008). Undoubtedly, one of the most notable topics is nanotechnology, which has the ability to address practically every element of microbial diseases (Cavalieri et al., 2014; Zhu et al., 2014). It was claimed that distinct NPs have a repressive impact against biofilm and plankton cells. NPs could be classified as organic and inorganic NPs.

- **Organic NPs:** Nanocarriers could be produced from a diverse range of organic and inorganic substances, including lipids and polymers (Rukavina & Vanić, 2016; Wolfmeier et al., 2018). Due to their special attributes, organic NPs have demonstrated higher potential than other approaches in resolving the issues caused by microbial biofilms (Li, Shen, Xia, & Shi, 2021; Li, Chen, & Xie, 2021)
  - i. **Silica NPs:** In particular, antimicrobial and melanoma therapies were discovered to massively benefit from the use of silica nanoparticles (Si NPs). Si NPs could be applied to bone cement and oral matrices as an antimicrobial ingredient (Letchmanan et al., 2017). Si NPs were also anticipated to destroy the biofilm when combined with nitric oxide (NO) without causing antibiotic resistance. Burroughs and coworkers in 2020 (Burroughs et al., 2020) reported that silicone catheters coated with polyacrylate films could be used to provide antimicrobial efficacy without generating deleterious immune reactions. In another study, silver-containing silica nanorattles ($Ag{-}SiO_2$) were developed, and its antibacterial effect against both *S. aureus and E. coli* was assessed. Additionally, their cytotoxic effect on immune cells was examined. According to the findings, there was significant bactericidal efficacy but no immune toxicity or proinflammatory reaction (Fulaz et al., 2019).
  - ii. **Peptides:** AMPs have widely been deployed as broad-spectrum antimicrobials for the eradication of both fungal and bacterial biofilm communities (Di Somma et al., 2020; Pletzer & Hancock, 2016). They engage through electrostatic interaction with the phospholipid membrane of bacterial cell membranes before being inserted into the membrane, thereby destroying bacteria (Selvarajan et al., 2020). Numerous AMPs could be used to inhibit biofilm formation. Mansour et al. (2015) stated that (p)ppGpp, a key signaling molecule in biofilm growth in *S. aureus* and multiple other bacteria, could be inhibited by treating with a synthetic peptide termed as 1018. It was very challenging for bacteria to acquire AMP tolerance because of the intricate nature of their antimicrobial properties, solving one of the main obstacles in classical antibiotics investigation (Yeaman & Yount, 2003). There have been instances suggesting that AMPs have worked in tandem with antimicrobial substances and have blocked a variety of biofilm-forming molecular pathways. AMPs could work as an alternative to conventional antibiotics (Hirt et al., 2018).
  - iii. **Hydrogels:** The three-dimensional cross-linked polymers known as hydrogels could react to shifts in natural conditions such as temperature, pH, electric potential, and the existence of enzymes by expanding or contracting appropriately without dissolving and producing any colloidal solution (Ahmed et al., 2013). Ramasamy & Lee, 2016 (Ramasamy & Lee, 2016) reported hydrogels coupled with vitamin E displayed

antimicrobial properties (Lee et al., 2013; Nevius et al., 2012). Au- and Ag-based NPs could be combined with hydrogels to form heterostructures. Ag NPs could be added to hydrogels, to enhance the antimicrobial properties and change their physical durability, stimulus ability to respond, and swelling ratio (Mekkawy et al., 2016; Zhao et al., 2015).

- **Inorganic NPs:** For millennia, inorganic metals such as silver, titanium, iron, copper, zinc, and gold have served as antiseptics and an alternative to standard antibiotic drugs. The surface topography of the biomaterials protected with the metals varies, depending on the filming process such as frittage (thermal method to partially liquefy tiny particles into a cohesive and solid form), anodizing, and abrasive blasting.

(i) **Silver NPs:** Silver oxynitrate (Ag $(Ag_3O_4)_2NO_3$ or Ag_7NsO_{11}) has already been discovered to be more effective toward bacterial biofilms than other combinations such as silver sulfadiazine (AgSD), silver sulfate (Ag_2SO_4), silver (II) oxide (AgO), silver nitrate ($AgNO_3$), and silver (I) oxide (Ag_2O) (Lemire et al., 2015). The use of silver NPs on titanium surfaces had been explored for application in oral care (Zhong et al., 2016), but the downside of this technique was that the antibiotic discharge from the implant material gradually ends. The superior metallurgical qualities, oxidation resistance, and biocompatibility of titanium and its blends were frequently utilized in orthopaedical grafts (Kargupta et al., 2014).

In contrast to synthetically manufactured silver NPs employed as a control, Salunke and coworkers in 2014 demonstrated bacterial proliferation could be suppressed by antimicrobial NPs (silver NPs) with a low concentration of inhibition. A new price-effective silver NP was recently developed by Gurunathan et al. (2014) by mixing Ag ions with *Allophylus cobbe* leaf extract. When coupled with antibiotics such as ampicillin and vancomycin, these Ag NPs have demonstrated antibiofilm properties against *P. aeruginosa* and *S. aureus.*

Moreover, Habash and his coworkers in an effort to enhance the antibacterial action of NPs discovered that capped 10-nm Ag NPs coupled with aztreonam damaged the biofilm framework of a *S. aureus* strain and *P. aeruginosa* (Habash et al., 2014).

(ii) **Iron NPs:** Iron (Fe) is a vital nanoelement that could be essential for the proper operation of structures and functions. Due to its versatility, magnetic NPs have been intensively researched for utilization in the medical sciences. Magnetic NPs could be affordable and biocompatible.

Iron oxide NPs have demonstrated a wide range of practice in biomedical fields such as cancer therapy, visualization, and magnetic hyperthermia, among others (Gudkov et al., 2021). Iron oxide NPs, unlike free ions, have no major detrimental impact on animal cells (Sihem et al., 2020; Li, Shen, Xia, & Shi, 2021). Anghel et al. (2012) suggested that catheter surfaces coated with iron have exhibited tolerance toward *P. aeruginosa* and *S. aureus* biofilms. Both Durmus et al. and Taylor et al. employed superparamagnetic iron oxide NPs to address methicillin-resistant *S. aureus* (MRSA) (Durmus et al., 2013; Taylor et al., 2012).

(iii) **Gold NPs:** Gold (Au), unlike silver (Ag), was not recognized for having inherent antibacterial characteristics. Nevertheless, the features of nanoscale gold (Au) enabled strong functionalization of particles, and scientists have looked into the use of gold

NPs in biofilm therapy. The antimicrobial properties of Au–Ag NPs have been previously demonstrated by Ding et al. (2017). They constructed NPs with gold in the center and silver as the crust and then used them to destroy *S. aureus*. Au NPs functionalized with zwitterion on surfaces were found to be efficient for both gram-positive and gram-negative bacteria (Hu et al., 2017). Furthermore, Au NPs were found to be photothermal (emitted a significant amount of thermal energy) in a confined area when exposed to near-infrared (NIR) radiations. These NIR radiations activated the Au NPs to emit heat, which destroyed the bacteria (Huo et al., 2016).

(iv) Copper NP: The antibacterial effects of copper coatings were mostly caused by the release of Cu^{2+}ions from metal substrates, although direct interaction between microbes and surfaces could also be essential (Mathews et al., 2013). The mechanism of activity of Cu and Cu-based surfaces could change depending on external factors such as pH, moisture, temperature, and the existence of any other metallic materials on the surface (Vincent et al., 2018). To prevent the growth of biofilm communities, Cu NPs are a less expensive substitute for Ag NPs (Gomes et al., 2020).

Cu NPs were used as protective materials on biomaterials to limit bacterial colonization. Cu NPs could be also employed as commercial antifungals in water filtration and as antifouling substances (Punniyakotti et al., 2020). Cu NPs prevented the biofilm growth at emerging phases in mature specimens (LewisOscar et al., 2015). Instead of eradicating biofilms, Cu NPs could be used to prevent biofilm formation.

- **Other polymeric materials:** It has already been known that bacteria within the established biofilm are more resistant to antimicrobial drugs in comparison with planktonic cells. Utilizing the polymeric substances could disturb the three-dimensional biofilm structure to interfere with the cell signaling network and render it more susceptible to antimicrobics.

One way of destructing the biofilm structure could be by using biological enzymes, such as dornase alfa was authorized for effectively treating biofilm-related infections within the lungs of cystic fibrosis sufferers, which seemed to reduce infection severity (Frederiksen et al., 2006). The second approach involved surgical or manual elimination of biofilm by brushing or flossing teeth. In an in vivo pig skin colonization paradigm, a combination of two extracellular polymeric matrix-degrading enzymes, i.e., dispersin B and DNase I, was reported to decrease staphylococcal epidermal colonization, remove the prior attached *S. aureus* cells from the epidermis, and increase povidone–iodine sensitivity (Kaplan et al., 2018). According to Tan et al. (2018), positively charged nanomaterials (DNase I and oxacillin) were utilized to address *S. aureus* biofilms, where the polymer loaded nanoparticles could destroy 70% and 100% of mature *S. aureus* biofilms (Tran et al., 2020).

Moreover, a sodium dodecyl sulfate (SDS)–based nanoporous polymeric substance and levofloxacin–coupled PLGA-poly(lactic-co-glycolic) acid were discovered to exhibit strong antibiofilm ability against *E. coli*. Furthermore, Elbi et al. in 2017 constructed chitosan-based NPs having fucoidan and ciprofloxacin coating to boost the antibacterial effect and the material's endocytosis into macrophages to eradicate intracellular microbes.

Previously, multiple quorum quenching enzymes have also been reported to partially eradicate the biofilms. Bjarnsholt et al. (2013) reported that genetic ablation of rhamnolipid synthesis enabled the elimination of biofilms formed by *P. aeruginosa* (Bjarnsholt et al., 2013; Gennip et al., 2012). Apart from the drug azithromycin, which was used to treat cystic fibrosis caused by *P. aeruginosa,* no QS inhibitors have been studied in clinical studies. Glycoprotein layers such as collagen, vitronectin, and fibrinogen could be used for prompt coating and treatment of the biomaterial devices, such as chlorhexidine-silver sulfaziadine is used for coating the surface of catheters (Berra et al., 2008).

Besides the aforementioned strategies, adequate dental hygiene strategies could incorporate routine brushing, mouth rinsing with an antibacterial solution, and flossing. This might have led to the treatment or avoidance of the linked consequences, such as preventing the onset of periodontal diseases as well as managing their potential influence on particular clinical manifestations. Innovative methods could be researched to prevent the innate tolerance of bacterial communities to antimicrobials and to reduce their pathogenic effects. The antibiofilm agents have been represented in Table 18.1.

8. Discussion

Microbial biofilms have been linked with many problems, which also include biofilm formation in medical devices. Medical devices used for implants such as pacemakers, bone implants, etc. are chiefly unprotected by microbial biofilm formation and disease. These surfaces are very prone to infection and thus serve as a prime focus for the microorganisms for causing biofilms.

These fastened implants are regularly affected with biofilms with lethal conditions. Therefore, several microorganisms share the customary capability to construct biofilms and occupy human tissues in the presence and absence of foreign bodies (Reid, 1999).

Methods that could be implemented to prevent or control biofilm-based medical device–related infections include strategies that are based on (1) prevention of adhesion of microbes to the surface and colonization, (2) biofilm development signaling molecules disrupt or interfere with those, and (3) disaggregation of the biofilm matrix.

(1) Inhibition of adhesion of microbes to device surfaces (Francolini & Donelli, 2010):

Hydrophilic coating: *S. epidermidis* adhesion was clearly reduced when polyurethane catheters and silicon shunts were covered with hydrophobic polymers such as hyaluronic acid (Cassinelli et al., 2000) and poly-*N*-vinylpyrrolidone (Boelens et al., 2000).

Antimicrobial agents: Surface coating of the device with one or more types of antimicrobial substances and entrapment method used to adhere the antimicrobial within the device are most common methods undertaken by research and development teams to formulate devices with different spectra of microbes and duration of antimicrobial effect (Darouiche, 2008; Zilberman & Elsner, 2008).

Orthopedic implants coated with antimicrobials: Artificial knee and hip orthopedic implants are most prone to infections despite antibiotic prophylaxis and strict hygienic protocols. An approach made developed in the past few decades for prevention and treatment of infections due to orthopedic bone implant infection by antibiotic-releasing polymethylmethacrylate bone cements and spacers. A broad study demonstrated 80% inhibition of *S. aureus* biofilm formation by widely used gentamicin-loaded polymethylmethacrylate

TABLE 18.1 Antibiofilm agents.

Sr. no.	Title	Inference	Author and references
1	Fatty Acids as Antibiofilm and Antivirulence Agents	Fatty acids could specifically prevent the biofilm production by various microbial pathogens such as *Pseudomonas aeruginosa, Serratia marcescens*, and *Staphylococcus aureus* by diminishing their attachment, flagellar actions, pathogenicity, and swarming motility.	Kumar et al. (2020)
2	Surfactants of Microbial Origin as Antibiofilm Agents	Due to less toxicity, biosurfactants could be regarded as ecologically friendly and biocompatible in nature. Owing to antimicrobial actions, they could be used against microbial biofilms.	Paraszkiewicz et al. (2021)
3	Antibiofilm Peptides: Potential as Broad-Spectrum Agents	Synthetic peptides have antibiofilm actions that destroyed multiple bacterial species, influenced bacterial activities by focusing on their stress responses, and has impeded matured biofilms.	Pletzer & Hancock (2016)
4	Microbial Glycoside Hydrolases as Antibiofilm Agents with Cross-Kingdom Activity	Glycoside hydrolases boost the hyphal infiltration of antifungal agents and have the potential to destabilize fungal biofilms by diminishing their pathogenicity.	Snarr et al. (2017)
5	Bioinspired Carbon Quantum Dots: An Antibiofilm Agents	Carbon dots (C-dots) are extremely small carbon particles, which were found to suppress adhesion and biofilm development by *Candida albicans* MTCC 227 on polystyrene surfaces.	Shaikh et al. (2019)
6	Hordenine: A Novel Quorum Sensing Inhibitor and Antibiofilm Agent against *Pseudomonas aeruginosa*	*P. aeruginosa* PAO1 when treated with hordenine (from budding barley) decreased the production of quorum sensing–associated extracellular pathogenic factors. Moreover, quorum sensing genes rhlR, lasR, rhlI, and lasI were considerably repressed.	Zhou et al. (2018)

beads with a novel biodegradable gentamicin-releasing poly-(trimethylene carbonate) (Neut et al., 2009).

CVCs (central venous catheters) coated with antimicrobials: CVCs are the main source of blood stream infections; minocycline-rifampicin (MR) antimicrobial-coated CVCs were found to exhibit high efficacy, prevented infection toward numerous gram-positive and gram-negative organisms, but were not effective to prevent *P. aeruginosa* or *Candida* spp. in vitro or in vivo (Sampath et al., 2001).

CVCs treated with chlorhexidine/silver sulfadiazine and benzalkonium chloride and rifampicin-miconazole were studied in the Casey review. Firstly, taken into account a clinical trial, CVCs impregnated with MR were observed to reduce colonization of catheterization for a period of 7.5 days; it has a reduction from 36% of patients to only 5% (Yücel et al., 2004).

Bacteria growing in the sessile mode have shown to have decreased antibiotic susceptibility; experiments were conducted with *S. aureus*- and *S. epidermidis*-grown biofilms on untreated or dispersin B-treated polyurethanes. It demonstrated the enzyme could promote antimicrobial and antibiofilm activity of cefamandole nafate (Francolini & Donelli, 2010) and triclosan (Darouiche et al., 2009); dispersin B is a β-*N*-acetylglucosaminidase that had been shown to dissolve the staphylococcal polysaccharide matrix (Kaplan et al., 2003)

Prevention of biofilm formation: Bacteria use small diffusible signaling molecules to communicate among themselves; the process is known as QS. Gram-positive bacteria used an oligopeptide, a family of autoinducer (AI) compounds known as AI-2− in both the bacteria, *N*-acyl homoserine lactones, in gram-negative bacteria. Potential targeting points in the process are (1) the generated signal, (2) the signal molecule, and (3) receptor for the signaling molecule. QS can be prevented by analogous signal molecule blocking the receptors (González & Keshavan, 2006; Rasmussen & Givskov, 2006).

The promoted microbial killing of established biofilm: Formation of multispecies biofilm formation and increase in antimicrobial resistance need better target sites to attack the organisms of the biofilm. It was demonstrated that when linezolid or vancomycin uses along with rifampicin, better biofilm killing was observed (Raad et al., 2007).

Groups working with magnetic NPs such as iron oxide to specifically deliver drug at infected site. It prevents both gram-positive and gram-negative bacteria after them with silver-containing polymer and ciprofloxacin (Francolini & Donelli, 2010).

Hwang et al. (2019) have developed catalytic antimicrobial robots also known as CARs that could destroy, dissolve, and eliminate biofilms by extraordinary precision and control. These magnetically powered robots could execute complicated duties in biotic environments while causing minimum damage. Catalytic antimicrobial robots could use NPs containing iron oxide (Fe_2O_3) with double catalytic properties that produced bactericidal free radicals and could thereby degrade the exopolysaccharide network of biofilms. The researchers have prepared two different CAR stages. In the first stage, the biohybrid CAR was developed using biofilm degradation compounds and nanomaterials that could completely eradicate biomass in a regulated way when powered by an externally applied magnetic field, thus preventing biofilm regrowth. In the second stage, the three-dimensional molded CAR was a smooth polymer robot constructed from three-dimensional components that could execute specified functions in confined areas (Hwang et al., 2019).

Biofilms had been important in terms of a medical device−associated infection due to their ability to resist and tolerate various antibiotics. The biofilm could be prevented by a number of techniques—mainly by using devices with antimicrobial coating and inhibitors (Khatoon et al., 2018).

Future research had been needed further to clarify the interaction between microorganisms and the implant surface. As we know, the rough surface of the implant contributes to the irreversible attachment of the bacteria to the implant; thus, the surface of the implants should be uniform and smooth (Dror et al., 2009).

The reappearance of infection even after proper treatment was an issue because of the microorganisms being resistant to phagocytic cells as well as antibiotics. There were a varied number of limitations that were yet to be worked upon and needed further dedicative research, which would enable more scopes in potential therapies for clinical trials. If the therapies could be a success, then there was a possibility of releasing the product in the market as well (Khatoon et al., 2018).

9. Conclusion

9.1 Significance

The increasingly grim situation caused by biofilm infection had undoubtedly become the challenge in the area of clinical safety. Public health has always been at risk from medical device–associated illnesses. Multiple pathogenic microorganisms such as yeasts, molds, *E. coli*, *Bacillus* spp., *Salmonella* spp., and *Pseudomonas* spp., among others, retained the capability to attach on various surfaces and equipment in the medical sectors where they produced biofilms and caused infections.

9.2 Prevention

The greatest challenge faced by biofilm was its ability to resist most of the currently existing antibiotics. Due to the protective effect of the biofilm structure, the microorganisms present within the biofilms showed increased resistance to antibacterial treatment compared with those in planktonic state. Biofilm removal could be possible by maintaining a systematic cleaning regimen. Cold plasma, a nonheat treatment technology, could be used in the medical industry to successfully remove biofilms from equipment and various surfaces. Traditional biofilm detection approaches, such as agar plating, were found to be ineffective as many bacteria such as *Listeria monocytogenes* attained a viable but not culturable state with reduced metabolic activity. Once the microcolony of biofilm becomes matured, then its eradication becomes difficult (EPS acts are a protective covering), so to prevent the proliferation of microbial cells, their initial attachment could be prevented. The biofilm could be prevented by a number of techniques such as by filming the device surfaces with antimicrobial coating such as the organic and inorganic NPs such as AMPs, glycoproteins, copper nanomaterials, and silver nanomaterials, among others.

Future research is needed further to clarify the interaction between microorganisms and the implant surface. As we know the rough surface of the implant contributes to the irreversible attachment of the bacteria to the implant, the surface of the implants should be uniform and smooth (Dror et al., 2009).

The reappearance of infection even after proper treatment is an issue because the microorganisms are resistant to phagocytic cells as well as antibiotics. There are a varied number of limitations that are yet to be worked upon, and further dedicative research is required, which will enable more scopes of potential therapies for clinical trials. If the therapies can be a success, then there is a possibility of releasing the product in the market as well (Khatoon et al., 2018).

Despite the fact that several antibiofilm agents have been identified as alternatives to conventional sanitizers, biocontrol activity demands attention, and many more innovative sanitization strategies are desired in the near term. Likewise, future research should focus on the pathogenic genes of bacteria responsible for biofilm development, as well as strategies for inhibiting them.

References

Ahmed, E. M., Aggor, F. S., Awad, A. M., & El-Aref, A. T. (January 16, 2013). An innovative method for preparation of nanometal hydroxide superabsorbent hydrogel. *Carbohydrate Polymers, 91*(2), 693–698. https://doi.org/10.1016/j.carbpol.2012.08.056. Epub 2012 Sep 4. PMID: 23121966.

Anghel, I., Limban, C., Grumezescu, A. M., et al. (2012). *In vitro* evaluation of anti-pathogenic surface coating nanofluid, obtained by combining Fe_3O_4/C_{12} nanostructures and 2-((4-ethylphenoxy) methyl)-*N*-(substituted-phenylcarbamothioyl)-benzamides. *Nanoscale Research Letters, 7*, 513. https://doi.org/10.1186/1556-276X-7-513

Aronson, J. K., Heneghan, C., & Ferner, R. E. (February 2020). Medical devices: Definition, classification, and regulatory implications. *Drug Safety, 43*(2), 83–93. https://doi.org/10.1007/s40264-019-00878-3. PMID: 31845212.

Balaure, P. C., & Grumezescu, A. M. (2020). Recent advances in surface nanoengineering for biofilm prevention and control. Part I: Molecular basis of biofilm recalcitrance. Passive anti-biofouling nanocoatings. *Nanomaterials, 10*, 1230. https://doi.org/10.3390/nano10061230

Barchowsky, A. (2020). Systemic and immune toxicity of implanted materials. In *Biomaterials science* (pp. 791–799). Academic Press.

Bayley, H., Cazimoglu, I., & Hoskin, C. E. (2019). Synthetic tissues. *Emerging Topics in Life Sciences, 3*(5), 615–622.

Becker, W., Becker, B. E., Ricci, A., Bahat, O., Rosenberg, E., Rose, L. F., … Israelson, H. (2000). A prospective multicenter clinical trial comparing one-and two-stage titanium screw-shaped fixtures with one-stage plasma-sprayed solid-screw fixtures. *Clinical Implant Dentistry and Related Research, 2*(3), 159–165. https://doi.org/10.1111/j.1708-8208.2000.tb00007.x

Berra, L., Kolobow, T., Laquerriere, P., et al. (2008). Internally coated endotracheal tubes with silver sulfadiazine in polyurethane to prevent bacterial colonization: A clinical trial. *Intensive Care Medicine, 34*, 1030–1037. https://doi.org/10.1007/s00134-008-1100-1

Bitterman, N. (2011). Design of medical devices—a home perspective. *European Journal of Internal Medicine, 22*(1), 39–42. https://doi.org/10.1016/j.ejim.2010.09.017

Bjarnsholt, T., Ciofu, O., Molin, S., et al. (2013). Applying insights from biofilm biology to drug development — can a new approach be developed? *Nature Reviews Drug Discovery, 12*, 791–808. https://doi.org/10.1038/nrd4000

Boelens, J. J., Tan, W. F., Dankert, J., & Zaat, S. A. (2000). Antibacterial activity of antibiotic-soaked polyvinylpyrrolidone-grafted silicon elastomer hydrocephalus shunts. *Journal of Antimicrobial Chemotherapy, 45*(2), 221–224. https://doi.org/10.1093/jac/45.2.221

Burroughs, L., Ashraf, W., Singh, S., Martinez-Pomares, L., Bayston, R., & Hook, A. L. (2020). Development of dual anti-biofilm and anti-bacterial medical devices. *Biomaterials Science, 8*(14), 3926–3934. https://doi.org/10.1039/D0BM00709A

Cassinelli, C., Morra, M., Pavesio, A., & Renier, D. (2000). Evaluation of interfacial properties of hyaluronan coated poly (methylmethacrylate) intraocular lenses. *Journal of Biomaterials Science, Polymer Edition, 11*(9), 961–977. https://doi.org/10.1163/156856200744138

Cavalieri, F., Tortora, M., Stringaro, A., Colone, M., & Baldassarri, L. (2014). Nanomedicines for antimicrobial interventions. *Journal of Hospital Infection, 88*(4), 183–190. https://doi.org/10.1016/j.jhin.2014.09.009

Chua, A. G., Ding, J., Schoch, P. E., & Cunha, B. A. (1998). Pacemaker-induced endocarditis due to Propionibacterium acnes. *Clinical Infectious Diseases, 27*(6), 1541–1542.

Cirioni, O., Mocchegiani, F., Ghiselli, R., Silvestri, C., Gabrielli, E., Marchionni, E., … Giacometti, A. (2010). Daptomycin and rifampin alone and in combination prevent vascular graft biofilm formation and emergence of antibiotic resistance in a subcutaneous rat pouch model of staphylococcal infection. *European Journal of Vascular and Endovascular Surgery, 40*(6), 817–822.

Costerton, J. W., Montanaro, L., & Arciola, C. R. (2005). Biofilm in implant infections: Its production and regulation. *The International Journal of Artificial Organs, 28*(11), 1062–1068. https://doi.org/10.1177/039139880502801103

Darouiche, R. O. (2008). Prevention of infections associated with vascular catheters. *The International Journal of Artificial Organs, 31*(9), 810–819. https://doi.org/10.1177/039139880803100909

Darouiche, R. O., Mansouri, M. D., Gawande, P. V., & Madhyastha, S. (2009). Antimicrobial and antibiofilm efficacy of triclosan and DispersinB® combination. *Journal of Antimicrobial Chemotherapy, 64*(1), 88–93. https://doi.org/10.1093/jac/dkp158

Darveau, R. (2010). Periodontitis: A polymicrobial disruption of host homeostasis. *Nature Reviews Microbiology, 8*, 481–490. https://doi.org/10.1038/nrmicro2337

Das, S., Singh, S., Matchado, M. S., Srivastava, A., & Bajpai, A. (2019). Biofilms in human health. In S. Kumar, N. Chandra, L. Singh, M. Hashmi, & A. Varma (Eds.), *Biofilms in human diseases: Treatment and control*. Cham: Springer. https://doi.org/10.1007/978-3-030-30757-8_3

Davide, C., Montanaro, L., & Renata Arciola, C. (April 2006). *The significance of infection related to orthopedic devices and issues of antibiotic resistance* (pp. 2331–2339). https://doi.org/10.1016/j.biomaterials.2005.11.044

De Castro, M. R., da Silva Fernandes, M., Kabuki, D. Y., & Kuaye, A. Y. (2017). Biofilm formation on stainless steel as a function of time and temperature and control through sanitizers. *International Dairy Journal, 68*, 9–16. https://doi.org/10.1016/j.idairyj.2016.12.005

De Jong, W. H., & Borm, P. J. (2008). Drug delivery and nanoparticles: Applications and hazards. *International Journal of Nanomedicine, 3*(2), 133. https://doi.org/10.2147/2Fijn.s596

Di Ciccio, P., Vergara, A., Festino, A. R., Paludi, D., Zanardi, E., Ghidini, S., & Ianieri, A. (2015). Biofilm formation by *Staphylococcus aureus* on food contact surfaces: Relationship with temperature and cell surface hydrophobicity. *Food Control, 50*, 930–936. https://doi.org/10.1016/j.foodcont.2014.10.048

Di Somma, A., Moretta, A., Canè, C., Cirillo, A., & Duilio, A. (2020). Antimicrobial and antibiofilm peptides. *Biomolecules, 10*(4), 652. https://doi.org/10.3390/biom10040652

Ding, X., Yuan, P., Gao, N., Zhu, H., Yang, Y. Y., & Xu, Q.-H. (2017). Au-Ag core-shell nanoparticles for simultaneous bacterial imaging and synergistic antibacterial activity. *Nanomedicine: Nanotechnology, Biology and Medicine, 13*, 297–305. https://doi.org/10.1016/j.nano.2016.09.003

Döring, M., Richter, S., & Hindricks, G. (2018). The diagnosis and treatment of pacemaker-associated infection. *Deutsches Arzteblatt International, 115*(26), 445–452. https://doi.org/10.3238/arztebl.2018.0445

Dror, N., Mandel, M., Hazan, Z., & Lavie, G. (2009). Advances in microbial biofilm prevention on indwelling medical devices with emphasis on usage of acoustic energy. *Sensors, 9*, 2538–2554. https://doi.org/10.3390/s90402538

Durmus, N. G., Taylor, E. N., Kummer, K. M., & Webster, T. J. (2013). Enhanced efficacy of superparamagnetic iron oxide nanoparticles against antibiotic-resistant biofilms in the presence of metabolites. *Advanced Materials, 25*, 5706–5713. https://doi.org/10.1002/adma.201302627

Elbi, S., Nimal, T. R., Rajan, V. K., Baranwal, G., Biswas, R., Jayakumar, R., & Sathianarayanan, S. (2017). Fucoidan coated ciprofloxacin loaded chitosan nanoparticles for the treatment of intracellular and biofilm infections of Salmonella. *Colloids Surf. B Biointerfaces, 160*, 40–47. https://doi.org/10.1016/j.colsurfb.2017.09.003

Ercan, U. K., İbis, F., Dikyol, C., Horzum, N., Karaman, O., Yıldırım, Ç., et al. (2018). Prevention of bacterial colonization on non-thermal atmospheric plasma treated surgical sutures for control and prevention of surgical site infections. *PLoS One, 13*(9), e0202703. https://doi.org/10.1371/journal.pone.0202703

Francolini, I., & Donelli, G. (August 2010). Prevention and control of biofilm-based medical-device-related infections. *FEMS Immunology & Medical Microbiology, 59*(3), 227–238. https://doi.org/10.1111/j.1574-695X.2010.00665.x

Frederiksen, B., Pressler, T., Hansen, A., Koch, C., & Høiby, N. (2006). Effect of aerosolized rhDNase (Pulmozyme®) on pulmonary colonization in patients with cystic fibrosis. *Acta Paediatrica, 95*(9), 1070–1074. https://doi.org/10.1080/08035250600752466

Fulaz, S., Vitale, S., Quinn, L., & Casey, E. (2019). Nanoparticle–biofilm interactions: The role of the EPS matrix. *Trends in Microbiology, 27*(11), 915–926. https://doi.org/10.1016/j.tim.2019.07.004

van Gennip, M., Christensen, L. D., Alhede, M., Qvortrup, K., Jensen, P.Ø., Høiby, N., … Bjarnsholt, T. (2012). Interactions between polymorphonuclear leukocytes and Pseudomonas aeruginosa biofilms on silicone implants in vivo. *Infection and immunity, 80*(8), 2601–2607. https://doi.org/10.1128/IAI.06215-11

Gomes, I. B., Simões, M., & Simões, L. C. (2020). Copper surfaces in biofilm control. *Nanomaterials, 10*(12), 2491. https://doi.org/10.3390/nano10122491

González, J. E., & Keshavan, N. D. (2006). Messing with bacterial quorum sensing. *Microbiology and Molecular Biology Reviews, 70*(4), 859–875. https://doi.org/10.1128/MMBR.00002-06

von Gotz, F., Haussler, S., Jordan, D., Saravanamuthu, S. S., Wehmhoner, D., Strußmann, A., … Steinmetz, I. (2004). Expression analysis of a highly adherent and cytotoxic small colony variant of *Pseudomonas aeruginosa* isolated from a lung of a patient with cystic fibrosis. *Journal of Bacteriology, 186*(12), 3837–3847. https://doi.org/10.1128/JB.186.12.3837-3847.2004

Gudkov, S. V., Burmistrov, D. E., Serov, D. A., Rebezov, M. B., Semenova, A. A., & Lisitsyn, A. B. (2021). Do iron oxide nanoparticles have significant antibacterial properties? *Antibiotics, 10*(7), 884. https://doi.org/10.3390/antibiotics10070884

Guillaume, B. (2016). Dental implants: A review. *Morphologie, 100*(331), 189–198. https://doi.org/10.1016/j.morpho.2016.02.002

Gurunathan, S., Han, J. W., Kwon, D. N., et al. (2014). Enhanced antibacterial and anti-biofilm activities of silver nanoparticles against Gram-negative and Gram-positive bacteria. *Nanoscale Research Letters, 9*, 373. https://doi.org/10.1186/1556-276X-9-373

Habash, M. B., Park, A. J., Vis, E. C., Harris, R. J., & Khursigara, C. M. (2014). Synergy of silver nanoparticles and aztreonam against *Pseudomonas aeruginosa* PAO1 biofilms. *Antimicrobial Agents and Chemotherapy, 58*(10), 5818–5830. https://doi.org/10.1128/AAC.03170-14

Hahnel, S. (2017). Biofilms on dental implants. In *Biofilms and implantable medical devices* (pp. 117–140). Woodhead Publishing. https://doi.org/10.1016/B978-0-08-100382-4.00005-8

Halim, A. S., Khoo, T. L., & Yussof, S. J. M. (2010). Biologic and synthetic skin substitutes: An overview. *Indian Journal of Plastic Surgery, 43*(S 01), S23–S28.

Han, A., Tsoi, J. K., Rodrigues, F. P., Leprince, J. G., & Palin, W. M. (2016). Bacterial adhesion mechanisms on dental implant surfaces and the influencing factors. *International Journal of Adhesion and Adhesives, 69*, 58–71. https://doi.org/10.1016/j.ijadhadh.2016.03.022

Herten, M., Bisdas, T., Knaack, D., Becker, K., Osada, N., Torsello, G. B., & Idelevich, E. A. (2017). Rapid in vitro quantification of *S. aureus* biofilms on vascular graft surfaces. *Frontiers in Microbiology, 8*, 2333.

Hirt, H., Hall, J. W., Larson, E., & Gorr, S. U. (2018). A D-enantiomer of the antimicrobial peptide GL13K evades antimicrobial resistance in the Gram-positive bacteria *Enterococcus faecalis* and Streptococcus gordonii. *PLoS One, 13*(3), e0194900. https://doi.org/10.1371/journal.pone.0194900

Hu, D., Li, H., Wang, B., Ye, Z., Lei, W., Jia, F., Jin, Q., Ren, K. F., & Ji, J. (September 26, 2017). Surface-adaptive gold nanoparticles with effective adherence and enhanced photothermal ablation of methicillin-resistant *Staphylococcus aureus* biofilm. *ACS Nano, 11*(9), 9330–9339. https://doi.org/10.1021/acsnano.7b04731. Epub 2017 Aug 17. PMID: 28806528.

Huo, S., Jiang, Y., Gupta, A., Jiang, Z., Landis, R. F., Hou, S., … Rotello, V. M. (2016). Fully zwitterionic nanoparticle antimicrobial agents through tuning of core size and ligand structure. *ACS Nano, 10*(9), 8732–8737. https://doi.org/10.1021/acsnano.6b04207

Hwang, G., Paula, A. J., Hunter, E. E., Liu, Y., Babeer, A., Karabucak, B., Stebe, K., Kumar, V., Steager, E., & Koo, H. (2019). Catalytic antimicrobial robots for biofilm eradication. *Science Robotics, 4*(29), eaaw2388. https://doi.org/10.1126/scirobotics.aaw2388

Jakubovics, N. S., Goodman, S. D., Mashburn-Warren, L., Stafford, G. P., & Cieplik, F. (2021). The dental plaque biofilm matrix. *Periodontology 2000, 86*(1), 32–56. https://doi.org/10.1111/prd.12361

Kaplan, J. B., Mlynek, K. D., Hettiarachchi, H., Alamneh, Y. A., Biggemann, L., Zurawski, D. V., … Granick, M. S. (2018). Extracellular polymeric substance (EPS)-degrading enzymes reduce staphylococcal surface attachment and biocide resistance on pig skin in vivo. *PLoS One, 13*(10), e0205526. https://doi.org/10.1371/journal.pone.020552

Kaplan, J. B., Ragunath, C., Ramasubbu, N., & Fine, D. H. (2003). Detachment of Actinobacillus actinomycetemcomitans biofilm cells by an endogenous β-hexosaminidase activity. *Journal of Bacteriology, 185*(16), 4693–4698. https://doi.org/10.1128/JB.185.16.4693-4698.2003

Kargupta, R., Bok, S., Darr, C. M., Crist, B. D., Gangopadhyay, K., Gangopadhyay, S., & Sengupta, S. (2014). Coatings and surface modifications imparting antimicrobial activity to orthopedic implants. *Wiley Interdisciplinary Reviews: Nanomedicine and Nanobiotechnology, 6*(5), 475–495. https://doi.org/10.1002/wnan.1273

Khatoon, Z., McTiernan, C. D., Suuronen, E. J., Mah, T. F., & Alarcon, E. I. (2018). Bacterial biofilm formation on implantable devices and approaches to its treatment and prevention. *Heliyon, 4*(12), e01067. https://doi.org/10.1016/j.heliyon.2018.e01067

Kornman, K. S., Newman, M. G., Alvarado, R., Flemmig, T. F., Nachnani, S., & Tumbusch, J. (1991). Clinical and microbiological patterns of adults with periodontitis. *Journal of Periodontology, 62*(10), 634–642. https://doi.org/10.1902/jop.1991.62.10.634

Kumar, P., Lee, J. H., Beyenal, H., & Lee, J. (2020). Fatty acids as antibiofilm and antivirulence agents. *Trends in Microbiology, 28*(9), 753–768. https://doi.org/10.1016/j.tim.2020.03.014

Kurtz, S., Ong, K., Lau, E., Mowat, F., & Halpern, M. (April 2007). Projections of primary and revision hip and knee arthroplasty in the United States from 2005 to 2030. *The Journal of Bone & Joint Surgery, 89*(4), 780–785. https://doi.org/10.2106/JBJS.F.00222

Lang, N. P., Bosshardt, D. D., & Lulic, M. (2011). Do mucositis lesions around implants differ from gingivitis lesions around teeth? *Journal of Clinical Periodontology, 38*, 182–187. https://doi.org/10.1111/j.1600-051X.2010.01667.x

Lascelles, A., & Claringbold, P. (1960). The effects of suture materials and suture techniques on the healing of wounds in the SKIN OF the sheep. *Australian Journal of Experimental Biology and Medical Science, 38*, 111–115. https://doi.org/10.1038/icb.1960.12

Lasserre, J. F., Brecx, M. C., & Toma, S. (2018). Oral microbes, biofilms and their role in periodontal and peri-implant diseases. *Materials, 11*(10), 1802. https://doi.org/10.3390/ma11101802

Lee, A. L., Ng, V. W., Wang, W., Hedrick, J. L., & Yang, Y. Y. (2013). Block copolymer mixtures as antimicrobial hydrogels for biofilm eradication. *Biomaterials, 34*(38), 10278–10286. https://doi.org/10.1016/j.biomaterials.2013.09.029

LeGeros, R. Z. (1992). Materials for bone repair, augmentation, and implant coatings. In S. Niwa, S. M. Perren, & T. Hattori (Eds.), *Biomechanics in orthopedics*. Tokyo: Springer. https://doi.org/10.1007/978-4-431-68216-5_8

Lemire, J. A., Kalan, L., Bradu, A., & Turner, R. J. (2015). Silver oxynitrate, an unexplored silver compound with antimicrobial and antibiofilm activity. *Antimicrobial Agents and Chemotherapy, 59*(7), 4031–4039. https://doi.org/10.1128/AAC.05177-14

Letchmanan, K., Shen, S. C., Ng, W. K., Kingshuk, P., Shi, Z., Wang, W., & Tan, R. B. (2017). Mechanical properties and antibiotic release characteristics of poly (methyl methacrylate)-based bone cement formulated with mesoporous silica nanoparticles. *Journal of the Mechanical Behavior of Biomedical Materials, 72*, 163–170. https://doi.org/10.1016/j.jmbbm.2017.05.003

Lewis Oscar, F., Mubarak Ali, D., Nithya, C., Priyanka, R., Gopinath, V., Alharbi, N. S., & Thajuddin, N. (2015). One pot synthesis and anti-biofilm potential of copper nanoparticles (CuNPs) against clinical strains of *Pseudomonas aeruginosa*. *Biofouling, 31*, 379–391. https://doi.org/10.1080/08927014.2015.1048686

Li, X., Chen, D., & Xie, S. (2021). Current progress and prospects of organic nanoparticles against bacterial biofilm. *Advances in Colloid and Interface Science, 294*, 102475. https://doi.org/10.1016/j.cis.2021.102475

Li, D., Shen, M., Xia, J., & Shi, X. (2021). Recent developments of cancer nanomedicines based on ultrasmall iron oxide nanoparticles and nanoclusters. *Nanomedicine, 16*(8), 609–612. https://doi.org/10.2217/nnm-2021-0033

Maderazo, E. G., Judson, S., & Pasternak, H. (April 1988). Late infections of total joint prostheses: A review and recommendations for prevention. *Clinical Orthopaedics and Related Research, 229*, 131–142. https://doi.org/10.1097/00003086-198804000-00017

Mahesh, L., Kumar, V. R., Jain, A., Shukla, S., Aragoneses, J. M., Martínez González, J. M., Fernández-Domínguez, M., & Calvo-Guirado, J. L. (2019). Bacterial adherence around sutures of different material at grafted site: A microbiological analysis. *Materials, 12*, 2848. https://doi.org/10.3390/ma12182848

Mansour, S. C., de la Fuente-Núñez, C., & Hancock, R. E. (2015). Peptide IDR-1018: Modulating the immune system and targeting bacterial biofilms to treat antibiotic-resistant bacterial infections. *Journal of Peptide Science, 21*(5), 323–329. https://doi.org/10.1002/psc.2708

Marrie, T. J., Nelligan, J., & Costerton, J. W. (1982). A scanning and transmission electron microscopic study of an infected endocardial pacemaker lead. *Circulation, 66*(6), 1339–1341. https://doi.org/10.1161/01.CIR.66.6.1339

Mathews, S., Hans, M., Mücklich, F., & Solioz, M. (2013). Contact killing of bacteria on copper is suppressed if bacterial-metal contact is prevented and is induced on iron by copper ions. *Applied and Environmental Microbiology, 79*(8), 2605–2611. https://doi.org/10.1128/AEM.03608-12

McConoughey, S. J., Howlin, R., Granger, J. F., Manring, M. M., Calhoun, J. H., Shirtliff, M., Kathju, S., & Paul, S. (2014). Biofilms in periprosthetic orthopedic infections. *Future Microbiology, 9*(8), 987–1007. https://doi.org/10.2217/fmb.14.64

McDougald, D., Rice, S. A., Barraud, N., Steinberg, P. D., & Kjelleberg, S. (2012). Should we stay or should we go: Mechanisms and ecological consequences for biofilm dispersal. *Nature Reviews Microbiology, 10*(1), 39–50. https://doi.org/10.1038/nrmicro2695

Mekkawy, A. I., El-Mokhtar, M. A., El-Shanawany, S. M., & Ibrahim, E. H. (2016). Silver nanoparticles-loaded hydrogels, a potential treatment for resistant bacterial infection and wound healing: A review. *Journal of Pharmaceutical Research International, 14*, 1–19. https://doi.org/10.9734/BJPR/2016/30525

Minkiewicz-Zochniak, A., Jarzynka, S., Iwańska, A., Strom, K., Iwańczyk, B., Bartel, M., … Oledzka, G. (2021). Biofilm formation on dental implant biomaterials by *Staphylococcus aureus* strains isolated from patients with cystic fibrosis. *Materials, 14*(8), 2030. https://doi.org/10.3390/ma14082030

Mohd Noor, S. N. A., & Mahmud, J. (2014). A review on synthetic skin: Materials investigation, experimentation and simulation. *Advanced Materials Research, 915*, 858–866.

Mombelli, A., & Décaillet, F. (2011). The characteristics of biofilms in peri-implant disease. *Journal of Clinical Periodontology, 38*, 203–213. https://doi.org/10.1111/j.1600-051X.2010.01666.x

Montanaro, L., Speziale, P., Campoccia, D., Ravaioli, S., Cangini, I., Pietrocola, G., Giannini, S., & Arciola, C. R. (2011). Scenery of Staphylococcus implant infections in orthopedics. *Future Microbiology, 6*(11), 1329–1349. https://doi.org/10.2217/fmb.11.117

Nandakumar, V., Chittaranjan, S., Kurian, V., et al. (2013). Characteristics of bacterial biofilm associated with implant material in clinical practice. *Polymer Journal, 45*, 137–152. https://doi.org/10.1038/pj.2012.130

Neut, D., Kluin, O. S., Crielaard, B. J., van der Mei, H. C., Busscher, H. J., & Grijpma, D. W. (2009). A biodegradable antibiotic delivery system based on poly-(trimethylene carbonate) for the treatment of osteomyelitis. *Acta Orthopaedica, 80*(5), 514–519. https://doi.org/10.3109/17453670903350040

Nevius, B. A., Chen, Y. P., Ferry, J. L., et al. (2012). Surface-functionalization effects on uptake of fluorescent polystyrene nanoparticles by model biofilms. *Ecotoxicology, 21*, 2205–2213. https://doi.org/10.1007/s10646-012-0975-3

Okuda, K. I., Nagahori, R., Yamada, S., Sugimoto, S., Sato, C., Sato, M., … Mizunoe, Y. (2018). The composition and structure of biofilms developed by Propionibacterium acnes isolated from cardiac pacemaker devices. *Frontiers in Microbiology, 9*, 182. https://doi.org/10.3389/fmicb.2018.00182

Ong, K., Yun, M., & White, J. (2015). New biomaterials for orthopedic implants. *Orthopedic Research and Reviews, 7*, 107–130. https://doi.org/10.2147/ORR.S63437

Orell, A., Schopf, S., Randau, L., & Vera, M. (2017). Biofilm lifestyle of thermophile and acidophile archaea. In *Biocommunication of archaea* (pp. 133–146). Cham: Springer. https://doi.org/10.1007/978-3-319-65536-9_9

Paraszkiewicz, K., Moryl, M., Płaza, G., Bhagat, D., Satpute, S., K., & Bernat, P. (2021). Surfactants of microbial origin as antibiofilm agents. *International Journal of Environmental Health Research, 31*(4), 401–420. https://doi.org/10.1080/09603123.2019.1664729

Pletzer, D., & Hancock, R. E. (2016). Antibiofilm peptides: Potential as broad-spectrum agents. *Journal of Bacteriology, 198*(19), 2572–2578. https://doi.org/10.1128/JB.00017-16

Punniyakotti, P., Panneerselvam, P., Perumal, D., et al. (2020). Anti-bacterial andanti-biofilm properties of green synthesized copper nanoparticles from *Cardiospermum halicacabum* leaf extract. *Bioprocess and Biosystems Engineering, 43*, 1649–1657. https://doi.org/10.1007/s00449-020-02357-x

Raad, I., Hanna, H., Jiang, Y., Dvorak, T., Reitzel, R., Chaiban, G., … Hachem, R. (2007). Comparative activities of daptomycin, linezolid, and tigecycline against catheter-related methicillin-resistant Staphylococcus bacteremic isolates embedded in biofilm. *Antimicrobialagents and Chemotherapy, 51*(5), 1656–1660. https://doi.org/10.1128/AAC.00350-06

Ramasamy, M., & Lee, J. (2016). Recent nanotechnology approaches for prevention and treatment of biofilm-associated infections on medical devices. *BioMed Research International, 2016*. https://doi.org/10.1155/2016/1851242

Rasmussen, T. B., & Givskov, M. (2006). Quorum-sensing inhibitors as anti-pathogenic drugs. *International Journal of Medical Microbiology, 296*(2–3), 149–161. https://doi.org/10.1016/j.ijmm.2006.02.005

Reid, G. (1999). Biofilms in infectious disease and on medical devices. *International Journal of Antimicrobial Agents, 11*(3–4), 223–226.

Reinbold, J., Uhde, A.-K., Müller, I., Weindl, T., Geis-Gerstorfer, J., Schlensak, C., Wendel, H.-P., & Krajewski, S. (2017). Preventing surgical site infections using a natural, biodegradable, antibacterial coating on surgical sutures. *Molecules, 22*, 1570. https://doi.org/10.3390/molecules22091570

Ribeiro, M., Monteiro, F. J., & Ferraz, M. P. (2012). Infection of orthopedic implants with emphasis on bacterial adhesion process and techniques used in studying bacterial-material interactions. *Biomatter*, 2(4), 176–194. https://doi.org/10.4161/biom.22905

Roy, S., Santra, S., Das, A., Dixith, S., Sinha, M., Ghatak, S., … Sen, C. K. (2020). *Staphylococcus aureus* biofilm infection compromises wound healing by causing deficiencies in granulation tissue collagen. *Annals of Surgery, 271*(6), 1174.

Rukavina, Z., & Vanić, Ž. (2016). Current trends in development of liposomes for targeting bacterial biofilms. *Pharmaceutics, 8*(2), 18. https://doi.org/10.3390/pharmaceutics8020018

Rzhepishevska, O., Hakobyan, S., Ruhal, R., Gautrot, J., Barbero, D., & Ramstedt, M. (2013). The surface charge of anti-bacterial coatings alters motility and biofilm architecture. *Biomaterials Science, 1*(6), 589–602. https://doi.org/10.1039/C3BM00197K

Salunke, G. R., Ghosh, S., Kumar, R. S., Khade, S., Vashisth, P., Kale, T., … Chopade, B. A. (2014). Rapid efficient synthesis and characterization of silver, gold, and bimetallic nanoparticles from the medicinal plant Plumbago zeylanica and their application in biofilm control. *International Journal of Nanomedicine, 9*, 2635. https://doi.org/10.2147/IJN.S59834

Sampath, L. A., Tambe, S. M., & Modak, S. M. (2001). In vitro and in vivo efficacy of catheters impregnated with antiseptics or antibiotics: Evaluation of the risk of bacterial resistance to the antimicrobials in the catheters. *Infection Control & Hospital Epidemiology, 22*(10), 640–646. https://doi.org/10.1086/501836

Schaldach, M. (1992). Materials in pacemaker technology. In *Electrotherapy of the heart*. Berlin, Heidelberg: Springer. https://doi.org/10.1007/978-3-642-50209-5_10

Selvarajan, V., Obuobi, S., & Ee, P. L. R. (2020). Silica nanoparticles—a versatile tool for the treatment of bacterial infections. *Frontiers in Chemistry, 8*, 602. https://doi.org/10.3389/fchem.2020.00602

Shah, A., & Goyal, R. (2008). Current status of the regulation for medical devices. *Indian Journal of Pharmaceutical Sciences, 70*, 695–700. https://doi.org/10.4103/0250-474X.49085

Shaikh, A. F., Tamboli, M. S., Patil, R. H., Bhan, A., Ambekar, J. D., & Kale, B. B. (2019). Bioinspired carbon quantum dots: An antibiofilm agents. *Journal of Nanoscience and Nanotechnology, 19*(4), 2339–2345. https://doi.org/10.1166/jnn.2019.16537

Shuai, C., Li, S., Peng, S., Pei, F., Lai, Y., & Gao, C. (2019). *Biodegradable Metallic Bone Implants, 3*, 544–562. https://doi.org/10.1039/C8QM00507A

Sihem, L., Hanine, D., & Faiza, B. (2020). Antibacterial activity of α-Fe_2O_3 and α-Fe_2O_3@Ag nanoparticles prepared by *Urtica* leaf extract. *Nanotechnology in Russia, 15*, 198–203. https://doi.org/10.1134/S1995078020020135

Simonetti, O., Rizzetto, G., Radi, G., Molinelli, E., Cirioni, O., Giacometti, A., & Offidani, A. (2021). New perspectives on old and new therapies of staphylococcal skin infections: The role of biofilm targeting in wound healing. *Antibiotics, 10*(11), 1377.

Siqueira, W. L., Custodio, W., & McDonald, E. E. (2012). New insights into the composition and functions of the acquired enamel pellicle. *Journal of Dental Research, 91*(12), 1110–1118. https://doi.org/10.1177/0022034512462578

Snarr, B. D., Baker, P., Bamford, N. C., Sato, Y., Liu, H., Lehoux, M., … Sheppard, D. C. (2017). Microbial glycoside hydrolases as antibiofilm agents with cross-kingdom activity. *Proceedings of the National Academy of Sciences, 114*(27), 7124–7129. https://doi.org/10.1073/pnas.1702798114

Song, F., Koo, H., & Ren, D. (2015). Effects of material properties on bacterial adhesion and biofilm formation. *Journal of Dental Research, 94*(8), 1027–1034. https://doi.org/10.1177/0022034515587690

Srey, S., Jahid, I. K., & Ha, S. D. (2013). Biofilm formation in food industries: A food safety concern. *Food Control, 31*(2), 572–585. https://doi.org/10.1016/j.foodcont.2012.12.001

Stewart, P. S., & William Costerton, J. (2001). Antibiotic resistance of bacteria in biofilms. *The Lancet, 358*(9276), 135–138. https://doi.org/10.1016/S0140-6736(01)05321-1

Tan, Y., Ma, S., Leonhard, M., Moser, D., Haselmann, G. M., Wang, J., Eder, D., & Schneider-Stickler, B. (November 15, 2018). Enhancing antibiofilm activity with functional chitosan nanoparticles targeting biofilm cells and biofilm matrix. *Carbohydrate Polymers, 200*, 35–42. https://doi.org/10.1016/j.carbpol.2018.07.072. Epub 2018 Jul 25. PMID: 30177175.

Taylor, E. N., Kummer, K. M., Durmus, N. G., Leuba, K., Tarquinio, K. M., & Webster, T. J. (2012). Superparamagnetic iron oxide nanoparticles (SPION) for the treatment of antibiotic-resistant biofilms. *Small, 8*(19), 3016–3027. https://doi.org/10.1002/smll.201200575

Teughels, W., Van Assche, N., Sliepen, I., & Quirynen, M. (2006). Effect of material characteristics and/or surface topography on biofilm development. *Clinical Oral Implants Research, 17*(S2), 68–81. https://doi.org/10.1111/j.1600-0501.2006.01353.x

Tolker-Nielsen, T. (2015). Biofilm development. *Microbiology Spectrum, 3*(2). https://doi.org/10.1128/microbiolspec.MB-0001-2014, 3-2.

Toyofuku, M., Inaba, T., Kiyokawa, T., Obana, N., Yawata, Y., & Nomura, N. (2016). Environmental factors that shape biofilm formation. *Bioscience, Biotechnology, and Biochemistry, 80*(1), 7–12. https://doi.org/10.1080/09168451.2015.1058701

Trampuz, A., et al. (2006). Diagnosis and treatment of infections associated with fracture-fixation devices. *Injury, 37*(2), S59–S66. https://doi.org/10.1016/j.injury.2006.04.010

Tran, H. M., Tran, H., Booth, M. A., Fox, K. E., Nguyen, T. H., Tran, N., & Tran, P. A. (2020). *Nanomaterials for treating bacterial biofilms on implantable medical devices*. https://doi.org/10.3390/nano10112253. ORCID Received: 20 October 2020/Revised: 31 October 2020/Accepted: 11 November 2020/Published: November 13, 2020.

Van Eck, C. F., Chen, A. F., Klatt, B. A., D'Antonio, J., & Fu, F. (2009). The classification of implants: Class I, II, III. *Journal of Long-Term Effects of Medical Implants, 19*(3). https://doi.org/10.1615/JlongTermEffMedImplants.v19.i3.30

Varshini, K. A., Charan, K. M., & Kumar, M. S. (2021). Concept design and analysis of self sustainable triboelectric pacemaker. In *, 2115(1)*. Journal of Physics: Conference Series (p. 012050). IOP Publishing. http://www.10.1088/1742-6596/2115/1/012050.

Veerachamy, S., Yarlagadda, T., Manivasagam, G., & Yarlagadda, P. K. (2014). Bacterial adherence and biofilm formation on medical implants: A review. *Proceedings of the Institution of Mechanical Engineers, Part H: Journal of Engineering in Medicine, 228*(10), 1083–1099.

Vieira, D., Angel, S. N., Honjol, Y., et al. (2022). Engineering surgical stitches to prevent bacterial infection. *Scientific Reports, 12*, 834. https://doi.org/10.1038/s41598-022-04925-5

Vincent, M., Duval, R. E., Hartemann, P., & Engels-Deutsch, M. (2018). Contact killing and antimicrobial properties of copper. *Journal of Applied Microbiology*. https://doi.org/10.1111/jam.13681

Wi, Y. M., & Patel, R. (2018). Understanding biofilms and novel approaches to the diagnosis, prevention, and treatment of medical device-associated infections. *Infectious Disease Clinics, 32*(4), 915–929. https://doi.org/10.1016/j.idc.2018.06.009

Wolfmeier, H., Pletzer, D., Mansour, S. C., & Hancock, R. E. (2018). New perspectives in biofilm eradication. *ACS Infectious Diseases, 4*(2), 93–106. https://doi.org/10.1021/acsinfecdis.7b00170

Xing, R., Lyngstadaas, S. P., Ellingsen, J. E., Taxt-Lamolle, S., & Haugen, H. J. (2015). The influence of surface nanoroughness, texture and chemistry of TiZr implant abutment on oral biofilm accumulation. *Clinical Oral Implants Research, 26*(6), 649–656. https://doi.org/10.1111/clr.12354

Yadav, S. K., & Sanyal, S. (2019). Biofilms: The good and the bad. In Biofilms in Human Diseases: Treatment and Control (pp. 13–26). Cham: Springer.

Yeaman, M. R., & Yount, N. Y. (2003). Mechanisms of antimicrobial peptide action and resistance. *Pharmacological Reviews, 55*(1), 27–55. https://doi.org/10.1124/pr.55.1.2

Yücel, N., Lefering, R., Maegele, M., Max, M., Rossaint, R., Koch, A., … Neugebauer, E. A. (2004). Reduced colonization and infection with miconazole–rifampicin modified central venous catheters: A randomized controlled clinical trial. *Journal of Antimicrobial Chemotherapy, 54*(6), 1109–1115. https://doi.org/10.1093/jac/dkh483

Zedtwitz-Liebenstein, K., Gabriel, H., & Graninger, W. (2003). Pacemaker endocarditis due to Propionibacterium acnes. *Infection, 31*(3), 184–185. https://doi.org/10.1007/s15010-002-2193-z

Zhao, F., Yao, D., Guo, R., Deng, L., Dong, A., & Zhang, J. (2015). Composites of polymer hydrogels and nanoparticulate systems for biomedical and pharmaceutical applications. *Nanomaterials (Basel, Switzerland), 5*(4), 2054–2130. https://doi.org/10.3390/nano5042054

Zhong, X., Song, Y., Yang, P., Wang, Y., Jiang, S., Zhang, X., & Li, C. (2016). Titanium surface priming with phase-transited lysozyme to establish a silver nanoparticle-loaded chitosan/hyaluronic acid antibacterial multilayer via layer-by-layer self-assembly. *PLoS One, 11*(1), e0146957. https://doi.org/10.1371/journal.pone.0146957

Zhou, J. W., Luo, H. Z., Jiang, H., Jian, T. K., Chen, Z. Q., & Jia, A. Q. (2018). Hordenine: A novel quorum sensing inhibitor and antibiofilm agent against *Pseudomonas aeruginosa*. *Journal of Agricultural and Food Chemistry, 66*(7), 1620–1628. https://doi.org/10.1021/acs.jafc.7b05035

Zhu, X., Radovic-Moreno, A. F., Wu, J., Langer, R., & Shi, J. (2014). Nanomedicine in the management of microbial infection–overview and perspectives. *Nano Today, 9*(4), 478–498. https://doi.org/10.1016/j.nantod.2014.06.003

Zilberman, M., & Elsner, J. J. (2008). Antibiotic-eluting medical devices for various applications. *Journal of Controlled Release, 130*(3), 202–215. https://doi.org/10.1016/j.jconrel.2008.05.02

CHAPTER

19

Targeting microbial biofilms using genomics-guided drug discovery

Damayanti Chakravarty[1], *Dipabarna Bhattacharya*[2], *Sayak Ganguli*[3] and *Upal Das Ghosh*[4]

[1]Cell and Molecular Biology Laboratory, The University of Southern Mississippi, Hattiesburg, MS, United States; [2]University of Helsinki, Helsinki, Finland; [3]Department of Biotechnology, St. Xavier's College, Kolkata, West Bengal, India; [4]Department of Microbiology, Bidhannagar College, Kolkata, West Bengal, India

1. Introduction

Biofilms can be defined as heterogenous assembly of microbial cells confined within a self-produced matrix. It is predicted that almost 80% of bacterial and archaeal cells reside in biofilms (Burmolle et al., 2010). Controlling the growth of biofilms has been one of the most researched avenues over the past few years, as it has been evidenced by numerous workers that the formation of biofilms generally tends to have a negative impact on the industrial, infrastructure, and medical fields. The matrix is made up of microbial biopolymers, which includes proteins, exopolysaccharides, and extracellular DNA, forging a distinct microenvironment (Chiang et al., 2013). Thus, for many microbes (including bacteria and archaea, as well as unicellular eukaryotes such as amoeba, flagellates, diatoms, and unicellular algae), biofilm formation is an adaptive measure that offers them protection from environmental perturbations (Costerton et al., 1995). The occurrence of biofilms is almost ubiquitous, and their expanse varies depending upon the environment in which they survive. They serve a myriad of functions such as ecosystem functioning, biogeochemical cycling, nutrient exchange between communities, and bioremediation in specific cases. Some workers also consider them to be part of the holobiont where they dwell closely with human, animal, and plant hosts. The flip side of this assemblage is that they contribute in addition to 80% to all bacterial infections in various settings, both human and industry (Fuqua et al., 2019). Thus, there is an urgent need to devise agents as well as identify targets, which could prevent the formation of such biofilms in anthropogenic and commercial settings.

Microbial Biofilms
https://doi.org/10.1016/B978-0-323-95715-1.00020-0

Over the years, our knowledge base has steadily increased, and now we have some clear indications toward the biology of biofilms, though gaps remain in context to their structure and function (Koo et al., 2017). High-throughput methods have been randomly used for functional studies in many disease cohorts and thus, offer rapid and efficient identification of the key regulators in any process under study. It has become clear that quorum sensing (QS) triggers and regulates the initiation and progression of biofilm formation aided by alterations in hydrodynamic conditions, cellular cross talk, signaling messengers, nutrient bioavailability, and other environmental fluctuations (Stewart et al., 2001). Microbial biofilms affect the society in a number of ways, of which antibiotic resistance is one of the most prominent issues. Apart from this, recalcitrant infections, spoilage of food, and deaths associated with biofilms are also contributory threats. Thus the focus of any biofilm study may be concentrated upon the following key features:

1. Explore the structure and the microbial diversity of biofilms.
2. Discover and further analyze the persistent and distinct functional cores.
3. Study the microbe–host interactions and community succession in them.
4. How to diagnose and treat various biofilm infections.

In this treatise, through a series of case studies, we shall provide some insights into how the genomic, transcriptomic, and proteomic methods have been successfully used and then conglomerate the veracity of the data to propose a pipeline for genomic-guided drug design.

2. Genomic approaches to study biofilm formation

Listeria monocytogenes is rapidly being identified as a major causal agent for food contamination. Such infections are on the rise as we become consume more and more canned and processed food. The pathogen's ability toward biofilm production is considered to be the prime mechanism by which it increases its persistence and also becomes more and more resistant. In a recent study (Lee et al., 2019), the food-processing environments (FPEs)–associated isolates were investigated, which included some unique genotypes having biofilm-forming abilities, which included the two most important characteristic features of adhesion and sessile biomass production under diverse environmental conditions. Sessile biomass quantity varied dependent upon growth conditions, serotype, lineage, and genotype, but cold temperature was associated with the formation of biofilm by the clonal complex (CC) 26 genotype. No uniform pattern of biofilm formation was noted across the genotypes under study, and different growth conditions revealed a wide diversity in the ability to form biofilms. The experimental conditions did not reveal any close associations of biofilm formation efficiency and persistent or prevalent genotypes. Loss of nutrients or deprivation was found to instigate the enhanced cellular adhesion. If the nutrient deficient condition was persisted with and prolonged, then it was found that biofilm maturation was getting affected. Another important observation was that addition of salt increased production of biofilms in both normal and nutrient-deprived environments. This Pan-genome-wide association study (Pan-GWAS) is first of its kind, which was used to assess the genetic composition in context of biofilm phenotypes. An interesting observation was the lack of common genes at various stages of growth and maturation, across the genotypes that were studied. However, a broad overview of the ontology contents revealed common functional patterns, which were independent of environmental perturbations. The enriched functions ubiquitously displayed by the genotypes were found to be associated with transformation/competence and surface proteins including internalins.

One of the most common biofilm-forming bacteria that dwell in different soil niches is *Bacillus cereus*. Despite several years of study, the exact mechanismsof biofilm formation was unelucidated until two genome wide approaches—transcriptome analysis and gene identification using transposon insertion mutagenesis were employed to test out the expression of the genes involved in biofilm formation. Transcriptome-based approach was instrumental in identifying around 500 genes, which exhibited differential expression under biofilm-promoting environment, and gene annotation successfully identified the following pathways and genes—ATP-dependent protease, transcription regulators nucleotide biosynthesis, iron salvage, and antibiotic production. Some of these biomolecules have been reported to be involved in volatile compound production, and thus, their overexpression may result in robust biofilms.

3. Transcriptomic approaches to study biofilm formation

Transcriptome-based analysis of biofilm-forming ability of the conditional pathogen *Haemophilus parasuis* (*H. parasuis*) was carried out since there are several reports of this pathogen escaping drug treatment resulting in severe chronic infections, which leads to significant commercial loss in the pig industry. Results indicated that the extracellular matrix of the organisms' biofilm was composed majorly of proteins and DNA. This study also reported a timescale of events for biofilm formation and progression where the initial 48 h exhibited rapid growth and then stable growth was observed up to 60 h. GO and KEGG analysis revealed that artM, artQ, ssrS, pflA, and HutX genes were differentially expressed. These genes have been reported to be involved in bacterial colonization and adhesion and thus can form the integral coterie of genes that affect biofilm formation. Metabolic pathways such as biosynthesis of secondary metabolites, ATP-binding cassette (ABC) transporters, starch and sucrose metabolism, etc. were found to be the most enriched. One of the most studied organism in context to biofilm formation is *Staphylococcus aureus*. These are richly abundant as foodborne pathogen and are under the WHO risk category of priority pathogens as indiscriminate use of antibiotics in the livestock industry has resulted in a large number of resistant strains of *S. aureus*.

It is unclear what is the effect of antibiotic pressure on the formation of biofilm by *S. aureus*. The aim of this study was to figure out what is the plausible mechanism by which ampicillin, in low concentration, regulates the growth of *S. aureus* biofilm formation. The study detailed the viability and biomass of biofilm formed with and without treatment with 1/4 MIC ampicillin for 8 h as determined by XTT and crystal violet straining assays, respectively (Hughes and Webber, 2017).

Quite a large number of differentially expressed genes were identified using comparative transcriptomics, KEGG and GO annotations, between samples treated with ampicillin and controls (noninduced). 530 genes were found to be either over (167) or underexpressed (363) (Huang et al., 2009).

GO functional enrichment analysis identified 183, 252, and 21 specific GO terms in biological process, molecular function and cellular component, respectively. Of the eight KEGG pathways identified, significantly enriched ones included "microbial metabolism in diverse environments," "S. aureus infection," and "monobactam biosynthesis." Additionally, "beta-lactam resistance" also identified as highly enriched pathway. Positive response of *S. aureus* to ampicillin was recorded as the following proteins exhibited upregulation of genes such as—efflux pump AbcA, penicillin-binding proteins PBP1, PBP1a/2, and PBP3, and antimicrobial resistance proteins VraF, VraG, Dlt, and Aur. With the advent of comparative genomics, there has been great advancement in identification of features that are either unique to or shared between cells or tissues. This in turn allows for discovery of novel disease biomarkers. Proteomics facilitates generation of high-throughput quantitative maps of protein expression; comparative proteomics, on the other hand, has been mostly restricted to the comparison of single cell lines or mutant strains. In a recent study using a data independent acquisition (DIA) technique, which allows global protein quantification of large sample cohorts, the proteome profiles of 27 clinical isolates from the opportunistic pathogen *Pseudomonas aeruginosa* were recorded. Analysis of the proteome of these 27 isolates, which were previously whole-genome sequenced and transcriptionally profiled, when grown under planktonic and biofilm growth conditions facilitated the identification of a core biofilm–associated signature. Moreover, when compared between different *P. aeruginosa* strains, the protein-to-mRNA ratio was found to correlate well, thus corroborating the fact that post-translational regulation of biofilm formation conforms to conserved proteomic patterns (Alhede et al., 2009). Such studies, which allow us to discover core regulatory pathways, which are quintessential to formation of biofilms in bacterial pathogens and their analogous antibiotic tolerance, hold promise as therapeutic potential by elucidating interactions between pathogenic species and their environment and in turn allowing development of clinical interventions that can impede associated infections.

4. Proteomic approaches to study biofilms

One of the most potent virulent pathogens prevalent in nosocomial infections of polymicrobial nature is *Enterococcus faecalis*, which also exhibits some resistance properties. They form biofilms on suitable surfaces, and their potency can be correlated to their biofilm-forming ability. In a study conducted by the *E. faecalis* SK460 strain isolated from a patient suffering from chronic diabetic foot ulcer, a label-free proteomic approach was used to quantify differential protein expression at the planktonic and biofilm stages (He et al., 2016). This strain of the pathogen is known as an esp (enterococcal surface protein) and fsr (two-component signal transduction system) negative non–gelatinase-producing strong biofilm former. Gene Ontology and Kyoto Encyclopedia of Genes and Genomes (KEGG) were the two servers used in the enrichment analyses to validate the proteomic data, and the results obtained indicated the overexpression of proteins associated with glycolysis, amino acid biosynthesis, biosynthesis of secondary metabolites, microbial metabolism in diverse environments, and stress response factors. Apart from these, stress and environmental response pathways, overexpression of LuxS-mediated QS, arginine metabolism, rhamnose biosynthesis, pheromone, and adhesion-associated proteins were noted during the stages of transition of the biofilm from the planktonic stages (Suryaletha et al., 2018). The expressions were cross-validated using quantitative real-time PCR. Computational prediction of interactions of the upregulated proteins revealed that the upregulated proteins identified through proteomic approaches formed a cohesive interaction network, which possessed regulatory functions (Szklarczyk et al., 2015). These validated biofilm determinants can serve as important targets for preventing or regulating the pathogenic cascade of enterococcal infections.

5. Drug discovery

A pathogenic microbial biofilm can be described as a collective of microbial cells, which is protected a self-produced polymer. These biofilms are inherently resistant to antibiotics and hence considered a global challenge. Living, as it does, in a community of microbial organisms in a clinical situation, makes it responsible for severe and dangerous cases of infection. Mounting an immune response often fails, which necessitates external high antibiotic doses for a prolonged time. Often, such attempts fail and lead to persistent infections, highlighting the limitations of therapeutic potential in these cases.

Over the years, several workers have pointed out to the fact that microbial cells depend on the degree of bacterial aggregation and multicellular communities to exhibit strong resistance in biofilms. The EPS matrix is also thought to play a pivotal role in the process (Mah & O'Toole, 2001; Mulcahy et al., 2010). Following the formation of biofilms, the dense aggregation of cells and the EPS matrix provide together impart strong resistance by eliciting genetic exchange, channelizing QS signals, developing persisters, chelating antimicrobials, and providing overall physical protection (Williams and Cámara, 2009; van Kessel, 2019). Thus, any disruption in these structures may directly lead to the loss of the resistance mechanism cross talks and restore the effect of antimicrobials.

Since their identification as potent mechanisms of drug resistance and disease incidence, biofilm inhibition has been investigated worldwide, which has led to the elucidation of two leading mechanisms by which treatment regimes can be planned (Table 19.1). One is to prevent their formation, and the other is to destroy the already-formed biofilm structures and aggregations. To achieve the former, planktonic cell attachment to the surfaces followed by maturation of early microcolonies to fully structured biofilms must be prevented. The initial attachment has been prevented using two approaches—the first involving the modification of the surface properties where biofilms are to be attached, and the second being the targeting of matrix proteins or molecules that mediate attachment. Special 3D patterns imprinted on surfaces have been found to alter physicochemical properties such as surface hydrophobicity leading to reduced attachment (Tang et al., 2011; Pogodin et al., 2013; Kim et al., 2015).

Another potent approach for preventing the formation of biofilms is to target QS inhibitors. This has been particularly investigated in *P. aeruginosa*, which is an opportunistic

TABLE 19.1 Summary of Major types of Antibiofilm agents.

Agents which directly inhibit biofilms	Mechanism of inhibition	Reference
QS inhibitors **i)** Azithromycin **ii)** Lactonases and acylases **iii)** Most AHL analogs, bergamottins, cyclic sulfur compounds, etc.	Majorly act by disrupting the signal generation of quorum sensing or by its degradation or competitive inhibition	Parsek et al. (1999), Favre-Bonté et al. (2003), Ishida et al. (2007)
Matrix-degrading enzymes **i)** Dispersion B, alpha-amylase, Alginate lyase **ii)** DNase I, thermonuclease, NucB, DNase 1L2 **iii)** Subtilisins, lysostaphin, trypsin, protease K, etc.	Main function is the degradation of polysaccharides, or eDNAs or proteinaceous components of the biofilm matrix	Schindler & Schuhardt (1964), Alkawash et al. (2006), Chaignon et al. (2007), Eckhart et al. (2007), Qin et al. (2007), Leroy et al. (2008), Mann et al. (2009), Nijland et al. (2010), Craigen et al. (2011)
Antimicrobial peptides **i)** LL-37 **ii)** Peptide 1018, DJK5, DJK6 **iii)** Peptide AS10	Mode of action relies on the reduction in attachment, promotion of twitching motility, and influencing QS systems. May also reduce intracellular PpGpp levels or use some currently undeciphered mechanism	Overhage et al. (2008), de la Fuente-Nunez et al. (2014, 2015)
Metal chelators		
EDTA, EGTA, Dequest 2006, lactoferrin	Increasing sensitivity of cells to antibacterial agents	Gray & Wilkinson (1965), Root et al. (1988), Gil et al. (1994), Raad et al. (2003), Banin et al. (2005)

gram-negative bacterium reported to be the pathogen in acute and chronic nosocomial infections contributing to increased patient mortality (Erdmann et al., 2019; Kamal et al., 2017; Favre-Bonte et al., 2003). Such infections thrive due to the formation of biofilms, which renders the properties of multidrug resistance, thus enhancing pathogenesis properties of the bacterium. A rational drug design strategy has been explored to inhibit the quorum signaling system of *P. aeruginosa* in which potent inhibitory lead molecules were designed against anthranilate-CoA ligase enzyme a product of the pqsA gene. The function of this enzyme is to produce autoinducers for cell-to-cell communication, which initiates and regulates biofilm formation, leading to an increased virulence of *P. aeruginosa*. A library of potential drug molecules was generated by employing a ligand-based virtual screening method aimed against a set of enzyme inhibitors.

Following this, structure-based virtual screening was performed to locate the compounds having the most stable binding conformation with the target enzyme. The compounds identified mainly formed numerous short-range hydrophilic and hydrophobic interactions with the enzyme. Confirmation of the interactions was performed using molecular dynamic simulation to calculate the affinity and stability of the hit compounds with the target enzyme.

6. Genomics-guided drug discovery

For over 20 years, genomics has aided in the acceleration of drug development. Target identification, target prioritization tractability, and validation have been made easier with the advent of high-throughput sequencing methods. The prediction of outcomes from pharmacological interventions comprising the field of pharmacogenomics has paved the way for personalized medicine. Population genomics and genome-wide association studies have enabled the sequencing and genotyping of hundreds and thousands on individuals, for an in-depth understanding of disease and drug interactions at the tissue and single cell levels, which can be evaluated using transcriptome analysis. CRISPR technologies have promised to enhance the mutation screening paradigm and extend it to the genome level, which should enable us to enhance our intervention capacities in complex environments such as biofilms (Ford et al., 2019).

The impact of genomics-guided drug discovery has been successfully performed for several human diseases. For example, the drug "herceptin" has been formulated to target HER2 gene, which was identified using functional genomics studies in a family of individuals (Hicks et al., 2019). Survival and longevity of patients under docetaxel chemotherapy were found to be enhanced when Perjeta and Herceptin were added specifically in cases of untreated HER2-positive metastatic breast cancer [Genentech. Herceptin Plus Chemotherapy Improved Disease-Free Survival and Overall Survival in Adjuvant Setting for Early-Stage Her2-Positive Breast Cancer Patients. May 13, 2005]. Thus, with the advancement of the genomic, transcriptomic, and proteomic repertoires and application of deep learning methods in genomics, it is now possible to understand the position, variation, evolution, expression, function, and interaction of a specific gene (Jørgensen et al., 2014; Li and Lee, 2017; Lim and Pavlidis, 2021; Long et al., 2017). Biofilms explored in a longitudinal framework using a timescale analysis would enable us to formulate suitable targets and design and test efficacious small molecules, which may serve to inhibit them.

References

Alhede, M., Bjarnsholt, T., Jensen, P.Ø., Phipps, R. K., Moser, C., Christophersen, L., Christensen, L. D., van Gennip, M., Parsek, M., Høiby, N., Rasmussen, T. B., & Givskov, M. (2009). *Pseudomonas aeruginosa* recognizes and responds aggressively to the presence of polymorphonuclear leukocytes. *Microbiology (Reading, England), 155*(Pt 11), 3500–3508. https://doi.org/10.1099/mic.0.031443-0

Alkawash, M. A., Soothill, J. S., & Schiller, N. L. (2006). Alginate lyase enhances antibiotic killing of mucoid *Pseudomonas aeruginosa* in biofilms. *APMIS: acta pathologica, microbiologica, et immunologica Scandinavica, 114*(2), 131–138. https://doi.org/10.1111/j.1600-0463.2006.apm_356.x

Banin, E., Vasil, M. L., & Greenberg, E. P. (2005). Iron and *Pseudomonas aeruginosa* biofilm formation. *Proceedings of the National Academy of Sciences of the United States of America, 102*(31), 11076–11081. https://doi.org/10.1073/pnas.0504266102

Burmølle, M., Thomsen, T. R., Fazli, M., Dige, I., Christensen, L., Homøe, P., Tvede, M., Nyvad, B., Tolker-Nielsen, T., Givskov, M., Moser, C., Kirketerp-Møller, K., Johansen, H. K., Høiby, N., Jensen, P.Ø., Sørensen, S. J., & Bjarnsholt, T. (2010). Biofilms in chronic infections - a matter of opportunity - monospecies biofilms in multispecies infections. *FEMS Immunology and Medical Microbiology, 59*(3), 324–336. https://doi.org/10.1111/j.1574-695X.2010.00714.x

Chaignon, P., Sadovskaya, I., Ragunah, C.h., Ramasubbu, N., Kaplan, J. B., & Jabbouri, S. (2007). Susceptibility of staphylococcal biofilms to enzymatic treatments depends on their chemical composition. *Applied Microbiology and Biotechnology, 75*(1), 125–132. https://doi.org/10.1007/s00253-006-0790-y

Chiang, W. C., Nilsson, M., Jensen, P.Ø., Høiby, N., Nielsen, T. E., Givskov, M., & Tolker-Nielsen, T. (2013). Extracellular DNA shields against aminoglycosides in *Pseudomonas aeruginosa* biofilms. *Antimicrobial Agents and Chemotherapy, 57*(5), 2352–2361. https://doi.org/10.1128/AAC.00001-13

Costerton, J. W., Lewandowski, Z., Caldwell, D. E., Korber, D. R., & Lappin-Scott, H. M. (1995). Microbial biofilms. *Annual Review of Microbiology, 49*, 711–745. https://doi.org/10.1146/annurev.mi.49.100195.003431

Craigen, B., Dashiff, A., & Kadouri, D. E. (2011). The use of commercially available alpha-amylase compounds to inhibit and remove *Staphylococcus aureus* biofilms. *The Open Microbiology Journal, 5*, 21–31. https://doi.org/10.2174/1874285801105010021

Eckhart, L., Fischer, H., Barken, K. B., Tolker-Nielsen, T., & Tschachler, E. (2007). DNase1L2 suppresses biofilm formation by *Pseudomonas aeruginosa* and *Staphylococcus aureus*. *The British Journal of Dermatology, 156*(6), 1342–1345. https://doi.org/10.1111/j.1365-2133.2007.07886.x

Erdmann, J., Thöming, J. G., Pohl, S., Pich, A., Lenz, C., & Häussler, S. (2019). The core proteome of biofilm-grown clinical *Pseudomonas aeruginosa* isolates. *Cells, 8*(10), 1129. https://doi.org/10.3390/cells8101129

Favre-Bonté, S., Kohler, T., & Van Delden, C. (2003). Environment-driven changes of mRNA and protein levels in *Pseudomonas aeruginosa*. Biofilm formation by *Pseudomonas aeruginosa*: Role of the C4-HSL cell- to-cell signal and inhibition by azithromycin. *Journal of Antimicrobial Chemotherapy, 52*, 598–604.

de la Fuente-Núñez, C., Reffuveille, F., Haney, E. F., Straus, S. K., & Hancock, R. E. (2014). Broad-spectrum anti-biofilm peptide that targets a cellular stress response. *PLoS Pathogens, 10*(5), e1004152. https://doi.org/10.1371/journal.ppat.1004152

de la Fuente-Núñez, C., Reffuveille, F., Mansour, S. C., Reckseidler-Zenteno, S. L., Hernández, D., Brackman, G., Coenye, T., & Hancock, R. E. (2015). D-enantiomeric peptides that eradicate wild-type and multidrug-resistant biofilms and protect against lethal *Pseudomonas aeruginosa* infections. *Chemistry & Biology, 22*(2), 196–205. https://doi.org/10.1016/j.chembiol.2015.01.002

Favre-Bonte, S., Köhler, T., & Van Delden, C (2003). Biofilm for- mation by pseudomonas aeruginosa: role of the C4-HSL cell- to-cell signal and inhibition by azithromycin. *Journal of Antimicrobial Chemotherapy, 52*(4), 598–604. https://doi.org/10.1093/jac/dkg397

Ford, K., McDonald, D., & Mali, P. (2019). Functional genomics via CRISPR-Cas. *Journal of Molecular Biology, 431*(1), 48–65. https://doi.org/10.1016/j.jmb.2018.06.034

Fuqua, C., Filloux, A., Ghigo, J. M., & Visick, K. L. (2019). Biofilms 2018: A diversity of microbes and mechanisms. *Journal of Bacteriology, 201*(18), e00118–e00119. Advance online publication. https://doi.org/10.1128/JB.00118-19

Gil, M. L., Casanova, M., & Martínez, J. P. (1994). Changes in the cell wall glycoprotein composition of Candida albicans associated to the inhibition of germ tube formation by EDTA. *Archives of Microbiology, 161*(6), 489–494. https://doi.org/10.1007/BF00307769

Gray, G. W., & Wilkinson, S. G. (1965). The effect of ethylenediaminetetra-acetic acid on the cell walls of some gram-negative bacteria. *Journal of General Microbiology, 39*(3), 385–399. https://doi.org/10.1099/00221287-39-3-385

He, Z., Liang, J., Zhou, W., Xie, Q., Tang, Z., Ma, R., & Huang, Z. (2016). Effect of the quorum-sensing luxS gene on biofilm formation by *Enterococcus faecalis*. *European Journal of Oral Sciences, 124*(3), 234–240. https://doi.org/10.1111/eos.12269

Hicks, M., Bartha, I., di Iulio, J., Venter, J. C., & Telenti, A. (2019). Functional characterization of 3D protein structures informed by human genetic diversity. *Proceedings of the National Academy of Sciences of the United States of America, 116*(18), 8960–8965. https://doi.org/10.1073/pnas.1820813116

Huang, D., Sherman, B. T., & Lempicki, R. A. (2009). Bioinformatics enrichment tools: Paths toward the comprehensive functional analysis of large gene lists. *Nucleic Acids Research, 37*(1), 1–13. https://doi.org/10.1093/nar/gkn923

Hughes, G., & Webber, M. A. (2017). Novel approaches to the treatment of bacterial biofilm infections. *British Journal of Pharmacology, 174*(14), 2237–2246. https://doi.org/10.1111/bph.13706

Ishida, T., Ikeda, T., Takiguchi, N., Kuroda, A., Ohtake, H., & Kato, J. (2007). Inhibition of quorum sensing in *Pseudomonas aeruginosa* by N-acyl cyclopentylamides. *Applied and Environmental Microbiology, 73*(10), 3183–3188. https://doi.org/10.1128/AEM.02233-06

Jørgensen, A. B., Frikke-Schmidt, R., Nordestgaard, B. G., & Tybjærg-Hansen, A. (2014). Loss-of-function mutations in APOC3 and risk of ischemic vascular disease. *The New England Journal of Medicine, 371*(1), 32–41. https://doi.org/10.1056/NEJMoa1308027

Kamal, A. A., Maurer, C. K., Allegretta, G., Haupenthal, J., Empting, M., & Hartmann, R. W. (2017). Quorum sensing inhibitors as path blockers for *Pseudomonas aeruginosa* infections: A new concept in anti-infective drug discovery. *Topics in Medicinal Chemistry, 26*, 185–210. https://doi.org/10.1007/7355_2017_17

Kim, S., Jung, U. T., Kim, S. K., Lee, J. H., Choi, H. S., Kim, C. S., & Jeong, M. Y. (2015). Nanostructured multifunctional surface with antireflective and antimicrobial characteristics. *ACS Applied Materials & Interfaces, 7*, 326–331.

Koo, H., Allan, R. N., Howlin, R. P., Stoodley, P., & Hall-Stoodley, L. (2017). Targeting microbial biofilms: Current and prospective therapeutic strategies. *Nature Reviews. Microbiology, 15*(12), 740–755. https://doi.org/10.1038/nrmicro.2017.99

Lee, B. H., Cole, S., Badel-Berchoux S, Guillier L, Felix B, Krezdorn N, S., Guillier, L., Felix, B., Krezdorn, N., et al. (2019). Biofilm formation of *Listeria monocytogenes* strains under food processing environments and pan-genome-wide association study. *Frontiers in Microbiology, 10*, 2698. https://doi.org/10.3389/fmicb.2019.02698

Leroy, C., Delbarre, C., Ghillebaert, F., Compere, C., & Combes, D. (2008). Effects of commercial enzymes on the adhesion of a marine biofilm-forming bacterium. *Biofouling, 24*(1), 11–22. https://doi.org/10.1080/08927010701784912

Lim, N., & Pavlidis, P. (2021). Evaluation of connectivity map shows limited reproducibility in drug repositioning. *Scientific Reports, 11*(1), 17624. https://doi.org/10.1038/s41598-021-97005-z

Li, X. H., & Lee, J. H. (2017). Antibiofilm agents: A new perspective for antimicrobial strategy. *Journal of Microbiology (Seoul, Korea), 55*(10), 753–766. https://doi.org/10.1007/s12275-017-7274-x

Long, T., Hicks, M., Yu, H. C., Biggs, W. H., Kirkness, E. F., Menni, C., Zierer, J., Small, K. S., Mangino, M., Messier, H., Brewerton, S., Turpaz, Y., Perkins, B. A., Evans, A. M., Miller, L. A., Guo, L., Caskey, C. T., Schork, N. J., Garner, C., Spector, T. D., … Telenti, A. (2017). Whole-genome sequencing identifies common-to-rare variants associated with human blood metabolites. *Nature Genetics, 49*(4), 568–578. https://doi.org/10.1038/ng.3809

Mah, T. F., & O'Toole, G. A. (2001). Mechanisms of biofilm resistance to antimicrobial agents. *Trends in Microbiology, 9*(1), 34–39.

Mann, E. E., Rice, K. C., Boles, B. R., Endres, J. L., Ranjit, D., Chandramohan, L., Tsang, L. H., Smeltzer, M. S., Horswill, A. R., & Bayles, K. W. (2009). Modulation of eDNA release and degradation affects *Staphylococcus aureus* biofilm maturation. *PloS One, 4*(6), e5822. https://doi.org/10.1371/journal.pone.0005822

Mulcahy, L. R., Burns, J. L., Lory, S., & Lewis, K (2010). Emergence of pseudomonas aeruginosa strains producing high levels of persister cells in patients with cystic fibrosis. *Journal of Bacteriology, 192*(23), 6191–6199. https://doi.org/10.1128/JB.01651-09

Nijland, R., Hall, M. J., & Burgess, J. G. (2010). Dispersal of biofilms by secreted, matrix degrading, bacterial DNase. *PloS One, 5*(12), e15668. https://doi.org/10.1371/journal.pone.0015668

Overhage, J., Campisano, A., Bains, M., Torfs, E. C., Rehm, B. H., & Hancock, R. E. (2008). Human host defense peptide LL-37 prevents bacterial biofilm formation. *Infection and Immunity, 76*(9), 4176–4182. https://doi.org/10.1128/IAI.00318-08

Parsek, M. R., Val, D. L., Hanzelka, B. L., Cronan, J. E., Jr., & Greenberg, E. P. (1999). Acyl homoserine-lactone quorum-sensing signal generation. *Proceedings of the National Academy of Sciences of the United States of America, 96*(8), 4360–4365. https://doi.org/10.1073/pnas.96.8.4360

Pogodin, ., Hasan, J., Baulin, V. A., Webb, H. K., Truong, V. K., Nguyen, T. H. P., Boshkovikj, V., et al. (2013). Biophysical model of bacterial cell interactions with nanopatterned cicada wing surfaces. *Biophysical Journal, 104*, 835–840.

Qin, Z., Ou, Y., Yang, L., Zhu, Y., Tolker-Nielsen, T., Molin, S., & Qu, D. (2007). Role of autolysin-mediated DNA release in biofilm formation of Staphylococcus epidermidis. *Microbiology (Reading, England), 153*(Pt 7), 2083–2092. https://doi.org/10.1099/mic.0.2007/006031-0

Raad, I., Chatzinikolaou, I., Chaiban, G., Hanna, H., Hachem, R., Dvorak, T., Cook, G., & Costerton, W. (2003). In vitro and ex vivo activities of minocycline and EDTA against microorganisms embedded in biofilm on catheter surfaces. *Antimicrobial Agents and Chemotherapy, 47*(11), 3580–3585. https://doi.org/10.1128/AAC.47.11.3580-3585.2003

Root, J. L., McIntyre, O. R., Jacobs, N. J., & Daghlian, C. P. (1988). Inhibitory effect of disodium EDTA upon the growth of Staphylococcus epidermidis in vitro: Relation to infection prophylaxis of hickman catheters. *Antimicrobial Agents and Chemotherapy, 32*(11), 1627–1631. https://doi.org/10.1128/AAC.32.11.1627

Schindler, C. A., & Schuhardt, V. T. (1964). Lysostaphin: A new bacteriolytic agent for the staphylococcus. *Proceedings of the National Academy of Sciences of the United States of America, 51*(3), 414–421. https://doi.org/10.1073/pnas.51.3.414

Stewart, P. S., & Costerton, J. W. (2001). Antibiotic resistance of bacteria in biofilms. *Lancet (London, England), 358*(9276), 135–138. https://doi.org/10.1016/s0140-6736(01)05321-1

Suryaletha, K., John, J., Radhakrishnan, M. P., George, S., & Thomas, S. (2018). Metataxonomic approach to decipher the polymicrobial burden in diabetic foot ulcer and its biofilm mode of infection. *International Wound Journal, 15*(3), 473–481. https://doi.org/10.1111/iwj.12888

Szklarczyk, D., Franceschini, A., Wyder, S., Forslund, K., Heller, D., Huerta-Cepas, J., Simonovic, M., Roth, A., Santos, A., Tsafou, K. P., Kuhn, M., Bork, P., Jensen, L. J., & von Mering, C. (2015). STRING v10: Protein-protein interaction networks, integrated over the tree of life. *Nucleic Acids Research, 43*, D447–D452. https://doi.org/10.1093/nar/gku1003

Tang, P. F., Zhang, W., Wang, Y., Zhang, B. X., Wang, H., Lin, C. J., & Zhang, L. H. (2011). Effect of superhydrophobic surface of titanium on *Staphylococcusaureus* adhesion. *Journal of Nanomaterials*. https://doi.org/10.1155/2011/178921

van Kessel, J. C. (2019). PQS signaling for more than a quorum: The collective stress response protects healthy *Pseudomonas aeruginosa* populations. *Journal of Bacteriology, 201*(23), 19–005688. https://doi.org/10.1128/JB.00568-19

Williams, P., & Cámara, M. (2009). Quorum sensing and environmental adaptation in *Pseudomonas aeruginosa*: A tale of regulatory networks and multifunctional signal molecules. *Current Opinion in Microbiology, 12*(2), 182–191. https://doi.org/10.1016/j.mib.2009.01.005

CHAPTER

20

Biofilms: a sociomicrobiological nexus

Lynn D'Lima and Manju Phadke

Department of Microbiology, S.I.E.S College of Arts, Science & Commerce (Autonomous), Mumbai, Maharashtra, India

1. Introduction to biofilms

Microorganisms throughout the history of microbiology have often been considered as free suspended cells growing in nutritious growth medium. However, microorganisms have been observed to form densely populated communities known as biofilm. Microorganisms attach themselves to surfaces of both biotic and abiotic entities. Later, these adherent cells are encased in a slimy layer known as the extracellular polymeric substances (EPS). The architecture of the established biofilm creates an environment for interaction between the cells, thereby affecting the processes occurring within the biofilm. Biofilms have proved to be both an advantage and a nuisance to human beings. Biofilms observed in water pipe networks compromise on the quality of water and lead to operational issues (Lang et al., 2016). They produce copious amounts of volatile organic compounds that include terpenoids, geosmin, pyrazines, and other compounds, which make freshwater unpalatable (Zhou et al., 2017). Food spoilage–causing bacteria can build biofilms within food-processing facilities, leading to food deterioration and risking consumer health (Galié et al., 2018). They have also caused much havoc in the medical field due to their growing insensitivity toward antibiotics (Estevez et al., 2020; Palanisamy et al., 2014). Biofilms have also been useful in microbial leaching, an extraction of valuable metals using microbial biofilms (Zannoni et al., 2021). Biofilms can assist in bioremediation of contaminated soil and groundwater (Farber et al., 2019). Although biofilms are known for their contribution to mankind, the biofilm itself is far more complex. Various positive and negative interactions occur within this complex structure. In this chapter, the events within the biofilm such as gene transfer, quorum sensing, chemical breakdown and utilization, symbiosis, predation, and competition will be discussed.

Microbial Biofilms
https://doi.org/10.1016/B978-0-323-95715-1.00018-2

2. Stages and complexity of a biofilm

A biofilm is a microbial system that is encased in a matrix that is primarily composed of polysaccharide attached to a surface. The biofilm matrix may also include materials such as soil particles, minerals, or components of blood depending on the environmental conditions suitable for the development of the biofilm (Donlan, 2002).

The attachment of the microbes is a result of the solid−liquid interface between a fluid such as water and a surface. Many factors such as hydrodynamics of the fluid, material, and the conditioning of the surface contribute to the attachment for biofilm initiation. It has been observed that an increase in microbial colonization is due to the increase in the surface roughness (Fletcher & Loeb, 1979). When substratum is placed in a fluid, the surfaces form a coat, primarily composed of protein, thus enabling attachment. This phenomenon is called conditioning of the surface. The characteristics of the microbial cell are equally important. The presence of a glycocalyx, fimbriae, pili, flagella, and the hydrophobicity of the cell may contribute to the attachment. These factors are required to establish contact and help overcome the repulsive forces present on surfaces of objects (Korber et al., 1989).

Furthermore, the attached cells experience complex gene regulation. Various genes are either upregulated or downregulated in this process. Genes for glycolysis or fermentation were upregulated postattachment. This could be due to the poor oxygen availability observed in a biofilm. Gene modulation is a result of the dynamic changes in the physicochemical factors surrounding the cell (Prigent-Combaret et al., 1999).

Biofilms are composed of numerous cells and the EPS layer. According to studies, the EPS layer can interact differently with bacterial cells depending on their gram nature. The EPS produced by gram-negative bacteria are anionic in nature, thereby attracting cationic ions and leading to biofilm development. In the case of gram-positive bacteria, the EPS is cationic (Hussain et al., 1993; Sutherland, 2001).

The EPS layer is a highly hydrated cross-linked structure. The level of cross-linking can result in its rigidity. The EPS layer acts as a protective layer. It reduces the adverse effects of desiccation and harmful chemicals such as antibiotics and metals (Sutherland, 2001). The EPS layer produced depends on numerous factors such as the age of the biofilm, the growth rate of the bacteria residing in the biofilm, and the availability of nutrients (Donlan, 2000).

The biofilms can be composed of either a mixed population or a single population. Within the biofilm, these are separated from each other through water channels or interstitial voids. The channels provide nutrients and oxygen as the fluid flows through it. The microcolonies also known as structural communities, observed in the biofilm, are surrounded by a microenvironment that ensures genetic transfer, quorum sensing, and nutrient gradient. The secluded microcolonies present in the biofilm can flow through the channels and migrate to different locations. This can be achieved through cell communication or quorum sensing. Microbial self-produced signals known as autoinducers consisting of quinolones, acyl homoserine lactones, and peptides are produced (Miller & Bassler, 2001; Zhang & Li, 2016). In addition, quorum sensing−dependent factors such as siderophores, lectins, and rhamnolipids are required for the formation of biofilm aggregates (Passos da Silva et al., 2017). Within the biofilm, ecological microcolony diversity is maintained. This helps in the cycling of nutrients such as carbon, nitrogen, sulfur, phosphorus, and other toxic compounds. Later, this leads to the disintegration of the biofilm and the further dissemination of the microcolonies, thus completing the cycle of the biofilm formation, establishment, and disintegration (Paula et al., 2020).

The spatial arrangement of the microcolonies can lead to various positive and negative interactions that include coordination, cooperation, competition, and predation. The cells in the proximity are the most affected. A balance between such interactions needs to be maintained to influence the macroscopic properties of the biofilm.

3. Competition in biofilms

During the formation of a biofilm, the primary colonizers are solitary and low in density. This behavior generally does not express social phenotypes, thus affecting the neighboring cells. At high population densities, the microcolonies turn genetically segregated. This could be due to the bottleneck effect experienced in nutrient-limited conditions. As several cell lineages interact with each other in nutrient-limited environment, competition is generally preferred. Competition can be mainly classified into exploitative and interference competition. The indirect interaction of cells that result in the prevention of resource access is called exploitative competition. On the other hand, interference competition leads to cellular damage of neighboring cells. Competition may also occur within kin due to adverse conditions (Rendueles & Ghigo, 2015). Competitive strategies include rapid growth, prompt resource acquisition, quick and effective attachment to surfaces, and most widely observed release of toxins, antibiotics, bacteriocins, and other inhibitory substances (Nadell et al., 2010). The susceptible cells are damaged, releasing their contents that are recycled as raw material by the neighbors. However, the cells producing toxins should be greater in number to have such detrimental effects on others (Foster & Bell, 2012; Hamilton, 1971; Mitri & Richard Foster, 2013). The mechanisms of these inhibitory molecules prove to be different. The toxins are secreted into neighboring cells, whereas antibiotics and bacteriocins travel to the extracellular space (Riley & Wertz, 2002; Wexler et al., 2016). Bacteriodes, a symbiont found in mammalian intestine, compete and thrive by a mechanism known as contact-dependent inhibition. *Proteus mirabilis* is a motile bacterium, which uses the type VI secreting mechanism as well as the motile equipment to eliminate competition (Borenstein et al., 2015). Lactobacilli create an acidic environment, thus preventing the growth of *Neisseria gonorrhoeae* (Vance, 2011). Toxic metabolic by-products can eliminate certain bacteria in the biofilm, which has been observed in the case of *Streptococcus pneumoniae* generating hydrogen peroxide against *Neisseria meningitides* (Pericone et al., 2000). Polymer production can prove to be advantageous over non–polymer-producing bacteria. Studies show that the survival rate of polymer producers is relatively low. The polymer production helps such cells compete by retarding the growth of neighboring nonpolymer producers (Xavier & Foster, 2007). A cellular communication mechanism known as quorum sensing is disrupted by degrading such molecules (Fan et al., 2017). Adhesion molecules are downregulated by competitors, as they secrete various enzymes and exopolysaccharides. A decrease in *Escherichia coli* biofilms has been observed due to the released exopolysaccharides, as they interfere with chemotaxis (Kim et al., 2009). Hence, competition has a strong part to play in microcolonies residing close to each other.

Biofilms also experience predation through phagocytosis, bacterial wolfpack hunting or cell hunting, and bacteriophage attack. *Dictyostelium discoideum*, a protozoan, performs phagocytosis. *Myxobacteria* attack cells through coordinating community movement. *Bdellovibrio bacteriovorus* uses a type IV pili to penetrate into its prey (Berleman & Kirby, 2009; Rendueles & Ghigo, 2015).

4. Coordination and cooperation in biofilms

Microbial aggregates are observed in a biofilm. When cells form such a social association, they can demonstrate tolerance to antibiotics, resistance to predation and desiccation, and enhanced reproductive rate (Butler et al., 2010; Gascoigne et al., 2009; Ruxton & Sherratt, 2006). Microbial genetics offers insights into the mechanisms of behavioral biofilm association. An ordered gene expression required for irreversible biofilm growth may contribute to a system of coordination and cooperation (Monds & O'Toole, 2009).

Microbes in biofilms are metabolically active and are a functionally integrated social system. Resources such as hydrogen, hydrogen sulfide, ammonia and several organic compounds, waste products, and electron acceptors such as NO_3^-, CO_2, O_2, etc. lead to molecular diffusion by chemical gradient formation (Berlanga, 2002; Sachs & Hollowell, 2012).

Individual cells and microcolonies perform cooperative activities involving the production and secretion of autoinducers vital for quorum sensing. Such activities aid in biofilm development, alteration of virulence factors, metabolic pathway regulation, and degradation of pollutants (Grandcl et al., 2016; Zhang & Li, 2016). Quorum sensing promotes the virulence factors of the pathogen with high population density and compromises the host defenses (Miller & Bassler, 2001). The Black Queen Hypothesis proposed by Morris et al. has two basic assumptions. The first assumption states that some bacteria produce resources that can be used publically. This facilitates mutualism. The interspecific cooperative relationship such as cross-protection and cross-feeding can contribute to mutualism. Mutualism has reciprocal benefits. It encourages microbial species diversity and ensures coexistence. According to the "by-product mutualism" model, an individual releases by-products that are ultimately utilized by the another individual for its growth and metabolism (Morris et al., 2012). The second assumption mentions that if an individual utilizes the by-products of the other, it will discontinue the use of its own metabolic pathway and will eventually delete it. This is known as adaptive genome streaming, often observed in symbionts and parasites. Such mutualistic behavior is a result of divergent resource utilization (Sachs et al., 2011). The release of such products includes catalase or peroxidase, iron chelators for enhanced iron solubility, and reduction of sulfur. An example of such an interaction is observed between *Prochlorococcus* and *Synechococcus*. *Synechococcus* produces catalase–peroxidase that breaks down hydrogen peroxide. *Prochlorococcus* is sensitive to hydrogen peroxide but grows with the help of the extracellular enzyme production of *Synechococcus* (Morris et al., 2011; Petasne & Zika, 1997). *Streptococcus gordonii* and *Actinomyces naeslundii* have a similar interaction involving arginine. *S. gordonii* unable to grow in the absence of arginine can only grow in the presence of *A. naeslundii*, when they coaggregate (Marchesi & Prosser, 2008). Such interactions turn complex due to resource exchange.

Biofilms can help in waste management, fermentation, and degradation of toxic compounds by bioremediation (Maksimova, 2014). In food industry, during the fermentation of natural olives, a shift in the microbial system is observed. Initially, gram-negative microbes grow and proliferate, and this is followed by lactic acid bacteria (LAB) and yeast. LAB or yeast utilizes the nutrients-producing acid, which drastically reduces the pH. Ethanol is also produced and is converted to products that impart flavor and odor to the olives (Botta & Cocolin, 2012). The oral cavity has also recorded such metabolic interactions. *Veillonella* sp. utilize the lactic acid produced from sugar fermentation by *Streptococcus oralis* (Kolenbrander et al., 2010).

Mixed biofilm species cooperate to survive in unfavorable conditions. Oxygen-sensitive anaerobic bacteria can thrive due to the presence of aerobic bacteria. Aerobes consume oxygen and thus contribute to anaerobic conditions for growth within the deeper layers of the mixed biofilm (Sbordone & Bortolaia, 2003).

Water treatment plants show species stratification. Aerobic conditions are maintained at the surface, thus facilitating the growth of aerobes. The microbes in the uppermost layer convert ammonium to nitrite. The layer below has microbes that oxidize nitrite to nitrate. The conversion of nitrate to elemental nitrogen is performed by anaerobes residing in the deeper layers. A similar observation has been made in the case of sulfate reduction (Lydmark et al., 2006; Ramsing et al., 1993).

Mutualism is required to maintain the stability of the community as it can promote coexistence. However, it has been observed that in the microbial society, some individuals gain maximum benefit and contribute negligibly. Such individuals are called cheaters in the biofilm hierarchy. They may destabilize and limit diversity in the microbial community (Popat et al., 2012; Smith & Schuster, 2019; Bronstein, 2001).

5. Role of mutations in biofilm

Mutations can prove either advantages or disadvantages for a cell. However, throughout evolution, mutations bring genetic diversity and complexity. According to the Black Queen Hypothesis, the following are a few predicted evolutionary events (Connor, 1986):

1. Selfish utilization of the by-products of another individual consequently reduces the production of one's own resources and leads to genome streamlining.
2. The evolution of cooperative traits develops to obtain maximum benefit from others. In such cases, loss of function of a gene through deletions will be prevalent in a population. This leads to dependence on others, and such interactions later are optimized.

Studies on metabolomes and transcriptomes showed that each phase of life has a unique transcriptional activity. Gene expression in planktonic and sessile cells varies as genes are upregulated and downregulated. Upregulation of genes for amino acid and carbohydrate transport, lipid, iron and sulfur metabolism, stress response, expression of adhesions, and synthesis of secondary metabolites have been observed (Nakamura et al., 2016). Overexpression of genes for iron acquisition has been observed in cells in the developed biofilm, but DNA repair genes are downregulated, thus giving rise to spontaneous mutants. Overexpression of the SOS response has been observed in planktonic cells dispersed from the biofilm (Nakamura et al., 2016; Guilhen et al., 2016). The SOS response is observed in cases with extensive DNA damage. It is an error-prone response responsible for the survival of the cell. The cell experiences extreme levels of mutagenesis. The regulation of SOS response is performed by Lex A and Rec A proteins. During normal conditions, Lex A acts as a transcriptional repressor for the SOS regulon genes. During excessive mutations, Rec A, an inducer of the SOS response, stimulates the self-cleavage and degradation of Lex A, thereby performing damage control (Maslowska et al., 2019). The SOS response proves advantageous as antibiotic resistance develops in cells within the biofilms (Ciofu & Tolker-Nielsen, 2019).

Vibrio fisheri showing bioluminescence has a symbiotic relationship with the bobtail squid. At high cell densities, the bioluminescence is controlled by the release of the *N*-acyl homoserine lactone inducer, thus activating the lux operon (Miyamoto et al., 2000). *V. fisheri* having null mutations lacks the lux gene (lux^-); hence, they are outnumbered by the wild type (lux^+) within the squid (Visick et al., 2000).

Mutations can also occur due to gene transfer mechanisms such as conjugation, transposition, transformation, and transduction. Conjugative plasmids carry gene for biofilm production, which are transferred to other cells. *Vibrio cholerae* becomes a potent toxin producer due to a phage infection, later infecting other cells (Williams, 1996). Transformation through DNA uptake occurs in competent sporulating *Bacillus subtilis* growing in nutrient-limited conditions (Chen et al., 2006). DNA exchange through transduction and conjugation increases the frequency of cooperation among cells (Smith, 2001; Wagner, 2005).

6. Genetic study of biofilm formation in vitro by the introduction of genes such as green fluorescent protein

Green fluorescent protein (GFP) is a fluorescent protein that was originally isolated from the luminous organ of the jellyfish *Aequorea victoria* (Ohba et al., 2013). GFP has been used as a marker to monitor the growth of viable cells, which can be observed under confocal scanning laser microscopy. GFP is a preferred marker as it is stable, requires no sample preparation and no substrate, and is compatible with numerous techniques (Martin Chalfie et al., 1993). The GFP is used as a reporter gene to detect gene expression and protein localization. Once introduced into the cell, it can be viewed microscopically. The major disadvantage is that it cannot be used under anaerobic conditions (Aspiras et al., 2000). The development of biofilm and adhesion can be studied using the gfp system. A plasmid coding for the GFP and a chloramphenicol resistance cassette was introduced into *Klebsiella pneumoniae* cells to study biofilm formation. The GFP production was confirmed under excitation by ultraviolet radiation. The confocal fluorescence images of the developed biofilms were observed using an argon ion laser (488 nm) (Wu et al., 2011). In a mixed species biofilm, *Pseudomonas putida* labeled with gfp showed colonization in the outer most layer, whereas *Acinetobacter* sp. was found closest to the substrate (Errampalli et al., 1999; Skillman et al., 1998). In another study, *E. coli* tagged with gfp showed the importance of hydrophobicity in bacterial adhesion. Biofilms were stained with both propidium iodide (PI) and gfp. The distribution of gfp-labeled cells was observed in green, whereas PI-stained cells were red. This study showed that surface-associated macromolecules are crucial for biofilm interactions (Errampalli et al., 1999). The survival of *P. putida* in activated sludge was studied using molecular gfp tags. Protozoans in the system grazed on the planktonic cells. The cells later attached themselves to the flocs in the sludge to evade predation (Eberl et al., 1997; Errampalli et al., 1999; Olofsson et al., 1998). It was also observed that the cells enter flocs through pores and channels.

Cells are known to transfer genes through plasmids. This was confirmed using a gfp-tagged TOL plasmid in *Pseudomonas* sp. transferred from donor to recipient (Christensen et al., 1996).

GFP-mediated genetic study has enabled scientists to understand the adhesion mechanism, special arrangement and location of cells, degradation of complex substrates, role in nutrient cycling, and bioremediation.

7. Conclusion

Biofilms are a complex microbial community. The initiation of the biofilms relies on several factors related to both the microorganism and the environment. The physiological complexity of biofilms suggests the occurrence of several positive as well as negative interactions. The interactions include antagonism, predation, mutualism, coordination, and cooperation. As the cells establish the biofilm, genetic changes are observed, and genes are either upregulated or downregulated to enhance their chances of survival and biofilm establishment. Mutations in biofilms can lead to ecological diversity. Cell communication or quorum sensing can further strengthen microbial associations. This may not be favorable for the growth of certain microbes as certain cells show antagonism, while other individuals benefit from such communication, as they can evade predation, avoid harsh environmental conditions, and utilize resources. Reliance of microbes for the utilization of another individual's by-products can lead to loss of function of genes as they evolve. Some microbes utilize the end products of other organisms for its growth. These are observed in the case of nutrient cycling, wastewater management, and bioremediation. Bacteria termed cheaters contribute to the bare minimum, and they threaten to disrupt the biofilm. The biofilms are studied with the aid of fluorescent microscopic techniques. However, studies on the socioevolution of the biofilms should be performed. The sociomicrobiological evolution observed in the biofilm is a result of the equilibrium between competition and cooperation, leading to the formation and development of biofilms.

References

Aspiras, M. B., Kazmerzak, K. M., Kolenbrander, P. E., McNab, R., Hardegen, N., & Jenkinson, H. F. (2000). Expression of green fluorescent protein in Streptococcus gordonii DL1 and its use as a species-specific marker in coadhesion with Streptococcus oralis 34 in saliva-conditioned biofilms in vitro. *Applied and Environmental Microbiology, 66*(9), 4074–4083. https://doi.org/10.1128/AEM.66.9.4074-4083.2000

Guerrero, R., Piqueras, M., & Berlanga, M. (2002). Microbial mats and the search for minimal ecosystems. *International Microbiology, 5*, 177–188. https://doi.org/10.1007/s10123-002-0094-8

Berleman, J. E., & Kirby, J. R. (2009). Deciphering the hunting strategy of a bacterial wolfpack. *FEMS Microbiology Reviews, 33*, 942–957. https://doi.org/10.1111/j.1574-6976.2009.00185.x

Borenstein, D. B., Ringel, P., Basler, M., & Wingreen, N. S. (2015). Established microbial colonies can survive type VI secretion assault. *PLOS Computational Biology, 11*, 1–16. https://doi.org/10.1371/journal.pcbi.1004520

Botta, C., & Cocolin, L. (2012). Microbial dynamics and biodiversity in table olive fermentation: Culture-dependent and independent approaches. *Frontiers in Microbiology, 3*, 1–10. https://doi.org/10.3389/fmicb.2012.00245, 245.

Bronstein, J., B. (2001). The exploitation of mutualisms. *Ecology Letters, 4*(3), 277–287. https://doi.org/10.1046/j.1461-0248.2001.00218.x

Butler, M. T., Wang, Q., & Harshey, R. M. (2010). Cell density and mobility protect swarming bacteria against antibiotics. *Proceedings of the National Academy of Sciences of the United States of America, 107*(8), 3776–3781. https://doi.org/10.1073/pnas.0910934107

Chen, I., Provvedi, R., & Dubnau, D. (2006). A macromolecular complex formed by a pilin-like protein in competent Bacillus subtilis. *Journal of Biological Chemistry, 281*(31), 21720–21727. https://doi.org/10.1074/jbc.M604071200

Christensen, B. B., Sternberg, C., & Molin, S. (1996). Bacterial plasmid conjugation on semi-solid surfaces monitored with the green fluorescent protein (GFP) from Aequorea victoria as a marker. *Gene*, 59–65. https://doi.org/10.1016/0378-1119(95)00707-5

Ciofu, O., & Tolker-Nielsen, T. (2019). Tolerance and resistance of *Pseudomonas aeruginosa* biofilms to antimicrobial agents-how *P. aeruginosa* can escape antibiotics. *Frontiers in Microbiology, 10*, 1–3. https://doi.org/10.3389/fmicb.2019.02164

Connor, R. C. (1986). Pseudo-reciprocity: Investing in mutualism. *Animal Behaviour, 34*(5), 1562–1566. https://doi.org/10.1016/S0003-3472(86)80225-1

Donlan, R. M. (2000). Role of biofilms in antimicrobial resistance. *ASAIO Journal, 46*(6), S47–S52. https://doi.org/10.1097/00002480-200011000-00037

Donlan, R. M. (2002). Biofilms: Microbial life on surfaces. *Emerging Disease Infections, 8*(9), 881–890. https://doi.org/10.3201/eid0809.020063

Eberl, L., Schulze, R., Ammendola, A., Geisenberger, O., Erhart, R., Sternberg, C., Molin, S., & Amann, R. (1997). Use of green fluorescent protein as a marker for ecological studies of active sludge communities. *FEMS Microbiology Letters, 149*(1), 77–83. https://doi.org/10.1016/S0378-1097(97)00058-X

Errampalli, D., Leung, K., Cassidy, M., B., Kostrzynska, M., Blears, M., Lee, H., & Trevors, J., T. (1999). Applications of the green fluorescent protein as a molecular marker in environmental microorganisms. *Journal of Microbiological Methods, 35*(3), 187–199. https://doi.org/10.1016/S0167-7012(99)00024-X

Estevez, M. B., Raffaelli, S., Mitchell, S. G., Faccio, R., & Alborés, S. (2020). Biofilm eradication using biogenic silver nanoparticles. *Molecules, 25*(9), 1–14. https://doi.org/10.3390/molecules25092023

Fan, X., Liang, M., Wang, L., Chen, R., Li, H., & Liu, X. (2017). Aii810, a novel cold-adapted N-acylhomoserine lactonase discovered in a metagenome, can strongly attenuate *Pseudomonas aeruginosa* virulence factors and biofilm formation. *Frontiers in Microbiology, 8*, 1–11. https://doi.org/10.3389/fmicb.2017.01950

Farber, R., Rosenberg, A., Rozenfeld, S., Benet, G., & Cahan, R. (2019). Bioremediation of artificial diesel-contaminated soil using bacterial consortium immobilized to plasma-pretreated wood waste. *Microorganisms, 7*(11), 1–17. https://doi.org/10.3390/microorganisms7110497

Fletcher, M., & Loeb, G. I. (1979). Influence of substratum characteristics on the attachment of a marine pseudomonad to solid surfaces. *Applied and Environmental Microbiology, 37*(1), 67–72. https://doi.org/10.1128/aem.37.1.67-72.1979

Foster, K. R., & Bell, T. (2012). Competition, not cooperation, dominates interactions among culturable microbial species. *Current Biology, 22*(19), 1845–1850. https://doi.org/10.1016/j.cub.2012.08.005

Galié, S., García-Gutiérrez, C., Miguélez, E. M., Villar, C. J., & Lombó, F. (2018). Biofilms in the food industry: Health aspects and control methods. *Frontiers in Microbiology, 9*, 1–18. https://doi.org/10.3389/fmicb.2018.00898

Gascoigne, J., Berec, L., Gregory, S., & Courchamp, F. (2009). Dangerously few liaisons: A review of mate-finding Allee effects. *Population Ecology, 51*(3), 355–372. https://doi.org/10.1007/s10144-009-0146-4

Grandcl, C., Mor, S., Dessaux, Y., & Faure, D. (2016). Quorum quenching: Role in nature and applied developments. *FEMS Microbiology Reviews, 40*(1), 86–116. https://doi.org/10.1093/femsre/fuv038

Guilhen, C., Charbonnel, N., Parisot, N., Gueguen, N., Iltis, A., Forestier, C., & Balestrino, D. (2016). Transcriptional profiling of *Klebsiella pneumoniae* defines signatures for planktonic , sessile and biofilm-dispersed cells. *BMC Genomics, 17*, 1–15. https://doi.org/10.1186/s12864-016-2557-x, 237.

Hamilton, W. D. (1971). The genetical evolution of social behavior. II. *Group selection* (1st, pp. 1–47). Routledge.

Hussain, M., Wilcox, M. H., & White, P. J. (1993). The slime of coagulase-negative staphylococci: Biochemistr y and relation to adherence. *FEMS Microbiology Letters, 104*(3-4), 191–208. https://doi.org/10.1111/j.1574-6968.1993.tb05867.x

Kim, Y., oh, S., & Kim, S. H. (2009). Released exopolysaccharide (r-EPS) produced from probiotic bacteria reduce biofilm formation of enterohemorrhagic *Escherichia coli* O157:H7. *Biochemical and Biophysical Research Communications, 379*(2), 324–329. https://doi.org/10.1016/j.bbrc.2008.12.053

Kolenbrander, P. E., Palmer, R. J., Periasamy, S., & Jakubovics, N. S. (2010). Oral multispecies biofilm development and the key role of cell-cell distance. *Nature Reviews Microbiology, 8*, 471–480. https://doi.org/10.1038/nrmicro2381

Korber, D. R., Lawrence, J. R., Sutton, B., & Caldwell, D. E. (1989). Effect of laminar flow velocity on the kinetics of surface recolonization by Mot+ and Mot-Pseudomonas fluorescens. *Microbial Ecology, 18*, 1–19. https://doi.org/10.1007/BF02011692

Lang, J. M., Erb, R., Pechal, J. L., Wallace, J. R., McEwan, R. W., & Benbow, M. E. (2016). Microbial biofilm community variation in flowing habitats: Potential utility as bioindicators of postmortem submersion intervals. *Microorganisms, 4*(1), 1–17. https://doi.org/10.3390/microorganisms4010001

Lydmark, P., Lind, M., Sörensson, F., & Hermansson, M. (2006). Vertical distribution of nitrifying populations in bacterial biofilms from a full-scale nitrifying trickling filter. *Environmental Microbiology, 8*(11), 2036–2049. https://doi.org/10.1111/j.1462-2920.2006.01085.x

Maksimova, Y. G. (2014). Microbial biofilms in biotechnological processes. *Applied Biochemistry and Microbiology, 50*, 750–760. https://doi.org/10.1134/S0003683814080043

Marchesi, J., & Prosser, J. I. (2008). Gut microbiology: The black box and beyond. *FEMS Microbiology Ecology, 66*(2), 485–486. https://doi.org/10.1111/j.1574-6941.2008.00607.x

Martin Chalfie, D. C. P., Tu, Y., Euskirchen, G., & Ward, W. W. (1993). Green fluorescent protein as a marker for gene expression. *Science, 263*(5148), 802–805. https://doi.org/10.1126/science.8303295

Maslowska, K. H., Makiela-Dzbenska, K., & Fijalkowska, I. J. (2019). The SOS system: A complex and tightly regulated response to DNA damage. *Environmental and Molecular Mutagenesis, 60*(4), 368–384. https://doi.org/10.1002/em.22267

Miller, M. B., & Bassler, B. L. (2001). Quorum sensing in bacteria. *Annual Review of Microbiology, 55*, 165–199. https://doi.org/10.1146/annurev.micro.55.1.165

Mitri, S., & Richard Foster, K. (2013). The genotypic view of social interactions in microbial communities. *Annual Review of Genetics, 47*, 247–273. https://doi.org/10.1146/annurev-genet-111212-133307

Miyamoto, C. M., Lin, Y. H., & Meighen, E. A. (2000). Control of bioluminescence in Vibrio fischeri by the LuxO signal response regulator. *Molecular Microbiology, 36*, 594–607, 3rd ed.

Monds, R. D., & O'Toole, G. A. (2009). The developmental model of microbial biofilms: Ten years of a paradigm up for review. *Trends in Microbiology, 17*(2), 73–87. https://doi.org/10.1016/J.TIM.2008.11.001

Morris, J. J., Johnson, Z. I., Szul, M. J., Keller, M., & Zinser, E. R. (2011). Dependence of the cyanobacterium Prochlorococcus on hydrogen peroxide scavenging microbes for growth at the ocean's surface. *PLoS One, 6*(2), 1–5. https://doi.org/10.1371/journal.pone.0016805

Morris, J. J., Lenski, R. E., & Zinser, E. R. (2012). The black queen hypothesis: Evolution of dependencies through adaptive gene loss. *MBio, 3*(2). https://doi.org/10.1128/mBio.00036-12

Nadell, C. D., Foster, K. R., & Xavier, J. B. (2010). Emergence of spatial structure in cell groups and the evolution of cooperation. *PLOS Computational Biology, 6*(3), 1–9. https://doi.org/10.1371/journal.pcbi.1000716

Nakamura, Y., Yamamoto, N., Kino, Y., Yamamoto, N., Kamei, S., Kurokawa, K., Nakashima, N., & Ken Kurokawa 1 2, Nobutaka Nakashima. (2016). Establishment of a multi-species biofilm model and metatranscriptomic analysis of biofilm and planktonic cell communities. *Applied Microbiology and Biotechnology, 100*(16), 7263–7279. https://doi.org/10.1007/s00253-016-7532-6

Ohba, Y., Fujioka, Y., Nakada, S., & Tsuda, M. (2013). Fluorescent protein-based biosensors and their clinical applications. *Progress in Molecular Biology and Translational Science, 113*, 313–348. https://doi.org/10.1016/B978-0-12-386932-6.00008-9

Olofsson, A. C., Zita, A., & Hermansson, M. (1998). Floc stability and adhesion of green-fluorescent-protein-marked bacteria to flocs in activated sludge. *Microbiology, 144*, 519–528. https://doi.org/10.1099/00221287-144-2-519

Palanisamy, N. K., Ferina, N., Amirulhusni, A. N., Mohd-Zain, Z., Hussaini, J., Ping, L. J., & Durairaj, R. (2014). Antibiofilm properties of chemically synthesized silver nanoparticles found against *Pseudomonas aeruginosa. Journal of Nanobiotechnology., 12*, 1–7. https://doi.org/10.1186/1477-3155-12-2

Passos da Silva, D., Schofield, M. C., Parsek, M. R., & Tseng, B. S. (2017). An update on the sociomicrobiology of quorum sensing in gram-negative biofilm development. *Pathogens, 6*(4), 1–15. https://doi.org/10.3390/pathogens6040051

Paula, A. J., Hwang, G., & Koo, H. (2020). Dynamics of bacterial population growth in biofilms resemble spatial and structural aspects of urbanization. *Nature Communications, 11*, 1–6. https://doi.org/10.1038/s41467-020-15165-4, 1354.

Pericone, C. D., Overweg, K., Hermans, P. W. M., & Weiser, J. N. (2000). Inhibitory and bactericidal effects of hydrogen peroxide production by Streptococcus pneumoniae on other inhabitants of the upper respiratory tract. *Infection and Immunity, 68*(7), 3990–3997. https://doi.org/10.1128/IAI.68.7.3990-3997.2000

Petasne, R. G., & Zika, R. G. (1997). Hydrogen peroxide lifetimes in south Florida coastal and offshore waters. *Marine Chemistry, 56*(3-4), 215–225. https://doi.org/10.1016/S0304-4203(96)00072-2

Popat, R., Crusz, S. A., Messina, M., Williams, P., West, S. A., & Diggle, S. P. (2012). Quorum-sensing and cheating in bacterial biofilms. *Proceedings of the Royal Society B: Biological Sciences, 279*(1748), 4765–4771. https://doi.org/10.1098/rspb.2012.1976

Prigent-Combaret, C., Vidal, O., Dorel, C., & Lejeune, P. (1999). Abiotic surface sensing and biofilm-dependent regulation of gene expression in *Escherichia coli. Journal of Bacteriology, 181*(19), 5993–6002. https://doi.org/10.1128/jb.181.19.5993-6002.1999

Ramsing, N. B., Kuhl, M., & Jorgensen, B., B. (1993). Distribution of sulfate-reducing bacteria, 02, and H2S in photosynthetic biofilms determined by oligonucleotide probes and microelectrodes. *Applied and Environmental Microbiology, 59*(11), 3840–3849. https://doi.org/10.1128/aem.59.11.3840-3849.1993

Rendueles, O., & Ghigo, J. (2015). Mechanisms of competition in biofilm communities. *Microbiology Spectrum, 3*(3), 1–18. https://doi.org/10.1128/microbiolspec.MB-0009-2014.f1

Riley, M. A., & Wertz, J. E. (2002). Bacteriocins: Evolution, ecology, and application. *Annual Review of Microbiology, 56*, 117–137. https://doi.org/10.1146/annurev.micro.56.012302.161024

Ruxton, G. D., & Sherratt, T. N. (2006). Aggregation, defence and warning signals: The evolutionary relationship. *Proceedings of the Royal Society B: Biological Sciences, 273*(1600), 2417–2424. https://doi.org/10.1098/rspb.2006.3570

Sachs, J. L., & Hollowell, A. C. (2012). The origins of cooperative bacterial communities the origins of cooperative bacterial communities. *mBio, 3*(3), 1–2. https://doi.org/10.1128/mBio.00099-12.Updated

Sachs, J. L., Skophammer, R. G., & Regus, J. U. (2011). Evolutionary transitions in bacterial symbiosis. *PNAS*, 10800–10807. https://doi.org/10.1073/pnas.1100304108

Sbordone, L., & Bortolaia, C. (2003). Oral microbial biofilms and plaque-related diseases: Microbial communities and their role in the shift from oral health to disease. *Clinical Oral Investigations, 7*(4), 181–188. https://doi.org/10.1007/s00784-003-0236-1

Skillman, L. C., Sutherland, I. W., & Jones, M. V. (1998). The role of exopolysaccharides in dual species biofilm development. *Journal of Applied Microbiology Symposium Supplement, 85*(1), 13–18. https://doi.org/10.1111/j.1365-2672.1998.tb05278.x

Smith, J. (2001). The social evolution of bacterial pathogenesis. *Proceedings of the Royal Society B: Biological Sciences, 268*(1462), 61–69. https://doi.org/10.1098/rspb.2000.1330

Smith, P., & Schuster, M. (2019). Public goods and cheating in microbes. *Current Biology, 29*, R442–R447. https://doi.org/10.1016/J.CUB.2019.03.001

Sutherland, I. W. (2001). Biofilm exopolysaccharides: A strong and sticky framework. *Microbiology, 147*, 3–9. https://doi.org/10.1099/00221287-147-1-3

Vance, R. R. (2011). Interference competition and the coexistence of two competitors on a single limiting resource. *Ecology, 65*(5), 1349–1357. https://doi.org/10.2307/1939115

Visick, K. L., Foster, J., Doino, J., McFall-Ngai, M., & Ruby, E. G. (2000). Vibrio fischeri lux genes play an important role in colonization and development of the host light organ. *Journal of Bacteriology, 182*(16), 4578–4586. https://doi.org/10.1128/JB.182.16.4578-4586.2000

Wagner, A. (2005). Periodic extinctions of transposable elements in bacterial lineages: Evidence from intragenomic variation in multiple genomes. *Molecular Biology and Evolution, 23*(4), 723–733. https://doi.org/10.1093/molbev/msj085

Wexler, A. G., Bao, Y., Whitney, J. C., Bobay, L. M., Xavier, J. B., Schofield, W. B., Barry, N. A., Russell, A. B., Tran, B. Q., Goo, Y. A., Goodlett, D. R., Ochman, H., Mougous, J. D., Goodman, A. L., & Isberg, R. R. (2016). Human symbionts inject and neutralize antibacterial toxins to persist in the gut. *Proceedings of the National Academy of Sciences of the United States of America, 113*(13), 3639–3644. https://doi.org/10.1073/pnas.1525637113

Williams, N. (1996). Phage transfer: A new player turns up in cholera infection. *Science, 272*(5270), 1869–1870. https://doi.org/10.1126/science.272.5270.1869

Wu, M. C., Lin, T. L., Hsieh, P. F., Yang, H. C., & Wang, J. T. (2011). Isolation of genes involved in biofilm formation of a klebsiella pneumoniae strain causing pyogenic liver abscess. *PLoS One, 6*(8), 1–11. https://doi.org/10.1371/journal.pone.0023500

Xavier, J. B., & Foster, K. R. (2007). Cooperation and conflict in microbial biofilms. *Proceedings of the National Academy of Sciences of the United States of America, 104*(3), 876–881. https://doi.org/10.1073/pnas.0607651104

Zannoni, D., Turner, R. J., Tuovinen, O. H., Poblete-Castro, I., & Bobadilla-Fazzini, R. A. (2021). Biofilm formation is crucial for efficient copper bioleaching from bornite under mesophilic conditions: Unveiling the lifestyle and catalytic role of sulfur-oxidizing bacteria bobadilla-fazzini and poblete-castro biofilm crucial for copper recovery from B. *Frontiers in Microbiology, 12*, 1–10. https://doi.org/10.3389/fmicb.2021.761997. www.frontiersin.org.

Zhang, W., & Li, C. (2016). Exploiting quorum sensing interfering strategies in gram-negative bacteria for the enhancement of environmental applications. *Frontiers in Microbiology, 6*, 1535. https://doi.org/10.3389/fmicb.2015.01535

Zhou, X., Zhang, K., Zhang, T., Li, C., & Mao, X. (2017). An ignored and potential source of taste and odor (T&O) issues—biofilms in drinking water distribution system (DWDS). *Applied Microbiology and Biotechnology, 101*(9), 3537–3550. https://doi.org/10.1007/s00253-017-8223-7

CHAPTER

21

In silico and in vivo methods for designing antibiofilm agents against *Pseudomonas aeruginosa* and *Staphylococcus aureus*

Saichandana Tamba[1,a], *Aravind Panicker*[1,a], *Renitta Jobby*[1], *Satish Kumar Rajasekharan*[2], *Karthi Shanmugam*[3] and *Vinothkannan Ravichandran*[4]

[1]Amity Institute of Biotechnology, Amity University Maharashtra, Mumbai, Maharashtra, India; [2]Department of Food Sciences, Institute of Postharvest Technology and Food Sciences, Agricultural Research Organization (ARO), The Volcani Center, Rishon LeZion, Israel; [3]School of Chemical and Biotechnology, SASTRA University, Thanjavur, Tamil Nadu, India; [4]Centre for Drug Discovery and Development (CD3), Amity Institute of Biotechnology (AIB), Amity University Maharashtra, Mumbai, Maharashtra, India

1. Introduction

Biofilm is a term used to describe microbes that are accumulated on a surface and have a polymer-encased community. After the microbial cells adhere to the surfaces, biofilm development is initiated. Suspended counterparts are differentiated from the biofilm associated cells by the up/down gene regulation, extracellular polymeric matrix generation, and reduction in the growth rate. Some basic requirements needed to be fulfilled for the biofilm formation are transcription of a variety of genes different from the planktonic cell and the intracellular signaling (Donlan, 2002; Kokare et al., 2009). Embodiment of surfaces for the attachment of microorganisms results in the substantial enhancement of bacterial activity and growth (Bottle effect) (Kokare et al., 2009).

[a] These authors contributed equally to this work.

Microbial Biofilms
https://doi.org/10.1016/B978-0-323-95715-1.00017-0

Biofilms are liable for a plethora of diseases with high levels of difficulty in effective treatment because of their resistance to host defense systems and antibiotics (Wilson, 2001). In case of oral infection, the oral biofilm contains a complex concoction of microorganisms, which constitutes a community with a suitable composition required to remain stable (termed a climax community) because of both antagonistic and advantageous variety of interactions such as coaggregation, bacteriocin production, competition for nutrients, and food webs between the constituent species (Marsh & Bradshaw, 1997). Apart from oral biofilms, biofilm-associated human diseases usually occur when the host defense systems display some sort of impairment like in the case of cystic fibrosis patients with pulmonary infections or due to the existence of some medical device implant such as joint prostheses, catheters, etc. (Costerton et al., 1999).

Due to the problems caused by the biofilms, prevention or proper eradication of biofilms is necessary, which leads to the formulation of antibiofilm strategies. Complete knowledge of the biofilm formation process is necessary for the designing of antibiofilm activity. So to fight against the biofilms, two strategies have been proposed after studying the biofilm formation process—first strategy comprises the eradication of biofilm cells and the second strategy is more of a prophylactic strategy, which aims toward the prevention of bacterial cell adhesion to the surface. Also at the same time, to limit the nosocomial infections, some preventive hygiene protocols such as cleaning the hands, rooms, or apparatus in a hospital environment have been formulated and followed (Jones et al., 1995; Thebault et al., 2013).

After the formulation and execution of antibiofilm strategy, it is studied and compared among various biofilm-forming microorganisms to understand the effectiveness of the eradicative or preventive strategy used.

2. Biofilms

A biofilm can be defined as a systematically participating, properly coordinating association of microorganisms. Biofilm formation is a cyclic step by step process comprising—conditioning layer formation, bacterial adhesion followed by bacterial growth, expansion of biofilm, and at last active dispersal (Fig. 21.1). Biofilms can be seen prevailing on almost all types of surfaces such as food products, materials for medical implantation, tissue, various soil particles, glass, wooden, metallic and plastic surfaces, etc. (Kokare et al., 2009). A bridge between the conditioning film and the bacteria is formed, which is mediated by the locomotive organelle of bacteria such as pili, flagella, and fimbriae to help in the production of extracellular polymeric substance (EPS) by the accretion of multilayered cells. Microbial biofilms naturally have a high level of organization with some special characteristics that they can develop a 3D structure or a single layer and may exist in communities of multiple or single species (Form, 2002; Kumar & Anand, 1998) (Table 21.1).

3. Biofilms and human diseases

The formation of biofilms not only makes bacteria less susceptible to antibacterial agents but also kills the host's immune effector mechanism, creating a new niche that provides a protective mechanism that allows pathogens to survive and enable them in hostile environments,

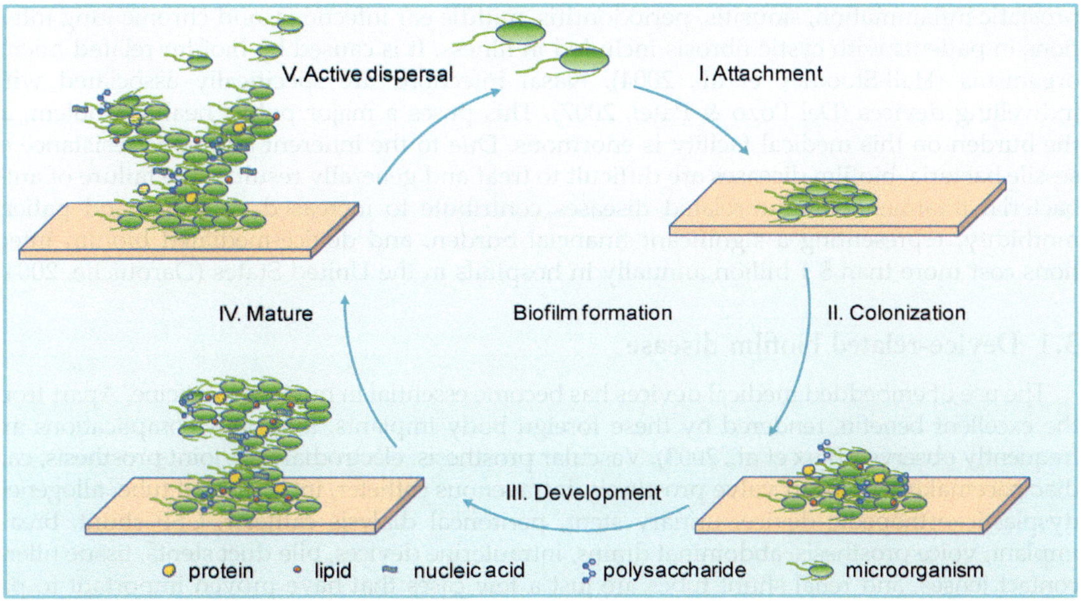

FIGURE 21.1 Cyclic step by step process of microbial biofilm formation consisting of five specific stages: I. Attachment: Reversible adhesion of microbial cells on the surface through weak forces (such as Van der Waals). II. Colonization: Irreversible attachment of microbial cells to the surface via locomotive organelle such as pilli, flagella, and fimbriae. III. Development: Production of EPS by the accumulation of multilayered cells. IV. Mature: Formation of a 3D stable community. V. Active dispersal: Return of microorganisms to the planktonic state after their dissemination from the accumulated biofilm (Yin et al., 2019).

TABLE 21.1 Biofilm-forming organisms and their respective site (Wilson, 2001).

Sr. No.	Microorganism	Biofilm formation site
1.	*Staphylococcus aureus*	Implantable medical devices
2.	*Staphylococcus epidermidis* and other coagulase-negative staphylococci	Implantable medical devices
3.	*Pseudomonas aeruginosa*	Lungs of cystic fibrosis patients
4.	*Escherichia coli* and other enterobacteria	Urinary catheters
5.	*E. coli*	Intestinal tract
6.	*Streptococcus* spp.	Teeth
7.	*Actinomyces* spp.	Teeth
8.	*Lactobacillus* spp.	Vagina, teeth

and allows the formation of colonies. Biofilm diseases include device-mediated infections, medical device malfunctions, and even chronic nonforeign body infections. Many infections in developed countries include myelitis, endocarditis, urinary tract infections, chronic prostatic inflammation, sinusitis, periodontitis, middle ear infections, and chronic lung infections in patients with cystic fibrosis included in illness. It is caused by biofilm-related microorganisms (Hall-Stoodley et al., 2004). Nasal infections are specifically associated with indwelling devices (Del Pozo & Patel, 2007). This poses a major public health problem, as the burden on this medical facility is enormous. Due to the inherent antibiotic resistance of sessile bacteria, biofilm diseases are difficult to treat and generally result in the failure of antibacterial treatment. Biofilm-related diseases contribute to increased mortality and patient morbidity, representing a significant financial burden, and device-mediated biofilm infections cost more than $ 1 billion annually in hospitals in the United States (Darouiche, 2004).

3.1 Device-related biofilm disease

The use of embedded medical devices has become essential in modern medicine. Apart from the excellent benefits rendered by these foreign body implants, infectious complications are frequently observed (Fux et al., 2003). Vascular prosthesis, electrodialyzer, joint prosthesis, cardiac pacemaker and heart valve prosthesis, intravenous catheter, intratracheal tube, allogeneic dysplastic orthopedic device, urinary stent, peritoneal dialysis catheter, CSF shunt, breast implant, voice prosthesis, abdominal drains, intrauterine devices, bile duct stents, tissue fillers, contact lenses, and renal shunt tubes are just a few cases that have proven important to patients. However, these devices are also associated with an increased risk of biofilm-mediated infection. These infections can be fatal and are often caused by bacterial species that are ubiquitous in soil, air, water, or human skin. Bacterial colonization on these devices can occur within 24 h and can be enhanced by host-produced conditioning films such as platelets, tissue proteins, and plasma (RaadBodey, 1992). Medical devices not only serve the main purpose of mimicking the normal functioning of vital organs, but also serve as a competitive ground for the host cells and microorganisms (Gristina et al., 1988). Device-related nosocomial infections are initiated by colonization of the material surface of medical equipment, but the origin of microbial infections is either endemic in vivo or extrinsic in vitro, followed by biofilm formation (Rutala & Weber, 1997).

3.2 Chronic biofilm diseases not related to devices

Cystic fibrosis, chronic obstructive pulmonary disease, tuberculosis, chronic wound infection, chronic sinusitis, osteomyelitis, tooth rot, bile duct infection, bacterial prostatic inflammation, endocarditis, periodontitis, and otitis media are a part of a biofilm disease that is not associated with implantable devices (Luis Del Pozo, 2017). It has been found that biofilm microflora present in soft tissues can be regularly exposed to antibiotics below minimal inhibitory concentrations. It affects the physiology of bacteria, causes phenotypic and genotypic changes in biofilms, and accelerates the spread of antibiotic-resistant strains from biofilms (Andersson & Hughes, 2014).

3.3 Malfunction of biofilm-related devices

The final stage of the inflammatory and wound healing response after transplantation of medical devices is the foreign body reaction between macrophages and foreign body giant cells. During the first 2–4 weeks after implantation of a medical device, the surface properties of the biomaterial play an important role in regulating the foreign body reaction (Glage et al., 2017a). Biofilm microorganisms that grow in a foreign body reaction environment are known to cause deterioration of biomaterials and lead to failure of clinical equipment (Del Pozo et al., 2009). Implant dysfunction presents with minor clinical symptoms such as slight soft tissue contraction and pain, as well as dysfunction due to negative inflammatory markers. The clinical manifestations of biofilm-related device malfunctions can be devastating. These include breast augmentation, capsule contraction, bile duct obstruction, endovascular catheter dysfunction, CSF shunt dysfunction, chemical deterioration and physical damage to pacemaker leads, crystalline attachment to urinary stents, and artificial joint dysfunction. It is included. All of this requires device removal, increases mortality and morbidity, and can result in additional hospital costs for the patient. Understanding these complex interactions between microorganisms and biofilms will lead to the emergence of new biomaterials and tissue engineering structures that can guide biological reactions.

4. Antibiofilm strategies

Two strategies are best known to fight against biofilms: the first one is to eliminate biofilm cells, and the second one is to prevent the formation of biofilm by blocking bacterial attachment. Simultaneously, novel hygiene protocols have come up to reduce nosocomial infections. These preventive hygiene protocols constitute cleaning of apparatus, hands, or rooms in the hospital arena. Jones et al. observed that the number of sessile staphylococci bacteria reduced by 50% when apparatus were cleaned on yearly basis as compared with uncleaned materials (Jones et al., 1995). Since some materials induce microbial contamination whereas others exhibit intrinsic antimicrobial properties, it has been proposed to use the "right materials" for the manufacturing of hospital equipment or fittings such as toilet seats, tap handles, ward entrance push plates, and door knobs. For instance, due to the antimicrobial activity of copper, a hospital in Ireland replaced door knobs in stainless steel by some in copper (Thebault and Lequeux).

4.1 Biofilm eradication

Elimination of biofilms requires high concentration of antibiotics or disinfectants, leading to severe environmental damages and emergence of multiresistance. Hence, some new strategies have been developed. One reliable strategy is the usage of quorum sensing (QS) inhibitors. The process by which bacteria achieve cell–cell communication is QS, which is mediated by small molecules secreted by microorganisms called self-inducing substances. By the inhibition of QS, the communication between the microorganisms gets disturbed. Gene modification is another strategy as seen in a study conducted by Jayaraman et al., where *lasl*, a structural gene for N-(3-oxododecanoyl) L-homoserine lactone (an autoinducer) synthetase in *Pseudomonas aeruginosa* was deleted (Jayaraman & Wood, 2008). This showed defects in biofilm formation due to loss of exopolysaccharide production. Another strategy is the usage of competitive molecules that interfere with autoinducer receptors (Lazar, in press). Nonclassical biocidal compounds can also be used. The mechanical strategy of removal of biofilm by simple friction can also be considered; however, it can be painful for the patient (Thebault and Lequeux). Enzymatic strategies could also be considered for the biofilm destruction, e.g., dispersin B (Fekete et al., 2011).

4.2 Biofilm formation prevention

This strategy constitutes the elaboration of antiadhesive and/or antimicrobial surfaces to prevent biofilm formation. It either aims to avoid bacterial adhesion or to kill cells that come in contact with the surface. To elaborate antiadhesive materials, the roughness is changed by physical modification (Mei et al., 2011) on surface energy (Churchley et al., 2008) or by immobilization of antiadhesive compounds such as polyethylene glycol (PEG) or polysaccharides (Dong et al., 2011). For the elaboration of biocidal surfaces, incorporation or covalent binding of antibiotics, silver, and quaternary ammonium is often used for immobilization. Another promising tool is the use of antimicrobial peptides (AMPs). This is because AMPs exhibit a wide range of antibacterial activity at very low concentrations and do not promote bacterial resistance. For instance, Humblot et al. immobilized gramicidin A on gold magainin I by self-assembled monolayer (SAM) methods. A reaction between the ester function of magainin and the amine function of gramicidin with the peptides has generated the antimicrobial surfaces. These surfaces showed antibacterial activity against gram-positive and gram-negative bacteria for about 6 months (Thebault and Lequeux).

5. Antibiofilm studies

Antibiofilm research suggests a good strategy to counter antibiotic research, as biofilms show increased resistance to most conventional antimicrobial agents prescribed by doctors. In recent years, the focus has been on the discovery of new bioactive compounds with desirable antibacterial properties. Recently, literature reports have pointed to the use of potential compounds that disrupt the biofilm-forming ability of microorganisms. Researchers have identified natural antibiotics due to the low efficacy of various treatments and the in vivo toxicity of available antibiotics (Cattò & Cappitelli, 2019). These approaches interfere with certain important steps associated with biofilm formation: (1) easy removal of microorganisms is possible prior to biofilm formation through a surface sensing process, keeping breakthrough cells in a floating form; (2) destruction of biofilm structure by attacking the matrix, or destruction of biofilm physical integrity by disrupting cell-cell communication; (3) helps spread biofilm by forcing a plankton state. Several natural and synthetic compounds, as well as matrix-targeting enzymes, are immobilized on the polymer surface based on the mechanism of action of prebiocide-free antibiofilms. This provides a promising bioinspired, environmentally friendly antibiofilm material that can replace or integrate with the currently mainstream biocide-based approaches (Cattò & Cappitelli, 2019).

For example, biocompatible poly(lactic-co-glycolic acid) coatings containing natural clove oil or eugenol exhibited efficient biofilm inhibition and less toxicity on solid surfaces against Enterohemorrhagic Escherichia coli O157:H7 (EHEC) (Cattò & Cappitelli, 2019). Natural compounds such as zosteric acid and salicylic acid were covalently grafted onto the surface of low density polyethylene, reducing *E. coli* adhesion and ultimately reducing biofilm formation by

73%. Phytochemicals are known to inhibit QS mechanisms by blocking QS inducers such as AHL and autoinducers (Ciric et al., 2019). Antiadhesion properties of *Psidium guajava* L. ethanol and acetone extracts have been studied (Razak & Rahim, 2003). Biosurfactants prevent the formation of biofilms by reducing the hydrophobicity of the cell surface, interfering with cell adhesion capacity through inhibition of the electron transport chain, membrane disruption, and controlling the energy demand of cells (Satpute et al., 2016). Analysis of the effects of the biosurfactants *Pediococcus acidilactici* and *Lactobacillus plantarum* on the QS signaling molecule revealed downregulation of biofilm-related genes such as autoinducer-2 and cidA, respectively (Yan et al., 2019). Antimicrobial Peptides (AMPs) esculentin-1a, esculentin-1a(1-21)NH_2 corresponding to the first 20 residues of esculentin-1a, as well as its diastereomer, Esc(1-21)-1c inhibits *P. aeruginosa* biofilm formation through its membrane-destroying activity (Casciaro et al., 2019). Efforts have been made to create specifically targeted multidomain AMPs containing species-specific peptides linked to broad-spectrum antibacterial killer peptide domains (Sztukowska et al., 2019). Researchers have set some of the limitations of natural antibiofilm drugs, including unexplored toxicity, degradation by host proteases, long amino acid sequences that boost production costs, and limited knowledge of the structure–function relationships of these compounds. Identified. Attempts to overcome the implementation (Fjell et al., 2012) include deletion and/or substitution of amino acid residues, cyclization, retro-inverso-peptide designs using D-enantiomer amino acids and hybrid construction, sequence truncation, or computational methods (Fuente-Núñez et al., 2016). The next step in this area is to confirm the activity of antibiofilm compounds in animal models of biofilm infection. Although some success has been achieved so far, much work must be done to establish these molecules as viable alternatives to antibiofilm agents and traditional antibiotics.

5.1 In silico

5.1.1 Molecular targets in **Pseudomonas aeruginosa**

An opportunistic pathogen *P. aeruginosa* is capable of performing group behaviors, including biofilm formation and swarm motility. HHQ (2-heptyl-4-hydroxyquinoline) and PQS (*Pseudomonas* quinolone signal) perform a compelling functions during biofilm provenance as well as in the virulence factor production and its regulation in *P. aeruginosa* (Storz et al., 2012). Transcriptional regulator PqsR is activated by PQS and HHQ, which enhances its own biosynthetic operon pqsABCDE (Wade et al., 2005). Production of signal molecules rapidly increases because of this autoinduction, and their diffusion into the environment leads to a more controlled and coordinated population behavior. In addition, HHQ and PQS show involvement in expressing various genes encoding for virulence factors such as rhamnolipids (a class of biosurfactants with sugar moiety being rhamnose), LasB (elastase B), HCN (hydrogen cyanide), LecA (lectin A), and pyocyanin (Déziel et al., 2005; Van Delden & Iglewski, 1998). Furthermore, PQS has some specific capability in regulating biofilms, even though its mechanism of action is not yet clearly understood (Diggle et al., 2003; Yang et al., 2009).

To identify the potential drug targets of *P. aeruginosa* (PAO1), Rajkumari et al. (2019) docked natural QS signaling molecules and 5-HMF to the transcriptional regulators of RhIR, LasR, and QS circuit. In this study, they used the crystal structure of *P. aeruginosa* regulatory protein LasR bound to its autoinducer solved at 1.80 Å (PDB ID: 2UV0). For RhIR, as the experimentally determined structure was unavailable, they have utilized their previously published homology modeled structure (Cordeiro et al., 2020a; Rajkumari et al., 2018, 2019) for docking studies. Using the AutoDock Vina program (https://vina.scripps.edu/), they docked the autoinducers C4-HSL, 3-oxo-C12HSL, and 5-HMF onto the binding site of RhIR and LasR proteins. LigPlot+ was used for analyzing hydrophobic and hydrogen bond interactions in between proteins and ligands. Based on docking scores and binding interactions, the best protein–ligand complex was chosen and further subjected to Molecular Dynamics Simulations in GROMACS 5.0.4 with Gromos force field. To further estimate the binding free energy and stability of the docked complexes, MM/PBSA method was used. Gibbs free energy was calculated, and the Visual Molecular Dynamics (VMD) tool was used to analyze MD trajectories (Rajkumari et al., 2019).

This in silico study clearly shows that the natural autoinducer molecules C4-HSL and 3-oxo-C12HSL exhibit a good binding affinity with the cognate receptors RhIR and LasR with the docking energies of −5.4 kcal/mol and −8.8 kcal/mol, respectively. In the catalytic site of RhIR, there is a formation of an H-bond between RhIR and Ser135 (2.99 Å). Similarly, in the catalytic site of LasR, four hydrogen bonds were formed by 3-oxo-C12HSL Asp73(2.6 Å), Ser129(2.5 Å), and Tyr56(2.8 Å, 3.1 Å). The RMSD (root mean square deviation) between natural ligands and biomolecular complexes was within 2 Å, indicating the docking accuracy and precise binding pocket. Molecular dynamics simulation output shows the stability of the LasR-HMF and LasR-3-oxo-C12HSL complexes. The binding free energy estimation from the molecular dynamics simulation trajectory shows that 5-HMF binds stronger with LasR than RhIR (Rajkumari et al., 2019).

5.1.2 Molecular targets in **Staphylococcus aureus**

Biofilm-related resistance of *Staphylococcus aureus* to persistent infections and conventional antibacterial drugs has elevated. Hence a demand is rapidly rising for discovery and creation of various advanced therapeutic alternatives as there is low availability of new antibacterial drugs.

To determine the antibacterial activity of myrtenol against *S. aureus*, Cordeiro et al. (2020a, 2020b) measured the MIC (minimum inhibitory concentration), MBC (minimum bactericidal concentration), and MBEC (minimum biofilm eradication concentration). In addition to this, they evaluated the effect of myrtenol in combination with existing FDA-approved antibacterial drugs to assess their synergistic activity against *S. aureus* biofilm formation. They also performed molecular docking and pharmacokinetic predictions to identify possible drug targets and drug likeness (Thomsen & Christensen, 2006).

They performed molecular docking to determine the possible interactions between myrtenol and PBP2 (penicillin-binding protein 2), a transpeptidase enzyme that plays a crucial role in the cell wall biosynthesis. Molecular redocking was performed to validate the efficiency of the docking algorithm in reproducing the crystal pose. Binding energy displayed by myrtenol with PBP2 was 52.3 kcal/mol, and on the data on the active site interactions of the enzyme, hydrogen bonding interactions were performed by the hydroxyl group of myrtenol with residues of Thr600 and Ser403, and with Lys406 van der Waals, hydrophobic interactions were observed. These set of interactions play a vital role as they are executed by the β-lactam drugs after they are get effectively anchored at the active site of PBP2 (Lim & Strynadka, 2002; https://vina.scripps.edu/).

The complete elucidation of the antimicrobial mechanism of action of myrtenol is not yet done. Here the results indicate that for myrtenol to act against *S. aureus*, PBP2 is a possible target. Thus, the substance would show interference in the bacterial cell wall synthesis, therefore leading to cell death (Lim & Strynadka, 2002; https://vina.scripps.edu/).

5.1.3 Molecular docking methods

To discover potential lead molecules and understand their possible drug targets against *P. aeruginosa* and *Staphylococcus*, several researchers used virtual screening and molecular docking as the primary in silico approach. Many different tools, algorithms, and software were utilized for this.

5.1.3.1 AutoDock Vina

AutoDock Vina is an open-source program developed by Scripps Research Institute for molecular docking purposes (https://www.ebi.ac.uk/thornton-srv/software/LIGPLOT/). It substantially elevates the moderate accuracy of predictions related to binding mode in comparison to AutoDock4. This software was successfully used for evaluating ligand molecules binding onto sites of LasR and RhlR proteins to predict the best possible docked conformations. The best-docked poses were analyzed based on their docking scores (Rajkumari et al., 2019).

5.1.3.2 LigPlot+

LigPlot+ is the successor to the LIGPLOT program. It is an open-source software used to study protein–ligand complex interactions. This algorithm automatically generates a 2D ligand–protein interaction diagram for the given protein–ligand docked complex. It mainly displays hydrogen bonds and their length and nonligand residues in hydrophobic contact(s). Hydrophilic contacts are characterized by spiked arc radiating toward contacted ligand atoms, whereas dashed lines in between involved atoms indicate hydrogen bonds. Spikes radiating back represent the contacted atoms (https://www.gromacs.org/About_Gromacs) (Fig. 21.2).

5.1.3.3 GROMACS

GROMACS is a package full of versatility and is mainly used to perform molecular dynamics simulations of proteins, nucleic acids, and lipids using Newtonian laws and equations of motion. It helps monitor the physical movements of atoms and molecules at specific time intervals. It also helps to study the behavior of systems such as protein–ligand complexes under a particular temperature, pressure, and pH (An Andy et al., 2021).

5.1.4 Recent advances

The current status of antibiofilm drug discovery significantly focuses on the process involved in regulating biofilm formation and inhibiting it. Hence an improved understanding of biofilm formation is essential for discovering various new potential drug targets during the development of antibiofilm agents. On the other hand, a detailed mechanism of action was not well studied for certain existing antibiofilm agents. Therefore, realizing the impact of these agents on biofilms can pave the way for discovering various new targets and pathways to be modulated by new agents.

The target genes, processes, and pathways necessary for biofilm formation can be well explored by omic approaches, including transcriptomics, genomics, and proteomics.

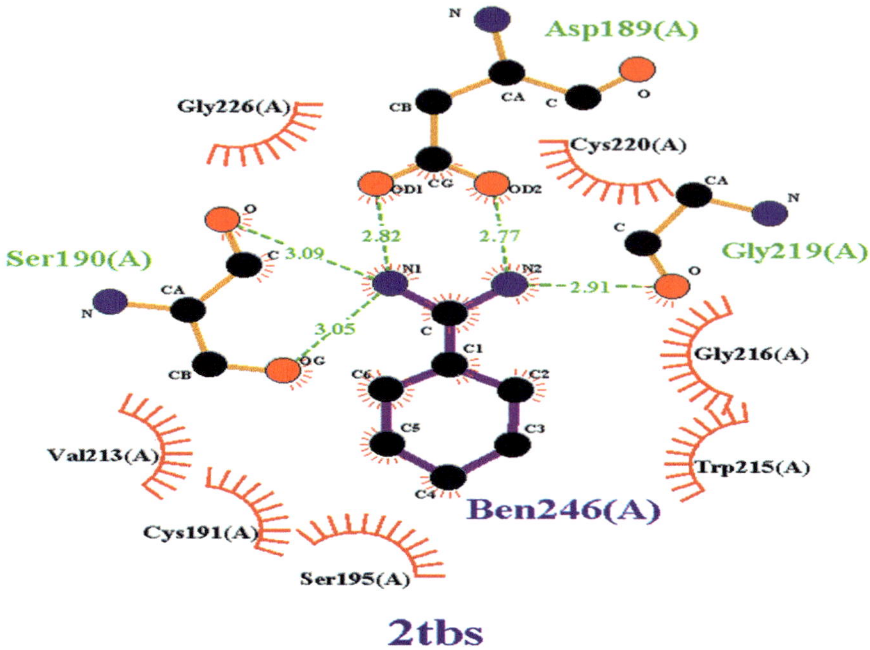

FIGURE 21.2 Ligplot of Benzamide in 2TBS crystral structure. *Courtesy: https://www.ebi.ac.uk/thornton-srv/software/LIGPLOT/gif/ligplot_2tbs.gif.*

Generally, these approaches focus on the abundance of different biological molecules such as metabolites, nucleic acids, and proteins in various conditions. Through the comparing molecular changes between diverse conditions in bacteria such as planktonic growth versus biofilm, ± an antibiofilm agent, mutant versus wild-type strains, it can be hypothesized that the changes observed indicate the treatment/condition and possibly affirm the details regarding potential causes behind the resistance and their mechanisms (Noble, 2006).

5.1.4.1 Machine learning

Machine learning is considered to be a more flexible approach as instead of focusing on the target to infer antibiofilm activity, it pays much attention to the properties of molecules. Machine learning is a set of systematic and powerful statistical methods used to make predictions in diverse scenarios, including the prediction of new antimicrobial agents specially targeted to biofilm infections. Usually, an algorithm is created using large significant data sets to learn a relationship between the characteristics expressing the data and the prediction task at hand. With respect to small molecule antibiofilm agents, these characteristics (also known as molecular descriptors) include lipophilicity, steric size, and 3D structure yet there are hundreds of physical–chemical parameters that can be used (Fig. 21.3). The aim of a perfect machine learning model is being unspecified to unseen examples; hence the precision of predictions is generally assessed on examples, which are not used for training (a validation set). Many algorithms have been developed to extricate complex linear and nonlinear

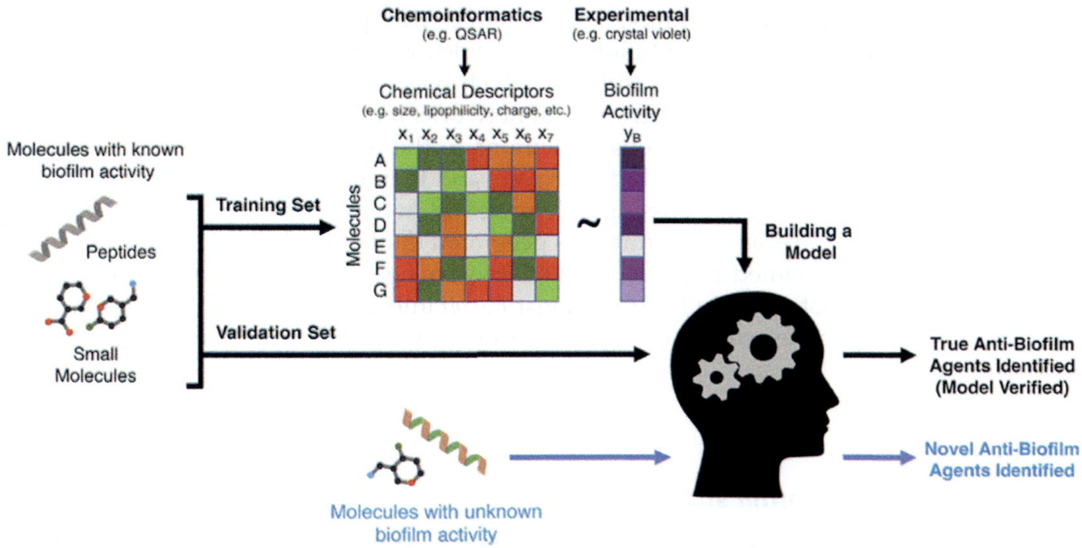

FIGURE 21.3 A schematic diagram of how machine learning can be used to discover new antibiofilm agents.

relationships and modify them into predictive models, including logistic regression, random forest, support vector machines (SVMs), and neural networks (Di Luca et al., 2015; LeCun et al., 2015). The Biofilm-active AMPs database (BaAMPs) was developed to provide access to researchers to a source of peptides to train machine learning models with antibiofilm activity (Gupta et al., 2016). In a recent study, BaAMPs was used to choose 178 antibiofilm peptides for training an SVM model; however, the non−antibiofilm set consisted of randomly produced peptides from all SwissProt database sequences (Sharma et al., 2016).

Likewise, BaAMPs database was used to select 80 antibiofilm peptides to train an SVM model, whereas only QS peptides with no antimicrobial effects were under nonbiofilm set (Haney et al., 2018b). Even though the model precisely predicted known antibiofilm peptides, its potential to predict unknown was not verified. However, it is essential to consider that such validations assume a reproducible standardized assay for evaluation, with in vitro MIC in contrast to biofilm inhibitory concentrations (BICs) and minimal biofilm inhibitory concentrations (MBICs) (Artini et al., 2018). Machine learning models that predicted the antibiofilm activity of naturally occurring essential oils were effectively implemented (Artini et al., 2018; Patsilinakos et al., 2019). The essential oils for two strains of *S. aureus* and two strains of *S. epidermidis* were assayed and utilized the results to train separate models for each strain. The value of training strain-specific machine learning models was highlighted by the fact that there is huge variation of antibiofilm activity of each essential oil (Lo et al., 2018).

The other approaches to represent molecules numerically for the application in machine learning are through fingerprinting and chemical graphs, which capture the atomic structure and connectivity of the molecule (Srivastava et al., 2020). These representations are commonly valid for small molecules instead of peptides with complex secondary structures.

In a study, a hybrid random forest model based on QSAR-type descriptors and chemical fingerprints to identify prospective antibiofilm molecules was trained (Stokes et al., 2020). A 10,208-unit chemical fingerprint was isolated, which was further combined with QSAR descriptors to produce a hybrid classifier. Neural networks are a category of machine learning models that impersonate the operations of neurons in the brain. Explicitly, they permit the models to learn the features through hidden layers and then utilize them to perform the prediction task. A directed message passing a deep neural network to learn a class of chemical fingerprints based on the graph structure was generated (Stokes et al., 2020; Yang et al., 2019). Even though it was not aimed to discover an antibiofilm agent, they predicted and validated the capability of antibiotic halicin for use against *E. coli* infections. This study also showcased the utilization of neural networks and feature learning, which can be applied to identify new antibiofilm agents.

5.2 In vivo

5.2.1 In vivo model biofilm studies

With regard to infection and pathogenesis, there is a continuous interaction between the host and the microorganism and between the microorganism itself. These interactions are complex and dynamic and can affect the fate of infection. Several approaches have been proposed to study this complexity of bacterial biofilms in vivo. Knowledge of physiology and biofilm formation gained through in vitro experiments is an important part of early research on biofilms. However, the urgent need to investigate biofilm-related infections and develop treatment strategies has led to a translational approach that allows the discovery of pathogenicity and pathogenicity determinants and the identification of new therapies. The in vitro biofilm model is a powerful tool in terms of easy treatment and reproducible testing of the effects of factors, but it is the interaction between the host's immune system and the bacteria that make up the biofilm. Therefore, in vitro and ex vivo studies are complemented by in vivo experiments to investigate the immune response to bacterial biofilm infections. Developing a robust in vivo model is the key to supporting in vitro results and is an important step in testing of new therapies and devices on the road to clinical implementation.

The earliest example of an in vivo model for analyzing biofilms was a rabbit for studying osteomyelitis (Scheman et al., 1941). Since then, much research effort has been devoted to the development and exploration of in vivo models. In vivo models of biofilm formation associated with tissue infections, device-mediated infections, and systemic infections are currently being found (Labeaux et al., 2013). Pulmonary infections, chronic wounds, ear, nose, and throat infections are notable tissue infections studied using an in vivo model. These models are also used in orthopedic implants, subcutaneous injection devices, and device-induced infections associated with the urethral catheter. Mammalian and nonmammalian models have been developed for this purpose. The nonmammalian model eliminates certain practical problems such as low cost associated with the breeding and maintenance of small and large animals (Ziegler et al., 2016) but fails to show a complex immune response and optimal growth temperature. It is narrow and has a short lifespan, making it inappropriate for the investigation of relevant chronic infections. On the other hand, an in vivo mammalian model consisting of rats, mice, rabbits, and pigs is used as a representative presenter of human pathology through similar anatomy, healing process, and immune response (Barre-Sinoussi & Montagutelli, 2015). However, there are situations where nonmammalian models are ideal such as high-throughput screening study (Letamendia et al., 2012).

5.2.2 *Animal models for* **Pseudomonas aeruginosa** *infections*

P. aeruginosa is an important opportunistic pathogen that can cause acute respiratory infections in immunocompromised and cystic fibrosis patients. These bacteria are encapsulated in a biofilm structure that can withstand a variety of environmental conditions such as immune response and antibacterial therapy. Therefore, *P. aeruginosa* is ubiquitous not only in the environment but also in the in-hospital environment.

5.2.2.1 Invertebrate models

Invertebrate models are inexpensive and relatively easy to set up and maintain. Although the invertebrate model lacks an adaptive immune response, it can be considered a simplified model of innate immunity and is highly responsive to infectious pathogens. The three major invertebrate models of *P. aeruginosa* are described in the following (Fig. 21.4).

5.2.2.2 Nematodes

Caenorhabditis elegans are nematodes that live in the soil. Easy adaptation to laboratory conditions, simple body organization, and a male–female way of life make it one of the broadest subjects of study for cell biologists and geneticists. Since nematodes mainly eat bacteria, they may ingest pathogens such as *P. aeruginosa*. A secondary metabolite called piocyanin is the main virulence factor for *P. aeruginosa*. It has been widely studied in C. elegans. Elevated levels of pyocyanin released into the environment have been shown to be positively correlated with clinical lung deterioration (Mowat et al., 2011). The use of *C. elegans* to study the effects of pyocyanin on the host was reported by Tan et al. (1999).

Infected *C. elegans* respond to pyocyanin with mitophagy, a lysosomal degradation of damaged mitochondria (Kirienko et al., 2015). This mechanism allows the organism to survive and resist infection. Perhaps bacterial killing does not correlate directly with autophagy but plays a role in the destruction of necrotic cells from infected intestines caused by bacterial infection. Autophagy is induced by bacteria in the gut, resulting in upregulation of the EGF

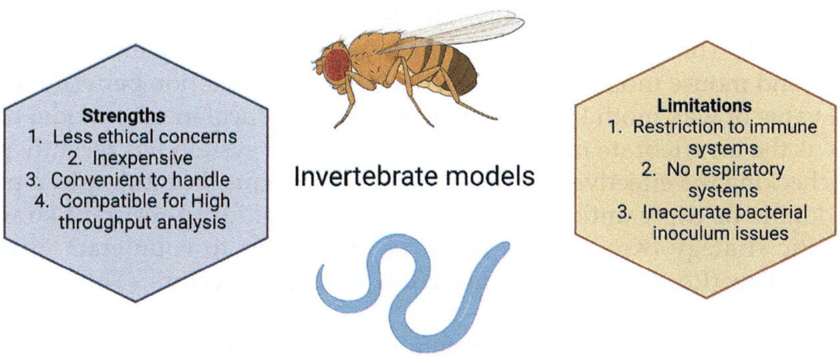

FIGURE 21.4 Overview of strengths and limitations using invertebrate animal models for the study of *Pseudomonas aeruginosa* infections (Lorenz et al., 2016).

homologous LIN3. This activates the MAP kinase-ERK cascade via the EFGR homolog LET2. The intestine is easily destroyed by bacteria, and without this mechanism, it can lead to a systemic infection. It emphasizes the value of considering elegance as a model of infection. *C. elegans* can be used to test the potential for newly induced antibiotic-resistant *P. aeruginosa* (Cabot et al., 2016). Therefore, *C. elegans* can be seen as a bridge between in vitro systems such as cell culture and the most complex in vivo vertebrate models. This helps in the evaluation of new compounds and leads to the establishment of translation potential. Metabolic experiments at *C. elegans* also show the potential of this model system.

While the *C. elegans* model has proven to be a powerful tool for infection research, it also has some drawbacks. First, worms feed on *P. aeruginosa*, but because the number of bacteria that can infect a worm is limited, it cannot affect the absorption of bacterial pathogens. For example, experiments using a high-density transposon mutant library for efficient genome-wide screening of infection-related genes can cause serious bottleneck problems. Also *C. elegans* lack important immune cells and appear unable to signal through the TLR/NF-κB pathway, a cellular signaling pathway that is highly valued in both the plant and animal kingdoms (Irazoqui et al., 2010).

5.2.2.3 Insects

Insects such as *Galleria mellonella* and *Drosophila melanogaster* have certain defense mechanisms that nematodes do not have. This is advantageous for use as an infection model for invertebrates. Pathogen phagocytosis, nodule formation, and encapsulation are characteristic cellular responses of insects to bacterial infections. In addition, the humoral response of insects forms the induction of coagulation, melanization, and antibacterial peptides. Insect blood cells, such as mammalian neutrophils and macrophages, can function as professional phagocytes that can ingest and kill bacteria (Kounatidis & Ligoxygakis, 2012).

Greater wax moths and wax moths belong to the order Lepidoptera. It is also known as a comb moth because female wax moths lay eggs in their nest boxes, eggs hatch, and larvae eat pollen and wax on combs. They have proven to be notable model organisms and are increasingly used in bacterial infection studies. They can be cultivated at 37°C, a natural mammalian host temperature. Caterpillars are available at feed stores and do not require special breeding or care in the laboratory. Caterpillar melanization can be used as a parameter to indicate a fatal infection (Koch et al., 2014). The potential of the *Galleria* larvae model to test *P. aeruginosa* virulence factor, which also affects mammals, has been previously evaluated (Jander et al., 2000). Thirty-two variants of pathogenicity-related genes were compared between *Galleria* and mouse models and showed a positive correlation between them. This emphasizes the value of wax moth larvae as a simple and practical in vivo model that is as close to mammals as the invertebrate model. The *Galleria* in vivo system has recently gained popularity by prechecking the effectiveness of antibiotic ingredients. For example, a potential combination of double or triple antibiotic therapy against multidrug-resistant strain revealed a new therapeutic strategy (Krezdorn et al., 2014). Strategies to counteract virulence factors such as pyoverdine (Ross-Gillespie et al., 2014) and QS effectors (Thomann et al., 2016) were investigated in the *Galleria* model.

D. melanogaster is a favorite subject of genetics and developmental research. Also known as *Drosophila*, its potential use as a model organism has not been fully investigated and is used in the study of bacterial infections. There are two main models of infection: (1) a feeding model in which a fly ingests a pathogen from the intestine and the etiology can be studied for several days, and (2) a septic injury model in which a fly is stung. They carry bacteria with needles. The power of *Drosophila* as a model for *P. aeruginosa* infection is demonstrated by coinfection experiments using QS variants (Schuster et al., 2013; Lutter et al., 2012). Isolates showing such mutations were compared to wild-type (WT) *P. aeruginosa*. Variants of the nicking system behaved almost like Wt, but in the feeding system, all mutants are nontoxic and require the presence of both QS systems to infect flies. This shows the diversity of the *Drosophila* in vivo system (Lutter et al., 2012).

The *D. melanogaster* model facilitates comparison of multiple genes involved in antioxidant reactions. Recently, FldP, a long-chain flavodoxin from *P. aeruginosa*, was discovered and characterized. It is presented as a virulence factor because it provides bacterial survival to the fly's infectious system and is required for most fly mortality (Moyano et al., 2014). In summary, all these examples demonstrate the importance of fruit fly and wax moth larvae as models of infection. However, the vertebrate model system is essential for more detailed characterization of molecular interactions between the host and bacteria (Fig. 21.4).

5.2.2.4 Vertebrate models

In the arms race between the host and the pathogen, some pathogenic microorganisms can overcome the barriers of the innate immune system. In primitive animals, host death is caused by the defeat of the innate immune system, whereas in higher animals, the adaptive immune system efficiently handles infection under these conditions.

5.2.2.5 Zebrafish

Adult zebrafish Daniel Orerio has both an innate and adaptive immune system. However, depending on the stage of fish development, there is a temporary disruption between the maturation of the two immune systems. Innate immune system or adaptive system "adult" is involved only in the immune response elicited by *Pseudomonas*, depending on the stage of development in which the infection occurred. Zebrafish have gills instead of lungs and are not directly associated with CF-related diseases, but their traceability and genetic characteristics are of importance. This shows that Rerio is a valuable in vivo model system. Zebrafish-fertilized eggs are transparent and fast-growing, so they can be used in combination with appropriate fluorescent markers to monitor the progression of infection in real time. It works by lockers. He emphasized that the bacteria can be injected intravenously in the case of systemic infections and into specific sites such as the posterior ventricle in the case of local infections (Rocker et al., 2015). Bacterial microcolonization, which represents the early stages of biofilm development, was observed in *P. aeruginosa* introduced into the posterior ventricle of the larva to avoid systemic spread and rapid death of the larva. The diversity of zebrafish as an infection model has been demonstrated by Peter et al. SOD2 detected by *P. aeruginosa* zebrafish infection has been shown to be upregulated in hematopoietic and myeloid hematopoietic organs (Peterman et al., 2015).

Detoxification of ROS in these cells has been shown to be important by removing SOD2 and infecting larvae at an early stage of development where only macrophages exist as professional immune cells. These larvae were susceptible to systemic infections and increased mortality. At the same time, zebrafish can be used to investigate the defense mechanisms present in pathogens against ROS. For example, the *P. aeruginosa* transcription factor PA2206 is activated by

ROS stress and also triggers 58 genes that successfully infect zebrafish (Reen et al., 2013). In addition, zebrafish can be used to characterize new concepts and alternative treatment strategies for *P. aeruginosa* infections. Due to the increasing complexity of vertebrate models, zebrafish raise ethical issues related to animal welfare. Apart from that, the temperature of the zebrafish shell is 28°C, which is different from the temperature of the human body.

5.2.2.6 Mammalian models

Mammalian models are specifically used to study a variety of pathogenic symptoms of *P. aeruginosa* infections, including biofilm-mediated infections. Various types of mammals have been used as model systems for studying infectious diseases from chinchillas to pigs and nonhuman primates.

5.2.2.7 Murine models (mouse, rat)

A small rodent model, primarily a mouse, was used to obtain most information about *P. aeruginosa* infection. Bacteria can be administered intranasally, intratracheally, or through the oropharyngeal route of administration. This is a factor that influences final lung colonization (Fisher et al., 2014), infection level, bacterial strain, and outcome of infection. It has little effect on the genetic background of the affected host. Systematic examination of various mouse strains revealed a significant difference between the strains tested and the immune response they exhibited (De Simone et al., 2014). Molecular interactions between pathogens and hosts can be studied in detail by mapping and characterizing these genetic differences.

The role of bacterial factors has been widely characterized in vitro and confirmed using the acute mouse infection model. For example, hemolytic phospholipase C (PlcH), which is involved in the degradation of phosphatidylcholine, a component of pulmonary surfactant, has been mentioned in a recent study (Jackson et al., 2013). It is emphasized that metabolites such as choline regulate the ANR-dependent promoter induced under hypoxic conditions. By the way, PlcH and its regulator GbdR were involved in biofilm formation of lung epithelial cells and colonization of the mouse respiratory tract in acute lung infections. Recently, a new two-component system BfmS/BfmR was discovered that controls quorum detection (Cao et al., 2014). The response regulator BfmR is negatively regulated by the sensory kinase BfmS. Removal of BfmS causes upregulation of BfmR and then rhlR, leading to biofilm maturation, which, along with reduced cytotoxicity to mouse lung epithelial cell lines, is a model of acute mouse infection. It showed a decrease in fitness.

Mammalian models are specifically used to study a variety of pathogenic symptoms of *P. aeruginosa* infections, including biofilm-mediated infections. Various types of mammals have been used as model systems for studying infectious diseases from chinchillas to pigs and nonhuman primates.

The effects of host factors on the establishment and progression of infection can be studied in mouse models. For example, in mice lacking lung-specific vascular endothelial growth factor (VEGF), severe infections and increased inflammatory responses were observed (Breen et al., 2013). The development of antibacterial strategies by discovering new bacterial etiological pathways and metabolic adaptations can be tested in an in vivo model system. Migiyama and others convincingly demonstrated AHL-lactase AiiMP in a mouse model of acute pneumonia. Expression of *P. aeruginosa* by bacterial strains facilitates the investigation of the in vivo efficacy of QS inhibitors (Migiyama et al., 2013). Benefits that mice offer as a test system include economic factors such as acquisition, care, housing, or genetic variation or alteration. Despite all these aspects, the small size can complicate certain experimental procedures. For example, the rat lung function test is cheaper than the mouse due to its large body size (Fig. 21.5).

5.2.2.8 Combination of different animal models

Collaborative research on various animal models has proven useful. This highlights the possibility of combining different in vivo infection models for an integrated view of the potential pathogenicity of *P. aeruginosa*. The nonvertebrate model is used to screen for new antibacterial compounds that provide information on efficacy and toxicity. You can then advance promising candidates to more complex mammalian models. A comparative study was conducted by Dubern et al. to understand the pathogenicity of C. elegans, and P. aeruginosa transposon mutants in various animal models such as melanogistor, mouse models. In some mutants, the pathogenicity profiles between the models were consistent, while in other mutants, variability was observed (Dubern et al., 2015). This highlights the possibility of combining different in vivo infection models for an integrated view of the potential pathogenicity of *P. aeruginosa*.

5.2.3 *Animal models for* Staphylococcus aureus *infections*

S. aureus is an important human pathogen that causes a variety of clinical infections. It is the leading cause of bacterialemia and infective endocarditis, skin and soft tissue infections, and chest lung and device infections. It is a symbiotic bacterium and a human pathogen. It is known that 30% of the population is colonized by *S. aureus*.

5.2.4 *Advantages and limitations of animal models*

The use of animals for research has been a long-standing practice and is the subject of ongoing debate in the scientific community. Due to the striking similarities between humans

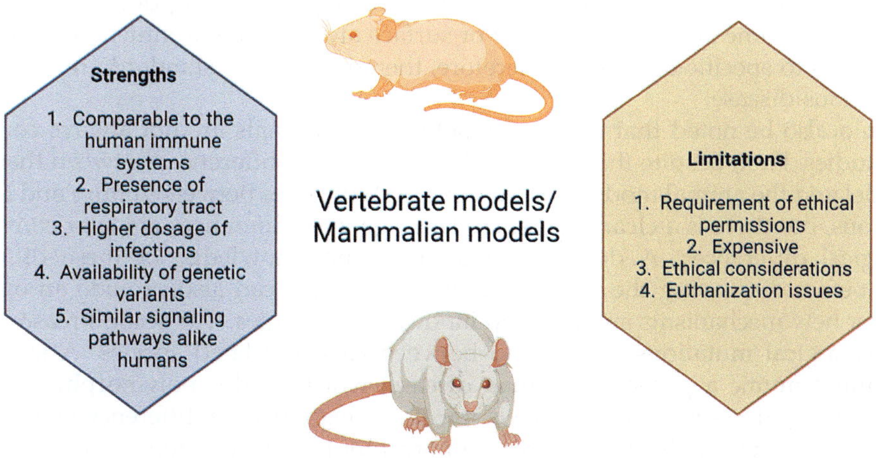

FIGURE 21.5 Overview of strengths and limitations of vertebrate and mammalian in vivo systems in *Pseudomonas aeruginosa* infection research (Lorenz et al., 2016).

and animals, especially mammals, both anatomically and physiologically, researchers have introduced new treatments for animals before testing them in humans. However, not all results obtained in animals can be extrapolated directly to humans. The right to use animals for human benefit, for example, in research, is still controversial as it can harm animals.

In vivo models have been used to answer a variety of scientific questions, from preliminary science to the development and evaluation of new vaccines or therapies. Animal species are also based on being susceptible to human illnesses such as type 1 diabetes, allergies, cancer, high blood pressure, myopathy, and epilepsy, in addition to infectious diseases. Due to the similar mechanisms involved in these diseases, 90% of veterinary drugs are very similar to those used in human treatment. Trials conducted solely on in vivo models are the reason for the great progress in medical research. Banting and McLoed won the 1921 Nobel Prize for establishing a treatment for canine type 1 diabetes. Cell therapies that use stem cells for tissue regeneration have been developed and tested in animals. After being tested in animals, even various surgical techniques have been developed and improved.

Complex organisms, such as humans and other mammals, have organs that have performed different physiological functions. Relationships between organs depend on hormones, circulatory factors, cell networks, and signal transduction between cells in all compartments. To explain and understand the mechanism, it is necessary to investigate at all levels such as molecules, cells, and organs. All three organizational levels can be examined in the in vitro model. However, the whole organism needs to explore the physiological functions and systemic interactions between organs. Hypotheses and models from in vitro study need to be tested and evaluated throughout the organism. If not, it remains speculative. Despite discussions by the Animal Rights Commission, there is still a long way to go to completely replace animal models in vivo.

Although some models of biofilm-related infections are lacking, the use of various in vivo models provides extensive information on most biofilm-related infections over the past two decades. Multiple models per infection to answer questions about initial adhesion, various surface assessments, and prophylactic approaches are shown in the multiple models used in the study of CVC-related infections. An in vivo model is being developed. Depending on the host's immune system, device size or surface area, and environment, each model provides answers to specific questions. Therefore, there is no "gold standard" model specific to each infectious disease.

It should also be noted that the results obtained in animals are not always confirmed in human studies. First, despite the great similarity, there are differences between the given human model and the animal model. For example, 95% of genes between mouse and human are homologous, but there is a clear difference. These genetic differences are later converted into physiological differences. A deep understanding and knowledge of these differences is required before disproving the value of animal models. It can also provide an opportunity to discover new mechanisms and develop innovative strategies. The second reason is genetic and physiological mutations within and between closely related species. A single animal model cannot mimic a particular human disease, which is also polymorphic between patients, but the strain–species difference is novel and explores the difference between disease onset and host response. It offers a great opportunity to develop step-by-step treatments.

The in vivo model has certain limitations. First, the immune system of animals, especially small rodents, is not comparable with the human immune system. Special equipment is required to evaluate biofilms in vivo to prevent the possibility of infection from other nearby colonies. The use of in vivo models is also considered expensive. In general, these studies are short-term and are conducted with a small sample size. In addition, regulatory in vivo models represent healthy animals that differ significantly from those observed in patients requiring implants or intervention procedures. For example, while the prevention of bacterial biofilms (*P. aeruginosa*) in skin repair is being studied, diabetic mice show impaired wound healing and have recently been reported to be more susceptible to bacterial infections than nondiabetic pairs (Lazurko et al., 2020). Another important parameter that can be considered a disadvantage of biofilm in vivo models is based on various methods for validating and quantifying biofilm formation.

Finally, we must also recognize the ethical limitations of developing an in vivo model that accurately represents human disease. The European Directive has created a regulatory framework for all animal testing. First, if a non–animal-based experimental approach is available, there is no need to use animal models. Second, we need to minimize the number of animals we use. Third, all necessary steps must be taken to reduce the harm done to animals throughout the research process. All projects, including the use of animals, are evaluated against these standard principles known as the Three Rs (Replacement, Reduction and Refinement).

Therefore, research into relevant, carefully designed, well-characterized, and controlled animal models is an essential step in basic discoveries, testing of biological-level hypotheses, and evaluation of human data. Animal welfare is taken into account as more reliable and informative in vivo models need to be improved. These two goals need to be balanced to produce high-quality science (Barre-Sinoussi & Montagutelli, 2015).

5.2.5 *Alternate models*

Since experiments on animal models are considered to be immoral and time-consuming and most of the time do not resemble human physiology, the researchers are now coming up with new models and animal-free methods, which are more relatively more relevant to human health to study diseases and to test products. These alternatives to in vivo models include advanced tests such as in vitro tests using human cells and tissues and sophisticated computer-modeling techniques popularly known as in silico models. These methods are quick, and research is not hampered by species differences.

5.2.5.1 In vitro models

A simplified in vitro model is important for answering questions about biofilm structure and physiology. Low cost, ease of setup, and high-throughput screening trends are the advantages offered by in vitro models. Based on the restoration of growth medium and the nutrients, these models are classified into static or dynamic type. Static models are generally used to substantiate the formation of biofilm and biomass accumulation (Barre-Sinoussi & Montagutelli, 2015).

Since the medium is not changed during the culture period, the nutritional supply of the static model is limited. These models are simple, cheap, and reproducible. In addition, they are often used to study the growth dynamics of biofilm and antibiofilm compounds (Salli & Ouwehand, 2015). The dynamic model creates an environment for biofilm growth that mimics the natural environment. This is made possible by a constant supply of nutrient throughout the assay. These models are used for long-term analysis of biofilm growth dynamics as the medium is continually updated and metabolites are discarded. The use of in vivo models has many advantages, such as nutrients for biofilm growth, which can lead to its incomplete maturation, and the lack of an immune response is another limitation of these models (Salli & Ouwehand, 2015; Tolker-Nielsen & Sternberg, 2014).

Currently, there are several commercially available in vitro models for testing biofilms. Microtiter plate-based system with 12-, 24-, or 96-well plates are the commonly used one for studying biofilms (Tolker-Nielsen & Sternberg, 2014). The cultured biofilm grows on the bottom and sides of the microtiter plate, or on specific surfaces placed in the wells of the plate. This system acts as a static model because the medium does not flow in or out (Roberts et al., 2015). The main drawback of these models is that they oversimplify the in vivo environment without in vivo simulation conditions during the assay procedure. The environment in the well changes over time, as nutrients gradually decrease and signal molecules accumulate. Due to the advantages of these plate-based methods, they are regularly used in research. This plate-based system is cost-effective because it uses only a small amount of reagents. It can be used to run many tests at the same time, demonstrating its usefulness in screening tests for further research (Labeaux et al., 2013). Therefore, these models can be used to explore new antimicrobials, disinfectants, and other compounds such as QS modulators with antibiofilm activity against plant extracts and biofilms (Heersink et al., 2003). The method of quantifying biofilms by the microtiter plate-based method is variable and relies on indirect measurements. Crystal violet staining, an XTT reduction assay used to check biofilm viability, is another major drawback of this method (Niu & Gilbert, 2004).

In vitro microplate-based methods can be supported on multiple devices such as biofilm ring tests and Calgary plates (Peeters et al., 2008; Sidrim et al., 2019). The biofilming test is performed on a microtiter plate (Goeres et al., 2009), and an assessment of biofilm growth is performed by quantifying the immobilization of paramagnetic beads by adherent cells and biofilm polymer matrix. This device can be used to study the dynamics of biofilm formation by a variety of bacteria (Goeres et al., 2009). It is also used to determine the effect of matrix components on biofilm growth (Rudney et al., 2012), to detect QS molecules (Chavant et al., 2007), and to compare biofilm-forming abilities of different bacterial species (Badel et al., 2008). Calgary biofilm devices are recognized as a new tool for rapid and reproducible antibacterial biofilm susceptibility testing. This is a plate-based model with a 96-well plate and a fixed lid (plastic invasion) inoculated with bacterial inoculum into wells containing a biofilm growth medium. This forms a biofilm on the surface of the vortex. This method makes it easy to transfer the biofilm from one plate to another, and the availability of fresh medium makes the study model relatively dynamic. It also reduces interference with biofilm growth in media containing metabolites and regulatory molecules (Huang et al., 2009). Due to differences in adhesive surfaces, nutritional support, and molecular interactions, the in vitro model cannot completely replace the in vivo model. However, this limitation can be overcome to some extent by associating these in vitro models with specific medical devices.

5.2.5.2 Microcosm-based biofilm models

It has been recommended to handle microcosms to represent the consumption of nutrient-rich sources or areas similar to location of infection to create a host-like microenvironment, thereby creating environmental conditions similar to infection (Sulaeman et al., 2010). The main goal of these models is to bring samples with minute modifications comprising clinical and fluid-based samples from various anatomical microenvironments such as reproductive and gastrointestinal tract as well as oral cavity to laboratory. Attempts to create an in vitro condition. Microcosms are broadly differentiated into three types—natural, mixed, and artificial microcosms. Natural microcosm is composed of natural liquids such as saliva and plaque, as well as the biochemical properties of microorganisms and the microenvironment (Goeres et al., 2009). The artificial microcosm contains a medium simulating necessary organic conditions (such as Lubbock's chronic wound model) and is essentially fully synthesized (Junka et al., 2016). In contrast, mixed microcosms contain samples from anatomical sites but are cultured in artificial medium (Sun et al., 2008). Oral microbial biofilms are easy to collect and exhibit a wide variety of microbial species, including those that cannot be cultivated in vitro, so microcosms are primarily developed through them. For primary development of microcosms, oral biofilms are used as they can be easily collected and also display a plethora of microbial species, comprising species that are unable to cultivate by in vitro methods. Moreover, dealing with the microcosm is difficult because it requires skilled human resources, special equipment, and experimental methods to ensure that the microenvironment of that particular anatomical site remains in a state similar to that of the microenvironment (Verardi et al., 2016).

Studies are underway to create reproducible oral microcosm models and to validate existing models because of certain inadequacy of specific methods and diversification within bacterial species (Sun et al., 2008). For example, Fernandez et al. (2017) (Sissons et al., 2007) created a reproducible subgingival biofilm model. They extracted subgingival plaque samples at four depths and used glass slides on the substrate for biofilm growth in patients with periodontitis. Biofilms were cultured under anaerobic conditions and analyzed every 2 weeks. This model is considered simple and effective, producing complicated yet powerful biofilms of various compositions.

Also, the producing and evaluating models of microcosm biofilm are important because research on in vitro biofilm models usually involves a single species that is far from reality. For example, the oral microbial flora contains hundreds of microorganisms that do not form a single biofilm (Fernandez y Mostajo et al., 2017). Microcosms are essentially more representative as they bring a complex microbial environment to the laboratory setting. For example, plaque and saliva infusion of human teeth extracted from the same individual can be performed to monitor tooth biofilm formation (Goeres et al., 2009), which can be used with antibiofilm compounds and more accurate evaluation of compounds with anticariogenic activity (Braga et al., 2018).

To evaluate the oral microcosm experiments and its reproducibility, various studies were conducted. For example, Rudney et al. (2012) to generate a dynamic state biofilm used oral microcosm (CDC biofilm reactor) including human saliva and plaque (Peeters et al., 2008) to assess whether saliva and dental specimen inoculations from the same donor have a similar composition when harvested on different days compared to the composition of inoculations between different donors. It was observed that in the first condition, the biofilm did not possess varied composition but in the second case where inoculum was collected from different individuals presented a different composition. Since, all the samples produced a thick biofilm which implied that they are reproducible (Peeters et al., 2008).

But there are only a few standard models for microcosm of anatomical sites other than the oral cavity. By observing the physicochemical, microbiological, and nutritional aspects of this environment, an improved understanding and studying of microorganisms and their anatomical insights on interaction with the environment can lead to the various concepts and uses proposed. Some researchers to discover an alternative for genesis of microcosms that can mimic the required environmental conditions for microorganisms are using the provided information on the anatomical sites regarding the biochemical and microbial composition (Woodworth et al., 2008). To study chronic sinusitis, an in vitro model of microcosm of *P. aeruginosa* was prepared by culturing cells of rat respiratory epithelium. Through this study, it was concluded that this model was dependable for assessing interactions among bacterial biofilms and rat respiratory epithelium (Woodworth et al., 2008). In another study, cells of human respiratory epithelium were taken for monitoring the biofilm formation *Aspergillus fumigatus* in patients undergoing EGD (esophagogastroduodenoscopy). *A. fumigatus* was found to have a complicated yet powerful biofilm formed on cells of epithelium. This helps in the further investigation of fungus and its pathogenic characteristics in chronic sinusitis (Singhal et al., 2011).

Cordeiro et al. (2020a, 2020b) using vaginal discharge from women suffering from candidiasis, proposed a microcosm-based model of biofilm to evaluate the infection (Singhal et al., 2011). The secretions were incubated for about 72 h in a medium, i.e., vaginal discharge simulator medium, which resulted in the production of a biofilm with the largest value of cellular density; also a biomass with resistance to fluconazole was obtained after incubating for 48 h. Metagenomic analysis of these biofilms showed that they were more sensitive to fluconazole when they carried high levels of *Lactobacillus* spp. compared with biofilms lacking this genus of bacteria. This model helped study strategies for treating and controlling vulvar vaginal candidiasis. A study was conducted by adding collagen and artificial skin for stimulating an environment similar to that of a protracting injury (Junka et al., 2016). To create a chronic wound environment, Sun et al. developed a Lubbock's chronic wound model using a variety of bacterial species such as *S. aureus*, *Enterococcus faecalis*, and *P. aeruginosa* for identifying the main components in protracting injuries. Therefore, a culture medium containing bovine plasma, meat hydrolyzate, and horse erythrocytes is used to produce plasma, erythrocytes, and necrotic tissue. Due to its inclusion of the main components of chronic wounds, this model is suitable because it allows the study of multiple biofilms along with anaerobic biofilms and the evaluation of antimicrobial agents against microbial biofilms. It also helps improve treatment strategies for these infections (Junka et al., 2016).

Using cell culture as a substrate, these models, which resemble body fluids, allow the interaction between bacteria and adhesive surfaces and the assessment of performance during the formation of microbial biofilm (provided peculiar nutritional support). It is important to be aware of what you do. However, it is not possible to analyze the interaction between symbiotic organisms or the effect of physicochemical properties of the anatomical environment on biofilm growth.

5.2.5.3 Ex vivo models

Investigation and evaluation of infectious processes associated with biofilm formation is possible through the development of many methods (Cordeiro et al., 2020b). To fill this gap, the ex vivo model was created as in vitro models cannot be considered reliable as it lacks valuable biological data on host (Eslami et al., 2019). For creating host-like environmental conditions, ex vivo models are used. Therefore, with certain modifications, an in vitro method can be used for stimulating an environment with in vivo conditions (Gabrilska & Rumbaugh, 2015) and used by the tissue itself as a substrate to generate a really naturally occurring sticky community in biofilm-related infections of common bacteria and fungi such as infective endocarditis, caries, periodontitis, osteomyelitis. The ex vivo model is used for biofilm research because it is likely to be (Cordeiro et al., 2020b; Eslami et al., 2019). Actually, teeth from various animals and humans were widely implemented for studying various dental ex vivo biofilm models on related diseases such as cavities (Vila et al., 2015), endodontic biofilms, and pulpitis (109). With the exception of pulp, teeth are primarily composed of calcified tissue, which makes them less prone to deterioration (Huang et al., 2019) and easier to sterilize, making them useful for ex vivo biofilm research. The reliability of the results obtained is improved by assigning them to the microcosm of the oral cavity. By creating an environment similar to the in vivo model, the ex vivo model allows large-scale experimental assays to be performed under a variety of conditions and at a relatively lower cost than the in vivo model (Eslami et al., 2019). It can be said as follows: The usefulness of the ex vivo model has provided diverse opportunities for studying multiple bacterial—host interactions. These studies can be performed for proper in vivo guidance and optimization of various assays while animal lives are saved (Carrasco et al., 2017). Therefore, the main advantages of using these ex vivo models are for significantly reducing the number of animals used in the study, addressing ethical concerns related to animal welfare, and reducing the costs associated with raising laboratory animals. These models should also consider using slaughterhouse by-products for maximizing the animal usage that humans will consume. However, from ex vivo point of view, these models have a major shortcoming, and that is the absence of interaction between biofilm structure and the host immune system, the potential lack of normal microbial flora at the studied anatomical site, and the in vivo model which is lack of fluid flow (Yang et al., 2013).

6. Recent advances

Many tissue-borne infections and commonly used medical devices have established animal models that allow biofilms to study disease progression and test new drugs, treatments, materials, and more. There was a push to create a search for a new model that exists to fill the gap where there is no model. A mouse model was developed by Saraswathi and Beuerman in 2015 to study *P. aeruginosa* biofilms in the process of corneal keratitis. They were able to track the progression of *P. aeruginosa*. A corneal biofilm derived from microcolonization of *P. aeruginosa* suspension in a corneal injury model using 7- to 8-week-old C57BL/6 mice. They can visualize and treat biofilm formation on the corneal surface from standard infections when mature biofilms are a common component of keratitis and treatment resistance is observed by Masu (Mao et al., 2018). This will be followed by a study by Ponce-Angulo et al. in 2020, we studied biofilm formation in an immunopreserved mouse model of mixed keratitis. A mixed bacterial and fungal biofilm was identified and characterized by coinfection with *S. aureus* and falciforme (Chuang-Smith et al., 2010) by inoculating a micropocket incision at the edge of the sclera of the cornea. Recently, Prezer and his group have developed a simple mouse abscess model for studying subcutaneous chronic gram-negative bacterial infections. Unlike other chronic infection models, this model is technically feasible and can be pursued without sacrificing animals. The formation of a visible abscess is caused by a subcutaneous injection of gram-negative bacteria, the progression of which is visually monitored and measured with a caliper. In addition, the course of the disease is visualized with a noninvasive in vivo imaging system by inoculation with luminescent bacteria. Subsequent homogenization and extraction of abscesses by plating enables quantitative analysis of colony-forming units. In addition, postmortem analysis of surrounding organs can be performed to determine the spread of the infection (Plezer et al., 2017). Zhang reported in 2018 on the first animal model of biofilm-related meningitis that investigated the role of *S. aureus*. A biofilm native to Switzerland that causes meningitis has been identified. After infection, quantitative culture and light microscopy of postsacrificial histopathological specimens confirmed the formation of bacterial colonies. This model was used to study the mechanism of bacterial meningococci and the efficacy of new drugs. Furthermore, the destruction of the blood–brain barrier and blood–cerebrospinal fluid barrier by the above streptococci has been shown to be an important step in the development of meningitis (Zhang et al., 2018).

In addition to these tissue-borne infections, several new biofilm models related to device-borne infections have recently been developed. These include new models of biofilm formation in cochlear implants, surgical screws, and central nervous system catheters. The guinea pig model was published by Cevizci et al. in 2015 by studying biofilm-borne infections from transplanted cochlear devices and testing QS inhibitors (Cevizci et al., 2015). This model was developed by injecting a cochlear implant through the eardrum into the pneumonia middle ear and implanting the cochlear implant under the skin behind the pinna. Visualization of biofilm formation on explanted device components using scanning electron microscopy demonstrates that these novel QS inhibitors can prevent biofilm formation on implants. For neurological devices, a rat model was developed to test whether intraoperative infection of the intracranial screw with *S. aureus* causes biofilm formation (Glage et al., 2017b).

6.1 Integrative biofilm models

In live organisms, the formation of bacterial infection models that are functional has been made possible by extensive research in the field of biofilms and infectious diseases over the past 50 years. A wide range of tailorable alternatives comprising high-throughput abilities can be performed for biofilm in vitro modeling at comparatively little cost. A mechanically accurate assessment of the development and progression of biofilms in sophisticated organisms is possible only through In vivo models, despite current limitations in their use.

Biologically relevant in vitro bacterial colonies can also be obtained from ex vivo models. In case of a corneal structure, immune response elicitation is relatively difficult; thus, cornea can be stated as an organ with immunity privileges. While assessing the potential antimicrobial agents in bacterial and fungal keratitis instead of the in vitro model, some alternatives were used such as corneas of porcine and rabbit (Pinnock et al., 2017; Snowden et al., 2012). However, when modeling endocarditis, physiological temperature, shear conditions, volume, and flow should be considered when dealing with perfused tissue such as the heart (Goeres et al., 2009). Rapid advances in microfluidics and organ-on-a-chips (commonly known as body-on-a-chips) bring the capacities for high-throughput analysis and integrated endogenously triggered immune response closer together during biofilm research. It may be fully possible in the future. Tip (Lauten et al., 2020). Such techniques, for example, in combination with the creation of 3D bioprinting structures, have the ability to mimic the complex structures of living organisms.

6.2 Challenges and future perspectives

The next frontier for generating a more accurate in vivo model of biofilm is the humanized animal model, but ethical issues need to be carefully considered. These humanized animal models should have comorbidities such as diabetes, cardiovascular diseases, or obesity, which can firmly mimic the exact conditions similar to that of human patients. Also it is hard to anticipate the upcoming generations of advanced biofilm models (Dehne et al., 2020).

References

An Andy, Y., Choi, K. Y. G., Baghela Arjun, S., & Hancock Robert, E. W. (2021). An overview of biological and computational methods for designing mechanism-informed anti-biofilm agents. *Frontiers in Microbiology, 12*. https://doi.org/10.3389/fmicb.2021.640787. https://www.frontiersin.org/article/10.3389/fmicb.2021.640787

Andersson, D. I., & Hughes, D. (2014). Microbiological effects of sublethal levels of antibiotics. *Nature Reviews Microbiology, 12*(7), 465–478. https://doi.org/10.1038/nrmicro3270. PubMed PMID: 24861036.

Artini, M., Patsilinakos, A., Papa, R., Božoviæ, M., Sabatino, M., Garzoli, S., et al. (2018). Antimicrobial and antibiofilm activity and machine learning classification analysis of essential oils from different Mediterranean plants against *Pseudomonas aeruginosa. Molecules Basel Switz, 23*, 482. https://doi.org/10.3390/molecules23020482

Badel, S., Laroche, C., Gardarin, C., Bernardi, T., & Michaud, P. (2008). New method showing the influence of matrix components in Leuconostoc mesenteroides biofilm formation. *Applied Biochemistry and Biotechnology, 151*, 364–370. https://doi.org/10.1007/s12010-008-8199-y

Barre- Sinoussi, F., & Montagutelli, X. (2015). Animal models are essential to biological research; issues and perspectives. *Future Science OA, 1*(4), Fso63.

Braga, A. S., Pires, J. G., & Magalh∼aes, A. C. (2018). Effect of a mouthrinse containing Malva sylvestris on the viability and activity of microcosm biofilm and on enamel demineralization compared to known antimicrobial mouthrinses. *Biofouling, 34*, 252–261. https://doi.org/10.1080/08927014.2018.1428957

Breen, E. C., Malloy, J. L., Tang, K., Xia, F., Fu, Z., Hancock, R. E. W., Overhage, J., Wagner, P. D., & Spragg, R. G. (2013). Impaired pulmonary defense against *Pseudomonas aeruginosa* in VEGF gene inactivated mouse lung. *Journal of Cellular Physiology, 228*, 371–379.

Cabot, G., Zamorano, L., Moyà, B., Juan, C., Navas, A., Blázquez, J., & Oliver, A. (2016). Evolution of *Pseudomonas aeruginosa* antimicrobial resistance and fitness under low and high mutation supply rates. *Antimicrobial Agents and Chemotherapy, 60*, 1767–1778.

Cao, Q., Wang, Y., Chen, F., Xia, Y., Lou, J., Zhang, X., Yang, N., Sun, X., Zhang, Q., Zhuo, C., et al. (2014). A novel signal transduction pathway that modulates rhl quorum sensing and bacterial virulence in *Pseudomonas aeruginosa. PLoS Pathogens, 10*, e1004340.

Carrasco, P. A., Brizuela, C. I., Rodriguez, I. A., Munoz, S., Godoy ∼, M. E., & Inostroza, C. (2017). Histological transformations of the dental pulp as a possible indicator of post mortem interval: A pilot study. *Forensic Science International, 279*, 251–257. https://doi.org/10.1016/j.forsciint.2017.09.001

Casciaro, B., Cappiello, F., Loffredo, M. R., Ghirg, F., & Mangoni, M. L. (2019). The potential of frog skin peptides for anti-infective therapies: The case of esculentin-1a (1-21) NH2. *Current Medicinal Chemistry, 27*, 1405–1419. https://doi.org/10.2174/0929867326666190722095408

Cattò, C., & Cappitelli, F. (2019). Testing anti-biofilm polymeric surfaces: Where to start? *International Journal of Molecular Sciences, 20*(15), 3794. https://doi.org/10.3390/ijms20153794

Cevizci, R., Duzlu, M., Dundar, Y., Noyanalpan, N., Sultan, N., Tutar, H., & Bayazit, Y. A. (2015). Preliminary results of a novel quorum sensing inhibitor against pneumococcal infection and biofilm formation with special interest to otitis media and cochlear implantation. *European Archives of Oto-Rhino-Laryngology, 272*, 1389–1393.

Chavant, P., Gaillard-Martinie, B., Talon, R., Hebraud, M., & Bernardi, T. (2007). A new device for rapid evaluation of biofilm formation potential by bacteria. *Journal of Microbiological Methods, 68*, 605–612. https://doi.org/10.1016/j.mimet.2006.11.010

Chuang-Smith, O. N., Wells, C. L., Henry-Stanley, M. J., & Dunny, G. M. (2010). Acceleration of *Enterococcus faecalis* biofilm formation by aggregation substance expression in an ex vivo model of cardiac valve colonization. *PloS One, 5*, e15798. https://doi.org/10.1371/journal.pone.0015798

Churchley, D., Rees, G. D., Barbu, E., Nevell, T. G., & Tsibouklis, J. (2008). Fluoropolymers as low-surface energy tooth coatings for oral care. *International Journal of Pharmaceutics, 352*, 44–49.

Ciric, A. D., Petrovic, J. D., Glamoclija, J. M., Smiljkovic, M. S., Nikolic, M. M., Stojkovic, D. S., et al. (2019). Natural products as biofilm formation antagonists and regulators of quorum sensing functions: A comprehensive review update and future trends. *South African Journal of Botany, 120*, 65–80. https://doi.org/10.1016/j.sajb.2018.09.010

Cordeiro, R. D. A., de Andrade, A. R. C., Portela, F. V. M., Pereira, L. M. G., Moura, S. G. B., Sampaio, M. D., Pereira, E. M. A., de Melo Guedes, G. M., Bandeira, S. P., de Lima-Neto, R. G., et al. (2020b). Proposal for a microcosm biofilm model for the study of vulvovaginal candidiasis. *Biofouling, 36*, 610–611. https://doi.org/10.1080/08927014.2020.1785435

Cordeiro, L., Figueiredo, P., Souza, H., Sousa, A., Andrade-Júnior, F., Barbosa-Filho, J., & Lima, E. (2020a). Antibacterial and antibiofilm activity of myrtenol against *Staphylococcus aureus*. *Pharmaceuticals, 13*(6), 133.

Costerton, J. W., Stewart, P. S., & Greenberg, E. P. (1999). Bacterial biofilms: A common cause of persistent infections. *Science, 284*(5418), 1318–1322.

Darouiche, R. O. (2004). Treatment of infections associated with surgical implants. *The New England Journal of Medicine, 350*(14), 1422–1429. https://doi.org/10.1056/NEJMra035415. PubMed PMID: 15070792.

De Simone, M., Spagnuolo, L., Lorè, N. I., Rossi, G., Cigana, C., De Fino, I., Iraqi, F. A., & Bragonzi, A. (2014). Host genetic background influences the response to the opportunistic *Pseudomonas aeruginosa* infection altering cell-mediated immunity and bacterial replication. *PLoS One, 9*, e106873.

Dehne, E. M., & Marx, U. (2020). Chapter 13- Human body-on-a-chip systems. In J. Hoeng, D. Bovard, & M. C. Peitsch (Eds.), *Organ-on-a-chip* (pp. 429–439). Academic Press.

Del Pozo, J. L., & Patel, R. (2007). The challenge of treating biofilm-associated bacterial infections. *Clinical Pharmacology & Therapeutics, 82*(2), 204–209. https://doi.org/10.1038/sj.clpt.6100247. PubMed PMID: 17538551.

Del Pozo, J. L., Tran, N. V., Petty, P. M., Johnson, C. H., Walsh, M. F., Bite, U., et al. (2009). Pilot study of association of bacteria on breast implants with capsular contracture. *Journal of Clinical Microbiology, 47*(5), 1333–1337. PubMed PMID: 19261794.

Déziel, E., Gopalan, S., Tampakaki, A. P., Lépine, F., Padfield, K. E., Saucier, M., … Rahme, L. G. (2005). The contribution of MvfR to *Pseudomonas aeruginosa* pathogenesis and quorum sensing circuitry regulation: Multiple quorum sensing-regulated genes are modulated without affecting lasRI, rhlRI or the production of N-acyl-L-homoserine lactones. *Molecular Microbiology, 55*(4), 998–1014.

Di Luca, M., Maccari, G., Maisetta, G., & Batoni, G. (2015). BaAMPs: The database of biofilm-active antimicrobial peptides. *Biofouling, 31*, 193–199. https://doi.org/10.1080/08927014.2015.1021340

Diggle, S. P., Winzer, K., Chhabra, S. R., Worrall, K. E., Cámara, M., & Williams, P. (2003). The *Pseudomonas aeruginosa* quinolone signal molecule overcomes the cell density-dependency of the quorum sensing hierarchy, regulates rhl-dependent genes at the onset of stationary phase and can be produced in the absence of LasR. *Molecular Microbiology, 50*(1), 29–43.

Dong, B., Manolache, S., Wong, A. C. L., & Denes, F. S. (2011). Antifouling ability of polyethylene glycol of different molecular weights grafted onto polyester surfaces by cold plasma. *Polymer Bulletin, 66*, 517–528.

Donlan, R. M. (2002). Biofilms: Microbial life on surfaces. *Emerging Infectious Diseases, 8*(9), 881.

Dubern, J., Cigana, C., De Simone, M., Lazenby, J., Juhas, M., Schwager, S., Bianconi, I., Döring, G., Eberl, L., Williams, P., et al. (2015). Integrated whole genome screening for *Pseudomonas aeruginosa* virulence genes using multiple disease models reveals that pathogenicity is host specific. *Environmental Microbiology, 17*, 4379–4397.

Eslami, L. M., Vatanpour, M., Aminzadeh, N., Mehrvarzfar, P., & Taheri, S. (2019). The comparison of intracanal medicaments, diode laser and photodynamic therapy on removing the biofilm of *Enterococcus faecalis* and Candida albicans in the root canal system (ex-vivo study). *Photodiagnosis and Photodynamic Therapy, 26*, 157–161. https://doi.org/10.1016/j.pdpdt.2019.01.033

Fekete, A., Borbas, A., Gyémant, G., Kandra, L., Fazekas, E., Ramasubbu, N., & Antus, S. (2011). Synthesis of β-(1-6)-linked N-acetyl-D-glucosamine oligosaccharide substrates and their hydrolysis by Dispersin B. *Carbohydrate Research, 346*, 1445–1453.

Fernandez y Mostajo, M., Exterkate, R. A. M., Buijs, M. J., Beertsen, W., Van Der Weijden, G. A., Zaura, E., & Crielaard, W. (2017). A reproducible microcosm biofilm model of subgingival microbial communities. *J Periodont Res, 52*, 1021–1031. https://doi.org/10.1111/jre.12473

Fisher, M. L., Sun, W., & Curtiss, R. (2014). The route less taken: Pulmonary models of enteric gram-negative infection. *Pathog Dis, 70*, 99–109.

Fjell, C. D., Hiss, J. A., Hancock, R. E. W., & Schneider, G. (2012). Designing antimicrobial peptides: Form follows function. *Nat. Rev. Drug Discov., 11*, 37–51.

Form, H. D. T. (2002). *Biofilms: Forming a defense strategy for the food plant. Order.*

de la Fuente-Núñez, C., Cardoso, M. H., de Souza Cândido, E., Franco, O. L., & Hancock, R. E. W. (2016). Synthetic antibiofilm peptides. *Biochimica et Biophysica Acta (BBA) - Biomembranes, 1858*(Issue 5), 1061–1069. https://doi.org/10.1016/j.bbamem.2015.12.015. ISSN 0005-2736.

Fux, C. A., Stoodley, P., Hall-Stoodley, L., & Costerton, J. W. (2003). Bacterial biofilms: A diagnostic and therapeutic challenge. *Expert Rev Anti Infect Ther, 1*(4), 667–683. PubMed PMID: 15482163.

Gabrilska, R. A., & Rumbaugh, K. P. (2015). Biofilm models of polymicrobial infection. *Future Microbiol, 10*, 1997–2015. https://doi.org/10.2217/fmb.15.109

Glage, S., Paret, S., Winkel, A., Stiesch, M., Bleich, A., Krauss, J. K., et al. (2017). A new model for biofilm formation and inflammatory tissue reaction: Intraoperative infection of a cranial implant with *Staphylococcus aureus* in rats. *Acta Neurochir (Wien)*. https://doi.org/10.1007/s00701-017-3244-7. PubMed PMID: 28647798.

Glage, S., Paret, S., Winkel, A., Stiesch, M., Bleich, A., Krauss, J. K., & Schwabe, K. (2017). A new model for biofilm formation and inflammatory tissue reaction: Intraoperative infection of a cranial implant with *Staphylococcus aureus* in rats. *Acta Neurochir, 159*, 1747–1756.

Goeres, D. M., Hamilton, M. A., Beck, N. A., Buckingham-Meyer, K., Hilyard, J. D., Loetterle, L. R., Lorenz, L. A., Walker, D. K., & Stewart, P. S. (2009). A method for growing a biofilm under low shear at the air-liquid interface using the drip flow biofilm reactor. *Nat Protoc, 4*, 783–788. https://doi.org/10.1038/nprot.2009.59

Gristina, A. G., Naylor, P., & Myrvik, Q. (1988). Infections from biomaterials and implants: A race for the surface. *Med Prog Technol, 14*(3–4), 205–224. PubMed PMID: 2978593.**First report of the so called race for the surface.

Gupta, S., Sharma, A. K., Jaiswal, S. K., & Sharma, V. K. (2016). Prediction of biofilm inhibiting peptides: An in silico approach. *Front. Microbiol., 7*, 949. https://doi.org/10.3389/fmicb.2016.00949

Hall-Stoodley, L., Costerton, J. W., & Stoodley, P. (2004). Bacterial biofilms: From the natural environment to infectious diseases. *Nat Rev Microbiol, 2*(2), 95–108. https://doi.org/10.1038/nrmicro821. PubMed PMID: 15040259.

Haney, E. F., Trimble, M. J., Cheng, J. T., Vallé, Q., & Hancock, R. E. W. (2018b). Critical assessment of methods to quantify biofilm growth and evaluate antibiofilm activity of host defence peptides. *Biomolecules, 8*, 29. https://doi.org/10.3390/biom8020029

Heersink, J., & Goeres, D. (2003). Reactor design considerations. In M. Hamilton, J. Heersink, J. Buckingham-Meyer, & D. Goeres (Eds.), *The biofilm laboratory: Step-by-step protocols for experimental design, analysis, and data interpretation* (pp. 13–15). Bozeman: Cytergy Publishing.

Huang, Z., Meric, G., Liu, Z., Ma, R., Tang, Z., & Lejeune, P. (2009). LuxS-based quorum-sensing signaling affects biofilm formation in Streptococcus mutans. *J Mol Microbiol Biotechnol, 17*, 12–19. https://doi.org/10.1159/000159193

Huang, Y., Thompson, T., Wang, Y., Yu, Q., Zhu, L., Xu, X., Wen, Z. T., & Townsend, J. A. (2019). Analysis of cariogenic potential of alternative milk beverages by in vitro Streptococcus mutans biofilm model and ex vivo caries model. *Archives Oral Biol, 105*, 52–58. https://doi.org/10.1016/j.archoralbio. 2019.05.033

Irazoqui, J. E., Urbach, J. M., & Ausubel, F. M. (2010). Evolution of host innate defense: Insights from *Caenorhabditis elegans* and primitive invertebrates. *Nat Rev Immunol, 10*, 47–58.

Jackson, A. A., Gross, M. J., Daniels, E. F., Hampton, T. H., Hammond, J. H., Vallet-Gely, I., Dove, S. L., Stanton, B. A., & Hogan, D. A. (2013). Anr and its activation by PlcH activity in *Pseudomonas aeruginosa* host colonization and virulence. *J Bacteriol, 195*, 3093–3104.

Jander, G., Rahme, L. G., & Ausubel, F. M. (2000). Positive correlation between virulence of *Pseudomonas aeruginosa* mutants in mice and insects. *J Bacteriol, 182*, 3843–3845.

Jayaraman, A., & Wood, T. H. (2008). Bacterial quorum sensing signals: Signals, circuits, and implications for biofilm and disease. *Annu Rev Biomed Eng, 10*, 145–167.

Jones, J. S., Hoerle, D., & Riekse, R. (1995). Stethoscopes: A potential vector of infection? *Annals of Emergency Medicine, 26*(3), 296–299.

Junka, A. F., Szymczyk, P., Secewicz, A., Pawlak, A., Smutnicka, D., Ziołkowski, G., Bartoszewicz, M., & Chlebus, E. (2016). The chemical digestion of Ti6Al7Nb scaffolds produced by Selective Laser Melting significantly reduces the ability of *Pseudomonas aeruginosa* to form biofilm. *Acta Bioeng Biomech, 18*, 115–120.

Kirienko, N. V., Ausubel, F. M., & Ruvkun, G. (2015). Mitophagy confers resistance to siderophore-mediated killing by *Pseudomonas aeruginosa*. *Proc Natl Acad Sci USA, 112*, 1821–1826.

Koch, G., Nadal-Jimenez, P., Cool, R. H., & Quax, W. J. (2014). Assessing *Pseudomonas* virulence with nonmammalian host: *Galleria mellonella*. *Methods Mol Biol, 1149*, 681–688.

Kokare, C. R., Chakraborty, S., Khopade, A. N., & Mahadik, K. R. (2009). *Biofilm: Importance and applications*.

Kounatidis, I., & Ligoxygakis, P. (2012). *Drosophila* as a model system to unravel the layers of innate immunity to infection. *Open Biol, 2*, 120075.

Krezdorn, J., Adams, S., & Coote, P. J. (2014). A *Galleria mellonella* infection model reveals double and triple antibiotic combination therapies with enhanced efficacy versus a multidrug-resistant strain of *Pseudomonas aeruginosa*. *J Med Microbiol, 63*, 945–955.

Kumar, C. G., & Anand, S. K. (1998). Significance of microbial biofilms in food industry: A review. *International Journal of Food Microbiology, 42*(1–2), 9–27.

Labeaux, D., Chauhan, A., Rendueles, O., & Beloin, C. (2013). From in vitro to *In vivo* models of bacterial biofilm related infections. *Pathogens (Basel), 2*, 288–356. https://doi.org/10.3390/pathogens2020288

Lauten, A., Martinovic, M., Kursawe, L., Kikhney, J., Affeld, K., Kertzscher, U., Falk, V., & Moter, A. (2020). Bacterial biofilms in infective endocarditis: As in vitro model to investigate emerging technologies of antimicrobial cardiovascular device coatings. *Clin. Res. Cardiol., 110*, 323–331.

Lazar, V. (2011). Quorum sensing in biofilms – how to destroy the bacterial citadels or their cohesion/power? *Anaerobe, 17*, 280–285.

Lazurko, K., Khatoon, Z., Goel, K., Sedlakova, V., Eren Cimenci, C., Ahumada, M., Zhang, L., Mah, T. F., Franco, W., Suuronen, E. J., et al. (2020). Multifunctional nano and collagen based therapeutic materials for skin repair. *ACS Biomat. Sci. Eng., 6*, 1124–1134.

LeCun, Y., Bengio, Y., & Hinton, G. (2015). Deep learning. *Nature, 521*, 436–444. https://doi.org/10.1038/nature14539

Letamendia, A., Quevedo, C., Ibarbia, I., Virto, J. M., Holgado, O., Diez, M., Izpisua Belmonte, J. C., & Callol Massot, C. (2012). Development and validation of an automated high throughput system for zebrafish *In vivo* screenings. *PLoS One, 7*, e36690.

Lim, D., & Strynadka, N. C. (2002). Structural basis for the β lactam resistance of PBP2a from methicillin-resistant *Staphylococcus aureus*. *Nature Structural Biology, 9*(11), 870–876.

Lo, Y.-C., Rensi, S. E., Torng, W., & Altman, R. B. (2018). Machine learning in chemoinformatics and drug discovery. *Drug Discov. Today, 23*, 1538–1546. https://doi.org/10.1016/j.drudis.2018.05.010

Lorenz, A., Pawar, V., Häussler, S., & Weiss, S. (2016). Insights into host–pathogen interactions from state-of-the-art animal models of respiratory *Pseudomonas aeruginosa* infections. *FEBS Lett, 590*, 3941–3959. https://doi.org/10.1002/1873-3468.12454

Luis Del Pozo, J. (2017). Biofilm-related disease. *Expert Review of Antiinfective Therapy*. https://doi.org/10.1080/14787210.2018.1417036

Lutter, E. I., Purighalla, S., Duong, J., & Storey, D. G. (2012). Lethality and cooperation of *Pseudomonas aeruginosa* quorum-sensing mutants in *Drosophila melanogaster* infection models. *Microbiology, 158*, 2125–2132.

Mao, Y., Singh-Varma, A., Hoffman, T., Dhall, S., Danilkovitch, A., & Kohn, J. (2018). The effect of cryopreserved human placental tissues on biofilm formation of wound-associated pathogens. *J Funct Biomater, 9*, 3. https://doi.org/10.3390/jfb9010003

Marsh, P. D., & Bradshaw, D. J. (1997). Physiological approaches to the control of oral biofilms. *Advances in Dental Research, 11*(1), 176–185.

Mei, L., Busscher, H. J., Van der Mei, H. C., & Ren, Y. (2011). Influence of surface roughness on streptococcal adhesion forces to composite resins. *Dental Materials, 27*, 770–778.

Migiyama, Y., Kaneko, Y., Yanagihara, K., Morohoshi, T., Morinaga, Y., Nakamura, S., Miyazaki, T., Hasegawa, H., Izumikawa, K., Kakeya, H., et al. (2013). Efficacy of AiiM, an N-acylhomoserine lactonase, against *Pseudomonas aeruginosa* in a mouse model of acute pneumonia. *Antimicrob Agents Chemother, 57*, 3653–3658.

Mowat, E., Paterson, S., Fothergill, J. L., Wright, E. A., Ledson, M. J., Walshaw, M. J., Brockhurst, M. A., & Winstanley, C. (2011). *Pseudomonas aeruginosa* population diversity and turnover in cystic fibrosis chronic infections. *Am J Respir Crit Care Med, 183*, 1674–1679.

Moyano, A. J., Tobares, R. A., Rizzi, Y. S., Krapp, A. R., Mondotte, J. A., Bocco, J. L., Saleh, M.-C., Carrillo, N., & Smania, A. M. (2014). A long-chain flavodoxin protects *Pseudomonas aeruginosa* from oxidative stress and host bacterial clearance. *PLoS Genet, 10*, e1004163.

Niu, C., & Gilbert, E. S. (2004). Colorimetric method for identifying plant essential oil components that affect biofilm formation and structure. *Applied and Environmental Microbiology, 70*, 6951–6956. https://doi.org/10.1128/AEM.70.12.6951-6956.2004

Noble, W. S. (2006). What is a support vector machine? *Nature Biotechnology, 24*, 1565–1567. https://doi.org/10.1038/nbt1206-1565

Patsilinakos, A., Artini, M., Papa, R., Sabatino, M., Božoviæ, M., Garzoli, S., et al. (2019). Machine learning analyses on data including essential oil chemical composition and in vitro experimental antibiofilm activities against Staphylococcus species. *Molecules Basel Switz., 24*, 890. https://doi.org/10.3390/molecules24050890

Peeters, E., Nelis, H. J., & Coenye, T. (2008). Comparison of multiple methods for quantification of microbial biofilms grown in microtiter plates. *Journal of Microbiological Methods, 72*, 157–165. https://doi.org/10.1016/j.mimet.2007.11.010

Peterman, E. M., Sullivan, C., Goody, M. F., Rodriguez-Nunez, I., Yoder, J. A., & Kim, C. H. (2015). Neutralization of mitochondrial superoxide by superoxide dismutase 2 promotes bacterial clearance and regulates phagocyte numbers in zebrafish. *Infection and Immunity, 83*, 430–440.

Pinnock, A., Shivshetty, N., Roy, S., Rimmer, S., Douglas, I., Macneil, S., & Grag, P. (2017). Ex vivo rabbit and human corneas as models for bacterial and fungal keratitis. *Graefe's Archive for Clinical and Experimental Ophthalmology, 255*, 333–342.

Plezer, D., Mansour, S. C., Wuerth, K., Rahanjam, N., & Hancock, R. E. W. (2017). New mouse model for chronic infections by gram negative bacteria by enabling the study of anti- infective efficacy and host microbe interactions. *mBio, 8*, e00140–17.

Ponce-Angulo, D. G., Bautista-Hernández, L. A., Calvillo-Medina, R. P., Castro-Tecorral, F. I., Aparicio-Ozores, G., López-Villegas, E. O., Ribas-Aparicio, R. M., & Bautista-de Lucio, V. M. (2020). Microscopic characterization of biofilm in mixed keratitis in a novel murine model. *Microbial Pathogenesis, 140*, 103953.

Raad, I. I., & Bodey, G. P. (1992). Infectious complications of indwelling vascular catheters. *Clinical Infectious Diseases, 15*(2), 197–208. PubMed PMID: 1520756.

Rajkumari, J., Borkotoky, S., Murali, A., Suchiang, K., Mohanty, S. K, & Busi, S. (2018). Cinnamic acid attenuates quorum sensing associated virulence factors and biofilm formation in Pseudomonas aeruginosa PAO1. *Biotechnology Letters, 40*(7). https://doi.org/10.1007/s10529-018-2557-9. In this issue.

Rajkumari, J., Borkotoky, S., Reddy, D., Mohanty, S. K., Kumavath, R., Murali, A., … Busi, S. (2019). Anti-quorum sensing and anti-biofilm activity of 5-hydroxymethylfurfural against *Pseudomonas aeruginosa* PAO1: Insights from in vitro, *In vivo* and in silico studies. *Microbiological Research, 226*, 19–26.

Razak, F. A., & Rahim, Z. H. (2003). The anti-adherence effect of *Piper betle* and *Psidium guajava* extracts on the adhesion of early settlers in dental plaque to saliva-coated glass surfaces. *Journal of Oral Science, 45*, 201–206. https://doi.org/10.2334/josnusd.45.201

Reen, F. J., Haynes, J. M., Mooij, M. J., & O'Gara, F. (2013). A non-classical LysR-type transcriptional regulator PA2206 is required for an effective oxidative stress response in *Pseudomonas aeruginosa*. *PLoS One, 8*, e54479.

Roberts, A. E., Kragh, K. N., Bjarnsholt, T., & Diggle, S. P. (2015). The limitations of in vitro experimentation in understanding biofilms and chronic infection. *Journal of Molecular Biology, 427*, 3646–3661. https://doi.org/10.1016/j.jmb.2015.09.002

Rocker, A. J., Weiss, A. R. E., Lam, J. S., Van Raay, T. J., & Khursigara, C. M. (2015). Visualizing and quantifying *Pseudomonas aeruginosa* infection in the hindbrain ventricle of zebrafish using confocal laser scanning microscopy. *Journal of Microbiological Methods, 117*, 85–94.

Ross-Gillespie, A., Weigert, M., Brown, S. P., & Kümmerli, R. (2014). Gallium-mediated siderophore quenching as an evolutionarily robust antibacterial treatment. *Evolution, Medicine, and Public Health, 2014*, 18–29.

Rudney, J. D., Chen, R., Lenton, P., Li, J., Li, Y., Jones, R. S., Reilly, C., Fok, A. S., & Aparicio, C. (2012). A reproducible oral microcosm biofilm model for testing dental materials. *Journal of Applied Microbiology, 113*, 1540–1553. https://doi.org/10.1111/j.1365-2672.2012.05439.x

Rutala, W. A., & Weber, D. J. (1997). Water as a reservoir of nosocomial pathogens. *Infection Control & Hospital Epidemiology, 18*(9), 609–616. PubMed PMID: 9309431.

Salli, K. M., & Ouwehand, A. C. (2015). The use of in vitro model systems to study dental biofilms associated with caries: A short review. *Journal of Oral Microbiology, 7*, 26149. https://doi.org/10.3402/jom. V7.26149

Saraswathi, P., & Beuerman, R. W. (2015). Corneal biofilms: from planktonic to microcolony formation in an experimental keratitis infection with Pseudomonas aeruginosa. *The ocular surface, 13*(4), 331–345.

Satpute, S. K., Banpurkar, A. G., Banat, I. M., Sangshetti, J. N., Patil, R. H., & Gade, W. N. (2016). Multiple roles of biosurfactants in biofilms. *Current Pharmaceutical Design, 22*, 1429–1448. https://doi.org/10.2174/1381612822666160120152704

Scheman, L., Janota, M., & Lewin, P. (1941). The production of experimental osteomyelitis; preliminary report. *The Journal of the American Medical Association, 117*, 1525–1529.

Schuster, M., Joseph Sexton, D., Diggle, S. P., & Peter Greenberg, E. (2013). Acyl-homoserine lactone quorum sensing: From evolution to application. *Annual Review of Microbiology, 67*, 43–63.

Sharma, A., Gupta, P., Kumar, R., & Bhardwaj, A. (2016). dPABBs: a novel in silico approach for predicting and designing anti-biofilm peptides. *Scientific Reports, 6*, 21839. https://doi.org/10.1038/srep21839

Sidrim, J. J., Amando, B. R., Gomes, F. I. F., do Amaral, S. M. S. G., de Sousa, P. C. P., Ocadaque, C. J., Brilhante, R. S. N., Cordeiro, R. A., Rocha, M. F. G., & Castelo-Branco, D. S. C. M. (2019). Chlorpromazine-impregnated catheters as a potential strategy to control biofilm- associated urinary tract infections. *Future Microbiology, 14*, 1023–1034. https://doi.org/10.2217/fmb2019-0092

Singhal, D., Baker, L., Wormald, P. J., & Tan, L. (2011). Aspergillus fumigatus biofilm on primary human sinonasal epithelial culture. *American Journal of Rhinology & Allergy, 25*, 219–225. https://doi.org/10.2500/ajra.2011.25.362

Sissons, C. H., Anderson, S. A., Wong, L., Coleman, M. J., & White, D. C. (2007). Microbiota of plaque microcosm biofilms: Effect of three times daily sucrose pulses in different simulated oral environments. *Caries Research, 41*, 413–422. https://doi.org/10.1159/000104801

Snowden, J. N., Beaver, M., Smeltzer, M. S., & Kielian, T. (2012). Biofilm- infected intracerebroventricular shunts elicit inflammation within the central nervous system. *Infection and Immunity, 80*, 3206–3214.

Srivastava, G. N., Malwe, A. S., Sharma, A. K., Shastri, V., Hibare, K., & Sharma, V. K. (2020). Molib: A machine learning based classification tool for the prediction of biofilm inhibitory molecules. *Genomics, 112*, 2823–2832. https://doi.org/10.1016/j.ygeno.2020.03.020

Stokes, J. M., Yang, K., Swanson, K., Jin, W., Cubillos-Ruiz, A., Donghia, N. M., et al. (2020). A deep learning approach to antibiotic discovery. *Cell, 180*, 688–702.e13. https://doi.org/10.1016/j.cell.2020.01.021

Storz, M. P., Maurer, C. K., Zimmer, C., Wagner, N., Brengel, C., de Jong, J. C., … Hartmann, R. W. (2012). Validation of PqsD as an anti-biofilm target in *Pseudomonas aeruginosa* by development of small-molecule inhibitors. *Journal of the American Chemical Society, 134*(39), 16143–16146.

Sulaeman, S., Le Bihan, G., Rossero, A., Federighi, M., De, E., & Tresse, O. (2010). Comparison between the biofilm initiation of Campylobacter jejuni and Campylobacter coli strains to an inert surface using BioFilm Ring Test (R). *Journal of Applied Microbiology, 108*, 1303–1312. https://doi.org/10.1111/j.1365-2672. 2009.04534.x

Sun, Y., Dowd, S. E., Smith, E., Rhoads, D. D., & Wolcott, R. D. (2008). In vitro multispecies Lubbock chronic wound biofilm model. *Wound Repair and Regeneration, 16*, 805–813. https://doi.org/10.1111/j. 1524-475X.2008.00434.x

Sztukowska, M. N., Roky, M., & Demuth, D. R. (2019). Peptide and non-peptide mimetics as potential therapeutics targeting oral bacteria and oral biofilms. *Molecular Oral Microbiology, 34*, 169–182. https://doi.org/10.1111/omi.12267

Tan, M. W., Rahme, L. G., Sternberg, J. A., Tompkins, R. G., & Ausubel, F. M. (1999). *Pseudomonas aeruginosa* killing of *Caenorhabditis elegans* used to identify *P. aeruginosa* virulence factors. *Proceedings of the National Academy of Sciences of the United States of America, 96*, 2408–2413.

Thebault, P., Lequeux, I., Jouenne T. CNRS UMR 6270, Laboratoire "Polymères, Biopolymères, Surfaces", Faculté des Sciences, Université de Rouen, France.

Thebault, P., Lequeux, I., & Jouenne, T. (2013). Antibiofilm strategies. *Journal of Wound Technology, 21*, 20–23.

Thomann, A., de Mello Martins, A. G. G., Brengel, C., Empting, M., & Hartmann, R. W. (2016). Application of dual inhibition concept within looped autoregulatory systems towards anti-virulence agents against *Pseudomonas aeruginosa* infections. *ACS Chemical Biology, 11*, 1279–1286.

Thomsen, R., & Christensen, M. H. (2006). MolDock: A new technique for high-accuracy molecular docking. *Journal of Medicinal Chemistry, 49*(11), 3315–3321.

Tolker-Nielsen, T., & Sternberg, C. (2014). Methods for studying biofilm formation: Flow cells and confocal laser scanning microscopy. *Methods in Molecular Biology, 1149*, 615–629. https://doi.org/10.1007/978-1-4939-0473-0_47

Van Delden, C., & Iglewski, B. H. (1998). Cell-to-cell signaling and *Pseudomonas aeruginosa* infections. *Emerging Infectious Diseases, 4*(4), 551.

Verardi, G., Cenci, M. S., Maske, T. T., Webber, B., & Santos, L. R. (2016). Antiseptics and microcosm biofilm formation on titanium surfaces. *Brazilian Oral Research, 30*, e30. https://doi.org/10.1590/1807-3107BOR-2016.vol30.0030

Vila, T. V., Rozental, S., & de Sa Guimar∼aes, C. M. (2015). A new model of in vitro fungal biofilms formed on human nail fragments allows reliable testing of laser and light therapies against onychomycosis. *Lasers Medical Science, 30*, 1031–1039. https://doi.org/10.1007/s10103-014-1689-y

Wade, D. S., Calfee, M. W., Rocha, E. R., Ling, E. A., Engstrom, E., Coleman, J. P., & Pesci, E. C. (2005). Regulation of Pseudomonas quinolone signal synthesis in *Pseudomonas aeruginosa*. *Journal of Bacteriology, 187*(13), 4372–4380.

Wilson, M. (2001). Bacterial biofilms and human disease. *Science Progress, 84*(3), 235–254.

Woodworth, B. A., Tamashiro, E., Bhargave, G., Cohen, N. A., & Palmer, J. N. (2008). An in vitro model of *Pseudomonas aeruginosa* biofilms on viable airway epithelial cell monolayers. *American Journal of Rhinology, 22*, 235–238. https://doi.org/10.2500/ajr.2008.22. 3178

Yang, L., Nilsson, M., Gjermansen, M., Givskov, M., & Tolker-Nielsen, T. (2009). Pyoverdine and PQS mediated subpopulation interactions involved in *Pseudomonas aeruginosa* biofilm formation. *Molecular Microbiology, 74*(6), 1380–1392.

Yang, Q., Phillips, P. L., Sampson, E. M., Progulske-Fox, A., Jin, S., Antonelli, P., & Schultz, G. S. (2013). Development of a novel ex vivo porcine skin explant model for the assessment of mature bacterial biofilms. *Wound Repair and Regeneration, 21*, 704–714. https://doi.org/10.1111/wrr.12074

Yang, K., Swanson, K., Jin, W., Coley, C., Eiden, P., Gao, H., et al. (2019). Analyzing learned molecular representations for property prediction. *Journal of Chemical Information and Modeling, 59*, 3370–3388. https://doi.org/10.1021/acs.jcim.9b00237

Yan, X., Gu, S., Cui, X., Shi, Y., Wen, S., Chen, H., et al. (2019). Antimicrobial, anti-adhesive and anti-biofilm potential of biosurfactants isolated from *Pediococcus acidilactici* and *Lactobacillus plantarum* against *Staphylococcus aureus* CMCC26003. *Microbial Pathogenesis, 127*, 12–20. https://doi.org/10.1016/j.micpath.2018.11.039

Yin, W., Wang, Y., Liu, L., & He, J. (2019). Biofilms: The microbial "protective clothing" in extreme environments. *International Journal of Molecular Sciences, 20*(14), 3423.

Zhang, S., Gao, X., Xiao, G., Lu, C., Yao, H., Fan, H., & Wu, Z. (2018). Intracranial subarachnoidal route of infection for investigating the roles of Streptococcus suis biofilms in meningitis in a mouse infection model. *Journal of Visualized Experiments, 137*, e57658.

Ziegler, A., Gonzalez, L., & Blikslager, A. (2016). Large animal models: The key to translational discovery in digestive disease research. *Cellular and Molecular Gastroenterology and Hepatology, 2*, 716–724.

https://vina.scripps.edu/.

https://www.ebi.ac.uk/thornton-srv/software/LIGPLOT/.

https://www.gromacs.org/About_Gromacs.

CHAPTER

22

Biofilm: a threat to medical devices

Jignasa H. Bera[1], Leyon Selvin Raj A.[1], Sishir Gang[2] and Dhara N. Patel[1]

[1]Department of Medical Laboratory Technology, Bapubhai Desaibhai Patel Institute of Paramedical Sciences (BDIPS), Charotar University of Science and Technology, Changa, Gujarat, India; [2]Department of Nephrology, Muljibhai Patel Urological Hospital, Nadiad, Gujarat, India

1. Introduction

Biofilms are bacterial communities that are complex cellular aggregates adhered to solid surfaces using cellular proteins and a highly hydrated extracellular polymeric matrix. The framework of biofilm contains polysaccharides, protein, extracellular deoxyribonucleic acid (eDNA), fatty acid, etc., produced by bacteria to protect themselves from the surrounding world (Shahrour et al., 2019). Above 90% of bacteria existing in nature prevail as biofilms rather than freely floating planktonic cells (Ilyina et al., 2004). Biofilms are known culprits in nosocomial infections and are found on the surfaces of medical implants and prosthetics, tubing, endoscopes, wounds, etc., making them incongruous in healthcare settings (Malheiro & Simões, 2017). The materials used in indwelling medical devices are vulnerable to microbial attachment, growth, and infection, making them the prime risk factors for acquiring a nosocomial infection (Malheiro & Simões, 2017; Percival et al., 2015). Propensity of bacteria to develop biofilms on or within medical devices is responsible for numerous medical device failures, resulting in chronic infections and implant-related complications (Yadav et al., 2020a). It is also reported that there are 2 million healthcare-related infections, out of which infections from indwelling medical devices account for 50%–70%, and the device-related mortality ranges from <5% to >25% depending upon the device, from dental implants to mechanical heart valves (Bryers, 2008; Darouiche, 2004; Li et al., 2021; VanEpps & Younger, 2016; Weinstein & Darouiche, 2001).

Medical device–associated infections result from the formation of biofilms, which is hard to eradicate, and treatment with antibiotics fails, ultimately leading to the removal of the device for the well-being of patients (Zander & Becker, 2018). Pathogenic biofilms on implanted devices (Fig. 22.1), such as orthopedic implants, cochlear implants, etc., are of significant

Microbial Biofilms
https://doi.org/10.1016/B978-0-323-95715-1.00010-8

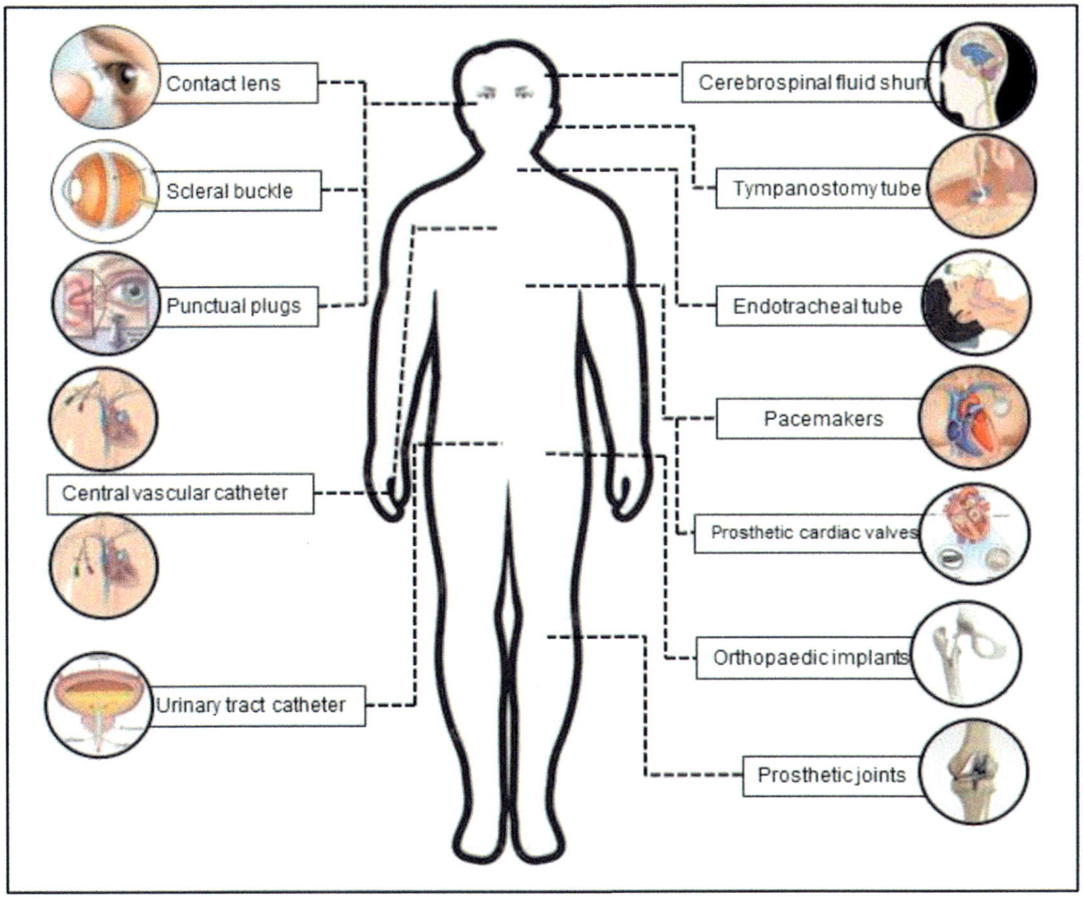

FIGURE 22.1 Commonly used medical devices and implants in the healthcare system associated with biofilm formation.

concern in medical device–associated infection (Veerachamy et al., 2014). Among all, tracheal and endotracheal tube aspiration isolates are the extensive biofilm producers (52.1%), followed by blood (17%) and urine (12.6%) (Gogoi et al., 2015). Microorganisms in a polymeric matrix cause chronic infection and persist locally at the site of infection, such as an implant surface (Xu et al., 2020). The most frequently reported pathogenic microbes in healthcare infections and biofilms of indwelling medical devices are *Escherichia coli, Staphylococcus aureus, Klebsiella* spp., *Pseudomonas aeruginosa, Enterococcus faecalis, coagulase-negative Staphylococcus* spp., *Enterococcus* spp, *Enterococcus faecium, Proteus* spp., *Candida albicans,* and others (Donlan, 2001; Weiner-Lastinger et al., 2020) (Fig. 22.2).

Biofilm in medical devices necessitates the removal and replacement of the contaminated implant with a new one, for the management of the infection, resulting in significant increases in mortality, morbidity, recovery period, and cost of therapy (Yadav et al., 2020a). Treatment of device-associated infections with antibiotics is challenging as bacteria become highly

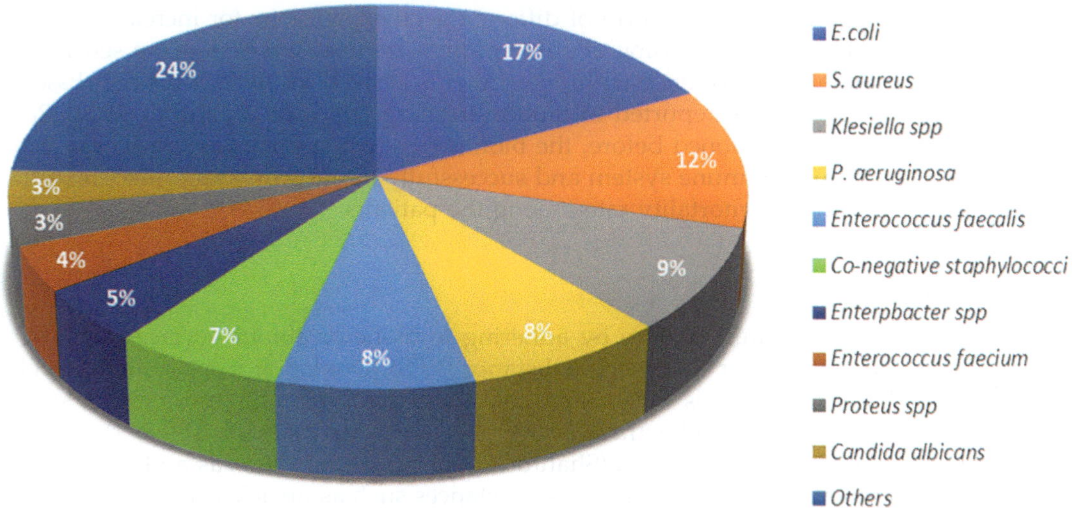

FIGURE 22.2 Infectious pathogens associated with healthcare medical devices.

resistant (even 1000 times more) to antibiotics, especially after biofilm formation (Mah, 2012; Yadav et al., 2020b). It contains a diverse cell population, including persisters or quiescent cells, which differentiate themselves into a metabolically inactive state highly tolerant to antibiotics (Shahrour et al., 2019). As a counteracting plan of action and to prevent microbial colonization on device surfaces, researchers have coated devices with antimicrobial substances, which mainly inhibits bacterial colonization and affirms the longevity of the implant in the patient (Zander & Becker, 2018).

To inhibit bacterial adhesion and biofilm formation, bactericidal compounds are coated onto the surface of medical devices; the bactericidal compounds include antibiotics, silver-based compounds, surfactants, antimicrobial peptides (AMPs), enzymes, and nanoparticles (NPs) (Atefyekta et al., 2019; Shahrour et al., 2021). Accumulation of host proteins on the surface of the foreign device implanted inside the body impairs its antimicrobial activity (Shahrour et al., 2021). Biofilm formation on medical devices reduces the bacterial susceptibility to conventional antibiotics while circumventing the human immune system, necessitating revision surgery and the use of antibiotics (Atefyekta et al., 2019). Treating biofilm and its consecutive infection is a challenge for public health. Hence this chapter addresses the role of biofilm on medical devices as a cause of secondary infections in humans and their effect on treatment.

2. Biofilm formation strategy on medical devices

Biofilm-forming cells have much higher suitability to colonize and persist inside a host than planktonic cells (Jamal et al., 2018). Furthermore, microbial colonization on medical devices depends on the physiochemical properties of the material, fluids to which the device is exposed, nutrient availability, and temperature (Percival et al., 2015). Induction of the general stress responds significantly in the form of different gene expressions for increased multiple drug resistance (MDR), overexpression of efflux pumps, modulation of quorum sensing systems (QS), and alteration of outer membrane proteins as well as surface charge. Essential modifications are extensively reported in biofilm-forming live cells (Shahrour et al., 2019). Following the strategy mentioned before, the biofilm-producing microorganisms can extensively escape the human immune system and successfully frame a dreadful infection, which elevates the morbidity and mortality rate among the patients.

2.1 Biofilm structure

Microbial colonies architect biofilm by adhering to biotic or abiotic surfaces and encase themselves in an extracellular polymeric substance (EPS) (Donlan & Costerton, 2002; Hall-Stoodley & Stoodley, 2005). This EPS matrix comprises a complex mixture of EPS, such as exopolysaccharides, proteins, and extracellular DNA, which accounts for 50%–90% of the total organic content of the biofilm mass (Shahrour et al., 2019). EPS acts as a physical barrier against external threats and traps exogenous substances such as nucleic acids, proteins, minerals, nutrients, and cell wall components found in the local environment (Shahrour et al., 2019). EPS can slow down or completely block the penetration of some antimicrobials into the biofilm cells. Fig. 22.3 shows one of the properties of bacterial biofilms, namely the physical and chemical gradients of nutrients, antibiotics, and oxygen from the peripheral surface to the bottom (Yadav et al., 2020b). Bacterial colony produces biofilm in response to stress as part of the survival mechanism and starts architecting EPS. The most common carbohydrates in the EPS matrix of *E. faecalis, S. aureus, Klebsiella pneumoniae, Acinetobacter baumannii, P. aeruginosa*, and *Enterobacter* spp. are mannose, galactose, and glucose, followed by *N*-acetylglucosamine, galacturonic acid, arabinose, fucose, rhamnose, and xylose. Microorganisms such as *E. coli* and *P. aeruginosa* have evolved a stage higher in biofilm production with the involvement of colanic acid and alginate in their EPS matrix (Bales et al., 2013; Di Somma et al., 2020; Rabin et al., 2015). Extracellular proteins are another essential element of the EPS matrix. In *P. aeruginosa*, biofilm formation and stabilization are aided by proteins linked to cell surfaces and exopolysaccharides such as alginate, Pel, and psl (Rabin et al., 2015). *Staphylococcus epidermidis* and *S. aureus* polysaccharide intracellular adhesion (PIA) antigen play a role in the initial attachment. Furthermore, PIA/poly-*N*-acetylglucosamine (PNAG)-dependent biofilms appear to be more robust and help the proliferating bacteria to bypass polymorphonuclear leukocytes and phagocytosis (Khatoon et al., 2018; Meskin, 1998). Glucan-binding proteins (Gbps) of *Staphylococcus mutans* play a crucial role in biofilm's architectural preservation. However, in *Pseudomonas* spp., the insoluble fibrous protein amyloids cause cell aggregation and enhance the biofilm architecture (Rabin et al., 2015).

Extracellular matrix enzymes such as dextranase and fructanase in *S. mutants* can break down biopolymers and release carbon and energy resources for the survival of the microbial cells present in the biofilm, especially during nutritional deprivation (Nguyen et al., 2012). Detachment and dispersal of biofilm also require enzymatic functions. EPS also contains extracellular DNAs (eDNAs), which not only derive from lysed cells but also are actively released, and eDNA is found to be critical for biofilm adhesion to the surface. eDNA can chelate Mg^{2+} and activate PhoPQ/PmrAB two-component systems, leading to antimicrobial peptide resistance in *P. aeruginosa, Salmonella enterica* serovar Typhimurium, and other gram-

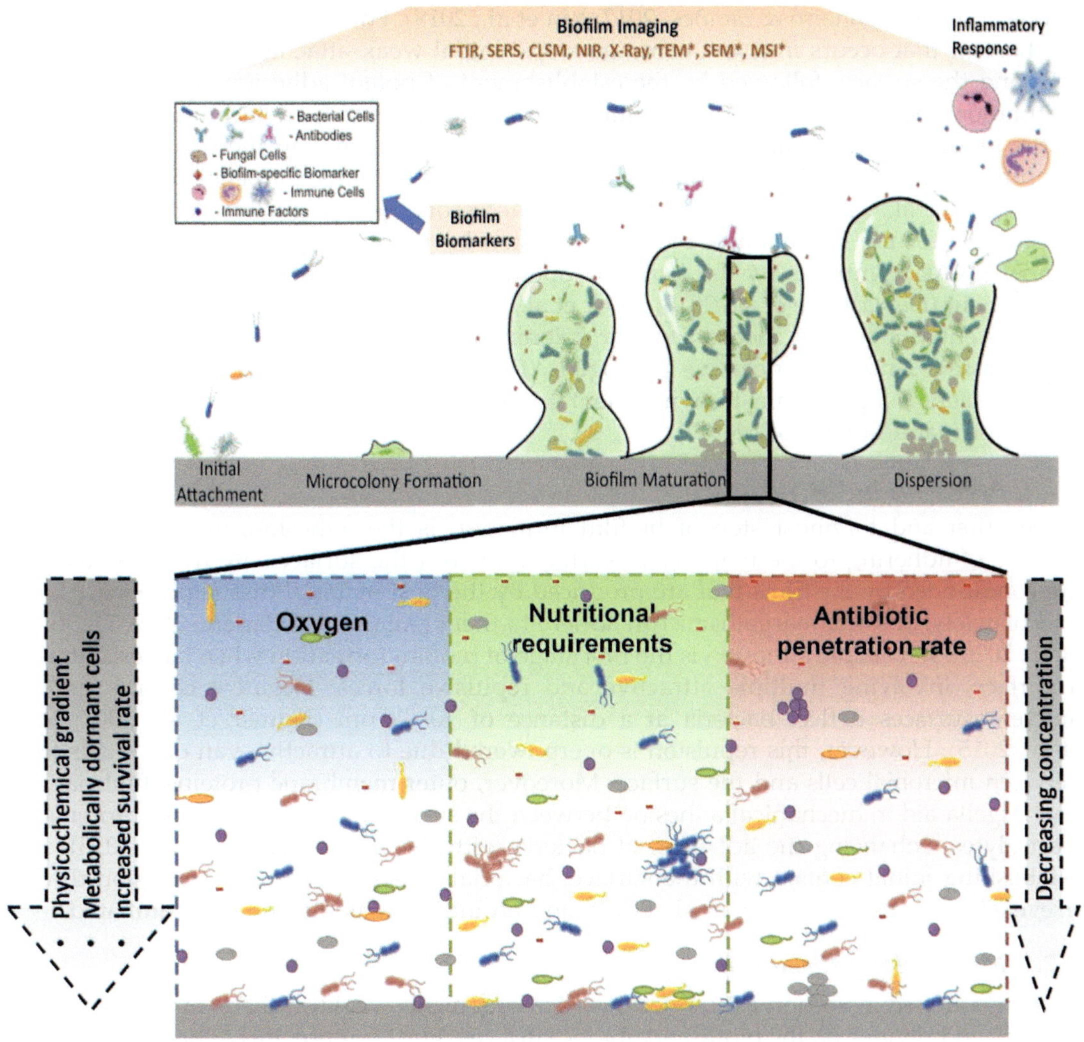

FIGURE 22.3 Schemetic representation of various stages of biofilm production on the abiotic surface (Xu et al., 2020) and biofilm micro-environment, depicting the physical and chemical gradient of various constituents from surface going towards the bottom.

negative bacteria (Johnson et al., 2013; Mulcahy et al., 2008). Components of the EPS matrix help bacteria survive in harsh conditions by managing nutritional sources and protecting against antimicrobial substances.

2.2 Biofilm genesis

Biofilm genesis is a microorganism's survival mechanism. Development of polysaccharide matrix structure allows them to withstand environmental stress while attaching to a surface or one another (Malheiro & Simões, 2017; Yin et al., 2019). Formation of biofilm is a complex mechanism that occurs in a few common steps: initial weak attachment between microbial cells and the surface, followed by the establishment of potent adhesion and microcolonies, maturation and formation of the biofilm architecture, and finally, detachment or dispersion of the biofilm (Jamal et al., 2018; Tandogan et al., 2017). Signaling mechanisms, including quorum sensing (QS) and nucleotide second messenger systems, influence biofilm development (Yin et al., 2021). In staphylococci, the accessory gene regulator Agr QS system is responsible for the induction of biofilm formation by upregulation of poly-*N*-acetylglucosamine (PNAG) synthesis by the *icaADBC* operon and also plays a role in maturation and detachment (Dewasthale et al., 2018; Mack et al., 2006). Nucleotide second messenger system is an essential signaling mechanism for colonization and adaptation and found in many gram-positive and gram-negative pathogens (Pletzer et al., 2016). Let us discuss how the mighty bacteria fight against environmental stress and successfully assemble the requirements for manifesting biofilm.

2.2.1 *Adhesion to the surface*

The first and foremost step of biofilm formation is the adhesion; microorganisms are capable of adhering to the majority of surfaces. Even if the surfaces are hydrophobic, they are conditioned by the films that are produced by the host material (tissue proteins, plasma, and platelets) and microorganisms can adhere to them (Agarwal & Radera, 2019; Di Somma et al., 2020). Reversible adhesion is the first stage of biofilm formation when bacteria attach to a surface, involving multiple attractive and repulsive forces. Negative charges on most ambient surfaces deflect bacteria at a distance of 10–20 nm (Palmer et al., 2007; Rabin et al., 2015). However, this repulsion is overpowered due to attractive van der Waals forces between microbial cells and the surface. Moreover, outer membrane proteins, fimbriae, pili, and flagella aid in mechanical adhesion between the surfaces in the presence of proteins and electrolytes, enhancing the adhesion of bacteria with the surface (Rabin et al., 2015). After establishing initial contact with the surface, bacterial cells reinforce their adhesion through irreversible bonds (Tandogan et al., 2017), and an initial weak attachment is initiated.

2.2.2 *Microcolony formation*

As mentioned in Section 2.2.1, once the microorganisms weakly adhere, microbial cell proliferates and divides on the biotic and abiotic surfaces. Furthermore, this adhesion helps them to form a stable microcolony through the production of EPS (Veerachamy et al., 2014). Following the initial attachment on the surface and microcolonization, many genes are activated and regulation and communications are maintained via QS molecules. This cellular activity continuously influences and manages the density of the microbial population. QS molecules further control genomic expression that regulates EPS production and promotes pathogenicity (Davies et al., 1998; Veerachamy et al., 2014). The microcolony grows and acquires a mushroom-like architecture and flourishes further to life as a matured biofilm (Di Somma et al., 2020).

2.2.3 *Maturation and architecture*

A mature biofilm is thought to be in a steady state, with balanced levels of nutrient transfer and biological activity. Microcolony may be composed of 10%–25% cells and 75%–90% EPS matrix. Mature biofilm has a stable three-dimensional (3D) structure that contains water channels to distribute nutrients that effectively exchange substrate, excretion of metabolic end products, and signaling molecules within the biofilm (Yin et al., 2019). The formation of a mature biofilm requires a time ranging from several hours to several weeks (Tandogan et al., 2017).

2.2.4 Detachment or dispersion of biofilm

Detachment or dispersal is the final phase of the biofilm formation cycle, where under suitable conditions, microbial cells can continue to settle on the surfaces and thus enter another biofilm life cycle. The detachment can occur either by shearing force of fluid flow or by enzymes such as deoxyribonuclease and dispersin B, which disintegrate the EPS and allow biofilm cells to move away from the mature biofilm and subsequent surface recolonization (Dewasthale et al., 2018; Di Somma et al., 2020; Tandogan et al., 2017). In addition, dispersal could be a way for bacteria to colonize new niches before space and nutrition resources become scarce (Hall-Stoodley & Stoodley, 2005). In the case of biofilm-associated medical device infection, the detachment can cause disseminated chronic infections that can lead to morbidity and mortality.

3. Biofilm and abiotic medical implants

Microbial biofilms on medical devices pose a serious threat to patient health and impair device function (Yadav et al., 2020a). Biofilms may be made up of single or multiple species of microorganisms, depending on the device and the length of time it is implanted in the patient (Donlan, 2001). Infections associated with devices placed in the vascular space certainly have a significant risk of bloodstream-related infections (VanEpps & Younger, 2016). Late-onset implant-associated infections are frequently associated with occult (subclinical) clinical manifestations that can delay disease detection by weeks, months, and sometimes years, with resolution often necessitating reoperation and replacement of the infected device (Edmiston et al., 2016).

3.1 Origin of microbial contamination on medical implant

The origin of the microbial contamination lies mainly in the skin microflora of the patients or healthcare professionals who come into contact with the medical devices during the insertion of the device (Donlan, 2001; Zander & Becker, 2018). After the insertion of medical devices, the surface of the devices is conditioned by the patient's proteins and other molecules in the body fluids. This helps the microbes attach to the surface and develop biofilm (Zander & Becker, 2018).

3.2 Orthopedic implant

Biofilm infections in orthopedic implants are a significant concern worldwide as they contribute to severe complications and mortality. Orthopedic implants are routinely used in the treatment such as fractures, correction of deformities, joint replacements, and soft tissue anchorage (Filipović et al., 2020; Ronin et al., 2022). Every year, more than 1 million hip and knee arthroplasties are performed in the United States (Ronin et al., 2022). Biofilm-associated infection of orthopedic implants can occur either as an exogenous process, with device contamination during surgery (or early in the postoperative period), or through bacteremia at any time after surgery; the major pathways of infections occur through perioperative inoculation, hematogenous inoculation, or contiguous (Edmiston et al., 2016; Trampuz & Zimmerli, 2005). Infections associated with orthopedic implants and prosthetic joint replacements can cause osteomyelitis, prosthetic joint infection, and septic arthritis (Dibartola et al., 2017). Most common causative agents of these infections are associated with the normal microflora of the skin, such as *S. aureus, S. epidermidis,* other coagulase-negative *staphylococci,* and *enterococci;* infections are also polymicrobial (Coraça-Huber et al., 2020; Dibartola et al., 2017; Moojen et al., 2007; Tunney et al., 1999).

3.3 Catheters

Catheters are used to treat various disorders; however, chronic use of catheters increases the likelihood of microbial biofilm formation (Sabir et al., 2017; Yadav et al., 2020b). Hemodialysis, parenteral feeding, and the delivery of chemotherapy or other treatments all require the use of catheters, such as central venous catheters. Patients are anticipated to have one episode of bacteremia every 1000 days of catheter use (Shahrour et al., 2019). Most bacteria growing on catheters are of the biofilm phenotype, and all the chronic infections are caused by biofilm (Høiby et al., 2015). In the case of acute infection, it is unclear whether they are caused by planktonic seeding or microscopic biofilm particles that have broken out into the bloodstream (Wolcott, 2020). It is crucial to make an early clinical diagnosis of sepsis and to confirm it with current microbiological approaches (Mermel et al., 2009). Commonly isolated microbes are *S. aureus, S. epidermidis, A. baumannii,* and *C. albicans* (Lin et al., 2019; Zander & Becker, 2018). *A. baumannii* is an emerging pathogen that causes catheter-related urinary tract infection, and *A. baumannii* biofilm produces a higher mortality rate than equivalent planktonic cells (Lin et al., 2019).

3.4 Cardiovascular implant

Interventional cardiologists and cardiac and vascular surgeons rely heavily on cardiovascular prostheses and medical devices, as well as a variety of complex devices available for patient management (Padera & Schoen, 2020). Infection of these intracardiac, arterial, and venous devices is frequently encountered in centers throughout the developed world (Baddour et al., 2003). 70% of the cardiovascular implantable electronic device infections are caused by coagulase-negative *Staphylococci* (42%) and *S. aureus* (29%), the remaining infections are caused by gram-negative (9%), other gram-positive (4%) organisms, and fungi (2%) (Palraj et al., 2017). This becomes a severe problem when it is a bridge-to-transplantation, it delays transplantation, and worse, it can lead to death. In patients undergoing bridge-to-transplant and destination therapy, infection is the leading cause of death (Padera, 2006). Infections in cardiovascular devices can lead to valvular endocarditis, and bloodstream infection with disseminated infections (Palraj et al., 2017).

3.5 Ocular implant

Biofilm formation imposes a constraint on the adoption and formulation of ocular devices, such as intraocular lenses, posterior contact lenses, conjunctival plugs, lacrimal intubation devices, scleral buckles, and orbital implants (Bispo et al., 2015). Many ocular infections are caused by microbial biofilms that form when such prosthetic device comes in contact with them and is implanted in the eye (Zegans et al., 2002). Biofilm on contact lenses is thought to contribute to the progression of keratitis (Behlau & Gilmore, 2008). *P. aeruginosa* is the predominant causative agent of lens-based microbial keratitis, and they form biofilm within 24 h once the bacteria adhered to the surface of the device (Cheng et al., 1999; Dutta et al., 2012; Houang et al., 2001; Lam et al., 2002). *Serratia marcescens, S. aureus,* and coagulase-negative staphylococci are the second most commonly found pathogens, and microorganisms associated with contact lens keratitis differ by geography, and in tropical climates, gram-negative bacteria are more common (Dutta et al., 2012; Lam et al., 2002).

4. Role of biofilm in disease

Medical implant surfaces provide a suitable environment for microorganisms to proliferate and produce biofilms (Yadav et al., 2020a). Microorganisms can colonize and form biofilms on or within the medical device, prostheses, and implants. Increasing evidence suggests that approximately 65% of chronic human infections are triggered by biofilms (Yadav et al., 2020b). Bacteria that produce biofilms have been shown to be crucial in many chronic infections, which presents a difficulty in therapeutic settings (Sabir et al., 2017). Infections such as chronic otitis media, chronic wound infections, endocarditis, pulmonary infections, chronic rhinosinusitis, endophthalmitis, catheter-associated urinary tract infection, keratitis, acute red eye, scleral buckle infection, lacrimal system, osteomyelitis, and other infections are just a few of the chronic or recurrent infections that biofilms can cause (Hall-Stoodley et al., 2012; Yadav et al., 2020b). The most prevalent pathogens associated with contaminating indwelling medical devices and growing biofilm on or within the medical devices are shown in Table 22.1, which also includes a list of some of these illnesses.

5. Prevention and control strategy for biofilm on medical devices

The increase in morbidity and mortality rate due to biofilm has led the scientific community to adopt various strategies to prevent the formation of biofilms on or within medical implants. Various technologies such as biocide coatings such as antibiotics, surfactants, AMPs, and NPs, as well as adhesion-resistant surfaces such as stainless steel, titanium, and certain hydrophobic surfaces, are used to manufacture different types of medical devices (Carvalho et al., 2020; Malheiro & Simões, 2017; Shahrour et al., 2021; Veerachamy et al., 2014). In this section, we will elaborate more on the various types of strategies utilized in the commercial world to control or limit the growth of biofilm on the implants.

TABLE 22.1 Medical devices—associated infections and their causative agent.

Medical devices	Infection	Biofilm producing microorganism	References
Cerebrospinal shunt	Cerebrospinal fluid shunt infection	*Staphylococcus aureus, coagulase-negative staphylococci, Citrobacter* spp., *Enterococcus* spp., *Pseudomonas* spp., *Klebsiella* spp.	Benachinmardi et al. (2017), Meskin (1998), Veerachamy et al. (2014)
Prosthetic heart valves	Endocarditis	*Streptococcus* spp., *S. aureus, Staphylococcus epidermidis, Enterococcus* spp., *Candida* spp.	Baddour et al. (2003), Donlan & Costerton (2002), Jamal et al. (2018)
Cardiac pacemakers	Endocarditis	*S. aureus, S. epidermidis, Streptococcus* spp., *Candida* spp.	Baddour et al. (2003), Yadav et al. (2020a)
Contact lenses	Keratitis	*Serratia* spp., *Pseudomonas aeruginosa, S. aureus,* other *Staphylococcus* spp.	Dutta et al. (2012), Jamal et al. (2018)
	Peripheral ulcer	*S. aureus*	Dutta et al. (2012)
Intraocular lens	Endophthalmitis	*S. epidermidis, Enterococcus* spp., *Propionibacterium* spp.	Garg et al. (2019)
Punctal plugs	Canaliculitis Dacryocystitis Conjunctivitis	*Staphylococcus haemolyticus* *Candida tropicalis*	Bispo et al. (2015), Yokoi et al. (2000)
	Lacrimal system infections	*Staphylococcus species*	
Scleral buckles	Scleral buckle infection	*Coagulase-negative staphylococci, Proteus mirabilis, Mycobacteroides chelonae*	Bispo et al. (2015), Garg et al. (2019)
Peritoneal dialysis catheters	Peritonitis	*Streptococci, staphylococci*	Agarwal & Radera (2019), VanEpps & Younger (2016)
Central venous catheters	Catheter-associated bloodstream infection	*S. aureus, coagulase-negative staphylococci, Streptococcus* spp., *Enterococcus* spp., *Klebsiella* spp., *Candida*	Magill et al. (2018)
Urinary catheters	Catheter-associated urinary tract infection	*Escherichia coli, P. aeruginosa, Klebsiella pneumoniae, Acinetobacter baumannii*	Percival et al. (2015)
Endotracheal tubes	Subglottic stenosis Ventilator-associated pneumonia Cystic fibrosis lung infection	*Streptococcus* spp., *Staphylococcus* spp., *E. coli, P. aeruginosa, Klebsiella* spp., *Enterobacter* spp., *Acinetobacter* spp.	Friedland et al. (2001), Hall-Stoodley & Stoodley (2005), Percival et al. (2015)
Orthopedic implant	Osteomyelitis	*S. aureus*	Hall-Stoodley et al. (2012)
Tympanostomy tube	Otitis media	*S. aureus, K. pneumoniae, Haemophilus influenza, Moraxella catarrhalis*	Pletzer et al. (2016)

5.1 Prevention of initial attachment

Initial attachment is a critical phase in biofilm formation, and limiting this can prove a successful strategy in preventing biofilm assembly on medical devices. Medical implants are covered with antimicrobial and antifouling substances such antibiotics, nitric oxide, furanone, chlorhexidine, and silver to prevent the growth of biofilm (Shahid et al., 2021). The material of implants is incorporated, coated, or covalently bonded with biocides that function by the gradual release of biocides, thereby killing the microbes coming in contact with it (Desrousseaux et al., 2013). In addition to this technology, antimicrobial release—based coating and nonreleasing coating are two more popular antimicrobial coating strategies for bactericidal action. Antimicrobial release—based coatings, as the name implies, releases antimicrobials after implantation, whereas nonreleasing-based coatings immobilize antimicrobial derivative on the device surface; both of them are proven strategies for bactericidal activity.

Moreover, antibiotics, antiseptics, metals, metal oxides, and furanones are all employed in release-based coatings. However, metal oxide NPs and cationic polymers are deployed as nonrelease-based coatings. Also, the cationic polymers employed are both natural and synthetic. That being said, there is no denying that while some antibacterial biocides such as silver compounds are very popular and have a broad spectrum of effects, they are toxic to human and animal cells (Ge et al., 2014; Kim et al., 2008).

On the contrary, certain molecular derivatives such as polymyxin B nanopeptides are less toxic (Duwe et al., 1986; Shahrour et al., 2021; Vaara, 2019). Other than this, antifouling coatings include superhydrophobic, poly-zwitterionic, PEG-based, and enzyme-based coatings (Shahid et al., 2021). In addition, it is also shown that by incorporating biocides into drug delivery vehicles such as degradable hydrogels, NPs, and liposomes, biocides can be released in a controlled manner, or that immobilizing biocides on the surface of medical devices improves stability and reduces toxicity (Atefyekta et al., 2019).

5.2 Disruption or eradication of established biofilm

Established biofilms are securely adhered to medical devices and are difficult to eliminate with typical antibiotic treatments (Zheng et al., 2018). To tackle biofilm infections, biofilm-dissolving tactics would be given top emphasis (Barraud et al., 2009). They can be eradicated by applying physical and chemical methods. In physical methods, the mechanical shear force is created by ultrasounds (Vyas et al., 2020), and magnetic fields (Li et al., 2019) are used to eradicate biofilms. When it comes to biochemical approaches, phage lysins, enzymatic treatment, and microbial secondary metabolites were used to eradicate biofilms (Yin et al., 2021). Phage lysins are peptidoglycan hydrolase enzymes produced by bacteriophages that cleave the peptidoglycan layer of the bacterial cell wall, which results in bacterial death (Schuch et al., 2017; Zhang et al., 2018). In the enzymatic treatment approach, the significant components of the biofilm materials are targeted using enzymes. Like in the case of *Candida auris* biofilm, they are rich in mannan-glucan, so applying mannosidase or glucanase will hydrolyze the biofilm matrix effectively (Dominguez et al., 2019).

Microbial secondary metabolites regulate gene expression by serving as intracellular signals, thus contributing to cellular processes, including biofilm (Dufour & Rao, 2011; Yang et al., 2012). Secondary metabolite called carolacton produced by *Sorangium cellulosum,* which is a soil bacterium, has high activity against *S. mutans* biofilm by causing cell morphology changes and cell death (Kunze et al., 2010). Likewise, rhamnolipid produced by *Pseudomonas* spp. showed activity against *S. aureus* biofilms (e Silva et al., 2017).

In *P. aeruginosa*, nitric oxide was previously recognized as a signal for biofilm dispersion. It has also been reported in the dispersion of pathogenic biofilm-forming microbes such as *E. coli, Vibrio cholerae, Neisseria gonorrhoeae, S. aureus,* and other bacterial multispecies and fungal species (Arora et al., 2015; Barraud et al., 2009; Schlag et al., 2007). The metabolites of probiotic bacteria such as *Lactobacillus* also can disrupt the preformed biofilms, which can also be used in the biofilm disruption/controlling strategy (Carvalho et al., 2020). However, traditional antibiotics should be used along with biofilm-controlling drugs to disperse and treat biofilm infections. Because most biofilm dispersing therapies do not kill pathogenic cells, combining biofilm dispersing agents with antibiotics may be beneficial (Rasmussen et al., 2005).

5.3 Biofilm signal disruption

Biofilm signal disruption is also an effective method to control biofilm, as biofilm formation is regulated by several signaling pathways, such as QS and nucleotide second messenger systems. Hence, interfering with the signaling pathway is one of the best methods to inhibit biofilm formation (Yin et al., 2021). It is evident that inference of QS signaling and antimicrobial treatment could be used successfully to treat biofilm infections in vivo (Christensen et al., 2012). QS signals to aid in the creation and maintenance of biofilm structures. However, if QS signal generation is prevented or hindered, biofilm structures become weak and unstructured (Abraham, 2016). To interrupt and regulate biofilm-forming pathogens, it is vital to disrupt and control the QS mechanism. This can be accomplished by limiting autoinducer synthesis by enzymatic breakdown of signaling molecules or by regulating their receptors via signal receptor blocking (Basavaraju et al., 2016; Gebreyohannes et al., 2019). Various bacterial and fungal species produce these kinds of quorum quenching molecules and enzymes (Chu et al., 2013; Hornby & Nickerson, 2004; Koul & Kalia, 2017).

6. Early detection of biofilm formation in medical implants

Adherence of pathogenic microbes and biofilm formation on medical device/implant surfaces causes chronic infection or systemic infection while persisting locally at the site of infection but eliciting a lesser inflammatory response than the planktonic cells (Loza-Correa & Ramírez-Arcos, 2017; Xu et al., 2020). Therefore, it is necessary to diagnose and prevent implant-associated infections. There are many methods to assess biofilm growth, which are limited, making it difficult to be consistent (Wilson et al., 2017). Due to the scarcity of biofilm-defined biomarkers and nondestructive imaging technologies, there are no specific identification methods to diagnose microbial biofilm in the early stage (Xu et al., 2020). Diagnosing a preformed biofilm in an implanted medical device in vivo is difficult because it is impossible to sample it without experiencing surgery and also challenging when cells are challenging to sample (Xu et al., 2020). Even if the cells can be harvested, slow-growing microbial variants and quiescent persister cells may not form colonies under conventional culturing conditions or culture-independent molecular techniques, and they can only capable of detecting planktonic cells, not sessile cells, consequently producing false results (Fisher et al., 2017; Shen et al., 2010). Early indications of biofilms in clinical specimens can be detected by routine light microscopy; the presence of bacteria with a surrounding matrix is the earliest indication of a biofilm-associated infection (Bjarnsholt et al., 2009). To diagnose medical device–associated biofilm, various methods are used, such as qualitative culture, elevated leukocytes count in blood and detection of antibodies, semiquantitative culture such as direct culture of device tip on the plate and culture of catheter fluid, and quantitative culture such as plate counting, in which sonication and vortex of suspected infected implants can improve culture positivity (Bose & Ghosh, 2015; Khatoon et al., 2018). There is a need to develop a suitable effective technique to overcome the difficulties. However, the diagnostic assays currently used to diagnose biofilms in healthcare settings are improved versions of conventional assays, which are routinely used to detect microbial infections (Yadav et al., 2020b). Confirmation of the presence of biofilm in biopsies of infected tissue can be identified by confocal microscopy and scanning electron microscopy, as well as species-specific identification of microbes through the application of hybridization probe techniques such as FISH (Malic et al., 2009; Xu et al., 2020). Furthermore, amplification and phylogenetic analysis of species specific genes, 16s rRNA gene and virulent genes can be utilized for identification of the pathogen (Kırmusaoğlu, 2019; Tunney et al., 1999).

7. Treatment approach for biofilm colonization on medical devices

Microbial adhesion and biofilm production on the surface of the medical implant hamper the therapeutic approach of the implants, and treatment of biofilm-associated infection is currently one of the significant challenges in medicine (Lee et al., 2021). Due to their protective mechanisms, such as EPS production, proliferation under stress, and quiescent nature, biofilms are highly tolerant to antimicrobial drugs and host defense mechanisms (Römling & Balsalobre, 2012). Because of this biofilm property, conventional treatment techniques are ineffective, thus posing severe treatment challenges for persistent biofilms and associated infections (Veerachamy et al., 2014). Adequate clearance of medical device–associated infections frequently necessitates two-stage revision surgeries and, in some circumstances, prolonged stays in critical care settings (Shahrour et al., 2021). The antibiotics used to combat biofilm infections have limited efficacy against infections associated with medical implants. Hence, further antibiotic treatment against these pathogens might result in the development of antimicrobial resistance (Zander & Becker, 2018). *S. aureus,* the most prevalent cause of device-associated infections, has exhibited significant antibiotic resistance after biofilm formation, resulting in significant treatment challenges (Hogan et al., 2016). Current antibiofilm approaches include systemic antibiotic prophylaxis during device insertion, antibiotic coating of implanted devices before insertion, early removal of an unnecessary device, surgical biofilm removal, and optimization of the antibiotic regimen against biofilm-induced infections (Girish et al., 2019). There are several antimicrobial drugs used to treat biofilm-forming *P. aeruginosa* strains, such as ciprofloxacin, which kills bacteria located on the biofilm surface.

In contrast, colistin kills bacteria located deep in the biofilm (Jamal et al., 2018). NPs, which are also used to treat biofilm-associated infections, such as therapy with inhalable NPs, could be effective against *P. aeruginosa* biofilms in patients with lung infection (Shahrour et al., 2019). Many other reliable options can be applied for the treatment of biofilm-associated infections, such as antibiotic lock therapy (ALT), antimicrobial peptides (AMPs), and phage therapy.

7.1 Antibiotic lock therapy

ALT is an alternative therapy to treat biofilm-associated medical device infections such as catheters (Rao et al., 2021). The ALT consists of filling the lumen of the catheter with a solution containing antimicrobial agents at extremely high concentrations to eradicate biofilms (Hogan et al., 2016). Antimicrobials are used at concentrations varying from 100 to 1000 times their planktonic MICs in the presence of an anticoagulant such as heparin at the site of infection and lead to a continuous release of the antimicrobial agent over a certain period (Shahrour et al., 2019). Several additives to the lock solution include ion chelators (citrate or ethylenediaminetetraacetic acid [EDTA]), anticoagulant (heparin), and antibiotics (Justo & Bookstaver, 2014). The administration of antibiotic lock solution (B-lock), a sterile liquid solution that contains trimethoprim, ethanol, and Ca-EDTA in a phosphate buffer, demonstrates the bactericidal/fungicidal activity against *Candida* spp., *E. faecalis*, methicillin-resistant *S. aureus*, methicillin-sensitive *S. aureus*, *P. aeruginosa*, and coagulase-negative *Staphylococcus* spp. (Ghannoum et al., 2011). The broad-spectrum activity for B-lock can be explained by its active ingredients, ethanol and trimethoprim. Ethanol disrupts the membrane structure, leading to a defective cell division process and facilitating the entry of trimethoprim into the cell. This prevents the enzyme dihydrofolate reductase, which plays a vital role in the synthesis of nucleic acid precursors (Ghannoum et al., 2011). ALT is an effective therapy for eradicating biofilm without removal of the device and can be carried out on an outpatient basis, reducing the adverse effects associated with a systemic antimicrobial treatment (Rao et al., 2021). There are potential risks associated with ALT, which are that the solution may expose the patient to an unnecessary high concentration of antibiotics associated with severe toxicities, the inappropriate concentration of anticoagulant promoting catheter blockage, and a low dosage of antibiotics that may increase the development of resistance (Hogan et al., 2016; Justo & Bookstaver, 2014). AMPs are a suitable choice for ALT due to their broad antimicrobial spectrum and biofilm disruption characteristics (Shahrour et al., 2019).

7.2 Antimicrobial peptides

AMPs are small amphiphilic oligopeptides with numerous properties that can be used to develop antibiofilm therapy. The AMP molecules have a broad spectrum of microbicidal activity, reduced propensity to induce resistance, the potential to kill quiescent cells, and synergistic activity with antibiotics combination (Shahrour et al., 2019). AMPs can interact with microbes and disrupt microbial membranes. Some AMPs have been reported to interrupt mechanisms necessary for biofilm formation, including EPS synthesis, maturation, regulation of QS signaling, and persister cell formation (Rao et al., 2021; Shahrour et al., 2019). Cathelicidin LL-37 is a well-studied human AMP that downregulates more than 50 QS-controlled genes of *P. aeruginosa* at nonlethal concentrations, inhibiting biofilm formation and initial adhesion of slime-producing *S. epidermidis* (Kazemzadeh-Narbat et al., 2021). The synthetic compound IDR-1018 binds to the second messenger ppGpp and stimulates its degradation within the cell. It is synthesized in response to harsh environmental conditions and plays an essential role in biofilm formation and the regulation of persister cell formation (De la Fuente-Núñez et al., 2014). The combined therapy of AMPs with antimicrobials and antibiofilm compounds is more effective in killing a broad range of microbes at low concentrations (Shahrour et al., 2019). However, AMP molecules are expensive to produce on an industrial scale because it requires complex extraction and purification steps (Shahrour et al., 2019).

7.3 Phage therapy

Bacteriophages are viruses that are the most abundant entities on earth that infect bacteria and have recently been explored for their hydrolytic effect to remove biofilms in medical implants (Connaughton et al., 2014). Numerous clinical trials are ongoing to study bacteriophages that may be used as an alternative to antibiotics (Agarwal & Radera, 2019). Phage encodes the enzymes called depolymerases, which break down bacterial capsular polysaccharides, allowing the phage to access its receptor on the surface of the cells (Pires et al., 2020). Phage therapy uses single phages or phage cocktails, phage-derived enzymes, phages in combination with antibiotics, and genetically engineered phages (Pires et al., 2020). Engineered phages that also express antibiofilm enzymes, such as engineered T7 bacteriophage, express dispersin, which hydrolyzes the β-1, 6-*N*-acetyl-D-glucosamine used to infect *E. coli* and kill the cells in the biofilm and also plays a role in the breakdown of the EPS matrix of the biofilm, and may be a promising alternative for biofilm eradication at the site of infection (Lu & Collins, 2007). A phage cocktail was formulated to treat catheter-related urinary tract infections caused by *Proteus mirabilis*, which shows potent biofilm destruction activity and prevents biofilm formation (Ferriol-González & Domingo-Calap, 2020). Phages can be exciting solutions to control the spreading step of biofilm infection; on the other hand, when phages are used as therapeutic tools in humans or animals, the disadvantage is that they can induce a rapid release of bacterial endotoxins, leading to unwanted inflammatory responses (Ferriol-González & Domingo-Calap, 2020).

8. Conclusion

In modern healthcare settings, the use of medical devices and prostheses is inevitable as most of the treatment and therapies involve them in one way or another. Despite all the advances and discoveries and developments of antibiotics, infections by microbes are inevitable. To escalate this, antimicrobial resistance makes treatment even more difficult. In addition, the ability of microbes to form a biofilm complicates the situation even further as the biofilm is much more challenging to manage and leads to treatment failure. In medical settings, biofilm-related illnesses increase complications and morbidity. Current knowledge of the role of the

EPS matrix component in biofilm maturation and persistence is limited, and more research is needed to understand its role in biofilm formation and to develop novel therapeutic strategies by targeting it. Understanding cell-to-cell communication allows for a better understanding of biofilm formation and more effective control strategies. To decrease the morbidity and death rates associated with acute and chronic infections, new therapeutic strategies other than antibiotics and recurrent surgeries are urgently needed. In this chapter, we examined many aspects of biofilm-related infections in medical devices, such as prevention, control, and treatment options. With the development of more treatment and prevention strategies, in the near future, they might be introduced into the application.

References

Abraham, W. R. (2016). Going beyond the control of quorum-sensing to combat biofilm infections. *Antibiotics, 5*(1), 3. https://doi.org/10.3390/ANTIBIOTICS5010003

Agarwal, J., & Radera, S. (2019). Biofilm-mediated urinary tract infections. In *Biofilms in human diseases: Treatment and control* (pp. 177–213). Cham: Springer. https://doi.org/10.1007/978-3-030-30757-8_13

Arora, D. P., Hossain, S., Xu, Y., & Boon, E. M. (2015). Nitric oxide regulation of bacterial biofilms. *Biochemistry, 54*(24), 3717–3728. https://doi.org/10.1021/BI501476N

Atefyekta, S., Pihl, M., Lindsay, C., Heilshorn, S. C., & Andersson, M. (2019). Antibiofilm elastin-like polypeptide coatings: Functionality, stability, and selectivity. *Acta Biomaterialia, 83*, 245–256. https://doi.org/10.1016/j.actbio.2018.10.039

Baddour, L. M., Bettmann, M. A., Bolger, A. F., Epstein, A. E., Ferrieri, P., Gerber, M. A., Gewitz, M. H., Jacobs, A. K., Levison, M. E., Newburger, J. W., Pallasch, T. J., Wilson, W. R., Baltimore, R. S., Falace, D. A., Shulman, S. T., Tani, L. Y., & Taubert, K. A. (2003). Nonvalvular cardiovascular device-related infections. *Circulation, 108*(16), 2015–2031. https://doi.org/10.1161/01.CIR.0000093201.57771.47

Bales, P. M., Renke, E. M., May, S. L., Shen, Y., & Nelson, D. C. (2013). Purification and characterization of biofilm-associated EPS exopolysaccharides from ESKAPE organisms and other pathogens. *PLoS One, 8*(6), e67950. https://doi.org/10.1371/journal.pone.0067950

Barraud, N., Storey, M. V., Moore, Z. P., Webb, J. S., Rice, S. A., & Kjelleberg, S. (2009). Nitric oxide-mediated dispersal in single- and multi-species biofilms of clinically and industrially relevant microorganisms. *Microbial Biotechnology, 2*(3), 370–378. https://doi.org/10.1111/j.1751-7915.2009.00098.x

Basavaraju, M., Sisnity, V. S., Palaparthy, R., & Addanki, P. K. (2016). Quorum quenching: Signal jamming in dental plaque biofilms. *Journal of Dental Sciences, 11*(4), 349–352. https://doi.org/10.1016/J.JDS.2016.02.002

Behlau, I., & Gilmore, M. S. (2008). Microbial biofilms in ophthalmology and infectious disease. *Archives of Ophthalmology, 126*(11), 1572–1581. https://doi.org/10.1001/archopht.126.11.1572

Benachinmardi, K. K., Ravikumar, R., & Indiradevi, B. (2017). Role of biofilm in cerebrospinal fluid shunt infections: A study at tertiary neurocare center from South India. *Journal of Neurosciences in Rural Practice, 8*(3), 335–341. https://doi.org/10.4103/jnrp.jnrp_22_17

Bispo, P. J. M., Haas, W., & Gilmore, M. S. (2015). Biofilms in infections of the eye. *Pathogens, 4*(1), 111–136. https://doi.org/10.3390/pathogens4010111

Bjarnsholt, T., Jensen, P.Ø., Fiandaca, M. J., Pedersen, J., Hansen, C. R., Andersen, C. B., Pressler, T., Givskov, M., & Høiby, N. (2009). *Pseudomonas aeruginosa* biofilms in the respiratory tract of cystic fibrosis patients. *Pediatric Pulmonology, 44*(6), 547–558. https://doi.org/10.1002/ppul.21011

Bose, S., & Ghosh, A. K. (2015). Diagnosis of biofilm-associated infections in medical devices. *Biomaterials and Medical Device - Associated Infections*, 71–80. https://doi.org/10.1533/9780857097224.1.71

Bryers, J. D. (2008). Medical biofilms. *Biotechnology and Bioengineering, 100*(1), 1–18. https://doi.org/10.1002/bit.21838

Carvalho, F. M., Teixeira-Santos, R., Mergulhão, F. J. M., & Gomes, L. C. (2020). The use of probiotics to fight biofilms in medical devices: A systematic review and meta-analysis. *Microorganisms, 9*(1), 27. https://doi.org/10.3390/microorganisms9010027

Cheng, K. H., Leung, S. L., Hoekman, H. W., Beekhuis, W. H., Mulder, P. G. H., Geerards, A. J. M., & Kijlstra, A. (1999). Incidence of contact-lens-associated microbial keratitis and its related morbidity. *Lancet, 354*(9174), 181–185. https://doi.org/10.1016/S0140-6736(98)09385-4

Christensen, L. D., Van Gennip, M., Jakobsen, T. H., Alhede, M., Hougen, H. P., Høiby, N., Bjarnsholt, T., & Givskov, M. (2012). Synergistic antibacterial efficacy of early combination treatment with tobramycin and quorum-sensing inhibitors against *Pseudomonas aeruginosa* in an intraperitoneal foreign-body infection mouse model. *The Journal of Antimicrobial Chemotherapy, 67*(5), 1198–1206. https://doi.org/10.1093/JAC/DKS002

Chu, Y. Y., Nega, M., Wölfle, M., Plener, L., Grond, S., Jung, K., & Götz, F. (2013). A new class of quorum quenching molecules from Staphylococcus species affects communication and growth of gram-negative bacteria. *PLoS Pathogens, 9*(9), e1003654. https://doi.org/10.1371/JOURNAL.PPAT.1003654

Connaughton, A., Childs, A., Dylewski, S., & Sabesan, V. J. (2014). Biofilm disrupting technology for orthopedic implants: What's on the horizon? *Frontiers in Medicine, 1*, 22. https://doi.org/10.3389/fmed.2014.00022

Coraça-Huber, D. C., Kreidl, L., Steixner, S., Hinz, M., Dammerer, D., & Fille, M. (2020). Identification and morphological characterization of biofilms formed by strains causing infection in orthopedic implants. *Pathogens, 9*(8), 649. https://doi.org/10.3390/pathogens9080649

Darouiche, R. O. (2004). Treatment of infections associated with surgical implants. *New England Journal of Medicine, 350*(14), 1422–1429. https://doi.org/10.1056/nejmra035415

Davies, D. G., Parsek, M. R., Pearson, J. P., Iglewski, B. H., Costerton, J. W., & Greenberg, E. P. (1998). The involvement of cell-to-cell signals in the development of a bacterial biofilm. *Science, 280*(5361), 295–298. https://doi.org/10.1126/science.280.5361.295

De la Fuente-Núñez, C., Reffuveille, F., Haney, E. F., Straus, S. K., & Hancock, R. E. W. (2014). Broad-spectrum antibiofilm peptide that targets a cellular stress response. *PLoS Pathogens, 10*(5), e1004152. https://doi.org/10.1371/journal.ppat.1004152

Desrousseaux, C., Sautou, V., Descamps, S., & Traoré, O. (2013). Modification of the surfaces of medical devices to prevent microbial adhesion and biofilm formation. *Journal of Hospital Infection, 85*(2), 87–93. https://doi.org/10.1016/J.JHIN.2013.06.015

Dewasthale, S., Mani, I., & Vasdev, K. (2018). Microbial biofilm: Current challenges in health care industry. *Journal of Applied Biotechnology & Bioengineering, 5*(3), 160–164. https://doi.org/10.15406/jabb.2018.05.00132

Di Somma, A., Moretta, A., Canè, C., Cirillo, A., & Duilio, A. (2020). Antimicrobial and antibiofilm peptides. *Biomolecules, 10*(4), 652. https://doi.org/10.3390/biom10040652

Dibartola, A. C., Swearingen, M. C., Granger, J. F., Stoodley, P., & Dusane, D. H. (2017). Biofilms in orthopedic infections: A review of laboratory methods. *Apmis, 125*(4), 418–428. https://doi.org/10.1111/apm.12671

Dominguez, E. G., Zarnowski, R., Choy, H. L., Zhao, M., Sanchez, H., Nett, J. E., & Andes, D. R. (2019). Conserved role for biofilm matrix polysaccharides in Candida auris drug resistance. *MSphere, 4*(1). https://doi.org/10.1128/MSPHEREDIRECT.00680-18, 006800-18.

Donlan, R. M. (2001). Biofilms and device-associated infections. *Emerging Infectious Diseases, 7*(2), 277–281. https://doi.org/10.3201/eid0702.010226

Donlan, R. M., & Costerton, J. W. (2002). Biofilms: Survival mechanisms of clinically relevant microorganisms. *Clinical Microbiology Reviews, 15*(2), 167–193. https://doi.org/10.1128/CMR.15.2.167-193.2002

Dufour, N., & Rao, R. P. (2011). Secondary metabolites and other small molecules as intercellular pathogenic signals. *FEMS Microbiology Letters, 314*(1), 10–17. https://doi.org/10.1111/J.1574-6968.2010.02154.X

Dutta, D., Cole, N., & Willcox, M. (2012). Factors influencing bacterial adhesion to contact lenses. *Molecular Vision, 18*, 14–21. https://www.ncbi.nlm.nih.gov/pmc/articles/PMC3258521/.

Duwe, A. K., Rupar, C. A., Horsman, G. B., & Vas, S. I. (1986). In vitro cytotoxicity and antibiotic activity of polymyxin B nonapeptide. *Antimicrobial Agents and Chemotherapy, 30*(2), 340–341. https://doi.org/10.1128/AAC.30.2.340

Edmiston, C. E., McBain, A. J., Kiernan, M., & Leaper, D. J. (2016). A narrative review of microbial biofilm in postoperative surgical site infections: Clinical presentation and treatment. *Journal of Wound Care, 25*(12), 693–702. https://doi.org/10.12968/jowc.2016.25.12.693

Ferriol-González, C., & Domingo-Calap, P. (2020). Phages for biofilm removal. *Antibiotics, 9*(5), 268. https://doi.org/10.3390/antibiotics9050268

Filipović, U., Dahmane, R. G., Ghannouchi, S., Zore, A., & Bohinc, K. (2020). Bacterial adhesion on orthopedic implants. *Advances in Colloid and Interface Science, 283*, 102228. https://doi.org/10.1016/J.CIS.2020.102228

Fisher, R. A., Gollan, B., & Helaine, S. (2017). Persistent bacterial infections and persister cells. *Nature Reviews Microbiology, 15*(8), 453–464. https://doi.org/10.1038/nrmicro.2017.42

Friedland, D. R., Rothschild, M. A., Delgado, M., Isenberg, H., & Holzman, L. (2001). Bacterial colonization of endotracheal tubes in intubated neonates. *Archives of Otolaryngology - Head and Neck Surgery, 127*(5), 525–528. https://doi.org/10.1001/archotol.127.5.525

Garg, P., Garg, R., & Raj, P. (2019). Biofilm-mediated diseases of the eye. In *Biofilms in human diseases: Treatment and control* (pp. 117–126). Cham: Springer. https://doi.org/10.1007/978-3-030-30757-8_8

Gebreyohannes, G., Nyerere, A., Bii, C., & Sbhatu, D. B. (2019). Challenges of intervention, treatment, and antibiotic resistance of biofilm-forming microorganisms. *Heliyon, 5*(8), e02192. https://doi.org/10.1016/j.heliyon.2019.e02192

Ge, L., Li, Q., Wang, M., Ouyang, J., Li, X., & Xing, M. M. Q. (2014). Nanosilver particles in medical applications: Synthesis, performance, and toxicity. *International Journal of Nanomedicine, 9*, 2399. https://doi.org/10.2147/IJN.S55015

Ghannoum, M. A., Isham, N., & Jacobs, M. R. (2011). Antimicrobial activity of B-lock against bacterial and Candida spp. causing catheter-related bloodstream infections. *Antimicrobial Agents and Chemotherapy, 55*(9), 4430–4431. https://doi.org/10.1128/AAC.00677-11

Girish, V. M., Liang, H., Aguilan, J. T., Nosanchuk, J. D., Friedman, J. M., & Nacharaju, P. (2019). Anti-biofilm activity of garlic extract loaded nanoparticles. *Nanomedicine: Nanotechnology, Biology, and Medicine, 20*, 102009. https://doi.org/10.1016/j.nano.2019.04.012

Gogoi, M., Sharma, A., & Hazarika, N. K. (2015). Biofilm formation by bacterial isolates from patients on indwelling medical devices. *Indian Journal of Medical Microbiology, 33*(2), 319–320. https://doi.org/10.4103/0255-0857.154896

Hall-Stoodley, L., & Stoodley, P. (2005). Biofilm formation and dispersal and the transmission of human pathogens. *Trends in Microbiology, 13*(1), 7–10. https://doi.org/10.1016/j.tim.2004.11.004

Hall-Stoodley, L., Stoodley, P., Kathju, S., Høiby, N., Moser, C., William Costerton, J., Moter, A., & Bjarnsholt, T. (2012). Towards diagnostic guidelines for biofilm-associated infections. *FEMS Immunology and Medical Microbiology, 65*(2), 127–145. https://doi.org/10.1111/j.1574-695X.2012.00968.x

Hogan, S., Zapotoczna, M., Stevens, N. T., Humphreys, H., O'Gara, J. P., & O'Neill, E. (2016). In vitro approach for identification of the most effective agents for antimicrobial lock therapy in the treatment of intravascular catheter-related infections caused by *Staphylococcus aureus*. *Antimicrobial Agents and Chemotherapy, 60*(5), 2923–2931. https://doi.org/10.1128/AAC.02885-15

Høiby, N., Bjarnsholt, T., Moser, C., Bassi, G. L., Coenye, T., Donelli, G., Hall-Stoodley, L., Holá, V., Imbert, C., Kirketerp-Møller, K., Lebeaux, D., Oliver, A., Ullmann, A. J., Williams, C., & ESCMID Study Group for Biofilms (ESGB), & Consulting External Expert Werner Zimmerli. (2015). ESCMID* guideline for the diagnosis and treatment of biofilm infections 2014. *Clinical Microbiology and Infection, 21*, S1–S25. https://doi.org/10.1016/j.cmi.2014.10.024

Hornby, J. M., & Nickerson, K. W. (2004). Enhanced production of farnesol by Candida albicans treated with four azoles. *Antimicrobial Agents and Chemotherapy, 48*(6), 2305. https://doi.org/10.1128/AAC.48.6.2305-2307.2004

Houang, E., Lam, D., Fan, D., & Seal, D. (2001). Microbial keratitis in Hong Kong: Relationship to climate, environment and contact-lens disinfection. *Transactions of the Royal Society of Tropical Medicine and Hygiene, 95*(4), 361–367. https://doi.org/10.1016/S0035-9203(01)90180-4

Ilyina, T. S., Romanova, Y. M., & Gintsburg, A. L. (2004). Biofilms as a mode of existence of bacteria in external environment and host body: The phenomenon , genetic control , and regulation systems of development. *Russian Journal of Genetics, 40*(11), 1189–1198. https://doi.org/10.1023/B:RUGE.0000048660.19958.C3

Jamal, M., Ahmad, W., Andleeb, S., Jalil, F., Imran, M., Nawaz, M. A., Hussain, T., Ali, M., Rafiq, M., & Kamil, M. A. (2018). Bacterial biofilm and associated infections. *Journal of the Chinese Medical Association, 81*(1), 7–11. https://doi.org/10.1016/j.jcma.2017.07.012

Johnson, L., Horsman, S. R., Charron-Mazenod, L., Turnbull, A. L., Mulcahy, H., Surette, M. G., & Lewenza, S. (2013). Extracellular DNA-induced antimicrobial peptide resistance in *Salmonella enterica* serovar Typhimurium. *BMC Microbiology, 13*(1), 1–8. https://doi.org/10.1186/1471-2180-13-115/FIGURES/4

Justo, J. A., & Bookstaver, P. B. (2014). Antibiotic lock therapy: Review of technique and logistical challenges. *Infection and Drug Resistance, 7*, 343–363. https://doi.org/10.2147/IDR.S51388

Kazemzadeh-Narbat, M., Cheng, H., Chabok, R., Alvarez, M. M., De la Fuente-Nunez, C., Phillips, K. S., & Khademhosseini, A. (2021). Strategies for antimicrobial peptide coatings on medical devices: A review and regulatory science perspective. *Critical Reviews in Biotechnology, 41*(1), 94–120. https://doi.org/10.1080/07388551.2020.1828810

Khatoon, Z., McTiernan, C. D., Suuronen, E. J., Mah, T. F., & Alarcon, E. I. (2018). Bacterial biofilm formation on implantable devices and approaches to its treatment and prevention. *Heliyon, 4*(12), e01067. https://doi.org/10.1016/j.heliyon.2018.e01067

Kim, Y. S., Kim, J. S., Cho, H. S., Rha, D. S., Kim, J. M., Park, J. D., Choi, B. S., Lim, R., Chang, H. K., Chung, Y. H., Kwon, I. H., Jeong, J., Han, B. S., & Yu, I. J. (2008). Twenty-eight-day oral toxicity, genotoxicity, and gender-related tissue distribution of silver nanoparticles in Sprague-Dawley rats. *Inhalation Toxicology, 20*(6), 575–583. https://doi.org/10.1080/08958370701874663

Kırmusaoğlu, S. (2019). The methods for detection of biofilm and screening antibiofilm activity of agents. *Antimicrobials, Antibiotic Resistance, Antibiofilm Strategies and Activity Methods*, 1–17. https://doi.org/10.5772/intechopen.84411

Koul, S., & Kalia, V. C. (2017). Multiplicity of quorum quenching enzymes: A potential mechanism to limit quorum sensing bacterial population. *Indian Journal of Microbiology, 57*(1), 100–108. https://doi.org/10.1007/S12088-016-0633-1

Kunze, B., Reck, M., Dötsch, A., Lemme, A., Schummer, D., Irschik, H., Steinmetz, H., & Wagner-Döbler, I. (2010). Damage of Streptococcus mutans biofilms by carolacton, a secondary metabolite from the myxobacterium Sorangium cellulosum. *BMC Microbiology, 10*(1), 1–13. https://doi.org/10.1186/1471-2180-10-199

Lam, D. S. C., Houang, E., Fan, D. S. P., Lyon, D., Seal, D., & Wong, E. (2002). Incidence and risk factors for microbial keratitis in Hong Kong: Comparison with Europe and North America. *Eye, 16*(5), 608–618. https://doi.org/10.1038/sj.eye.6700151

Lee, S. W., Phillips, K. S., Gu, H., Kazemzadeh-Narbat, M., & Ren, D. (2021). How microbes read the map: Effects of implant topography on bacterial adhesion and biofilm formation. *Biomaterials, 268*, 120595. https://doi.org/10.1016/j.biomaterials.2020.120595

Li, J., Nickel, R., Wu, J., Lin, F., Van Lierop, J., & Liu, S. (2019). A new tool to attack biofilms: Driving magnetic iron-oxide nanoparticles to disrupt the matrix. *Nanoscale, 11*(14), 6905–6915. https://doi.org/10.1039/C8NR09802F

Lin, M. F., Lin, Y. Y., & Lan, C. Y. (2019). A method to assess influence of different medical tubing on biofilm formation by Acinetobacter baumannii. *Journal of Microbiological Methods, 160*(52), 84–86. https://doi.org/10.1016/j.mimet.2019.03.023

Li, X., Sun, L., Zhang, P., & Wang, Y. (2021). Novel approaches to combat medical device-associated biofilms. *Coatings, 11*(3), 294. https://doi.org/10.3390/coatings11030294

Loza-Correa, M., & Ramírez-Arcos, S. (2017). Detection of bacterial adherence and biofilm formation on medical surfaces. In *Biofilms and implantable medical devices* (pp. 181–193). Woodhead Publishing. https://doi.org/10.1016/B978-0-08-100382-4.00008-3

Lu, T. K., & Collins, J. J. (2007). Dispersing biofilms with engineered enzymatic bacteriophage. *Proceedings of the National Academy of Sciences of the United States of America, 104*(27), 11197–11202. https://doi.org/10.1073/pnas.0704624104

Mack, D., Rohde, H., Harris, L. G., Davies, A. P., Horstkotte, M. A., & Knobloch, J. M. (2006). Biofilm formation in medical device-related infection. *The International Journal of Artificial Organs, 29*(4), 343–359. https://journals.sagepub.com/doi/abs/10.1177/039139880602900404.

Magill, S. S., O'Leary, E., Janelle, S. J., Thompson, D. L., Dumyati, G., Nadle, J., Wilson, L. E., Kainer, M. A., Lynfield, R., Greissman, S., Ray, S. M., Beldavs, Z., Gross, C., Bamberg, W., Sievers, M., Concannon, C., Buhr, N., Warnke, L., Maloney, M., … Edwards, J. R. (2018). Changes in prevalence of health care–associated infections in U.S. Hospitals. *New England Journal of Medicine, 379*(18), 1732–1744. https://doi.org/10.1056/nejmoa1801550

Mah, T. F. (2012). Biofilm-specific antibiotic resistance. *Future Microbiology, 7*(9), 1061–1072. https://doi.org/10.2217/fmb.12.76

Malheiro, J., & Simões, M. (2017). Antimicrobial resistance of biofilms in medical devices. In *Biofilms and implantable medical devices* (pp. 97–113). Woodhead Publishing. https://doi.org/10.1016/B978-0-08-100382-4.00004-6

Malic, S., Hill, K. E., Hayes, A., Percival, S. L., Thomas, D. W., & Williams, D. W. (2009). Detection and identification of specific bacteria in wound biofilms using peptide nucleic acid fluorescent in situ hybridization (PNA FISH). *Microbiology, 155*(8), 2603–2611. https://doi.org/10.1099/mic.0.028712-0

Mermel, L. A., Allon, M., Bouza, E., Craven, D. E., Flynn, P., O'Grady, N. P., Raad, I. I., Rijnders, B. J. A., Sherertz, R. J., & Warren, D. K. (2009). Clinical practice guidelines for the diagnosis and management of intravascular catheter-related infection: 2009 update by the infectious diseases society of America. *Clinical Infectious Diseases, 49*(1), 1–45. https://doi.org/10.1086/599376

Meskin, I. (1998). Staphylococcus epidermidis. *Pediatrics in Review/American Academy of Pediatrics, 19*(3), 105–106. https://doi.org/10.1542/pir.19-3-105

Moojen, D. J. F., Spijkers, S. N. M., Schot, C. S., Nijhof, M. W., Vogely, H. C., Fleer, A., Verbout, A. J., Castelein, R. M., Dhert, W. J. A., & Schouls, L. M. (2007). Identification of orthopaedic infections using broad-range polymerase chain reaction and reverse line blot hybridization. *Journal of Bone and Joint Surgery, 89*(6), 1298–1305. https://doi.org/10.2106/JBJS.F.00822

Mulcahy, H., Charron-Mazenod, L., & Lewenza, S. (2008). Extracellular DNA chelates cations and induces antibiotic resistance in *Pseudomonas aeruginosa* biofilms. *PLoS Pathogens, 4*(11), e1000213. https://doi.org/10.1371/JOURNAL.PPAT.1000213

Nguyen, T., Roddick, F. A., & Fan, L. (2012). Biofouling of water treatment membranes: A review of the underlying causes, monitoring techniques and control measures. *Membranes, 2*(4), 804–840. https://doi.org/10.3390/membranes2040804

Padera, R. F. (2006). Infection in ventricular assist devices: The role of biofilm. *Cardiovascular Pathology, 15*(5), 264–270. https://doi.org/10.1016/J.CARPATH.2006.04.008

Padera, R. F., & Schoen, F. J. (2020). Cardiovascular medical devices. In *Biomaterials science* (pp. 999–1032). Academic Press. https://doi.org/10.1016/b978-0-12-816137-1.00067-2

Palmer, J., Flint, S., & Brooks, J. (2007). Bacterial cell attachment, the beginning of a biofilm. *Journal of Industrial Microbiology and Biotechnology, 34*(9), 577–588. https://doi.org/10.1007/s10295-007-0234-4

Palraj, B. R., Farid, S., & Sohail, M. R. (2017). Strategies to prevent infections associated with cardiovascular implantable electronic devices. *Expert Review of Medical Devices, 14*(5), 371–381. https://doi.org/10.1080/17434440.2017.1322506

Percival, S. L., Suleman, L., Vuotto, C., & Donelli, G. (2015). Healthcare-Associated infections, medical devices and biofilms: Risk, tolerance and control. *Journal of Medical Microbiology, 64*(4), 323–334. https://doi.org/10.1099/jmm.0.000032

Pires, D. P., Costa, A. R., Pinto, G., Meneses, L., & Azeredo, J. (2020). Current challenges and future opportunities of phage therapy. *FEMS Microbiology Reviews, 44*(6), 684–700. https://doi.org/10.1093/femsre/fuaa017

Pletzer, D., Coleman, S. R., & Hancock, R. E. W. (2016). Anti-biofilm peptides as a new weapon in antimicrobial warfare. *Current Opinion in Microbiology, 33*, 35–40. https://doi.org/10.1016/j.mib.2016.05.016

Rabin, N., Zheng, Y., Opoku-Temeng, C., Du, Y., Bonsu, E., & Sintim, H. O. (2015). Biofilm formation mechanisms and targets for developing antibiofilm agents. *Future Medicinal Chemistry, 7*(4), 493–512. https://doi.org/10.4155/fmc.15.6

Rao, H., Choo, S., Mahalingam, S. R. R., Adisuri, D. S., Madhavan, P., Akim, A. M., & Chong, P. P. (2021). Approaches for mitigating microbial biofilm-related drug resistance: A focus on micro- and nanotechnologies. *Molecules, 26*(7), 1870. https://doi.org/10.3390/molecules26071870

Rasmussen, T. B., Skindersoe, M. E., Bjarnsholt, T., Phipps, R. K., Christensen, K. B., Jensen, P. O., Andersen, J. B., Koch, B., Larsen, T. O., Hentzer, M., Eberl, L., Hoiby, N., & Givskov, M. (2005). Identity and effects of quorum-sensing inhibitors produced by Penicillium species. *Microbiology, 151*(5), 1325–1340. https://doi.org/10.1099/MIC.0.27715-0

Römling, U., & Balsalobre, C. (2012). Biofilm infections, their resilience to therapy and innovative treatment strategies. *Journal of Internal Medicine, 272*(6), 541–561. https://doi.org/10.1111/joim.12004

Ronin, D., Boyer, J., Alban, N., Natoli, R. M., Johnson, A., & Kjellerup, B. V. (2022). Current and novel diagnostics for orthopedic implant biofilm infections: A review. *APMIS, 130*(2), 59–81. https://pubmed.ncbi.nlm.nih.gov/34862649/.

Sabir, N., Ikram, A., Zaman, G., Satti, L., Gardezi, A., Ahmed, A., & Ahmed, P. (2017). Bacterial biofilm-based catheter-associated urinary tract infections: Causative pathogens and antibiotic resistance. *American Journal of Infection Control, 45*(10), 1101–1105. https://doi.org/10.1016/j.ajic.2017.05.009

Schlag, S., Nerz, C., Birkenstock, T. A., Altenberend, F., & Götz, F. (2007). Inhibition of staphylococcal biofilm formation by nitrite. *Journal of Bacteriology, 189*(21), 7911–7919. https://doi.org/10.1128/JB.00598-07

Schuch, R., Khan, B. K., Raz, A., Rotolo, J. A., & Wittekind, M. (2017). Bacteriophage lysin CF-301, a potent antistaphylococcal biofilm agent. *Antimicrobial Agents and Chemotherapy, 61*(7). https://doi.org/10.1128/AAC.02666-16, 026666-16.

Shahid, A., Aslam, B., Muzammil, S., Aslam, N., Shahid, M., Almatroudi, A., Allemailem, K. S., Saqalein, M., Nisar, M. A., Rasool, M. H., & Khurshid, M. (2021). The prospects of antimicrobial coated medical implants. *Journal of Applied Biomaterials and Functional Materials, 19.* https://doi.org/10.1177/22808000211040304, 22808000211040304.

Shahrour, H., Dandache, I., Martínez-López, A. L., González-Gaitano, G., Chokr, A., & Martínez-de-Tejada, G. (2021). An antibiotic potentiator retains its activity after being immobilized on silicone and prevents growth of multidrug-resistant *Pseudomonas aeruginosa* biofilms. *Materials Science and Engineering C, 121,* 111876. https://doi.org/10.1016/j.msec.2021.111876

Shahrour, H., Ferrer-Espada, R., Dandache, I., Bárcena-Varela, S., Sánchez-Gómez, S., Chokr, A., & Martínez-de-Tejada, G. (2019). AMPs as anti-biofilm agents for human therapy and prophylaxis. *Advances in Experimental Medicine and Biology, 1117,* 257–279. https://doi.org/10.1007/978-981-13-3588-4_14

Shen, Y., Stojicic, S., & Haapasalo, M. (2010). Bacterial viability in starved and revitalized biofilms: Comparison of viability staining and direct culture. *Journal of Endodontics, 36*(11), 1820–1823. https://doi.org/10.1016/j.joen.2010.08.029

e Silva, S. S., Carvalho, J. W. P., Aires, C. P., & Nitschke, M. (2017). Disruption of *Staphylococcus aureus* biofilms using rhamnolipid biosurfactants. *Journal of Dairy Science, 100*(10), 7864–7873. https://doi.org/10.3168/jds.2017-13012

Tandogan, N., Abadian, P. N., Huo, B., & Goluch, E. D. (2017). Characterization of bacterial adhesion and biofilm formation. In *Antimicrobial coatings and modifications on medical devices* (pp. 67–95). Cham: Springer. https://doi.org/10.1007/978-3-319-57494-3_3

Trampuz, A., & Zimmerli, W. (2005). Prosthetic joint infections: Update in diagnosis and treatment. *Swiss Medical Weekly, 135*(17–18), 243–251. https://doi.org/10.4414/smw.2005.10934

Tunney, M. M., Patrick, S., Curran, M. D., Ramage, G., Hanna, D., Nixon, J. R., Gorman, S. P., Davis, R. I., & Anderson, N. (1999). Detection of prosthetic hip infection at revision arthroplasty by immunofluorescence microscopy and PCR amplification of the bacterial 16S rRNA gene. *Journal of Clinical Microbiology, 37*(10), 3281–3290. https://doi.org/10.1128/jcm.37.10.3281-3290.1999

Vaara, M. (2019). Polymyxin derivatives that sensitize Gram-negative bacteria to other antibiotics. *Molecules, 24*(2), 249. https://doi.org/10.3390/MOLECULES24020249

VanEpps, J. S., & Younger, J. G. (2016). Implantable device-related infection. *Shock (Augusta, Ga.), 46*(6), 597–608. https://doi.org/10.1097/SHK.0000000000000692

Veerachamy, S., Yarlagadda, T., Manivasagam, G., & Yarlagadda, P. K. (2014). Bacterial adherence and biofilm formation on medical implants: A review. *Proceedings of the Institution of Mechanical Engineers, Part H: Journal of Engineering in Medicine, 228*(10), 1083–1099. https://doi.org/10.1177/0954411914556137

Vyas, N., Wang, Q. X., Manmi, K. A., Sammons, R. L., Kuehne, S. A., & Walmsley, A. D. (2020). How does ultrasonic cavitation remove dental bacterial biofilm? *Ultrasonics Sonochemistry, 67,* 105112. https://doi.org/10.1016/J.ULTSONCH.2020.105112

Weiner-Lastinger, L. M., Abner, S., Edwards, J. R., Kallen, A. J., Karlsson, M., Magill, S. S., Pollock, D., See, I., Soe, M. M., Walters, M. S., & Dudeck, M. A. (2020). Antimicrobial-resistant pathogens associated with adult healthcare-associated infections: Summary of data reported to the National Healthcare Safety Network, 2015–2017. *Infection Control & Hospital Epidemiology, 41*(1), 1–18. https://doi.org/10.1017/ice.2019.296

Weinstein, R. A., & Darouiche, R. O. (2001). Device-associated infections: A macroproblem that starts with microadherence. *Clinical Infectious Diseases, 33*(9), 1567–1572. https://doi.org/10.1086/323130

Wilson, C., Lukowicz, R., Merchant, S., Valquier-Flynn, H., Caballero, J., Sandoval, J., … Holmes, A. E. (2017). Quantitative and qualitative assessment methods for biofilm growth: A mini-review. *Research and Reviews. Journal of Engineering and Technology, 6*(4).

Wolcott, R. (2020). Biofilm and catheter-related bloodstream infections. *British Journal of Nursing, 30*(8), S4–S9.

Xu, Y., Dhaouadi, Y., Stoodley, P., & Ren, D. (2020). Sensing the unreachable: Challenges and opportunities in biofilm detection. *Current Opinion in Biotechnology, 64,* 79–84. https://doi.org/10.1016/j.copbio.2019.10.009

Yadav, M. K., Song, J. J., Singh, B. P., & Vidal, J. E. (2020). Microbial biofilms and human disease: A concise review, *1. New and Future Developments in Microbial Biotechnology and Bioengineering: Microbial Biofilms Current Research and Future Trends in Microbial Biofilms* (pp. 1–13). Elsevier B.V. https://doi.org/10.1016/B978-0-444-64279-0.00001-3

Yadav, M. K., Vidal, J. E., & Song, J. J. (2020). Microbial biofilms on medical indwelling devices. In *New and future developments in microbial biotechnology and bioengineering: Microbial biofilms current research and future trends in microbial biofilms* (pp. 15–28). Elsevier. https://doi.org/10.1016/B978-0-444-64279-0.00002-5

Yang, L., Liu, Y., Wu, H., Song, Z., Høiby, N., Molin, S., & Givskov, M. (2012). Combating biofilms. *FEMS Immunology and Medical Microbiology, 65*(2), 146–157. https://doi.org/10.1111/J.1574-695X.2011.00858.X

Yin, W., Wang, Y., Liu, L., & He, J. (2019). Biofilms: The microbial "protective clothing" in extreme environments. *International Journal of Molecular Sciences, 20*(14), 3423. https://doi.org/10.3390/ijms20143423

Yin, W., Xu, S., Wang, Y., Zhang, Y., Chou, S. H., Galperin, M. Y., & He, J. (2021). Ways to control harmful biofilms: Prevention, inhibition, and eradication. *Critical Reviews in Microbiology, 47*(1), 57–78. https://doi.org/10.1080/1040841X.2020.1842325

Yokoi, N., Okada, K., Sugita, J., & Kinoshita, S. (2000). Acute conjunctivitis associated with biofilm formation on a punctal plug. *Japanese Journal of Ophthalmology, 44*(5), 559–560. https://doi.org/10.1016/S0021-5155(00)00214-8

Zander, Z. K., & Becker, M. L. (2018). Antimicrobial and antifouling strategies for polymeric medical devices. *ACS Macro Letters, 7*(1), 16–25. https://doi.org/10.1021/acsmacrolett.7b00879

Zegans, M. E., Becker, H. I., Budzik, J., & O'Toole, G. (2002). The role of bacterial biofilms in ocular infections. *DNA and Cell Biology, 21*(5–6), 415–420. https://doi.org/10.1089/10445490260099700

Zhang, Y., Cheng, M., Zhang, H., Dai, J., Guo, Z., Li, X., Ji, Y., Cai, R., Xi, H., Wang, X., Xue, Y., Sun, C., Feng, X., Lei, L., Han, W., & Gu, J. (2018). Antibacterial effects of phage lysin LysGH15 on planktonic cells and biofilms of diverse staphylococci. *Applied and Environmental Microbiology, 84*(15). https://doi.org/10.1128/AEM.00886-18, 008866-18.

Zheng, Y., He, L., Asiamah, T. K., & Otto, M. (2018). Colonization of medical devices by staphylococci. *Environmental Microbiology, 20*(9), 3141–3153. https://doi.org/10.1111/1462-2920.14129

CHAPTER

23

Biofilm formation in acute and chronic respiratory infections caused by nosocomial gram-negative bacteria

Shreya Karmakar[1], *Soumyadeep Mukherjee*[1], *Natasha Mary Joseph*[2], *Archana Jena Priyadarshini*[3], *Vinothkannan Ravichandran*[4], *Satish Kumar Rajasekharan*[5] and *Renitta Jobby*[6,7]

[1]Department of Botany, Institute of Science, Banaras Hindu University, Varanasi, Uttar Pradesh, India; [2]Department of Microbiology, School of Life Sciences, Pondicherry University, Puducherry, India; [3]Post Graduate Department of Microbiology, St. Xavier's College (Autonomous), Kolkata, West Bengal, India; [4]Centre for Drug Discovery and Development (CD3), Amity Institute of Biotechnology (AIB), Amity University Maharashtra, Mumbai, Maharashtra, India; [5]Department of Food Sciences, Institute of Postharvest Technology and Food Sciences, Agricultural Research Organization (ARO), The Volcani Center, Rishon LeZion, Israel; [6]Amity Institute of Biotechnology, Amity University Maharashtra, Mumbai, Maharashtra, India; [7]Amity Centre of Excellence in Astrobiology, Amity University Maharashtra, Mumbai, Maharashtra, India

1. Introduction

Bacteria have developed survival mechanisms to escape from antimicrobial treatment and also to find strategies that would shield them from the immune response of the host. Of the various survival strategies adopted by bacteria, one of the most crucial strategies is biofilm formation. Bacteria can exist as both freely floating planktonic forms and sessile aggregates

Microbial Biofilms
https://doi.org/10.1016/B978-0-323-95715-1.00008-X

in the form of biofilm. Biofilm development can be defined as the irreversible attachment and growth of microorganisms on a suitable surface. After attachment to the surface, the microorganism produces polymers of extracellular origin, which help them to attach to a surface irreversibly and the formation of a matrix. This results in an alteration in the phenotype expressed by the microorganism that alters its rate of growth and allows the transcription of genes that are not seen during the planktonic stage (Donlan, 2001). Thus, we can say that biofilms are highly organized three-dimensional structured bacterial communities that are seen to be growing on solid surfaces.

Biofilm formation by bacteria on surfaces such as implants and host tissues has caused the infections by bacteria to persist over longer periods and has increased their resistance to antibiotics (Dufour et al., 2010). Infections caused by biofilms are common in urinary tract infections, endocarditis, periodontitis, and osteomyelitis (Nandakumar et al., 2013). Because resistance shown by bacterial biofilms toward antibiotics may elude the immune system of the host, their presence in upper respiratory tract infections raises serious concerns. Chronic rhinosinusitis, chronic adenoiditis, recurrent acute tonsillitis, and acute and chronic middle ear disorders may all be linked to biofilms (Nazzari et al., 2015). The drugs that are used for treatment in the current scenario are effective only against planktonic forms, which can float freely in the medium but fail to act against those who are embedded deep in the matrix of biofilm made up of exopolysaccharides (Stewart & William Costerton, 2001). The large quantities of drug-resistant bacteria entrenched in the biofilm are blamed for the failure of antimicrobial treatment in general. When implanted in a biofilm, it is seen that the bacteria, which was sessile and was highly sensitive to a particular antibiotic developed 1000 times more resistance to that antibiotic. Various factors are responsible for this increased antibiotic resistance in biofilm. A few of these factors contributing to antibiotic resistance are the encapsulation of bacteria in the exopolysaccharide matrix, reduction in growth factors, activation of efflux pumps, and cell-to-cell communication existing among them in the form of quorum sensing (Tenke et al., 2004).

Bacterial biofilms are usually pathogenic when it is caused by gram-negative bacteria. These bacteria can cause hospital-acquired (nosocomial) infections and chronic infections. According to a current estimate by the National Institutes of Health (NIH), biofilms caused by bacteria are involved in 65% and 80% of microbiological and chronic infections, respectively (Jamal et al., 2018). Biofilm can induce diseases using mediums that are related to medical devices as well as through mediums that are not related to medical devices to spread infections, which raises health concerns. Device-related infection rates have been reported for a variety of devices, including 4% for mechanical heart valves account for 4% of cases, pacemakers and defibrillators account for 4%, breast implants account for 2%, 10% for ventricular shunts, 2% for joint prostheses, and over 40% for ventricular-assisted devices (Darouiche, 2004).

The relation of biofilm with diverse infectious diseases and its association with medical device–linked infections has increased the concern of the medical fraternity. Of all the causes of death found in underdeveloped countries, one major cause is death due to pulmonary infections. Death due to pulmonary infections has been ranked as the third-largest cause of death all over the world. Apart from high mortality, pulmonary infections have imposed a financial strain on the worldwide medical system. Viruses, bacteria, fungi, or combinations of these pathogens cause lung infections. Most bacterial infections can be cured by antibiotic treatment. But sometimes bacteria acquire through biofilm formation as a consequence giving rise to recurring infections including lung diseases such as cystic fibrosis (CF), which are chronic. As the immunological defense system of the lungs gets altered, infections in the respiratory tract are quite common with certain pulmonary illnesses such as chronic obstructive pulmonary disease, non-CF bronchiectasis, CF, and ventilator-associated pneumonia. Often, repetitive infections can become chronic when the bacteria gain the ability to produce biofilms. Antimicrobial medicines are difficult to use to treat chronic biofilm-based infections (Ding et al., 2021).

This chapter discusses the nosocomial and chronic infections caused by biofilm-forming gram-negative bacteria. As antibiotic resistance is emerging day by day, a greater understanding of biofilm formation can lead to discovering new drug therapy as well as improving patient management.

2. Biofilms: what are they?

Biofilms are formed when microorganisms aggregate with each other upon getting embedded in a matrix secreted by themselves. This matrix is made up of extracellular polysaccharides (EPS). This allows the attachment of bacterial cells to the surface and also to the cells that lie close to each other. Biofilm biogenesis appears to be an existential survival mechanism that enables microbes to survive in unfavorable environments (Dang & Lovell, 2016). The pattern of life cycle and survival in biofilms is completely distinct from their freely floating planktonic members. Biofilms have high complexities and often are composed of more than one species. The density of bacterial cells found in a biofilm ranges from 10^8 to 10^{11} cells per gram. Certain local conditions induce respective changes that make the cells in biofilm to get differentiated. This includes a life cycle where the expression of certain genes is coordinated at different stages (Konopka, 2009).

Biofilms are one of the most extensively spread and effective forms of life on earth (Hall-Stoodley & Stoodley, 2002). Biofilms have wide biotechnological applications where it is used to purify drinking water by filtering it, treatment of polluted water in wastewater treatment plants, and degradation of solid waste. All higher organisms, including humans, are colonized by microorganisms that are also able to produce biofilms. This production of biofilms has been linked to chronic infections in plants (Ramey et al., 2004) animals, and humans. They are also responsible for the contamination (Brady et al., 2008; Davies et al., 1998) on implants, catheters, prosthetic heart valves, teeth, and contact lenses (Bjarnsholt et al., 2009).

2.1 Why do bacteria form biofilms?

Many bacteria can alter between planktonic and biofilm stages. The cell growth and reproduction rates of planktonic bacteria are relatively high as compared with biofilms. The biofilm state, on the other hand, appears to be the bacteria's natural and dominating condition. The requirement of bacteria to form biofilms can be attributed to several reasons (Jefferson, 2004). The metamorphosis from a freely floating planktonic stage to a sessile phenotype is known to be triggered by changes in environmental factors. These environmental elements, such as pH, nutrition level, temperature, and ionic strength, can affect the development of biofilms. To protect themselves from dangerous host circumstances and to colonize a nutrient-rich location, bacteria form biofilms (Agarwal et al., 2011; Rossi et al., 2016). A bacterium is bestowed with a lot of advantages when it lives in cooperative association in the form of a biofilm. One of the benefits achieved due to biofilm formation is that bacteria can withstand new environments that are not favorable for their survival. To prevent from getting washed away by the bloodstream inside the body or by water, they hold on to substratum or tissues. Biofilms formed in the oral cavity are much more resistant to prolonged and continuous shear stresses. The resistance against antimicrobials increases 1000 times in biofilms as compared with the planktonic stage (Rasmussen & Givskov, 2006). The diffusion of antimicrobials gets highly reduced in biofilms, thus protecting the bacterial cells that lie in the inner layers of biofilm. As the movement of bacterial cells gets limited inside the biofilm, the density of cells also increases to a greater extent. This favors DNA (plasmid) exchange among the members. Often, antibiotic resistance genes are transferred among the members in this manner. The event of gene transfer occurs horizontally and is a frequent phenomenon seen in bacteria living inside biofilms than in freely floating bacterial cells (Hausner & Wuertz, 1999).

3. Biofilm biogenesis

The complex mechanism by which a biofilm is formed can be broadly classified into three basic steps, which include attachment of a planktonic bacteria to a surface, maturation and growth of biofilm, and evasion from the matrix (Kostakioti et al., 2013) (Fig. 23.1). The first stage of a biofilm's biogenesis is when bacterial cells attach themselves to the surfaces where they are newly introduced. The initial attachment of cells to the surface is reversible and depends upon various factors such as availability of nutrients, pH of the surrounding medium, temperature, and the properties of the bacterial cell surface. The effect of repulsive forces of the medium, as well as the fluid motion, is minimized by the bacteria, which use flagella for their locomotion. The attachment to the surface is also influenced by chemotaxis (Lemon et al., 2007; Toutain et al., 2007), and (Schmidt & Kirschning, 2012).

The bacterial cell, which has been reversibly attached, can be removed from the surface due to hydrodynamic forces (Dunne, 2002), or if the bacteria encounter any change in nutrient availability (Anderson et al., 2008; Banin et al., 2005; Wu & Wayne Outten, 2009). The bacterial cells, which can overcome such shearing forces, get irreversibly attached to the surface. *Pseudomonas aeruginosa* counters the shearing forces of the fluid medium by using the twitching motility of type IV pili and gets irreversibly attached to the surface (Klausen, Aaes-Jørgensen et al., 2003; Klausen, Heydorn, et al., 2003; O'Toole & Kolter, 1998). On the other hand, *Escherichia coli* curli fibers, antigen 43, and type I pilli are used to attach to the surfaces irreversibly and also for interacting with each other (Anderson et al., 2003; Beloin et al., 2008; Cegelski et al., 2009; Danese & Pratt, 2000).

The contact with the surface after irreversible attachment triggers changes at the gene expression level. This leads to the upregulation of certain factors that help in the formation of sessile aggregates (Inagaki et al., 2005; Otto & Silhavy, 2002). The bacteria residing in biofilm produce polysaccharides that allow them to form aggregates, adhere to the surface and

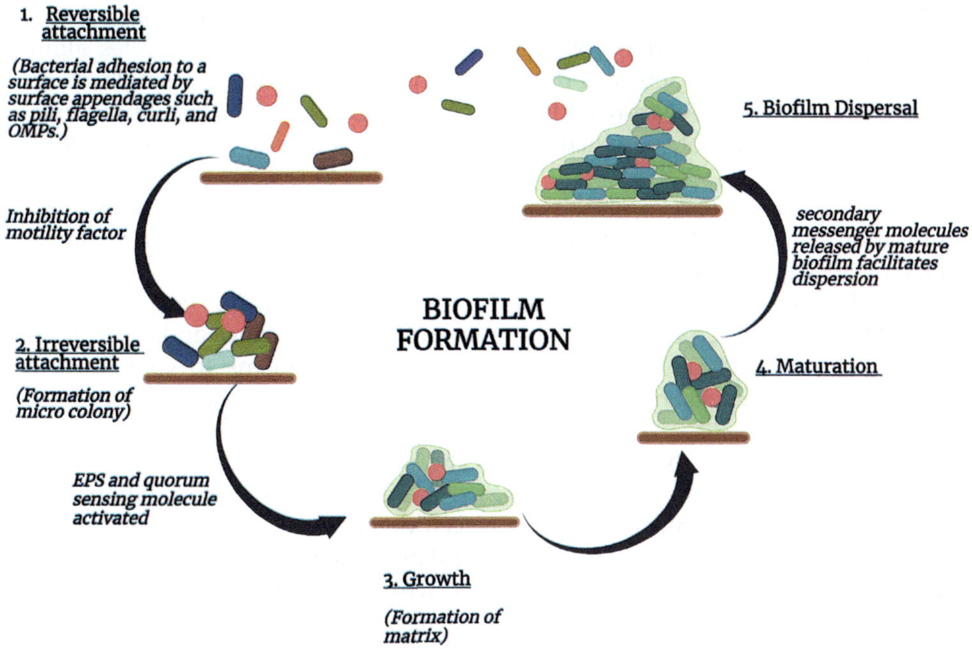

FIGURE 23.1 **Biofilm biogenesis:** (1) Planktonic cells move independently and cling to surfaces. These adhering cells begin to construct biofilms on the surface and get partially enclosed in exopolymeric material. (2) When adherent cells release an extracellular polymeric substance (EPS) and permanently adhere to the surface, cell aggregation and matrix formation take place. (3) As the biofilm grows, it forms microcolonies and water channel architecture, and it becomes substantially more layered. (4) Once completely established, the biofilm attains its highest possible cell density and can be described as a three-dimensional community. (5) The established biofilm disperses and releases microcolonies of cells, allowing them to migrate to other surfaces and spread the infection to new sites. *Created with BioRender.com.*

make them tolerant toward the surface. This helps the bacteria to successfully colonize the surface (Laue et al., 2006).

A microcolony that is stable is formed when the bacteria bind to the inert surface. Once the germs have stuck to the surface of the implants, bacterial cell growth and intercellular adhesion occur. Polysaccharide intercellular adhesion (PIA), an antigen composed of polysaccharides, is responsible for adhesion among the cells and the development of biofilm in *staphylococci*. The bacteria grow and form many-layered clusters of cells on the surface of the foreign material during this accumulation phase. Microcolonies become macrocolonies, which are surrounded by an extracellular polysaccharide matrix (Fernando et al., 2020). This phase produces extra polysaccharide substances (EPS), which are required for the binding of cells and adherence to the surface (Sutherland, 2001; Wilson et al., 2009). The EPS matrix functions as a shield, protecting the bacteria from the elements.

The polysaccharides secreted differ from species to species. The matrix of biofilm formed by *E. coli* is made up of polyglucosamine, cellulose, and colonic acid (Agladze et al., 2005; Zogaj et al., 2003). However, *P. aeruginosa* exhibits two different forms of polysaccharides. These two types of polysaccharides are capsular and aggregative. They help to form the extracellular matrix of *P. aeruginosa*. Aggregative polysaccharides provide structural stability to the matrix of the biofilm (Mann & Wozniak, 2012). *P. aeruginosa* secretes alginate as a capsular polysaccharide that acts as a dynamic polymer surrounding a single or more layer of cells (Davies et al., 1998). Apart from these, proteins, surfactants, DNA, lipids, and calcium ions are also found in the matrix, which adds to the structural properties of the biofilm (Soto, 2014).

Intercellular signaling, also known as quorum sensing (QS), occurs inside the EPS matrix. Communication between cells controls the processes and has an important role in which the cells attach or detach to a surface from QS (Daniels et al., 2004; Donlan et al., 2002), which is based on a phenomenon that gets autoinduced (Eberhard et al., 1981) and provides a mechanism that allows microbial cells to organize among themselves and control themselves (Parsek & Greenberg, 2005). The QS is used by microorganisms to coordinate activities among their communities, such as biofilm biogenesis, motility, and formation of EPS matrix (Xiong & Liu, 2010).

The Agr QS system, for example, controls the synthesis of lethal factors that improve adhesion to cells of the host, defense proteins that prevent host elimination, and factors that increase bacterial internalization and host cell death in *S. aureus.* Two components are involved mainly in this QS system: response regulator AgrA and the sensor histidine kinase AgrC. ArgC is autoinduced by certain peptides, which helps in the production of RNAIII. This regulatory RNA monitors the expression of virulence factors. When cells are densely packed, RNAIII switches from the synthesis of proteins used to evade immune response and binding to the matrix of the host to facilitate the development of many extracellular toxins, such as the α-hemolysin encoded by *hla* (Novick & Muir, 1999; Wesson et al., 1998). The action of acylated homoserine lactones facilitates cell communication in gram-negative bacteria (acyl-homoserine lactone [AHLs]) (Fuqua et al., 1996). AHL buildup in a forming biofilm leads individual cells to transition from planktonic to biofilm phenotype and coordinates their activity (Davies et al., 1998). In gram-negative bacteria, the LuxI/R operon acts as the universal system essential for the quorum sensing mechanism. In *P. aeruginosa,* four distinct QS systems are present, known as LasI/R, RhlI/r, PQS, and IQS. This system forms a hierarchy led by the LasI/R system. The AHL synthase produces small signaling molecules (C12-HSL for LasR, C4-HSL for RhlR, PQS for MvfR) that transport out in the extracellular environment. At high bacterial density, the number of AHL molecules increases and diffuses back to bind with its cognate receptor protein, LasR/RhlR/MvfR, activating the production of toxins, siderophore, toxic enzymes, HCN, and toxic secondary metabolites for increasing bacterial pathogenicity, whereas *E. coli* releases C6-HSL molecule that interacts with sdiA, receptor protein, and enhances antibiotic production, horizontal gene transfer, and toxin production, leading to biofilm development contributing to antibiotic resistance (Saeki et al., 2020).

After a biofilm attaches successfully, the second crucial step is the maturity of biofilms. They develop a three-dimensional structure. Extracellular matrix components, which are produced by themselves, are required for these three-dimensional formations with macrocolony shapes. To create these structures, adhesins, EPS, exopolysaccharides, and amyloid-forming proteins (all found in the biofilm matrix) are required, as are gradients of water, waste products, nutrients, and signaling compounds present along with the different areas of the biofilm, conditioning the cells' metabolism (Serra, Richter, and Hengge, 2013).

The third step that occurs in fully grown biofilms is detachment. Detachment permits cells to return to a planktonic condition, allowing them to build biofilm in new environments. Bacterial detachment might be triggered by both active and passive processes mediated by external pressures and erosion. Active mechanisms include enzymatic destruction of the matrix of biofilm and quorum sensing when change occurs in conditions of the external environment such as a decrease in nutrient concentration and oxygen levels (Davies et al., 1998; Kaplan, 2010; Karatan & Watnick, 2009; Hong, Wang, and Wood 2010; Rowe et al., 2010). Biofilm dispersion is a key stage that helps several bacterial species to transmit from the environment to the human, as well as among hosts (Kaplan, 2010). The significance of c-di-GMP concentration in the dispersion of biofilm has been discovered in recent times (Tamayo et al., 2007). It has been hypothesized that an inflated concentration of c-di-GMP boosts the bacteria's tendency to attach itself to a substratum, whereas low levels promote their motility. In *E. coli*, c-di-GMP concentration plays an important role in EPS synthesis, biofilm biogenesis, cell length, and swimming motility (Kostakioti et al., 2013; Méndez-Ortiz et al., 2006).

3.1 Risk of bacterial biofilms

Bacteria may colonize and create biofilms on vivid surfaces, which includes both living and nonliving objects (Hall-Stoodley and Stoodley, 2002; Sweet et al., 2011). It has been seen that the majority of human microbial diseases are due to their ability to form biofilms (Zottola & Sasahara, 1994). Infections acquired in hospitals rank as the fourth largest cause of death that occur in the United States (Wenzel, 2007). About 60%–70% of nosocomial infections are linked to a medical device that has been installed in the patient. Biofilms have been seen on surfaces of devices that are implanted inside the human body for medical purposes, dead tissues (such as dead bone tissue), and inside living tissues in healthcare settings (e.g., lung tissue, teeth surfaces) (Alav et al., 2018). Bacterial biofilms can also form in systems that are used to distribute water in hospitals. Biofilms formed by *P. aeruginosa* are found on the inner lining of metal pipes that are used to distribute water in hospitals (Loveday et al., 2014). A plethora of fatal diseases such as chronic obstructive pulmonary disease, cystic fibrosis (CF), noncystic fibrosis bronchiectasis (NCFBE) associated with chronic biofilm infection, and ventilator-associated pneumonia are caused by bacteria that can form biofilms (Southey-Pillig et al., 2005). In patients with CF, *P. aeruginosa* biofilm can cause serious lung infections (Southey-Pillig et al., 2005). More than 65% of the infections caused by microbes are due to organisms that are able to form biofilms. It is seen that they are tolerant to antimicrobial therapy and can evade both adaptive and innate immune responses of the host (Ciofu & Tolker-Nielsen, 2019; Jamal et al., 2018). Hence tackling biofilms in living as well as non living setting is a very difficult task.

4. Biofilms and nosocomial infections

The complex structure of biofilm has made it highly resistant to antibiotics and is one of the main culprits behind nosocomial infections brought on by indwelling devices (Divakar et al., 2019).

When bacteria develop biofilms, they are not only able to withstand changes in pH, a lack of nutrients, mechanical stress, and shear forces, but they are also protected against antibiotics and host immune cells, which has resulted in the creation of multidrug-resistant bacteria (Davies et al., 1998; Fux et al., 2005; McCarty et al., 2012).

Recent studies have showcased that mycobacterial species form biofilms in various environmental conditions and have been shown to cause outbreaks of mycobacterial infections. One such study done by Kohler P et al. showcased that biofilm has contaminated the heater–cooler units of surgery. These lead to what is called hospital-acquired infections (Kohler et al., 2015).

Infections that patients contract while receiving medical care in hospitals, long-term care facilities, and ambulatory settings are referred to as "hospital-acquired infections" (HAIs), also known as "nosocomial infections." These infections can also be contracted during the process of being discharged from these healthcare facilities. Gram-negative bacteria cause infections, which are now becoming a major concern as these organisms are extremely efficient in adapting themselves to combat against antibiotic resistance. This has led to the emergence of several gram-negative bacteria, which are resistant to several antibiotics, all the more so when antibiotic selection pressure is present (Peleg & Hooper, 2010).

There are various kinds of HAIs, and the Centre for Disease Control and Prevention has broadly classified the various types as follows:

1. Central line–associated bloodstream infections (CLABSI)
2. Ventilator-associated pneumonia (VAP)
3. Catheter-associated urinary tract infections (CAUTI)
4. Surgical site infections (SSI)

The most frequent pathogens in healthcare settings are bacteria, which also account for the majority of nosocomial infections. Fungi and viruses are next in line as the main culprits (Sikora et al., 2021).

Some of the common gram-negative bacteria include the following species (Jernigan et al., 2020; Vincent et al., 2009).

Enterobacteriaceae family, which include (*Proteus mirabilis, Klebsiella pneumoniae, E. coli, Klebsiella oxytoca* and *Enterobacter species*), *P. aeruginosa, Acinetobacter baumanii, Burkholderia cepacia.*

Several gram-negative bacteria, which are fairly common in hospital settings, are now resistant toward multiple drugs. Table 23.1 gives a list of the organisms and the antibiotics to which they have become resistant.

5. Role of biofilms in nosocomial infections with reference to chronic respiratory infections

Complex multifactorial infections or disorders of the airways and other lung structures which persist in the body for a long time are commonly referred to as chronic respiratory diseases or infections (Maniscalco et al., 2019). Patients who have weak immune systems or compromised lungs are more susceptible to respiratory infections caused by opportunistic pathogens, which can result in the development of biofilms in the respiratory tract.

TABLE 23.1 Gram-negative bacterial infection in patients in nosocomial settings, their transmission, and antibiotic resistance.

Gram-negative bacteria	Affected group	Transmission caused by	Antibiotics Used	Antibiotic: it is resistant toward	References
Pseudomonas aeruginosa	Patients with bronchiectasis can get *P. aeruginosa* infections. In cystic fibrosis, the infection can occur at an early stage.	Water or soil contaminated with these germs leads to their transmission among people in hospital settings. They can spread through contamination of hands surfaces and equipment present in hospital settings.	Combination of an antipseudomonal beta-lactam (e.g., penicillin or cephalosporin) Carbapenems (e.g., imipenem, meropenem) combined with antipseudomonal quinolones may be administered in conjunction with an aminoglycoside	Aminoglycosides, quinolones, and β-lactams	Aldallal et al. (2002), Pang et al. (2018)
Burkholderia cepacia	Devastating effects have emerged in the cystic fibrosis (CF) community. Pulmonary colonization can hasten the decline of lung function.	Infected people can transmit this pathogen to others by either direct contact or already-contaminated surfaces.	Trimethoprim-sulfamethoxazole (TMP-SMX) and ceftazidime as a first-line treatment Aztreonam, doripenem, tobramycin	Aminoglycosides and cephalosporins, and cotrimoxazole and fluoroquinolones. It is also resistant to colistin	Adair et al. (1999), Waters et al. (2013), Zobell et al. (2014), Waters et al. (2017), Meena et al. (2019)
Achromobacter xylosoxidans	*A. xylosoxidans* mostly occur in patients with some form of immunosuppression, mainly hematological malignancies. Patients with cystic fibrosis as their lungs are affected by bacterial airway infection	Contaminated liquids are the main source of transmission of this pathogen.	Trimethoprim-sulfamethoxazole, ceftazidime, piperacillin, and carbapenems	Aminoglycosides, antipseudomonal penicillins, antipseudomonal third-generation cephalosporins, and polymyxin B	Barragán (2018)
Haemophilus influenzae	Patients suffering from sickle cell disease, asplenia, HIV infection, and cancer patients require chemotherapy, and radiation therapy. Patients suffering from antibody deficiency syndromes and patients requiring bone marrow stem cell transplant are also susceptible to this pathogen.	Through respiratory droplets caused mainly by coughing or sneezing. If the pathogen is present in the nose and throat, then it can get transmitted from people who do not showcase any signs or symptoms.	Trimethoprim-sulfamethoxazole, cefuroxime axetil, cefixime, clarithromycin, azithromycin, and fluoroquinolones	Resistant toward most β-lactam antibiotics especially ampicillin	Agrawal & Murphy (2011)

(Continued)

TABLE 23.1 Gram-negative bacterial infection in patients in nosocomial settings, their transmission, and antibiotic resistance.—cont'd

Gram-negative bacteria	Affected group	Transmission caused by	Antibiotics Used	Antibiotic: it is resistant toward	References
Stenotrophomonas maltophilia	Most frequently discovered in patients suffering from cystic fibrosis in the respiratory system.	Transmitted through fluids used to wash wounds or ear canal or bladder and intravenous (IV) fluids. They are present in faucets, water, sink drains, and sponges of hospitals. Can also spread through the mucus of patients suffering from respiratory and urinary tract infections.	Levofloxacin, trimethoprim-sulfamethoxazole (SXT)	It is resistant to various antibiotics such as β-lactam antibiotics, cephalosporins, fluoroquinolones, TMP-SMX, aminoglycosides, carbapenems, chloramphenicol, macrolides, tetracyclines, and polymyxins	King et al. (2010), He et al. (2014)
Moraxella catarrhalis	Causes otitis media in upper respiratory tract of children and lower respiratory tract infection in elderly patients.	Contamination is caused due to secretion of droplets. Direct contact is the major transmission method.	Amoxicillin-clavulanate, trimethoprim-sulfamethoxazole	Penicillin, ampicillin, and amoxicillin	Verduin et al. (2002), Raveendran et al. (2020)
Proteus mirabilis	Patients undergoing long-term catheterization.	Direct contact with infected people. Contaminated food, objects, and surface also lead to their transmission.	Aminoglycosides, carbapenems (except imipenem), and third-generation cephalosporins	Resistant to polymyxins (colistin), nitrofurans, tigecycline, and tetracycline	Schaffer & Pearson (2016), Armbruster et al. (2018)
Porphyromonas gingivalis	Has been linked to adults suffering from severe forms of periodontitis.	Found in the oral cavity, and can be transmitted through the saliva of individuals.	Clindamycin, erythromycin, doxycycline, minocycline, hydrochloride tetracycline, amoxicillin, ampicillin metronidazole	Chlorhexidine, minocycline, metronidazole, amoxicillin, doxycycline, cefuroxime, ampicillin, and ofloxacin	Fiorillo et al. (2019)

According to some estimates, pulmonary infections are the third-leading cause of death worldwide and the primary killer in underdeveloped nations. Adding to its high morbidity and mortality rates, this infection has also resulted in an increasing financial load on the worldwide healthcare system. Bacteria, fungi, viruses, or combinations of these pathogenic organisms can cause respiratory infections (Ding et al., 2021).

Few lung diseases such as cystic fibrosis (CF), ventilator-associated pneumonia (VAP), noncystic fibrosis bronchiectasis, and chronic obstructive pulmonary diseases (COPD) are usually associated with chronic lung disorder due to the physiological alterations of the lung immune system.

We shall discuss them in detail.

5.1 Cystic fibrosis

CF is a progressive, deadly, inherited genetic anomaly and is most common in Caucasians. Cystic Fibrosis Transmembrane Conductance Regulator (CFTR) gene of chromosome 7 gets mutated and gives rise to this disease. It was discovered in the year 1989 (Davis, 2006). CFTR gene expresses CFTR protein, which can be found in all mucus-producing organs, such as lungs, pancreas, intestines, as well as sweat glands. Because of a reduction in the secretion of chloride ions by epithelium and a rise in the absorption of salts, this sticky mucus in the lungs becomes dehydrated. This dehydrated mucus becomes critical to get cleared by coughing, as they turn thick and sticky. Mucociliary clearance (MCC), the major natural defensive system of the airways, is compromised, resulting in mucus plugs and plaques that provide a breeding ground for bacteria (Matsui et al., 2006). Poor removal of bacteria that enter the airways by inhalation or aspiration causes sputum retention, infection, and inflammation (Lund-Palau et al., 2016).

5.1.1 Cystic fibrosis biofilm infections caused by gram-negative bacteria

P. aeruginosa is a bacterium that is gram-negative and is responsible for causing opportunistic infections in human beings. It can be found in the natural environment such as water, soil, vegetation, and even on the skin of a few healthy individuals. Although *P. aeruginosa* is rarely pathogenic in healthy people, it acts as a regular source of critical infections in cancer patients, burns, immunosuppressed individuals such as AIDS patients, and those who are on mechanical ventilation (Sadikot et al., 2005). It is one of the most often isolated and consequently clinically significant respiratory pathogens in CF patients, and it is responsible for faster development of the illness. The fascinating characteristic of *P. aeruginosa* strains reported in the CF lung sets them apart from bacteria discovered in other diseases: they can undergo hypermutation (Oliver et al., 2000). They have the potential to respond quickly to changes in surroundings by turning genes on or off and by raising the frequency of mutation events in genes. Conversion to mucoid phenotype, which is nearly pathognomonic for CF, is triggered by one such mutation event (Pedersen et al., 1992). Table 23.2 lists down the features of *P. aeruginosa* which makes it a persistant organism in CF infections.

It has been reported that inflammatory responses of the host cause conversion of *P. aeruginosa* to mucoid phenotype. Reactive oxygen species produced by activated polymorphonuclear leukocytes as a part of inflammatory response help in this conversion of

TABLE 23.2 Remarkable features of *Pseudomonas aeruginosa* resulting in persistence and virulence in the cystic fibrosis lungs.

Virulence factors of *Pseudomonas aeruginosa*	
1. Surface structures	• Flagella • Lipopolysaccharide • Pili
2. Extracellular products	• Pyocyanin • Hydrogen cyanide • Hemolysin • Proteases • Exotoxin A
3. Type 3 secretions	• ExoS • ExoU • ExoT • ExoY
4. Antibiotic resistance	Multidrug efflux pumps like PA1875-1877
5. Hypermutability	–
6. Mucoidy	–
7. Quorum sensing	–

phenotype (John et al., 1999). These reactive oxygen species have been shown to cause mutations in the mucA gene as well as excessive alginate production. The respiratory passages of patients affected with cystic fibrosis have sputum that comprises biofilms in the form of free aggregates (Moradali et al., 2017). *P. aeruginosa* has a diverse set of virulence characteristics that allow it to acclimate and survive in the CF airway. Lipopolysaccharide (LPS), type IV pili, and flagella are examples of surface features that donate to pathogenicity and stimulate host immunological activation. The immunogenic virulence factors, like LPS with structural variability in both the lipid A component and the O side chains, affect the immunity of the host (Pier, 2007). The predilection for particular penta- and hexa-acylated lipid A structures and the lack of LPS O-antigen side chains were shown to differentiate isolates of *P. aeruginosa* from long-term infected CF patients compared with environmental isolates, suggesting bacterial adaptability within the CF lung (Pier, 2007; Sap, 1999). Flagella and type IV pili are required for swimming, twitching motility, attachment to the host epithelium, and biofilm biogenesis (Feldman et al., 1998; O'Toole & Kolter, 1998). Toll-like receptor (TLR)-5 expressed on the host epithelium binds to flagellin, leading to activation of nuclear factor NF-κB and also stimulates the synthesis of tumor necrosis factor (TNF) and other proinflammatory cytokines (Hayashi et al., 2001). The effector protein determines the strain phenotype after infection, with ExoU-secreting strains inducing host cells to lyse rapidly and ExoS-secreting strains commencing a cell death sequence that mimics apoptosis (Hauser, 2009).

Recent reports have mentioned T6SS and its coregulation with biofilm biogenesis. Component proteins of T6SS have been discovered in the sputum isolated from cystic fibrosis patients having persistent *P. aeruginosa* infection (Mougous et al., 2006). Due to the thick mucus produced by CF patients, alginate can grow in conditions where the concentration of oxygen is low or when oxygen is completely absent (Worlitzsch et al., 2002). The accurate series of events that occurred when a tiny inoculum of *Pseudomonas* was inhaled into the CF airway is uncertain. However, adhesion (either to airway secretions or directly to the cell surface) and resistance to innate defenses are thought to be essential components in the initial stages, like most other bacterial infections (Davies & Bilton, 2009). Host tissue is degraded, and the inflammatory response is disrupted by hemolysins, secreted proteases (elastase and alkaline protease), and toxins. Elastase can break down host immunoglobulins, surfactant proteins A and D, and products of complement activation, further impairing host defenses (Mariencheck et al., 2003). The phenazine-derivative pyocyanin (which turns the cultures of *P. aeruginosa* blue-green) is a redox-active chemical that can produce reactive oxygen species and cause oxidative stress in host cells (Rada & Leto, 2013). Pyocyanin and hydrogen cyanide, both released by *P. aeruginosa*, have shown reduction in the beat frequency of cilia, thereby impairing innate airway defense mechanisms further (Davies et al., 2006; Nair et al., 2014). One of *P. aeruginosa*'s most intricate secretion systems is the type III secretion system (T3SS), where harmful effector proteins are injected into the host by using a syringe of macromolecular dimension. The T3SS has been linked to over 36 genes that code for proteins involved in pore development and needle assembly in eukaryotic cells, and effector activities (Filloux, 2011). These four effector proteins have been identified in *P. aeruginosa*: ExoU, ExoT, ExoY, and ExoS. Along with ExoU, ExoS protein shows the strongest cytotoxic action. Recent reports have mentioned T6SS and its coregulation with biofilm biogenesis. Component proteins of T6SS have been discovered in the sputum isolated from cystic fibrosis patients having persistent *P. aeruginosa* infection (Mougous et al., 2006). The bacteria produce some molecules, which can diffuse freely like acyl homoserine lactones (AHLs) that can easily diffuse across the bacterial membrane. By using freely diffusible chemicals such as AHLs, bacteria can use quorum sensing to detect the presence of other similar bacteria nearby. Genes that control the formation of biofilm are expressed once these chemicals are present in high concentrations within the organisms. The biofilm matrix in this form shields the microcolonies of bacteria from phagocytic cells and antibiotic penetration. Moreover, *P. aeruginosa* has several multidrug efflux pumps, including the biofilm-specific PA1875-1877 (Zhang & Fah Mah, 2008) and MexAB-OprM pumps (Poole et al., 2011).

Burkholderia. cepacia complex (Bcc) are opportunistic pathogens in CF patients linked with some distinctive unfavorable clinical results such as systemic spread (Davies & Bilton, 2009). Four different species of this bacteria are present in patients infected with CF: 46% of *Burkholderia cenocepacia* (genomovar III), 38% of *Burkholderia multivorans* (genmovar II), 5% of *Burkholderia vietnamiensis* (genmovar V), and the remaining *Burkholderia dolosa* (genmovar VI). The formation of biofilm has also been observed with Bcc. Different species express various multidrug efflux pumps that allow bacterial biofilms to become drug resistant. The RND efflux pumps BCAL1672-1676 (RND-3) were crucial for biofilm resistance to both ciprofloxacin and tobramycin, whereas the RND efflux pumps BCAM0925-0927 (RND-8) and BCAM1945-1947 (RND-9) protected *B. cepacia* complex biofilms against tobramycin. The coexistence of *P. aeruginosa* and Bcc in biofilms as the mixed form has been found in patients suffering from cystic fibrosis (Saiman et al., 1990; Sajjan et al., 2000).

Achromobacter xylosoxidans are usually found in the sputum of CF patients (Igra-Siegman et al., 1980). Although the medicinal relevance of *A. xylosoxidans* infection in CF patients is unknown, it can develop antibiotic resistance and form biofilms in the course of the patient's chronic infection of the lungs (Macfarlane et al., 2011; Sibley & Surette, 2011; Xiao et al., 2011).

Haemophilus influenzae is nonmotile, gram-negative in nature, which usually infects the respiratory pathways of pediatric CF patients (Hiller et al., 1990). *H. influenzae* usually forms biofilms to hinder host immune responses.

Stenotrophomonas maltophilia is a bacillus that is gram-negative in nature, requires aerobic conditions for growth, and is nonfermentative. CF patients infected with *S. maltophilia*-related chronic pulmonary infection are vulnerable to a high risk of pulmonary exacerbation and death (Waters et al., 2013). *S. maltophilia* persists in the CF lungs due to the biofilm formation and associated groups that enable them to gain resistance against antibiotics and the immune system of the host (Pompilio et al., 2010).

5.2 Chronic biofilm infection associated with noncystic fibrosis bronchiectasis

Noncystic fibrosis bronchiectasis (NCFBE) can be defined as a chronic inflammatory progressive respiratory condition that is characterized by permanent dilatation of the bronchi with symptoms of copious sputum production, persistent cough, and repeated infective exacerbations. There are three most commonly identified bacterial pathogens found in the sputum of patients suffering from NCFBE, which are *P. aeruginosa, H. influenzae,* and *Moraxella catarrhalis*. Similar to CF, severe NCFBE patients experience persistent *P. aeruginosa* infection, which is what causes the sharp fall in function of the lungs in patients suffering from NCFBE (Davies et al., 2006). The biofilm of *P. aeruginosa in* NCFBE patients is nearly identical to that in CF patients. Low antimicrobial concentrations in the respiratory tract increase *P. aeruginosa*, shifts from nonmucoid to mucoid strains, and increasing mutation frequency, which results in biofilm development (Lee et al., 2005).

5.3 Chronic obstructive pulmonary disease (COPD)—associated with chronic infection

Gradual malfunction of the respiratory system and parenchymal tissue of the lungs characterize COPD as a disease. The most frequently found pathogens linked with COPD include nontypeable *H. influenzae* (NTHi) and *M. catarrhalis*. Gram-negative bacteria, *H. parainfluenzae,* and *Enterobacteriaceae,* are also isolated from COPD patients. Unlike asthma, which has reversible airflow obstruction, COPD causes irreversible obstruction in the flow of air (Celli et al., 2004). The pulmonary blood vessels, parenchyma tissue of lungs, peripheral respiratory passages, and central respiratory passages all undergo structural alterations in COPD patients as a result of activated innate immune cells. Respiratory tract inflammation activates innate immune cells (Davidson & Bai, 2005). Additionally, a significant factor causing the metamorphosis of the pathogenic bacterium from the free-floating form to the biofilm phase is the fact that markers of inflammation reactions occurring in respiratory passages (such as inflammatory cytokines or chemokines such as interleukin, TNF, etc.) are greater in patients of COPD patients than in physically fit individuals. This is because of the high level of oxidative stress (Barnes et al., 2003). Impaired mucociliary clearance is another cause of biofilm development. COPD patients, such as CF patients, have inadequate mucociliary clearance in the respiratory pathway due to faulty cilia, and their respiratory epithelial cells undergo squamous metaplasia (Hogg & Timens, 2009).

It has been demonstrated that NTHi isolates found in the sputum COPD patients can form biofilms outside the body. Peroxiredoxin-glutaredoxin NTHi antigen, which is expressed solely at the time of biofilm biogenesis, was also found in the sera of these patients. This concludes that NTHi biofilm is also present in patients suffering from COPD (Juneau et al., 2015; Murphy et al., 2005). The biofilm generation of NTHi is aided by the addition of sialic acid residues to lipooligosaccharides (LOS) (Swords et al., 2004). However, the requirement of sialic acid in biofilm biogenesis is debatable. According to Bakaletz and Jurcisek, eDNA can help NTHi biofilm production, whether it comes from the bacterium or the host (Jurcisek & Bakaletz, 2007). In vivo, the neutrophils of the host system were discovered in biofilms by them. The neutrophil extracellular traps (NETs) allow these neutrophils to capture microbes by causing extrusion of their own genomic DNA (Brinkmann et al., 2004). Pili proteins are found in vitro to sustain bacterial adherence and biofilm development in NTHi biofilms (Carruthers et al., 2012).

5.4 Chronic infection associated with ventilator-associated pneumonia

When hospitalized patients are kept in intensive care units (ICUs), they are often affected by nosocomial pulmonary infections. The disease-causing bacteria reach the respiratory tract of the patient through the endotracheal tube when it is used for a minimum of 48 h to assist in ventilation mechanically in ICU (Sarda et al., 2019). The lining of the endotracheal tubes provides a suitable place for pathogenic bacteria to adhere, which acts as a storage for pathogens capable of forming biofilms. It has been shown that the endotracheal tube biofilm interacts with the lung in 70% VAP patients, suggesting that the biofilm serves as a critical and incessant source of harmful microorganisms (Adair et al., 1999). It is still unknown how the development of VAP in patients is related to the biofilm found on the endotracheal tube. Some researchers have proposed that the formation of VAP is linked to the fourth stage (i.e., detachment or apoptosis), which is independent of intubation time (Wilson et al., 2009).

5.5 Gram-negative bacteria responsible for ventilator-associated pneumonia

Certain gram-negative bacteria such as *P. aeruginosa*, *Acinetobacter baumannii*, *Porphyromonas gingivalis*, *K. pneumoniae*, and *P. mirabilis* have been isolated frequently in most patients suffering from VAP (Sarda et al., 2019). *P. gingivalis* and *P. mirabilis* biofilm are formed on endotracheal tubes in VAP patients. *A. baumannii* form biofilms on a broad range of surfaces that include epithelial cells of the host, and abiotic surfaces such as polypropylene and stainless steel. Generally, various virulence factors are responsible for bacterial cell adherence to the surfaces, but in the case of *A. baumannii*, the plasticity found in its genome has contributed to developing strain-specific biofilm variations. The two most conserved genes in *A. baumannii* are CsuE and OmpA, which are remarkable biofilm-associated genes (Colquhoun & Rather, 2020). The most vital virulence factors that are responsible for the pathogenesis of infections caused by *K. pneumoniae* are type 1, type 3 pili, and capsular polysaccharides. These factors have significant contributions to biofilm biogenesis (Seifi et al., 2016).

6. Current inhalation treatments for chronic infections caused by pulmonary biofilms

It is not easy to treat biofilm-based infections, which are chronic with antimicrobials. Currently available treatments that are delivered to the lungs by inhalation are mentioned in Table 23.3.

However, inhalation therapy can cause antibiotic resistance with prolonged use of excessive amounts of inhalable antimicrobials. Thus, novel antimicrobial agents that can defeat bacterial tolerance are required. AMPs (antimicrobial peptides), NO (nitric oxide), and Ga-based medicines are a few examples of therapies that have been produced or are currently being explored to combat biofilms (Ding et al., 2021).

TABLE 23.3 List of inhalable antibiotics used in clinical studies treating biofilm infections brought on by the pulmonary disease.

Classes of drugs	Inhalable antimicrobial agents	Commercialized name	Formulation	Use	Phase of trial	References
Fluoroquinolone	Levofloxacin	Aeroquin®	Aqueous nebulization solution	To treat any newly acquired bacterial strains by COPD patients. To reduce the bacterial burden	Phase 1b (NCT00752414) Phase 2 (NCT00739648)	Elborn et al. (2016)
Fluoroquinolone	Ciprofloxacin	Cipro®	For inhalation dry powder Aqueous nebulization solution	To treat infections caused by bacteria in COPD.	Phase 1 (NCT01052298) Phase 1 (NCT00961038) Phase 1 (NCT01072942) Phase 1 (NCT01168895)	Dorkin et al. (2015)

TABLE 23.3 List of inhalable antibiotics used in clinical studies treating biofilm infections brought on by the pulmonary disease.—cont'd

Classes of drugs	Inhalable antimicrobial agents	Commercialized name	Formulation	Use	Phase of trial	References
Monocyclic beta-lactam class of drug	Aztreonam	Cayston®	Aqueous nebulization solution	To treat infection caused by *Pseudomonas aeruginosa* in patients with cystic fibrosis	Phase 2 (NCT01055847) Phase 3 (NCT0012,8492) Phase 4 (NCT02894684)	Assael et al. (2013), McCoy et al. (2008), Quittner et al. (2009), McCoy et al. (2008), Tiddens et al. (2015), Wainwright et al. (2011)
Colistin	Colistimethate sodium (CMS), colistinsulfomethate	Promixin®	Aqueous nebulization solution	To treat chronic infection caused by *P. aeruginosa* in NCFBE patients	Phase 3 (NCT03460704)	Abdellatif et al. (2016)
Gas molecule	Nitric oxide (NO)	Thiolanox	Gaseous form	Patients suffering from chronic cystic fibrosis	Phase 1/2 (NCT00570349) Phase 1/2 (NCT01958944) Phase 2 (NCT02498535)	Deppisch et al. (2016)

More updates about the stage of trials can be found at http://clinicaltrials.gov.

7. Conclusion

The cure for diseases caused by gram-negative bacteria has been challenging due to increased tolerance to antibiotics among them. They are frequently associated with infections that are acquired from hospitals. Though the role of biofilms formed by them in conferring resistance to antimicrobial therapy in gram-negative bacteria is well established, their physiological role in the spread of fatal, nosocomial pulmonary infections needs thorough investigation. In addition, the treatment of biofilm is complex. Biofilms are clinically challenging to treat due to their great variability and multifactorial nature; as a result, combination therapies that target many elements of the biofilm milieu are necessary. The current treatment procedures are not fully effective in eradicating these infections and have made them recurrent. Studies should be conducted to find novel ways to fight against these infections and completely cure them. Apart from treatment, efforts should be made to identify diagnostic markers to detect early onset of such infections and prevent the occurrence of any fatal outcome.

Acknowledgments

We acknowledge the support of the MentX Internship program of the Young Academy of India (YAI).

References

Abdellatif, S., et al. (2016). Efficacy and toxicity of aerosolised colistin in ventilator-associated pneumonia: A prospective, randomised trial. *Annals of Intensive Care, 6*(1).

Adair, C. G., Gorman, S. P., Feron, B. M., Byers, L. M., Jones, D. S., Goldsmith, C. E., Moore, J. E., Kerr, J. R., Curran, M. D., Hogg, G., Webb, C. H., McCarthy, G. J., & Milligan, K. R. (1999). Implications of endotracheal tube biofilm for ventilator-associated pneumonia. *Intensive Care Medicine, 25*(10), 1072–1076. https://doi.org/10.1007/s001340051014.

Agarwal, R. K., Singh, S., Bhilegaonkar, K. N., & Singh, V. P. (2011). Optimization of microtitre plate assay for the testing of biofilm formation ability in different Salmonella serotypes. *International Food Research Journal, 18*(4), 1493–1498.

Agladze, K., Wang, X., & Romeo, T. (2005). Spatial periodicity of Eschenchia coli K-12 biofilm microstructure initiates during a reversible, polar attachment phase of development and requires the polysaccharide adhesin PGA. *Journal of Bacteriology, 187*(24), 8237–8246.

Agrawal, A., & Murphy, T. F. (2011). Haemophilus influenzae infections in the H. influenzae type b conjugate vaccine era. *Journal of Clinical Microbiology, 49*(11), 3728–3732.

Alav, I., Mark Sutton, J., & Miraz Rahman, K. (2018). Role of bacterial efflux pumps in biofilm formation. *Journal of Antimicrobial Chemotherapy, 73*(8), 2003–2020.

Aldallal, N., et al. (2002). Inflammatory response in airway epithelial cells isolated from patients with cystic fibrosis. *American Journal of Respiratory and Critical Care Medicine, 166*(9), 1248–1256.

Anderson, G. G., et al. (2003). Intracellular bacterial biofilm-like pods in urinary tract infections. *Science, 301*(5629), 105–107.

Anderson, G. G., Moreau-Marquis, S., Stanton, B. A., & O'Toole, G. A. (2008). In vitro analysis of tobramycin-treated Pseudomonas aeruginosa biofilms on cystic fibrosis-derived airway epithelial cells. *Infection and Immunity, 76*(4), 1423–1433.

Armbruster, C. E., et al. (2018). *2018互联网女皇报告_中文版.Pdf, 8*(1), 1–123.

Assael, B. M., et al. (2013). Inhaled aztreonam lysine vs. inhaled tobramycin in cystic fibrosis: A comparative efficacy trial. *Journal of Cystic Fibrosis, 12*(2), 130–140. https://doi.org/10.1016/j.jcf.2012.07.006

Banin, E., Vasil, M. L., & Peter Greenberg, E. (2005). Iron and Pseudomonas aeruginosa biofilm formation. *Proceedings of the National Academy of Sciences of the United States of America, 102*(31), 11076–11081.

Barnes, P. J., Shapiro, S. D., & Pauwels, R. A. (2003). Chronic obstructive pulmonary disease: Molecular and cellular mechanisms. *European Respiratory Journal, 22*(4), 672–688.

Barragán, E. P. (2018). *2018 span society Chem Achromobacter Xylosidans bacteremia 10 Yr case series, 28041*(3), 268–273.

Beloin, C., Roux, A., & Ghigo, J. (2008). *Escherichia coli biofilms - samenvatting* (pp. 249–289).

Bjarnsholt, T., et al. (2009). Pseudomonas aeruginosa biofilms in the respiratory tract of cystic fibrosis patients. *Pediatric Pulmonology, 44*(6), 547–558.

Brady, R. A., Calhoun, J. H., Leid, J. G., & Shirtliff, M. E. (2008). The role of biofilms in device-related infections. *Infections of orthopaedic implants and devices.*

Brinkmann, V., et al. (2004). Neutrophil extracellular traps kill bacteria. *Science, 303*, 1532–1535.

Carruthers, M. D., et al. (2012). Biological roles of nontypeable Haemophilus influenzae type IV pilus proteins encoded by the pil and com operons. *Journal of Bacteriology, 194*(8), 1927–1933.

Cegelski, L., et al. (2009). Small-molecule inhibitors target Escherichia coli amyloid biogenesis and biofilm formation. *Nature Chemical Biology, 5*(12), 913–919. https://doi.org/10.1038/nchembio.242

Celli, B. R., et al. (2004). Standards for the diagnosis and treatment of patients with COPD: A summary of the ATS/ERS position paper. *European Respiratory Journal, 23*(6), 932–946.

Ciofu, O., & Tolker-Nielsen, T. (2019). Tolerance and resistance of Pseudomonas aeruginosabiofilms to antimicrobial agents-how P. Aeruginosa can escape antibiotics. *Frontiers in Microbiology, 10*, 913.

Colquhoun, J. M., & Rather, P. N. (2020). Insights into mechanisms of biofilm formation in acinetobacter baumannii and implications for uropathogenesis. *Frontiers in Cellular and Infection Microbiology, 10*, 253.

Danese, P. N., & Pratt, L. A. (2000). Exopolysaccharide production is required for development of. *Society, 182*(12), 3593–3596.

Dang, H., & Lovell, C. R. (2016). Microbial surface colonization and biofilm development in marine environments. *Microbiology and Molecular Biology Reviews, 80*(1), 91–138.

Daniels, R., Vanderleyden, J., & Michiels, J. (2004). Quorum sensing and swarming migration in bacteria. *FEMS Microbiology Reviews, 28*(3), 261–289.

Davidson, W., & Bai, T. R. (2005). Lung structural changes in chronic obstructive pulmonary diseases. *Current Drug Targets: Inflammation and Allergy, 4*(6), 643–649.

Davies, J. C., & Bilton, D. (2009). Bugs, biofilms, and resistance in cystic fibrosis. *Respiratory Care, 54*(5), 628–638.

Davies, D. G., et al. (1998). The involvement of cell-to-cell signals in the development of a bacterial biofilm. *Science, 280*(5361), 295–298.

Davies, G., et al. (2006). The effect of Pseudomonas aeruginosa on pulmonary function in patients with bronchiectasis. *European Respiratory Journal, 28*(5), 974–979.

Davis, Pamela B. (2006). Cystic fibrosis since 1938. *American Journal of Respiratory and Critical Care Medicine, 173*(5), 475–482.

Deppisch, C., et al. (2016). Gaseous nitric oxide to treat antibiotic resistant bacterial and fungal lung infections in patients with cystic fibrosis: A phase I clinical study. *Infection, 44*(4), 513–520.

Ding, Li, et al. (2021). Pulmonary biofilm-based chronic infections and inhaled treatment strategies. *International Journal of Pharmaceutics, 604*, 120768. https://doi.org/10.1016/j.ijpharm.2021.120768

Divakar, S., Lama, M., & Kan Asad, U. (2019). Antibiotics versus biofilm: An emerging battleground in microbial communities | enhanced reader. *Antimicrobial Resistance and Infection Control, 8*(76), 1–10.

Donlan, R. M. (2001). Biofilms and device-associated infections. *Emerging Infectious Diseases, 7*(2), 277–281.

Donlan, R. M., Costerton, J. W., Rodney, M. D., & Costerton, J. W. (2002). Cmr.15.2.167-193.2002, *15*(2).

Dorkin, H. L., et al. (2015). Ciprofloxacin DPI: A randomised, placebo-controlled, phase IIb efficacy and safety study on cystic fibrosis. *BMJ Open Respiratory Research, 2*(1), 1–10.

Dufour, D., Leung, V., & Lévesque, C. M. (2010). Bacterial biofilm: Structure, function, and antimicrobial resistance. *Endodontic Topics, 22*(1), 2–16.

Dunne, W. M. (2002). Bacterial adhesion: Seen any good biofilms lately? *Clinical Microbiology Reviews, 15*(2), 155–166.

Eberhard, A., et al. (1981). Structural identification of autoinducer of photobacterium fischeri luciferase. *Biochemistry, 20*(9), 2444–2449.

Elborn, J. S., et al. (2016). Safety and efficacy of prolonged levofloxacin inhalation solution (APT-1026) treatment for cystic fibrosis and chronic Pseudomonas aeruginosa airway infection. *Journal of Cystic Fibrosis, 15*(5), 634–640. https://doi.org/10.1016/j.jcf.2016.01.005

Feldman, M., et al. (1998). Role of flagella in pathogenesis of Pseudomonas aeruginosa pulmonary infection. *Infection and Immunity, 66*(1), 43–51.

Fernando, S. M., et al. (2020). Diagnosis of ventilator-associated pneumonia in critically ill adult patients—a systematic review and meta-analysis. *Intensive Care Medicine, 46*(6), 1170–1179. https://doi.org/10.1007/s00134-020-06036-z

Filloux, A. (2011). Protein secretion systems in Pseudomonas aeruginosa: An essay on diversity, evolution, and function. *Frontiers in Microbiology, 2*, 1–21.

Fiorillo, L., et al. (2019). Porphyromonas gingivalis, periodontal and systemic implications: A systematic review. *Dentistry Journal, 7*(4), 1–15.

Fuqua, C., Winans, S. C., & Peter Greenberg, E. (1996). Census and consensus in bacterial ecosystems: The LuxR-LuxI family of quorum-sensing transcriptional regulators. *Annual Review of Microbiology, 50*, 727–751.

Fux, C. A., Costerton, J. W., Stewart, P. S., & Stoodley, P. (2005). Survival strategies of infectious biofilms. *Trends in Microbiology, 13*(1), 34–40.

Hall-Stoodley, L., & Paul, S. (2002). Developmental regulation of microbial biofilms. *Current Opinion in Biotechnology, 13*(3), 228–233.

Hauser, A. R. (2009). The type III secretion system of Pseudomonas aeruginosa: Infection by injection. *Nature Reviews Microbiology, 7*(9), 654–665. https://doi.org/10.1038/nrmicro2199

Hausner, M., & Wuertz, S. (1999). High rates of conjugation in bacterial biofilms as determined by quantitative in situ analysis. *Applied and Environmental Microbiology, 65*(8), 3710–3713.

Hayashi, F., et al. (2001). The innate immune response to bacterial flagellin is mediated by toll-like receptor 5. *Nature, 410*(6832), 1099–1103.

He, X., et al. (2014). Synergistic combination of marine oligosaccharides and azithromycin against Pseudomonas aeruginosa. *Microbiological Research, 169*(9–10), 759–767. https://doi.org/10.1016/j.micres.2014.01.001

Hiller, E. J., Ispahani, P., & Baker, M. (1990). *Haemophilus infection* (pp. 255–258).

Hogg, J. C., & Timens, W. (2009). The pathology of chronic obstructive pulmonary disease. *Annual Review of Pathology: Mechanisms of Disease, 4*, 435–459.

Hong, S. H., Wang, X., & Wood, T. K. (2010). Controlling biofilm formation, prophage excision and cell death by rewiring global regulator H-NS of Escherichia colim. *Microbial Biotechnology, 3*(3), 344–356.

Igra-Siegman, Y., Chmel, H., & Cobbs, C. (1980). Clinical and laboratory characteristics of achromobacter xylosoxidans infection. *Journal of Clinical Microbiology, 11*(2), 141–145.

Inagaki, S., Kuramitsu, H. K., & Sharma, A. (2005). Contact-dependent regulation of a tannerella forsythia virulence factor, BspA, in biofilms. *FEMS Microbiology Letters, 249*(2), 291–296.

Jamal, M., et al. (2018). Bacterial biofilm and associated infections. *Journal of the Chinese Medical Association, 81*(1), 7–11. https://doi.org/10.1016/j.jcma.2017.07.012

Jefferson, K. K. (2004). What drives bacteria to produce a biofilm? *FEMS Microbiology Letters, 236*(2), 163–173.

Jernigan, J. A., et al. (2020). Multidrug-resistant bacterial infections in U.S. Hospitalized patients, 2012–2017. *New England Journal of Medicine, 382*(14), 1309–1319.

John, A. H., et al. (1999). *Virulence activation in the cystic fibrosis lung.*

Juneau, R. A., et al. (2015). Peroxiredoxin-glutaredoxin and catalase promote resistance of nontypeable Haemophilus influenzae 86-028NP to oxidants and survival within neutrophil extracellular traps. *Infection and Immunity, 83*(1), 239–246.

Jurcisek, J. A., & Bakaletz, L. O. (2007). Biofilms formed by nontypeable haemophilus influenzae in vivo contain both double-stranded DNA and type IV pilin protein. *Journal of Bacteriology, 189*(10), 3868–3875.

Kaplan, J. B. (2010). Biofilm dispersal: Mechanisms, clinical implications, and potential therapeutic uses. *Journal of Dental Research, 89*(3), 205–218.

Karatan, E., & Watnick, P. (2009). Signals, regulatory networks, and materials that build and break bacterial biofilms. *Microbiology and Molecular Biology Reviews, 73*(2), 310–347.

King, P., et al. (2010). In vitro pharmacodynamics of levofloxacin and other aerosolized antibiotics under multiple conditions relevant to chronic pulmonary infection in cystic fibrosis. *Antimicrobial Agents and Chemotherapy, 54*(1), 143–148.

Klausen, M., Aaes-Jørgensen, A., Molin, S., & Tolker-Nielsen, T. (2003). Involvement of bacterial migration in the development of complex multicellular structures in Pseudomonas aeruginosa biofilms. *Molecular Microbiology, 50*(1), 61–68.

Klausen, M., Heydorn, A., et al. (2003). Biofilm Formation by Pseudomonas aeruginosa wild type, flagella and type IV pili mutants. *Molecular Microbiology, 48*(6), 1511–1524.

Kohler, P., et al. (2015). Healthcare-associated prosthetic heart valve, aortic vascular graft, and disseminated Mycobacterium chimaera infections subsequent to open heart surgery. *European Heart Journal, 36*(40), 2745–2753.

Konopka, A. (2009). What is microbial community ecology. *ISME Journal, 3*(11), 1223–1230.

Kostakioti, M., Hadjifrangiskou, M., & Hultgren, S. J. (2013). Bacterial biofilms: Development, dispersal, and therapeutic strategies in the dawn of the postantibiotic era. *Cold Spring Harbor Perspectives in Medicine, 3*(4), 1–23.

Laue, H., et al. (2006). Contribution of alginate and levan production to biofilm formation by Pseudomonas syringae. *Microbiology, 152*(10), 2909–2918.

Lee, B., et al. (2005). Heterogeneity of biofilms formed by nonmucoid Pseudomonas aeruginosa isolates from patients with cystic fibrosis. *Journal of Clinical Microbiology, 43*(10), 5247–5255.

Lemon, K. P., Higgins, D. E., & Kolter, R. (2007). Flagellar motility is critical for Listeria monocytogenes biofilm formation. *Journal of Bacteriology, 189*(12), 4418–4424.

Loveday, H. P., et al. (2014). Association between healthcare water systems and Pseudomonas aeruginosa infections: A rapid systematic review. *Journal of Hospital Infection, 86*(1), 7–15. https://doi.org/10.1016/j.jhin.2013.09.010

Lund-Palau, H., et al. (2016). Pseudomonas aeruginosa infection in cystic fibrosis: Pathophysiological mechanisms and therapeutic approaches. *Expert Review of Respiratory Medicine, 10*(6), 685–697.

Macfarlane, S., Bahrami, B., & Macfarlane, G. T. (2011). Mucosal biofilm Communities in the human intestinal tract (1st ed.), *75. Advances in applied microbiology*. Elsevier Inc. https://doi.org/10.1016/B978-0-12-387046-9.00005-0

Maniscalco, M., et al. (2019). Clinical metabolomics of exhaled breath condensate in chronic respiratory diseases (1st ed.), *88. Advances in clinical chemistry*. Elsevier Inc. https://doi.org/10.1016/bs.acc.2018.10.002

Mann, E. E., & Wozniak, D. J. (2012). Pseudomonas biofilm matrix composition and niche biology. *FEMS Microbiology Reviews, 36*(4), 893–916.

Mariencheck, W. I., Alcorn, J. F., Palmer, S. M., & Wright, J. R. (2003). Pseudomonas aeruginosa elastase degrades surfactant proteins A and D. *American Journal of Respiratory Cell and Molecular Biology, 28*(4), 528–537.

Matsui, H., et al. (2006). A physical linkage between cystic fibrosis airway surface dehydration and Pseudomonas aeruginosa biofilms. *Proceedings of the National Academy of Sciences of the United States of America, 103*(48), 18131–18136.

McCarty, S. M., Cochrane, C. A., Clegg, P. D., & Percival, S. L. (2012). The role of endogenous and exogenous enzymes in chronic wounds: A focus on the implications of aberrant levels of both host and bacterial proteases in wound healing. *Wound Repair and Regeneration, 20*(2), 125–136.

McCoy, K. S., et al. (2008). Inhaled aztreonam lysine for chronic airway Pseudomonas aeruginosa in cystic fibrosis. *American Journal of Respiratory and Critical Care Medicine, 178*(9), 921–928.

Meena, S., et al. (2019). Emergence of Burkholderia cepacia in ICU setting. *Indian Journal of Critical Care Medicine, 23*(9), 423–426.

Méndez-Ortiz, M. M., Hyodo, M., Hayakawa, Y., & Membrillo-Hernández, J. (2006). Genome-wide transcriptional profile of Escherichia coli in response to high levels of the second messenger 3′,5′-cyclic diguanylic acid. *Journal of Biological Chemistry, 281*(12), 8090–8099.

Moradali, M. F., Ghods, S., & Rehm, B. H. A. (2017). Pseudomonas aeruginosa lifestyle: A paradigm for adaptation, survival, and persistence. *Frontiers in Cellular and Infection Microbiology, 7*, 39.

Mougous, J. D., et al. (2006). A virulence locus of Pseudomonas aeruginosa encodes a protein secretion apparatus. *Science, 312*(5779), 1526–1530.

Murphy, T. F., Kirkham, C., Sethi, S., & Lesse, A. J. (2005). Expression of a peroxiredoxin-glutaredoxin by Haemophilus influenzae in biofilms and during human respiratory tract infection. *FEMS Immunology and Medical Microbiology, 44*(1), 81–89.

Nair, C., et al. (2014). Cyanide levels found in infected cystic fibrosis sputum inhibit airway ciliary function. *European Respiratory Journal, 44*(5), 1253–1261.

Nandakumar, V., Chittaranjan, S., Kurian, V. M., & Doble, M. (2013). Characteristics of bacterial biofilm associated with implant material in clinical practice. *Polymer Journal, 45*(2), 137–152.

Nazzari, E., et al. (2015). Role of biofilm in children with recurrent upper respiratory tract infections. *European Journal of Clinical Microbiology and Infectious Diseases, 34*(3), 421–429.

Novick, R. P., & Muir, T. W. (1999). Virulence gene regulation by peptides in staphylococci and other gram- positive bacteria. *Current Opinion in Microbiology, 2*(1), 40–45.

Oliver, A., et al. (2000). High frequency of hypermutable Pseudomonas aeruginosa in cystic fibrosis lung infection. *Science, 288*(5469), 1251–1253.

O'Toole, G. A., & Kolter, R. (1998). Flagellar and twitching motility are necessary for.Pdf. *Molecular Microbiology, 30*(2), 295–304.

Otto, K., & Silhavy, T. J. (2002). Surface sensing and adhesion of Escherichia coli controlled by the cpx-signaling pathway. *Proceedings of the National Academy of Sciences of the United States of America, 99*(4), 2287–2292.

Pang, Z., et al. (2018). PT US. *Biotechnology advances*. https://doi.org/10.1016/j.biotechadv.2018.11.013

Parsek, M. R., & Greenberg, E. P. (2005). Sociomicrobiology: The connections between quorum sensing and biofilms. *Trends in Microbiology, 13*(1), 27–33.

Pedersen, S. S., Hoiby, N., Espersen, F., & Koch, C. (1992). Role of alginate in infection with mucoid Pseudomonas aeruginosa in cystic fibrosis. *Thorax, 47*(1), 6–13.

Peleg, A. Y., & Hooper, D. C. (2010). Hospital-acquired infections due to gram-negative bacteria types of infections NIH public access. *The New England Journal of Medicine, 362*(19), 1804–1813.

Pier, G. B. (2007). Pseudomonas aeruginosa lipopolysaccharide: A major virulence factor, initiator of inflammation and target for effective immunity. *International Journal of Medical Microbiology, 297*(5), 277–295.

Pompilio, A., et al. (2010). Adhesion to and biofilm formation on IB3-1 bronchial cells by stenotrophomonas maltophilia isolates from cystic fibrosis patients. *BMC Microbiology, 10*, 1–15.

Poole, K., Hazlett, L. D., State, W., & Greenberg, E. P. (2011). Pseudomonas aeruginosa: Resistance to the max. *Frontiers in Microbiology, 2*, 1–13.

Quittner, A. L., et al. (2009). Determination of the minimal clinically important difference scores for the cystic fibrosis questionnaire-revised respiratory symptom scale in two populations of patients with cystic fibrosis and chronic Pseudomonas aeruginosa airway infection. *Chest, 135*(6), 1610–1618.

Darouiche, R. O. (2004). Treatment of infections associated with surgical implants. *The New England Journal of Medicine, 350,* 1422–1429. https://www.mendeley.com/research-papers/fulltext-120068747/?utm_source=desktop&utm_medium=1.17.13&utm_campaign=open_catalog&userDocumentId=%7B0f8c2161-b590-38d3-ab62-849241906d57%7D.

Rada, B., & Leto, T. L. (2013). Pyocyanin effects on respiratory epithelium: Relevance in Pseudomonas aeruginosa airway infections. *Trends in Microbiology, 21*(2), 73–81.

Ramey, B. E., Koutsoudis, M., Von Bodman, S. B., & Fuqua, C. (2004). Biofilm Formation in plant-microbe associations. *Current Opinion in Microbiology, 7*(6), 602–609.

Rasmussen, T. B., & Givskov, M. (2006). Quorum-sensing inhibitors as anti-pathogenic drugs. *International Journal of Medical Microbiology, 296*(2–3), 149–161.

Raveendran, S., Kumar, G., Sivanandan, R. N., & Dias, M. (2020). Moraxella catarrhalis: A cause of concern with emerging resistance and presence of BRO beta-lactamase gene - report from a tertiary care hospital in south India. *International Journal of Microbiology*, 1–5.

Rossi, C., et al. (2016). Influence of incubation conditions on biofilm formation by Pseudomonas fluorescens isolated from dairy products and dairy manufacturing plants. *Italian Journal of Food Safety, 5*(3).

Rowe, M. C., Withers, H. L., & Swift, S. (2010). Uropathogenic Escherichia coli forms biofilm aggregates under iron restriction that disperse upon the supply of iron. *FEMS Microbiology Letters, 307*(1), 102–109.

Sadikot, R. T., Blackwell, T. S., Christman, J. W., & Prince, A. S. (2005). Pathogen-host interactions in Pseudomonas aeruginosa pneumonia. *American Journal of Respiratory and Critical Care Medicine, 171*(11), 1209–1223.

Saeki, E. K., Kobayashi, R. K. T., & Nakazato, G. (2020). Quorum sensing system: Target to control the spread of bacterial infections. *Microbial Pathogenesis, 142*, 104068. https://doi.org/10.1016/j.micpath.2020.104068

Saiman, L., Cacalano, G., & Prince, A. (1990). Pseudomonas cepacia adherence to respiratory epithelial cells is enhanced by Pseudomonas aeruginosa. *Infection and Immunity, 58*(8), 2578–2584.

Sajjan, U., Wu, Y., Kent, G., & Forstner, J. (2000). Preferential adherence of cable-piliated Burkholderia cepacia to respiratory epithelia of CF knockout mice and human cystic fibrosis lung explants. *Journal of Medical Microbiology, 49*(10), 875–885.

Sap, M. (1999). *Ernst1999.*

Sarda, C., Fazal, F., & Rello, J. (2019). Management of ventilator-associated pneumonia (VAP) caused by resistant gram-negative bacteria: Which is the best strategy to treat? *Expert Review of Respiratory Medicine, 13*(8), 787–798. https://doi.org/10.1080/17476348.2019.1632195

Schaffer, J. N., & Pearson, M. M. (2016). Proteus mirabilis and urinary tract infections. *Urinary Tract Infections: Molecular Pathogenesis and Clinical Management, 3*(5), 383–433.

Schmidt, T., & Kirschning, A. (2012). Total synthesis of carolacton, a highly potent biofilm inhibitor. *Angewandte Chemie - International Edition, 51*(4), 1063–1066.

Seifi, K., et al. (2016). Evaluation of biofilm formation among Klebsiella pneumoniae isolates and molecular characterization by ERIC-PCR. *Jundishapur Journal of Microbiology, 9*(1), 2–7.

Serra, D. O., Richter, A. M., & Hengge, R. (2013). Cellulose as an architectural element in spatially structured Escherichia coli biofilms. *Journal of Bacteriology, 195*(24), 5540–5554.

Sibley, C. D., & Surette, M. G. (2011). The polymicrobial nature of airway infections in cystic fibrosis: Cangene gold medal lecture. *Canadian Journal of Microbiology, 57*(2), 69–77.

Sikora, M., et al. (2021). Clinical implications of intestinal barrier damage in psoriasis. *Journal of Inflammation Research, 14*, 237–243.

Soto, S. M. (2014). Importance of biofilms in urinary tract infections: New therapeutic approaches. *Advances in Biology, 2014*, 1–13.

Southey-Pillig, C. J., Davies, D. G., & Sauer, K. (2005). Characterization of temporal protein production in Pseudomonas aeruginosa biofilms. *Journal of Bacteriology, 187*(23), 8114–8126.

Stewart, P. S., & William Costerton, J. (2001). Antibiotic resistance of bacteria in biofilms. *Lancet, 358*(9276), 135–138.

Sutherland, I. W. (2001). The biofilm matrix - an immobilized but dynamic microbial environment. *Trends in Microbiology, 9*(5), 222–227.

Sweet, M. J., Croquer, A., & Bythell, J. C. (2011). Development of bacterial biofilms on artificial corals in comparison to surface-associated microbes of hard corals. *PLoS One, 6*(6).

Swords, W. E. (2004). Sialylation of lipooligosaccharides promotes biofilm formation by nontypeable *Haemophilus influenzae. Infection and Immunity, 72*(1), 106–113.

Tamayo, R., Pratt, J. T., & Camilli, A. (2007). Roles of cyclic diguanylate in the regulation of bacterial pathogenesis. *Annual Review of Microbiology, 61*, 131–148.

Tenke, P., et al. (2004). Bacterial biofilm formation on urologic devices and heparin coating as preventive strategy. *International Journal of Antimicrobial Agents, 23*(Suppl. 1), 67–74.

Tiddens, H. A. W. M., et al. (2015). Open label study of inhaled aztreonam for Pseudomonas eradication in children with cystic fibrosis: The ALPINE study. *Journal of Cystic Fibrosis, 14*(1), 111–119. https://doi.org/10.1016/j.jcf.2014.06.003

Toutain, C. M., Caizza, N. C., Zegans, M. E., & O'Toole, G. A. (2007). Roles for flagellar stators in biofilm formation by Pseudomonas aeruginosa. *Research in Microbiology, 158*(5), 471–477.

Verduin, C. M., et al. (2002). Moraxella catarrhalis: From emerging to established pathogen. *Clinical Microbiology Reviews, 15*(1), 125–144.

Vincent, J. L., Rello, J., Marshall, J., Silva, E., Anzueto, A., Martin, C. D., Moreno, R., Lipman, J., Gomersall, C., Sakr, Y., Reinhart, K., & EPIC II Group of Investigators. (2009). International study of the prevalence and outcomes of infection in intensive care units. *Jama, 302*(21), 2323–2329.

Wainwright, C. E., et al. (2011). Aztreonam for inhalation solution (AZLI) in patients with cystic fibrosis, mild lung impairment, and P. Aeruginosa. *Journal of Cystic Fibrosis, 10*(4), 234–242. https://doi.org/10.1016/j.jcf.2011.02.007

Waters, V., et al. (2013). Chronic stenotrophomonas maltophilia infection and mortality or lung transplantation in cystic fibrosis patients. *Journal of Cystic Fibrosis, 12*(5), 482–486. https://doi.org/10.1016/j.jcf.2012.12.006

Waters, V., et al. (2017). Pilot trial of tobramycin inhalation powder in cystic fibrosis patients with chronic Burkholderia cepacia complex infection. *Journal of Cystic Fibrosis, 16*(4), 492–495. https://doi.org/10.1016/j.jcf.2017.02.008

Wenzel, R. P. (2007). Health care-associated infections: Major issues in the early years of the 21st century. *Clinical Infectious Diseases, 45*(Suppl. 1), 85–88.

Wesson, C. A., et al. (1998). Staphylococcus aureus Agr and Sar global regulators influence internalization and induction of apoptosis. *Infection and Immunity, 66*(11), 5238–5243.

Wilson, A. M., Gray, D. M., & Thomas, J. G. (2009). Increases in endotracheal tube resistance are unpredictable relative to duration of intubation. *Chest, 136*(4), 1006–1013. https://doi.org/10.1378/chest.08-1938

Worlitzsch, D., et al. (2002). Effects of reduced mucus oxygen concentration in airway Pseudomonas infections of cystic fibrosis patients. *Journal of Clinical Investigation, 109*(3), 317–325.

Wu, Y., & Wayne Outten, F. (2009). IscR controls iron-dependent biofilm Formation in Escherichia coli by regulating type I fimbria expression. *Journal of Bacteriology, 191*(4), 1248–1257.

Xiao, X., Wu, Z. C., & Chou, K. C. (2011). A multi-label classifier for predicting the subcellular localization of gram-negative bacterial proteins with both single and multiple sites. *PLoS One, 6*(6), e20592.

Xiong, Y., & Liu, Y. (2010). Biological control of microbial attachment: A promising alternative for mitigating membrane biofouling. *Applied Microbiology and Biotechnology, 86*(3), 825–837.

Zhang, L., & Mah, T. F. (2008). Involvement of a novel efflux system in biofilm-specific resistance to antibiotics. *Journal of Bacteriology, 190*(13), 4447–4452.

Zobell, J. T., Kemper, A. L., & Young, D. C. (2014). The use of doripenem in pediatric cystic fibrosis patients in case of meropenem shortages. *Pediatric Pulmonology, 49*(3), 48–51.

Zogaj, X., Werner, B., Nimtz, M., & Römling, U. (2003). Production of cellulose and curli fimbriae by members of the family Enterobacteriaceae isolated from the human gastrointestinal tract. *Infection and Immunity, 71*(7), 4151–4158.

Zottola, E. A., & Sasahara, K. C. (1994). Microbial biofilms in the food processing industry-should they be a concern? *International Journal of Food Microbiology, 23*(2), 125–148.

Index

Note: 'Page numbers followed by "*f*" indicate figures and "*t*" indicate tables.'

D

H

K

L

M

9780323957151